Augsburger Schriften zur Mathematik, Physik und Informatik
Band 28

herausgegeben von:
Professor Dr. F. Pukelsheim
Professor Dr. B. Aulbach
Professor Dr. W. Reif
Professor Dr. B. Schmidt
Professor Dr. D. Vollhardt

Bibliografische Information der Deutschen Nationalbibliothek

Die Deutsche Nationalbibliothek verzeichnet diese Publikation in der Deutschen Nationalbibliografie; detaillierte bibliografische Daten sind im Internet über http://dnb.d-nb.de abrufbar.

ISBN 978-3-8325-4028-9
ISSN 1611-4256

Logos Verlag Berlin GmbH
Comeniushof, Gubener Str. 47,
10243 Berlin
Tel.: +49 030 42 85 10 90
Fax: +49 030 42 85 10 92
INTERNET: http://www.logos-verlag.de

Effective Theories for Brittle Materials

A Derivation of Cleavage Laws and Linearized Griffith Energies from Atomistic and Continuum Nonlinear Models

Dissertation

zur Erlangung des akademischen Grades

Dr. rer. nat.

eingereicht an der

Mathematisch-Naturwissenschaftlich-Technischen Fakultät

der Universität Augsburg

von

Manuel Friedrich

Augsburg, März 2015

Erstgutachter: Prof. Dr. Bernd Schmidt (Universität Augsburg)

Zweitgutachter: Prof. Dr. Marco Cicalese (Technische Universität München)

Drittgutachter: Univ.-Prof. Ulisse Stefanelli, PhD (Universität Wien)

Mündliche Prüfung: 09. Juni 2015

Contents

Introduction

The main focus of this thesis lies on the derivation of effective models for brittle materials in the simultaneous passage from discrete-to-continuum and nonlinear to linearized systems. Such materials show an elastic response to very small displacements and develop cracks already at moderately large strains. Typically there is no plastic regime in between the restorable elastic deformations and complete failure due to fracture. In spite of its importance in applications, a thorough understanding of the cleavage behavior of brittle crystals remains a challenging problem in theoretical mechanics. In particular, it is fundamental to identify critical loads for failure and to analyze the geometry of crack paths that occur in the fractured regime.

In variational fracture mechanics displacements and crack paths are determined from an energy minimization principle. Following the pioneering work of Griffith [52], Francfort and Marigo [44] have introduced an energy functional comprising elastic bulk contributions for the intact regions of the body and surface terms that assign energy contributions on the crack paths comparable to the size of the crack of codimension one. Subsequently, these models have been investigated and extended in various directions. Among the vast body of literature we only mention the work of Dal Maso and Toader [37]; Francfort and Larsen [43]; Dal Maso, Francfort and Toader [35] and refer to [11] for further references. Determining energy minimizers of such functionals leads to solving a free discontinuity problem in the language of Ambrosio and De Giorgi [38] as the crack path, i.e., the set of discontinuity of the diplacement field is not pre-assigned but has to be found as a solution to the variational problem. In particular, these models also lead to efficient numerical approximation schemes, cf., e.g., [7, 10, 58, 59, 66].

In the continuum setting many Griffith energies contain anisotropic surface terms (see e.g. [2, 22, 42, 58]) modeling the fact that due to the crystalline structure of the materials certain directions for the formation of cracks are energetically favored. Indeed, under tensile boundary loads fracture typically occurs in the form of cleavage along crystallographic planes of the atomic lattice. Ultimately, such a continuum model should be identified as an effective theory derived from atomistic interactions. One aim of this thesis is to provide a rigorous study of discrete systems for the validity and failure of crystal cleavage in the multidimensional framework with vector valued deformations.

In the engineering literature such discrete systems had been analyzed computationally in [53, 55] and formally by renormalization group techniques in [61]. Braides, Lew and Ortiz [18] then showed analytically that in the continuum limit the energy satisfies a certain cleavage law with a universal form independent of the specific choice of the interatomic potential. In all these models the crack geometry is pre-assigned and fracture may only occur along planes leading effectively to a one-dimensional problem which is much easier to analyze. However, in order to understand the physical and geometrical cause for cleavage in the fracture regime it is indispensable to examine vectorial problems in more than one space dimension.

In our model we assume that the macroscopic region occupied by the specimen is a cuboid subject to uniaxial tensile boundary conditions. Our focus on such boundary values corresponds to one of the basic experiments in determining e.g. the Poisson ratio and is naturally motivated by our main goal of analyzing cleavage behavior. We suppose that in our discrete model the atoms in the reference configuration are given by the portion of a Bravais lattice lying in that region. The interaction of the material points will be described by a general class of 'cell energies' including well known mass-spring models where the pair interaction of neighboring atoms is modeled by potentials of Lennard-Jones type.

We prove that under uniaxial tension in the continuum limit the energy satisfies a cleavage law of a universal form essentially only depending on the stiffness and toughness of the material which may be deduced from the cell energy. The limiting energy exhibits quadratic response to small boundary displacements followed by a sharp constant cut-off beyond some critical value. Similarly as in the one-dimesional seminal paper [18] we find that the most interesting regime for the elastic strains is given by $\sqrt{\varepsilon}$ (ε denotes the typical interatomic distance) as in this particular regime the elastic and the crack energy are of the same order. This is in accordance to the observation that brittle materials develop cracks already at moderately large strains.

Suitable test configurations then show that asymptotically optimal configurations are given by homogeneous elastic deformations for subcritical boundary values and by configurations cleaved along specific crystallographic hyperplanes beyond critical loading. It seems that our analysis provides a first multidimensional result for the validity of crystal cleavage. However, it turns out that there are non-generic models where also other, much more complex crack geometries are energetically optimal whence in the general case a full characterization of all asymptotically minimizing sequences seems currently out of reach.

Consequently, for a deeper investigation we restrict our analysis to a specific two-dimensional model problem where the discrete system is given by the portion of a triangular lattice in a rectangular strip interacting via next neighbor Lennard-Jones type potentials. Indeed, in this case we can show that any sequence of minimizers converges (up to subsequences) to a homogeneous continuum deformation for subcritical boundary values, while it converges to a continuum deformation which is cracked completely and does not store elastic energy in the supercritical case. We prove that in the generic case cleavage occurs along a unique crystallygraphic line, whereas for specific symmetric orientations of the crystal cleavage might fail. Nevertheless, also in these special cases we obtain a complete characterization of all possible limiting crack geometries. The model under investigation leads, in particular, to configurations respecting the Poisson effect, which would not be possible in scalar models. These results justify rigorously the aforementioned assumptions in the derivation of cleavage laws as, e.g., in [18].

Even though the uniaxial tension test is a natural set-up for investigating cleavage phenomena, it is desirable to also incorporate more general boundary conditions and to identify limiting continuum configurations and energies in the same energy regime which are not necessarily asymptotically energy minimizing.

While the passage from discrete systems to effective continuum models via Γ-convergence (see [12, 33]) is by now well understood for one-dimensional brittle chains, see e.g. [14, 15, 16], not much is known on discrete-to-continuum limits for models allowing for fracture in more than one dimension. The farthest reaching developments in that direction seem to be results for scalar valued models (see [17]) and approximations of vector valued free discontinuity problems where the elastic bulk part of the energy is characterized by linearized terms (see [2]) or by a quasiconvex stored energy density (see [42]).

However, in more than one dimension the energy density of discrete systems such as well-known mass spring models is in general not given in terms of a discretized continuum quasiconvex function. For large strains these lattices typically become even unstable, see e.g. the basic model discussed in [50]. Consequently, in the regime of finite elasticity it is a subtle question if minimizers for given boundary data exist at all. On the other hand, for sufficiently small strains one may expect the Cauchy-Born rule to apply so that individual atoms do in fact follow a macroscopic deformation gradient, see [50, 27]. In particular this applies to the regime of infinitesimal elastic strains. For purely elastic interactions this relation has also been obtained in the sense of Γ-convergence for a simultaneous passage from discrete to continuum and linearization process in [19, 65].

In the present context the investigation of cleavage laws already showed that the most interesting regime for the elastic strains is given by $\sqrt{\varepsilon}$. Consequently, a passage from discrete-to-continuum systems naturally involves a linearization process. Indeed, we will prove that in the small displacement regime the energies associated to the discrete energies under consideration can be related to a continuum Griffith energy functional with anisotropic surface contributions.

In our analysis we first make the simplifying assumption that we consider deformations lying $\sqrt{\varepsilon}$-close to the identity mapping. Indeed, we discuss physically interesting applications where such a smallness assumption can be justified rigorously, e.g. (1) a boundary value problem describing uniaxial extension and (2) fractured magnets in an external field. However, in general a small distance form the identity mapping can not be inferred from energy bounds and identifying all possible limiting continuum configurations and energies is a subtle task. The main challenge is to establish suitable rigidity estimates being essential in the passage from nonlinear to linearized theory (see [19, 65]).

One aim of this thesis is the derivation of such quantitative geometric rigidity estimates for brittle materials which allows to establish a general Γ-limit result in the passage from nonlinear to linearized energies in fracture mechanics without any a priori assumption on the deformation and the crack geometry. To avoid further complicacies of technical nature concerning the topological structure of cracks in higher dimensions and to concentrate on the essential difficulties arising from the frame indifference of the energy density, we will tackle this problem in a continuum setting in two dimensions with isotropic crack energies. However, we believe that our results can be extended to discrete systems with anisotropic surface terms whereby we can justify a posteriori the aforementioned smallness assumptions.

Clearly, such a derivation is not only important in the context of discrete systems but interesting on its own. Indeed, for many realistic models in fracture mechanics being genuinly nonlinear it is desirable to identify an effective linear theory and in this way to rigorously show that in the small displacement regime the neglection of effects arising from the non-linearities is a good approximation of the problem. In fact, the propagation of crack was studied in the framework of linearized elasticity since the seminal work of Griffith and led to a lot of realistic applications in engineering. Also from a mathematical point of view the theory is well developed (see [5, 8]) and adopted in many recent works in applied analysis (see e.g. [8, 26, 54, 66]) since, as discussed above, such models are often significantly easier to treat as their nonlinear counterparts due to the convexity of the bulk energy density.

In general, the analysis of fracture models is often very involved due to the two coexistent, competing energy forms, the elastic and the crack energy, showing different scaling properties. Therefore, as a first approach to the nonlinear-to-linear limit we consider each of the regimes separately and discuss the results which are available in the literature.

For elastic bodies not exhibiting cracks the passage from nonlinear to linearized models is by now well understood and was first rigorously derived by Dal Maso, Negri and Percivale in [36] in a continuum setting using Γ-convergence (cf. also [19, 65]). The main ingredient in their analysis is a quantitative geometric rigidity estimate by Friesecke, James and Müller [49] which allows to establish a compactness result for rescaled displacement fields. For configurations with small elastic energy the frame indifference of the energy density induces that the deformation gradient is pointwise near a rotation. The result states, loosely speaking, that then the deformation is globally near one single rigid motion.

In the framework of fracture mechanics one difficulty arises form the fact that global rigidity may fail if the body is disconnected by the jump set. Under the constraint that the material does not store elastic energy Chambolle, Giacomini and Ponsiglione [25] could show that the body behaves piecewise rigidly, i.e. the only possibility that global rigidity can fail is that the body is divided into various parts each of which subject to a different rigid motion.

The goal of our analysis is to combine the aforementioned results and to tackle the problem for general Griffith models where both energy forms are coexistent. As a first observation we see that, without passing to rescaled configurations, in the small strain limit the energies converge to a limiting functional which is finite for piecewise rigid motions and measures the segmentation energy which is necessary to disconnect the body. Consequently, in order to arrive at a limiting model showing coexistence of elastic in surface energy the strategy is to pass to appropriate rescaled displacement fields similarly as in [36].

The farthest reaching result in this direction seems to be a recent work by Negri and Toader [60] where a nonlinear-to-linear analysis is performed in the context of quasistatic evolution for a restricted class of admissible cracks. In particular, in their model the different components of the jump set are supposed to have a least positive distance rendering the problem considerably easier. In fact, one can essentially still employ the rigidity estimate [49] and the specimen cannot be separated into different parts effectively leading to a simple relation between the deformation and the rescaled displacement field. In the present work we do not presume any a priori assumptions on the jump set and thus treat the full free discontinuity problem.

The major challenge is the derivation of a – to the best of our knowledge – new kind of geometric rigidity result in the framework of geometric measure theory. We call this estimate an SBD-rigidity result as it is formulated in terms of *special functions of bounded deformation* (see [5, 8]). The derivation is very involved as among other things one has to face the problems that (1) the body might be disconnected by the jump set, (2) the body might be still connected but only in a small region where the elastic energy is possibly large, (3) the crack geometry might become extremely complex due to relaxation of the elastic energy by oscillating crack paths and infinite crack patterns occurring on different scales. The common difficulty of all these phenomena is the possible high irregularity of the jump set. Even if one can assume that the domain can be decomposed into different sets with Lipschitz boundary (e.g. by a density argument), there are no uniform bounds on the constants of several necessary inequalities such as the Poincaré and Korn inequality and the rigidity estimate [49].

The rigidity result provides the relation between the deformation of a material and the rescaled displacements, which measure the distance from piecewise rigid motions being constant on each connected component of the cracked body. It proves to be the fundamental ingredient for the derivation of linearized Griffith models via Γ-convergence in a small strain limit. As before the limiting configurations consist of a piecewise rigid motion with corresponding segmentation energy. Additionally, on each component of the partition there is an associated displacement field whose energy is of Griffith-type in the realm of linearized elasticity.

Finally, we discuss that the general Γ-limit result can be applied to solve boundary value problems of uniaxial compression which is as the uniaxial tension test a natural problem. Hereby we complete the picture about our derivation of cleavage laws in the passage from discrete-to-continuum systems.

Outline

We now give a more specific outline of the content of this thesis.

In Chapter 1 we introduce the discrete models and present our main results about the derivation of cleavage laws and the establishing of a discrete-to-continuum Γ-limit. Although the investigation of the general nonlinear-to-linear limit is a related problem, we prefer to postpone the presentation of the results to Chapter 6 as we base our analysis on a slightly different setting and wish to avoid confusion between the different models.

In Section 1.1 we introduce our multidimensional discrete model. We suppose that the atoms in the reference configuration are given by the portion $\varepsilon\mathcal{L} \cap \Omega$, where the macroscopic region $\Omega \subset \mathbb{R}^d$ occupied by the body is a cuboid and $\varepsilon\mathcal{L}$ is some Bravais lattice scaled by the typical interatomic distance $\varepsilon \ll 1$. The main structural assumption is that the energy of a deformation $y : \varepsilon\mathcal{L} \cap \Omega \to \mathbb{R}^d$ may be decomposed as a sum over cell energies, i.e.

$$E_\varepsilon(y) = \sum\nolimits_Q W_{\mathrm{cell}}(\bar{\nabla} y|_Q).$$

Here the sum runs over the scaled cells $Q \subset \Omega$ induced by $\varepsilon\mathcal{L}$ and the cell energy on its part depends on the discrete gradient $\bar{\nabla} y$ encoding all the relative displacements of atoms in a cell and satisfies some reasonable assumptions in the elastic regime (see e.g. [27]), particularly the frame indifference. As forces between well separated atoms are governed by dipole interactions we assume that for large deformation gradients the cell energy reduces to a pair interaction energy neglecting multiple point interactions.

In Section 1.2 we show by a heuristic argument that the most interesting regime of boundary values is given by $\sqrt{\varepsilon}$ as in this case the energies of typical elastic deformations and configurations with cleavage are of the same order. This scaling was first proposed by Nguyen and Ortiz [61], who investigated the problem with renormalization group techniques. In Section 1.3 we present our main cleavage law.

In Section 1.4 we introduce a specific two-dimensional model where the atoms in the reference configuration form a triangular lattice. Although being a model problem we emphasize that it contains the main features essential for our analysis. Indeed, it is (1) is frame indifferent in its vector-valued arguments in more than one dimension, (2) gives rise to non-degenerate elastic bulk terms and (3) leads to surface contributions sensitive to the crack geometry with competing crystallographic lines. Moreover, we will discuss an application to the stability of brittle nanotubes under interior expansive pressure and observe that such two-dimensional lattice surfaces naturally appear in the analysis of thin structures.

In Section 1.5 and Section 1.6 we state the main results about the characterization of minimizing sequences and the convergence of the variational problems, respectively. Here we study an application to fractured magnets in an external field and briefly discuss that the findings on crystal cleavage can be re-derived by investigating a limiting variational problem.

Chapter 2 is devoted to preliminaries. In Section 2.1 we first derive formulae for the essential constants appearing in the cleavage law characterizing the stiffness and the toughness of the material. Here we already see that it is optimal to cleave along a crystallographic hyperplane. In Section 2.2 we introduce interpolations both for the elastic regime following the methods in [65] and for the fracture regime being adapted for the application of slicing techniques. In Section 2.3 we state a short lemma about the length of Lipschitz curves in sets of bounded variation being substantially important in dimensions $d \geq 4$. Finally, in Section 2.4 we derive some elementary properties for the cell energy of the triangular lattice. In particular, we introduce a lower-bound comparison energy providing fine estimates on the discrete minimal energies.

Chapter 3 is entirely devoted to the derivation of cleavage laws for the limiting minimal energy. As a warm-up we first give the proof for the triangular lattice in Section 3.1 by reducing the problem to one-dimensional segments using projection and slicing arguments. Due to the isotropy of the linearized elastic energy and the planar geometry whereby cracks may not concentrate on lower dimensional structures the result can be established in a comparatively elementary way.

By way of contrast, the analogous result in arbitrary dimensions with general lattices and interaction potentials requires (1) new projection estimates for the size of cracks in the specimen, (2) a thorough analysis of all possible crack modes of a lattice unit cell and in particular (3) a full dimensional analysis of an auxiliary problem on a 'mesoscopic cell' whose size is carefully chosen between the microscopic scale ε and the macroscopic magnitude of the specimen.

On the one hand, by choosing this size of the mesoscopic cell small enough it is possible to separate the effects arising from the bulk elastic and the surface crack energy and to apply elaborated methods in the various regimes, including rigidity estimates [49] and slicing techniques for *special functions of bounded variation* (SBV) (see e.g. [6]). On the other hand, given that the size is large with respect to ε we can exploit the validity of the Cauchy-Born-rule for sufficiently small strains which means, loosely speaking, that every single atom follows the mesoscopic deformation gradient and atomistic oscillations are effectively excluded (see [27,

50]). More precisely and in mathematical terms, passing simultaneously from discrete to continuum theory and from finite to infinitesimal elasticity the discrete gradient of the atomic displacements reduces to a classical gradient leading to a simpler description of the stored elastic energy (cf. [65]). With the help of tailor-made interpolations depending on individual crack modes, it can be shown that the fracture energy consisting of all contributions from pair interactions of neighboring atoms reduces to a surface energy in the continuum limit which only depends on the crack geometry (cf. also [17]) and is minimized for a specific crystallographic hyperplane.

Section 3.2 contains the essential technical estimates providing a lower comparison potential for the energy of a 'cell of mesoscopic size' under given averaged boundary conditions. The proof is mainly divided into three parts each of which dealing with one particular regime: The elastic regime where we show that linear elasticity theory applies, the fracture regime where we use a slicing argument in the framework of SBV functions and an intermediate regime. Beyond that, in the case $d \geq 4$ an additional intermediate regime has to be introduced due to the fact that in higher dimensions it becomes more difficult to derive uniform bounds on the difference of boundary values.

Section 3.3 is devoted to the proof of the main theorem which relies on the application of the comparison energy derived in Section 3.2 and a slicing argument in the space direction were the tensile boundary conditions were imposed.

In Section 3.4 we give further examples of mass-spring models to which the aforementioned results apply and provide the limiting minimal energy as well as asymptotically optimal configurations. We first analyze the nearest and next-to-nearest neighbor interaction in a square lattice and see that in addition to the Poisson-effect elastic minimizers generically also show a shear effect due to the anisotropy of the linearized elastic energy. Whereas the energetically favorable crack line in the triangular lattice was exclusively determined by the geometry of the problem, we find that for the square lattice two competing crystallographic lines occur due to possible different microscopic structures of fracture. Afterwards, we apply our results to a general nearest and next-to-nearest neighbor model in 3d considered e.g. in [63, 65].

In Chapter 4 we reduce our analysis to the triangular lattice and provide a characterization of all minimizing sequences for the boundary value problem of uniaxial extension. As a preparation we first establish finer estimates on the discrete minima by deriving higher order terms (see Section 4.1). In particular, our proof illustrates the typical behavior of brittle materials already seen in the continuum cleavage law also in a discrete framework: There is essentially no plastic regime besides the elastic and the crack regime. More precisely, we see that for almost minimizers the deformation is either $\sqrt{\varepsilon}$-close to the identity mapping (representing elastic response) or springs between adjacent atoms are elongated by a factor scaling like $\frac{1}{\sqrt{\varepsilon}}$ (leading to fracture in the limit description).

Here we can already see that homogeneous deformations or cleavage along specific lines are asymptotically optimal.

We then proceed to show that, under appropriate assumptions, in terms of suitably rescaled displacement fields indeed all discrete energy minimizers converge strongly to such continuum deformations. As alluded to above, the main challenge is to establish a suitable compactness result which is typically based on geometric rigidity estimates. In this specific setting we do not employ our general SBD rigidity result, but pursue a straighter way which effectively leads to stronger results. The strategy is to provide a fine characterization of the crack, i.e. of the number and position of largely elongated springs (see Section 4.2). In the subcritical case the contribution of such springs is abitrarily small such that the purely elastic theory applies. In the generic case, for supercritical boundary values largely deformed springs lie in a small stripe in direction of the optimal cristallographic line in such a way that the two components on the right and on the left of the stripe essentially behave elastically whence the rigidity result by Friesecke, James and Müller [49] is applicable. This characterization is the key ingredient to prove strong discrete-to-continuum convergence results in the various regimes (see Section 4.3).

Chapter 5 is devoted to the derivation and investigation of a limiting variational problem. In Section 5.1 we establish a corresponding continuum energy functional via Γ-convergence. It turns out that this problem is an issue similar to those considered in [2, 17, 42]. Nevertheless, we believe that the present Γ-convergence result is interesting as (1) it gives rise to a limiting Griffith functional in the realm of linearized elasticity which can be explicitly investigated for cleavage, (2) there are applications to systems with small displacements for small energies and (3) to the best of our knowledge our approach to the problem differs from techniques which are predominantly used when treating discrete systems in the framework of fracture mechanics.

The reduction to one-dimensional sections using slicing properties for SBV functions turned out to be a useful tool not only to derive general properties of these function spaces but also to study discrete systems and variational approximation of free discontinuity problems. E.g., the original proofs of the main compactness and closure theorems in SBV (see [3]) as well as the Γ-convergence results in [17, 42] make use of this integral-geometric approach. Similar to the fact that there are simplified proofs of these compactness theorems being derived without the slicing technique (see [1]), we show that in our framework the lower bound of the Γ-limit can be achieved in a different way. In fact, we carefully construct the crack shapes of discrete configurations in an explicit way which allows us to directly appeal to lower semicontinuity results for SBV functions. The elastic part can then be treated similarly as in [49, 65].

In Section 5.2 we analyze the continuum problem under tensile boundary values. A similar problem has been studied recently by Mora-Corral in [57], where he investigates a rectangular bar of brittle, incompressible, homogeneous and isotropic material subject to uniaxial extension and shows that, depending on the loading, the minimizers are either given by purely elastic configurations or deformations with horizontal fracture. We extend these results to anisotropic and compressible materials and moreover re-derive in part the aforementioned convergence results of Chapter 4. A careful analysis of the anisotropic surface contribution shows that in the generic case there is a unique optimal direction for the formation of fracture, while in a symmetrically degenerate case cleavage fails and all energetically optimal crack geometries can be characterized by specific Lipschitz curves. As in [57] the proof makes use of a qualitative rigidity result for SBV functions (see [25]) and of the structure theorem on the boundary of sets of finite perimeter by Federer [41].

In Chapter 6 we present the main results about geometric rigidity in SBD and the derivation of linearized Griffith energies from nonlinear counterparts. We adopt a slightly different point of view and consider continuum fracture models of the form

$$E_\varepsilon(y) = \frac{1}{\varepsilon}\int_\Omega W(\nabla y(x))\,dx + \mathcal{H}^1(J_y),$$

where W is a frame-indifferent energy density and $\mathcal{H}^1(J_y)$ denotes the size of the jump set of the deformation y. We briefly note that the small parameter ε denotes not only the order of the elastic energy, but in models arising from discrete systems as considered in Chapter 1 it again represents the typical interatomic distance. In fact, the length scale ε plays an important role in our analysis as the system shows remarkably different behavior on scales smaller and larger than the atomistic unit. We also discuss that besides the main nonlinear rigidity and compactness theorems there can be established associated results in a linearized regime which are much easier to prove and interesting on its own (see Section 6.1 and Section 6.2). As particularly the proof of the rigidity result is very long and technical, we present a thorough overview and highlight the principal proof strategies for the convenience of the reader (see Section 6.4).

Chapter 7 is devoted to some preliminaries. We have already discussed in the first part of the introduction that the constants in certain inequalities crucially depend on the shape of the domain. We carry out a careful analysis for the constants of the geometric rigidity estimate and a Korn-Poincaré inequality in BD (see [62]). At this point we notice that easy counterexamples to rigidity estimates in SBD can be constructed if one does not admit a small modification of the deformation. Moreover, we establish a trace theorem in SBV which allows to control the L^2-norm of the functions on the boundary.

As a first approach to the proof of the SBD-rigidity result it is convenient to replace the nonlinear problem by a linearized version. In Chapter 8 we establish a Korn-Poincaré-type inequality in SBD which measures the distance of the displacement from an infinitesimal rigid motion in terms of ε. This problem is significantly easier as (1) the estimate only involves the function itself and not its derivative and (2) the set of infinitesimal rigid motions is a linear space in contrast to $SO(d)$. It turns out that this estimate will be one of the key ingredients to derive our main result which can be compared with the fact that in elasticity theory the linearized rigidity estimate, called Korn's inequality (see [29]), is one of the fundamental steps to establish the geometrically nonlinear result in [49]. To the best of our knowledge our Korn-Poincaré inequality differs from other inequalities of this type available in the SBV-setting as it is not based on a truncation argument for the configurations (see [39, 23]), but on a modification of the jump set (see Sections 8.1 - 8.4) and a subsequent determination of the jump heights (see Section 8.5). In particular, the set where the function and the modification differ has a rather simple geometry being the union of a finite number of rectangles. Consequently, in contrast to the recently established theorem in [24], the estimate is suitable for the application of compactness result for (G)SBD functions (see e.g. [34]). Although tailor-made for the applications in Chapter 9 and Chapter 10, we believe that this result is of independent interest and may contribute to solve related problems in the future, especially concerning fracture models in the realm of linearized elasticity which are related to problems in SBV where Poincaré inequalities (see [39]) have proved to be useful.

Chapter 9 is devoted to the proof of the SBD-rigidity estimate. It turns out that the result can only be established under the additional condition that we admit an arbitrarily small modification of the deformation. One essential point is to derive an inequality for the symmetric part of the gradient. We also see that in general it is not possible to gain control over the full gradient which is not surprising as there is no analogue of Korn's inequality for SBV functions. In addition, we provide an L^2-bound for the configurations. In contrast to the setting in elasticity theory this is highly nontrivial as Poincaré's inequality cannot be applied due to the possibly present complicated crack geometry.

The main strategy of the proof is to establish local rigidity results on cells of mesoscopic size (Section 9.2) which together with the Korn-Poincaré inequality allows to replace the deformation by a modification where the least length of the crack components has increased (Section 9.3). Repeating the arguments on various mesoscopic scales becoming gradually larger it is possible to show that the modified deformation behaves rigidly on each connected component of the domain (Section 9.4). The fact that we analyze the problem on different length scales is indispensable to understand specific size effects correctly such as the accumulation of crack patterns on certain scales. Moreover, we briefly note that similarly as in Section 3.2 a mesoscopic localization technique proves to be useful to tackle problems in the framework of brittle materials as hereby effects arising from the bulk and the surface contributions can be separated.

Basically, this is enough the establish the requirements for compactness results in the space of SBD functions. However, as we are interested in the derivation of effective linearized models we have to assure that we do not change the total energy of the deformation during the modification procedure. In particular, for the surface energy this is a subtle problem and in Section 9.5 a lot of effort is needed to show that the modified configurations can be constructed in a way such that the crack length does not increase substantially.

We observe that many arguments in the proof are valid also in dimension $d \geq 3$. The essential reason why we restrict ourselves to the two-dimensional framework is the derivation of the Korn-Poincaré-type inequality in Chapter 8 where a lot of technical difficulties concerning the topological structure of the crack geometry occur. Nevertheless, we believe that we provide the principal techniques which are necessary to prove the result in arbitrary space dimension. Moreover, we are confident that our methods, in particular the small modification of the deformation and the jump set, may contribute to the solution of related problems.

Finally, Chapter 10 is devoted to the identification of limiting configurations and to the derivation of linearized Griffith energies via Γ-convergence. In Section 10.1 we present the main compactness result showing that the configurations consist of piecewise rigid motions and corresponding displacement fields. As there is no uniform bound on the functions it turns out that the limiting displacements are generically not summable and we naturally end up in the space of GSBD functions (for the definition and basic properties we refer to [34]). We believe that our results are interesting also outside of this specific context as they allow to solve more general variational problems in fracture mechanics. Typically, for compactness results in function spaces as SBV and SBD one needs L^∞ or L^1 bounds on the functions (see [4, 8, 34]). However, in many applications, in particular for atomistic systems and for models dealing with rescaled deformations, such estimates cannot be inferred from energy bounds. Nevertheless, we are able to treat problems without any a priori bound by passing from the deformations to displacement fields whose distance from rigid motions can be controlled.

In Section 10.2 we discuss the properties of the limiting partition which is related to the piecewise rigid motion. We observe that an even finer segmentation may occur if on a connected component of the partition the jump set of the corresponding displacement further disconnects the domain. Here it becomes apparent that we treat a real multiscale model as the jump heights at the boundaries associated to the coarse partition are of order $\gg \sqrt{\varepsilon}$, whereas the jump heights of the finer partition are of order $\sqrt{\varepsilon}$. In particular, it is evident that the choice of the limiting partition is not unique. However, we propose a selection principle and show that the partition can always be chosen in a way such that a further coarsening is not possible.

In Section 10.3 we derive the main Γ-limit which is almost immediate due to the preparations in Section 10.1. Finally, in Section 10.4 we return to our discussion about cleavage laws and investigate a specific boundary value problem, where we essentially follow the proof presented in Section 5.2 (cf. also [57]). Whereas the strategy followed in Chapter 3, which basically relied on a slicing technique, was not appropriate to treat the case of compression, at least in a continuum setting for isotropic surface energies the full Γ-limit allows to extend the results obtained in Section 1.6 to the case of uniaxial compression. It turns out that in the linearized limit the behavior for compression and extension is virtually identical. We briefly note that to avoid unphysical effects such as self-penetrability further modeling assumptions would be necessary.

Acknowledgement

It is a great pleasure to thank my supervisor Prof. Bernd Schmidt for having proposed interesting topics for my thesis, and for his guidance, continuous advice and encouragement. I would also like to thank the whole group "Nichtlineare Analysis" in Augsburg for many interesting discussions and joint activities.

The financial support of this thesis by a scholarship from the Universität Bayern e.V. is gratefully acknowledged. Moreover, I would like to thank the Stiftung Maximilianeum in Munich for offering me much more than an accommodation during the years of my studies.

Finally, I am grateful to my family and friends, in particular to Michaela Hofmann, for their support, encouragement and patience in the last years.

Part I

An analysis of crystal cleavage in the passage from atomistic models to continuum theory

Chapter 1

The model and main results

1.1 The discrete model

Let $\Omega \subset \mathbb{R}^d$ be the macroscopic region occupied by the body under consideration. To simplify the exposition we assume that $\Omega = (0, l_1) \times \ldots \times (0, l_d)$ is rectangular, but remark that all our results extend without difficulty to more general geometries as $\Omega = (0, l_1) \times \omega$, $\omega \subset \mathbb{R}^{d-1}$ open, for which cleavage boundary values as discussed in Section 1.2 below may be imposed. Let $\mathcal{L}$ be some Bravais lattice in $\mathbb{R}^d$, i.e. there are linearly independent vectors $v_1, \ldots, v_d \in \mathbb{R}^d$ such that

$$\mathcal{L} = \{\lambda_1 v_1 + \ldots \lambda_d v_d : \lambda_1, \ldots, \lambda_d \in \mathbb{Z}\} = A\mathbb{Z}^d,$$

where A is the matrix $(v_1, \ldots, v_d)$. Without restriction we may assume that the vectors v_i are labeled such that $\det A > 0$. The portion of the scaled lattice $\mathcal{L}_\varepsilon = \varepsilon\mathcal{L}$ lying in Ω represents the positions of the specimen's atoms in the reference position. Here ε is a small parameter measuring the typical interatomic distance eventually tending to zero. Note that $\mathcal{L}_\varepsilon$ partitions $\mathbb{R}^d$ into cells of the form $\varepsilon A(\lambda + [0,1)^d)$ for $\lambda \in \mathbb{Z}^d$. The shifted lattice $\varepsilon A((\frac{1}{2}, \ldots, \frac{1}{2})^T + [0,1)^d)$ consisting of the midpoints of the cells is denoted by $\mathcal{L}'_\varepsilon$. For $x \in \mathbb{R}^d$ we denote by $\bar{x} = \bar{x}(x, \varepsilon)$ the center of the ε-cell containing the point x and set $Q_\varepsilon(x) = \bar{x}(x,\varepsilon) + \varepsilon A[-\frac{1}{2}, \frac{1}{2})$.

We choose a numbering $z_1, \ldots, z_{2^d}$ of the corners $A\left\{-\frac{1}{2}, \frac{1}{2}\right\}^d$ of the reference cell $A[-\frac{1}{2}, \frac{1}{2})^d$ and set

$$Z = (z_1, \ldots, z_{2^d}), \quad \mathcal{Z} = \{z_1, \ldots, z_{2^d}\}. \tag{1.1}$$

For subsets $U \subset \Omega$ we define the following lattice subsets with respect to the midpoints $\mathcal{L}'_\varepsilon$ and the corners $\mathcal{L}_\varepsilon$:

$$\mathcal{L}'_\varepsilon(U) = \{\bar{x} \in \mathcal{L}'_\varepsilon : Q_\varepsilon(\bar{x}) \cap U \neq \emptyset\}, \quad \mathcal{L}_\varepsilon(U) = \mathcal{L}'_\varepsilon(U) + \varepsilon\{z_1, \ldots, z_{2^d}\},$$
$$(\mathcal{L}'_\varepsilon(U))^\circ = \{\bar{x} \in \mathcal{L}'_\varepsilon : \overline{Q}_\varepsilon(\bar{x}) \subset U\}, \quad (\mathcal{L}_\varepsilon(U))^\circ = (\mathcal{L}'_\varepsilon(U))^\circ + \varepsilon\{z_1, \ldots, z_{2^d}\}.$$

We call $Q_\varepsilon(\bar{x})$ for $\bar{x} \in (\mathcal{L}'_\varepsilon(\Omega))^\circ$ an inner cell and set $\Omega_\varepsilon = \bigcup_{\bar{x} \in (\mathcal{L}'_\varepsilon(\Omega))^\circ} Q_\varepsilon(\bar{x})$.

The deformations of our system are mappings $y : \mathcal{L}_\varepsilon \cap \Omega \to \mathbb{R}^d$. Given $x \in \Omega_\varepsilon$ and the corresponding midpoint $\bar{x} \in (\mathcal{L}'_\varepsilon(\Omega))^\circ$ we denote the images of the atoms in $\overline{Q_\varepsilon(x)}$ by $y_i = y(\bar{x} + \varepsilon z_i)$ for $i = 1, \ldots, 2^d$ and view

$$Y(x) = (y_1, \ldots, y_{2^d}). \tag{1.2}$$

as elements of $\mathbb{R}^{d \times 2^d}$. We define the discrete gradient $\bar{\nabla} y(x) \in \mathbb{R}^{d \times 2^d}$ by

$$\bar{\nabla} y := \varepsilon^{-1}(y_1 - \bar{y}, \ldots, y_{2^d} - \bar{y}), \quad \bar{y} := \frac{1}{2^d} \sum_{i=1}^{2^d} y_i. \tag{1.3}$$

In particular, $\bar{\nabla} y$ is a function on Ω_ε, which is constant on each cube $Q_\varepsilon(\bar{x})$, $\bar{x} \in (\mathcal{L}'_\varepsilon(\Omega))^\circ$.

We also need to keep track of the atomic positions within subsets of $\mathcal{Z}$. Therefore, for a given matrix $G = (g_1, \ldots, g_{2^d}) \in \mathbb{R}^{d \times 2^d}$ and $\tilde{\mathcal{Z}} \subset \mathcal{Z}$ we define

$$G[\tilde{\mathcal{Z}}] = (g_j)_{z_j \in \tilde{\mathcal{Z}}} \in \mathbb{R}^{d \times \# \tilde{\mathcal{Z}}}. \tag{1.4}$$

In cells with large deformation it will be convenient to measure the distance of different subsets of the atoms forming the cell. For $G \in \mathbb{R}^{d \times 2^d}$ and $\mathcal{Z}_1, \mathcal{Z}_2 \subset \mathcal{Z}$ we set

$$d(G; \mathcal{Z}_1, \mathcal{Z}_2) := \min \left\{ |g_i - g_j| : z_i \in \mathcal{Z}_1, z_j \in \mathcal{Z}_2 \right\}. \tag{1.5}$$

We now define the set of interaction directions

$$\mathcal{V} = A\{-1, 0, 1\}^d \setminus \{0\}$$

and characterize the crystallographic hyperplanes spanned by the corners of a unit cell by their normal vectors. Let $S^{d-1} = \{\xi \in \mathbb{R}^d : |\xi| = 1\}$ and set

$$\mathcal{P} := \{\xi \in S^{d-1} : \exists u_1, \ldots, u_{d-1} \in \mathcal{V}, \ \mathrm{span}\{u_1, \ldots u_{d-1}\} = \xi^\perp\}.$$

Note that every hyperplane is represented twice in $\mathcal{P}$, by ξ and $-\xi$.

Our basic assumption is that the energy associated to deformations $y : \mathcal{L}_\varepsilon \cap \Omega \to \mathbb{R}^d$ can be written as a sum over cell energies $W_{\mathrm{cell}} : \mathbb{R}^{d \times 2^d} \to [0, \infty]$ in the form

$$E_\varepsilon(y) = \sum_{\bar{x} \in (\mathcal{L}'_\varepsilon(\Omega))^\circ} W_{\mathrm{cell}}(\bar{\nabla} y(\bar{x})). \tag{1.6}$$

For convenience the energy is defined as a sum over the inner cells only as the energy contribution of cells with midpoints lying in $\mathcal{L}'_\varepsilon(\Omega) \setminus (\mathcal{L}'_\varepsilon(\Omega))^\circ$ are negligible in our model for uniaxial extension. We briefly note that W_{cell} is of order one in atomic units and therefore we will have to consider a suitably scaled quantity of E_ε to arrive at macroscopic energy expressions for small ε. This will be discussed in the next section.

Remark 1.1.1. A decomposition as in (1.6) is, in particular, possible for many mass spring models, as will be exemplified in Section 1.4 and Section 3.4: The energy stored in an atomic bonds which lies on a face of more than one unit cell will then be equidistributed to the energy contribution of all adjacent cells. Moreover, energy functionals of the form (1.6) can also incorporate bond angle dependent energy terms.

We let
$$\bar{SO}(d) := \{\bar{R} = RZ : R \in SO(d)\} \subset \mathbb{R}^{d\times 2^d},$$
where Z is as defined in (1.1) and now describe the general assumptions on the cell energy W_{cell} in detail.

Assumption 1.1.2. *(i) $W_{\text{cell}} : \mathbb{R}^{d\times 2^d} \to [0,\infty]$ is invariant under translations and rotations, i.e. for $G \in \mathbb{R}^{d\times 2^d}$ we have*

$$W_{\text{cell}}(G) = W_{\text{cell}}(RG + (c, \ldots, c))$$

for all $R \in SO(d)$ and $c \in \mathbb{R}^d$.

(ii) $W_{\text{cell}}(G) = 0$ if and only if there exists $R \in SO(d)$ and $c \in \mathbb{R}^d$ such that

$$G = RZ + (c, \ldots, c).$$

(iii) W_{cell} is continuous and C^2 in a neighborhood of $\bar{SO}(d)$. The Hessian $Q_{\text{cell}} = D^2W_{\text{cell}}(Z)$ at the identity is positive definite on the complement of the subspace spanned by translations $(c, \ldots, c)$ and infinitesimal rotations HZ, with $H + H^T = 0$.

(iv) If there is a partition $\mathcal{Z} = \mathcal{Z}_1 \dot\cup \ldots \dot\cup \mathcal{Z}_n$ such that $\min_{1\le i<j\le n} d(G; \mathcal{Z}_i, \mathcal{Z}_j)$ is near infinity for $G \in \mathbb{R}^{d\times 2^d}$ then the energy W_{cell} decomposes, i.e. there are $W^{\mathcal{Z}_i} : \mathbb{R}^{d\times \#\mathcal{Z}_i} \to [0,\infty)$, $i = 1, \ldots, n$, and $\beta(z_s, z_t) = \beta(z_t, z_s) \ge 0$ for $z_s, z_t \in \mathcal{Z}$, $z_s \neq z_t$, such that

$$W_{\text{cell}}(G) = \sum_{i=1}^{n} W^{\mathcal{Z}_i}(G[\mathcal{Z}_i]) + \frac{1}{2} \sum_{\substack{1\le i,j\le n \\ i\neq j}} \sum_{z_s\in\mathcal{Z}_i} \sum_{z_t\in\mathcal{Z}_j} \beta(z_s, z_t) + o(1)$$

as $\min_{1\le i<j\le n} d(G; \mathcal{Z}_i, \mathcal{Z}_j) \to \infty$, where the triple sum on the right hand side is strictly positive unless $n = 1$. The components of the energy satisfy $W^{\mathcal{Z}_i}(H[\mathcal{Z}_i]) \le C$ for $H \in \mathbb{R}^{d\times 2^d}$ for some $C > 0$ and

$$W^{\mathcal{Z}_i}(H[\mathcal{Z}_i]) \le C W_{\text{cell}}(H) \tag{1.7}$$

in a neighborhood of $\bar{SO}(d)$.

Note that the above assumptions imply that the quadratic form Q_{cell} satisfies

$$Q_{\text{cell}}(c, \ldots, c) = 0, \quad Q_{\text{cell}}(HZ) = 0 \tag{1.8}$$

for all $c \in \mathbb{R}^d$ and $H \in \mathbb{R}^{d \times d}$ with $H + H^T = 0$. Moreover, there are constants $0 < C_1 \leq C_2$ such that

$$W_{\text{cell}} \leq C_2 \ \text{ and } \ C_1 \leq \liminf_{|G| \to \infty} W_{\text{cell}}(G). \tag{1.9}$$

Partitioning the set of interaction directions as $\mathcal{V} = \mathcal{V}_1 \cup \ldots \cup \mathcal{V}_d$ with

$$\mathcal{V}_k = \left\{ At : t \in \{-1, 0, 1\}^d, \#t = k \right\}$$

for $1 \leq k \leq d$, where $\#t$ denotes the number of non zero entries of the vector $t \in \{-1, 0, 1\}^d$, we may assume that for all $1 \leq k \leq d - 1$ and for all $\nu \in \mathcal{V}_k$ there are $\beta(\nu) \geq 0$ such that

$$\frac{1}{2} \beta(z_s, z_t) = 2^{k-d} \beta(\nu) \text{ for all } z_s, z_t \in \mathcal{Z} : z_s - z_t = \nu. \tag{1.10}$$

Indeed, a bond in ν-direction, $\nu \in \mathcal{V}_k$ is shared by 2^{d-k}, $1 \leq k \leq d - 1$ different cells. If these cells give different interaction energies $\beta^1, \ldots, \beta^{2^{d-k}}$ we would set $\beta(\nu) = \frac{1}{2}(\beta^1 + \ldots + \beta^{2^{d-k}})$ and would replace β^i by $2^{k-d+1}\beta(\nu)$ without affecting the energy. The additional factor of $\frac{1}{2}$ takes account of the fact that every atomic bond is represented twice in $\mathcal{V}$.

Remark 1.1.3. Pair interaction potentials fulfilling assumption (iv) are, e.g., potentials of 'Lennard-Jones-type'. They are characterized by the fact that for large distances the interaction energy is near a fixed positive value. We note that in the case of large deformation the energy reduces to a pair interaction energy neglecting multiple point interactions. This is meaningful as the forces between well separated atoms are governed by dipole interactions, while angle-dependent potentials are needed in order to describe the chemical binding effects for close atoms appropriately. Condition (1.7) is a compatibility condition and ensures that surface terms cannot dominate bulk terms (cf. [65]).

1.2 Boundary values and scaling

We are interested in the behavior of the specimen under uniaxial extension, say in $\mathbf{e}_1$-direction. In particular, we would like to investigate when and how the body breaks, i.e.,

(1) at which value of the boundary displacement energetic minimizers are no longer elastic deformations but exhibit cracks and

(2) if indeed it is most favorable for the cracks to separate the body along specific crystallographic hyperplanes.

In order to avoid geometric artefacts and complicated crack geometries, we will therefore assume that the specimen is 'long enough' so that it is possible for the body to completely break apart along crystallographic hyperplanes not passing through the boundary parts $B_1 = \{x \in \overline{\Omega} : x_1 = 0\}$ and $B_2 = \{x \in \overline{\Omega} : x_1 = l_1\}$.

More precisely, we choose $l_1 \geq L = L(\sqrt{A^T A}, W_{\mathrm{cell}}, l_2, \ldots, l_d)$ (see (2.2) below). In particular, we will see that L may be chosen independently of the orientation of the lattice. We briefly note that, under additional symmetry conditions on the cell energy, L is independent of W_{cell} and thus the minimum length only depends on the geometry of the problem. Such a symmetry condition is satisfied, e.g., if springs associated to the set $\mathcal{V}_k$ are of 'the same type', i.e. give the same interaction energy for large expansion (see the examples in Section 3.4). We mention that one may overcome this technical difficulty alternatively by imposing periodic boundary conditions or considering infinite crystals.

Due to the discreteness of the underlying atomic lattice the boundary conditions of uniaxial extension have to be imposed in atomistically small neighborhoods of B_1 and B_2 as otherwise unphysical boundary effects may occur, in particular cracks near the boundary might become energetically more favorable. Define

$$l_A = \sum_{j=1}^{d} |v_j \cdot \mathbf{e}_1| \tag{1.11}$$

and let $B_1^\varepsilon = \{x \in \overline{\Omega} : x_1 \leq 2 l_A \varepsilon\}$ and $B_2^\varepsilon = \{x \in \overline{\Omega} : x_1 \geq l_1 - 2 l_A \varepsilon\}$. For $a_\varepsilon > 0$ we set

$$\begin{aligned} \mathcal{A}(a_\varepsilon) = \{y = (y^1, \ldots, y^d)^T : \mathcal{L}_\varepsilon \cap \Omega \to \mathbb{R}^d : \\ y^1(x) = (1 + a_\varepsilon) x_1 \text{ for } x \in B_1^\varepsilon \cup B_2^\varepsilon\}. \end{aligned} \tag{1.12}$$

There is some arbitrariness in the implementation of boundary conditions. A possible alternative is, e.g.,

$$y^1(x) = x_1 \text{ for } x \in B_1^\varepsilon \quad \text{and} \quad y^1(x) = x_1 + a_\varepsilon l_1 \text{ for } x \in B_2^\varepsilon. \tag{1.13}$$

We remark that such different choices do not change the results of our analysis and we will say that a deformation satisfies the boundary condition if either (1.12) or (1.13) is satisfied. Note that there are no assumptions on the other $d-1$ components of the deformation y near the boundaries B_1 and B_2 of the boundary displacement, i.e. the atoms may 'slide along the boundary'.

There are two obvious choices for deformations satisfying the boundary conditions: The homogeneous elastic deformation $y^{\mathrm{el}}(x) = (1+a_\varepsilon)x$ and a cracked body deformation y^{cr}, which has the form $y^{\mathrm{cr}} = x\chi_{\Omega_1} + (x + a_\varepsilon l_1 \mathbf{e}_1)\chi_{\Omega_2}$, where the sets

Ω_1 and Ω_2 form a partition of Ω and are seperated by some hyperplane (or manifold) intersecting $\overline{\Omega}$ in the set $\overline{\Omega} \setminus (B_1^\varepsilon \cup B_2^\varepsilon)$. Noting that $W_{\text{cell}} \sim \operatorname{dist}^2(\cdot, \bar{SO}(d))$ in a neighborhood of $\bar{SO}(d)$ and recalling (1.9) it is not hard to see that for $\varepsilon \ll a_\varepsilon \ll 1$ we have

$$E_\varepsilon(y^{\text{el}}) \sim \varepsilon^{-d} a_\varepsilon^2, \qquad E_\varepsilon(y^{\text{cr}}) \sim \varepsilon^{1-d}.$$

We are particularly interested in the regime where both of these energy values are of the same order, i.e. $E_\varepsilon(y^{\text{el}}) \sim E_\varepsilon(y^{\text{cr}}) \sim \varepsilon^{1-d}$. This implies $a_\varepsilon \sim \sqrt{\varepsilon}$. As alluded to above in order to arrive at finite and nontrivial energies in the limit $\varepsilon \to 0$, we rescale E_ε to $\mathcal{E}_\varepsilon := \varepsilon^{d-1} E_\varepsilon$.

1.3 Limiting minimal energy and cleavage laws

We now state our main result about the limiting minimal energy as $\varepsilon \to 0$ when $a_\varepsilon/\sqrt{\varepsilon} \to a \in [0, \infty]$. In particular, we will see that the minimal energy is given by elastic deformations for a_ε up to some critical value a_{crit} of boundary displacements and by cleavage along a specific crystallographic hyperplane beyond this value. Before we state the theorem we introduce two constants occurring in the limiting minimal energy which describe the stiffness and toughness of the material. The constant in the fracture energy is given by

$$\beta_A = \min_{\xi \in \mathbb{R}^d \setminus \{0\}} \frac{\sum_{\nu \in \mathcal{V}} \beta(\nu) |\nu \cdot \xi|}{|\mathbf{e}_1 \cdot \xi|} \tag{1.14}$$

with $\beta(\nu)$ as in (1.10). In Lemma 2.1.3 below we will show that the minimum β_A is attained for some $\xi \in \mathcal{P}$. In particular, this means that cleavage along a crystallographic hyperplane is energetically favorable. Concerning the elastic regime we define a *reduced energy* for the quadratic form Q_{cell} by

$$\tilde{Q}(r) := \min \left\{ Q_{\text{cell}}(e(G) \cdot Z) : G \in \mathbb{R}^{d \times d}, g_{11} = r \right\} \tag{1.15}$$

for $r \in \mathbb{R}$, $e(G) = \frac{1}{2}(G + G^T)$. As the problem is quadratic with a linear constraint, it is not hard so see that $\tilde{Q}(r) = \alpha_A r^2$ for a specific $\alpha_A > 0$. We will see that

$$\alpha_A = \frac{\det(\mathcal{Q})}{\det(\hat{\mathcal{Q}})}, \tag{1.16}$$

where, roughly speaking, $\mathcal{Q}$ is the projection of Q_{cell} onto the linear subspace orthogonal to infinitesimal rotations and $\hat{\mathcal{Q}}$ arises from $\mathcal{Q}$ by cancellation of the first row and column. This will be stated more precisely in Lemma 2.1.2 below.

Theorem 1.3.1. *Let $l_1 \geq L(\sqrt{A^T A}, W_{\text{cell}}, l_2, \dots, l_d)$ and suppose $a_\varepsilon/\sqrt{\varepsilon} \to a \in [0, \infty]$. The limiting minimal energy is given by*

$$\mathcal{E}_{\text{lim}}(a) := \lim_{\varepsilon \to 0} \inf \{ \mathcal{E}_\varepsilon(y) : y \in \mathcal{A}(a_\varepsilon) \} = \frac{\prod_{j=2}^d l_j}{\det A} \min \left\{ \frac{1}{2} l_1 \alpha_A a^2, \beta_A \right\}.$$

A detailed proof of this result will be given in Section 3.3 and is content of the paper [47]. As discussed above, for $a \in \{0, \infty\}$ either the elastic or the fracture regime is energetically favorable. The more interesting case is $a \in (0, \infty)$ where both energies are of the same order. In particular, we are interested in the behavior of the specimen when a is near the critical value of boundary displacements

$$a_{\text{crit}} = \sqrt{\frac{2\beta_A}{l_1 \alpha_A}}. \tag{1.17}$$

We briefly indicate asymptotically optimal configurations. In the subcritical case $a \leq a_{\text{crit}}$ we consider the sequence of configurations

$$y_\varepsilon^{\text{el}}(x) = x + \bar{F}(a_\varepsilon)\, x, \quad x \in \mathcal{L}_\varepsilon \cap \Omega,$$

where $\bar{F}(a_\varepsilon)$ is the solution of the minimization problem (1.15) with $r = a_\varepsilon$ (see Lemma 2.1.2 below). The deformations behave purely elastically and as we will see in Corollary 1.5.2 and in the examples in Section 3.4 show elongation in $\mathbf{e}_1$-direction and contraction in the other space directions, a manifestation of the Poisson-effect. Moreover, the configurations illustrate the validity of the Cauchy-Born-rule in this regime as each individual atom follows the macroscopic deformation gradient. In the supercritical case $a \geq a_{\text{crit}}$ there is some $\xi \in \mathcal{P}$ and $c \in \mathbb{R}$ such that the hyperplane $\Pi = \{x \in \mathbb{R}^d : x \cdot \xi = c\}$ satisfies $\Pi \cap \overline{\Omega} \subset \overline{\Omega} \setminus (B_1^\varepsilon \cup B_2^\varepsilon)$ and the configurations

$$y_\varepsilon^{\text{cr}}(x) = \begin{cases} x, & x \cdot \xi < c, \\ x + l_1 a_\varepsilon \mathbf{e}_1, & x \cdot \xi > c, \end{cases} \quad x \in \mathcal{L}_\varepsilon \cap \Omega, \tag{1.18}$$

are asymptotically optimal. As $\xi \in \mathcal{P}$, we conclude that Π is a crystallographic hyperplane, as desired.

Let us also remark that our class of atomistic interactions is rich enough to model any non-degenerate linearly elastic energy density, respectively, any preferred cleavage normal in $\mathcal{P}$ (not perpendicular to $\mathbf{e}_1$) in the continuum limit: In the elastic regime, this has in fact been observed in [20, Prop. 1.10]. Now suppose that $\xi \in \mathcal{P}$ with $\xi \cdot \mathbf{e}_1 \neq 0$ is orthogonal to $\text{span}\{u_1, \ldots, u_{d-1}\}$ with $u_1, \ldots, u_{d-1} \in \mathcal{V}$. Then, if $\beta(u_1), \ldots, \beta(u_{d-1})$ are much larger than $\beta(\nu)$ for all $\nu \in \mathcal{V} \setminus \{u_1, \ldots, u_{d-1}\}$, it is elementary to see that the minimum of

$$\min_{\varsigma \in \mathcal{P}} \frac{\sum_{\nu \in \mathcal{V}} \beta(\nu) |\nu \cdot \varsigma|}{|\mathbf{e}_1 \cdot \varsigma|},$$

is attained at $\varsigma = \xi$, so that indeed ξ defines a crack normal for an asymptotically optimal configuration as in (1.18).

The main idea in the proof of Theorem 1.3.1 is based on the derivation of a lower comparison potential for a certain cell energy depending on the expansion in $\mathbf{e}_1$-direction and on the application of a slicing argument. Testing either with elastic deformations or configurations forming jumps along specific hyperplanes as given above, we will then see that this lower bound is sharp. Actually, it will turn out that the lower bound coincides with the reduced energy $\tilde{Q}$ in the regime of infinitesimal elasticity. In contrast to the local definition of $\tilde{Q}$, however, it is in general not convenient to optimize the energy W_{cell} of single cells individually as it is geometrically nonlinear and therefore, due to possible rotations, the corresponding minimizer for one cell might not be compatible with deformations defined on the whole domain. As a remedy we will introduce a mesoscopic localization technique and will consider 'large cells' defined on a mesoscopic scale $\varepsilon^{\frac{3d-1}{3d}}$. This main technical result is addressed in Section 3.2.

1.4 A specific model: The triangular lattice in two dimensions

We present a planar model where the atoms in the reference configuration are given by the portion of a triangular lattice and only interact with their nearest neighbors. In Section 3.4 we will see that the model is admissible in the sense of Assumption 1.1.2. It serves as the most basic non-trivial example to which our theory applies. In particular, it (1) is frame indifferent in its vector-valued arguments in more than one dimension, (2) gives rise to non-degenerate elastic bulk terms and (3) leads to surface contributions sensitive to the crack geometry with competing crystallographic lines. The goal is to perform a much more complete analysis of this model including a detailed characterization of low energy configurations and the identification of a limiting variational problem.

Let $\mathcal{L}$ denote the rotated triangular lattice

$$\mathcal{L} = R_{\mathcal{L}} \begin{pmatrix} 1 & \frac{1}{2} \\ 0 & \frac{\sqrt{3}}{2} \end{pmatrix} \mathbb{Z}^2 = \{\lambda_1 \mathbf{v}_1 + \lambda_2 \mathbf{v}_2 : \lambda_1, \lambda_2 \in \mathbb{Z}\},$$

where $R_{\mathcal{L}} \in SO(2)$ is some rotation and $\mathbf{v}_1$, $\mathbf{v}_2$ are the lattice vectors $\mathbf{v}_1 = R_{\mathcal{L}}\mathbf{e}_1$ and $\mathbf{v}_2 = R_{\mathcal{L}}(\frac{1}{2}\mathbf{e}_1 + \frac{\sqrt{3}}{2}\mathbf{e}_2)$, respectively. Without loss of generality we assume that $R_{\mathcal{L}} = \begin{pmatrix} \cos\phi & -\sin\phi \\ \sin\phi & \cos\phi \end{pmatrix}$ for $\phi \in [0, \frac{\pi}{3})$. We collect the basic lattice vectors in the set $\mathcal{V} = \{\mathbf{v}_1, \mathbf{v}_2, \mathbf{v}_3\}$, where $\mathbf{v}_3 = \mathbf{v}_2 - \mathbf{v}_1$. The region $\Omega = (0, l) \times (0, 1) \subset \mathbb{R}^2$, $l > 0$, is considered the macroscopic region occupied by the body under investigation, where for the sake of simplicity we set $l_1 = l$ and $l_2 = 1$. As before, to avoid geometric artefacts, we will assume that $l > \frac{1}{\sqrt{3}}$, so that it is possible for the body to completely break apart along lines parallel to $\mathbb{R}\mathbf{v}_1$, $\mathbb{R}\mathbf{v}_2$ or $\mathbb{R}\mathbf{v}_3$ not passing through the left or right boundaries.

In the reference configuration the positions of the specimen's atoms are given by the points of the scaled lattice $\mathcal{L}_\varepsilon = \varepsilon\mathcal{L}$ that lie within Ω. The deformations of our system are mappings $y : \mathcal{L}_\varepsilon \cap \Omega \to \mathbb{R}^2$ and the energy associated to such a deformation y is assumed to be given by nearest neighbor interactions as

$$E_\varepsilon(y) = \frac{1}{2} \sum_{\substack{x,x' \in \mathcal{L}_\varepsilon \cap \Omega \\ |x-x'|=\varepsilon}} W\left(\frac{|y(x) - y(x')|}{\varepsilon}\right). \tag{1.19}$$

Note that the scaling factor $\frac{1}{\varepsilon}$ in the argument of W takes account of the scaling of the interatomic distances with ε. The pair interaction potential $W : [0,\infty) \to [0,\infty]$ is supposed to be of 'Lennard-Jones-type' (cf. Remark 1.1.3):

(i) $W \geq 0$ and $W(r) = 0$ if and only if $r = 1$.

(ii) W is continuous on $[0,\infty)$ and C^2 in a neighborhood of 1 with $\alpha := W''(1) > 0$.

(iii) $\lim_{r\to\infty} W(r) = \beta > 0$.

In order to obtain fine estimates on limiting energies and configurations we will also consider the following stronger versions of hypotheses (ii) and (iii):

(ii') W is continuous on $[0,\infty)$ and C^4 in a neighborhood of 1 with $\alpha := W''(1) > 0$ and arbitrary $\alpha' := W'''(1)$.

(iii') $W(r) = \beta + O(r^{-2})$ as $r \to \infty$,

which is still satisfied, e.g., by the classical Lennard-Jones potential.

In order to analyze the passage to the limit as $\varepsilon \to 0$ it will be useful to interpolate and rewrite the energy as an integral functional. Let $\mathcal{C}_\varepsilon$ be the set of equilateral triangles $\triangle \subset \Omega$ of sidelength ε with vertices in $\mathcal{L}_\varepsilon$ and define $\Omega_\varepsilon = \bigcup_{\triangle \in \mathcal{C}_\varepsilon} \triangle$. By $\tilde{y} : \Omega_\varepsilon \to \mathbb{R}^2$ we denote the interpolation of y, which is affine on each $\triangle \in \mathcal{C}_\varepsilon$. The derivative of $\tilde{y}$ is denoted by $\nabla\tilde{y}$, whereas we write $(y)_\triangle$ for the (constant) value of the derivative on a triangle $\triangle \in \mathcal{C}_\varepsilon$. Then (1.19) can be rewritten as

$$\begin{aligned} E_\varepsilon(y) &= \sum_{\triangle \in \mathcal{C}_\varepsilon} W_\triangle((\tilde{y})_\triangle) + E_\varepsilon^{\text{boundary}}(y) \\ &= \frac{4}{\sqrt{3}\varepsilon^2} \int_{\Omega_\varepsilon} W_\triangle(\nabla\tilde{y})\, dx + E_\varepsilon^{\text{boundary}}(y), \end{aligned} \tag{1.20}$$

where

$$W_\triangle(F) = \frac{1}{2}\Big(W(|F\mathbf{v}_1|) + W(|F\mathbf{v}_2|) + W(|F(\mathbf{v}_3)|)\Big). \tag{1.21}$$

(Note that $|\triangle| = \sqrt{3}\varepsilon^2/4$.) Here the boundary term is the sum of pair interaction energies $\frac{1}{4}W(\frac{|y(x)-y(x')|}{\varepsilon})$ or $\frac{1}{2}W(\frac{|y(x)-y(x')|}{\varepsilon})$ over nearest neighbor pairs which form the side of only one or non triangle in $\mathcal{C}_\varepsilon$, respectively. As above, in order to obtain finite and nontrivial energies in the limit $\varepsilon \to 0$, we accordingly rescale E_ε to $\mathcal{E}_\varepsilon := \varepsilon E_\varepsilon$.

We impose the same boundary conditions of uniaxial extension with possible alternatives as described in (1.12) and (1.13), where we set $B_1^\varepsilon = \{x \in \overline{\Omega} : x_1 \le \varepsilon\}$ and $B_2^\varepsilon = \{x \in \overline{\Omega} : x_1 \ge l - \varepsilon\}$. In the special case $\phi = 0$ we will in addition assume that there is an upper bound R_0 on the elongation of every atomic bond in a small $\psi(\varepsilon)$-neighborhood of the lateral boundaries of width $\psi(\varepsilon) > 0$ with $\varepsilon \ll \psi(\varepsilon) \ll 1$:

$$|y(x) - y(x')| \le R_0 \varepsilon \text{ if } x_1, x_1' \le \psi(\varepsilon) \text{ or } x_1, x_1' \ge l - \psi(\varepsilon). \tag{1.22}$$

Without such an assumption, in the general low energy regime to be considered later, for $\phi = 0$ the boundary values are not strong enough to prevent the specimen from breaking on the boundary into a large amount of completely separated components, rendering the system too sensitive to unphysical boundary effects.

Conceivable alternative implementations of the boundary conditions as alluded to above will then result in energy changes of order $O(\varepsilon)$. We will account for all such possibilities by characterizing not only energy minimizing configurations, but more generally all configurations which are energy minimizing up to an error term of order $O(\varepsilon)$.

Besides describing a basic experiment on elastic bodies, the assumption on the boundary values allows for a direct application of our results to the stability analysis of nanotubes: If the rotation $R_\mathcal{L}$ and the length l are such that for a sequence $\varepsilon_k \to 0$ the translated lattice $\varepsilon_k \mathcal{L} + (l, 0)$ concides with the original lattice $\varepsilon_k \mathcal{L}$, we may view the system as an atomistic nanotube with macroscopic region $\frac{l}{2\pi} S^1 \times (0,1)$. (Note that for small ε_k the bending energy contributions when rolling up $(0,l) \times (0,1)$ into a cylinder are negligible as this mapping is an isometric immersion and thus infinitesimally rigid.) Imposing periodic boundary conditions, for arbitrary $l > 0$ our system then models deformations of a nanotube subject to expansion of the diameter.

We first provide the basic energetic cleavage law for the triangular lattice.

Theorem 1.4.1. *Let $l > \frac{1}{\sqrt{3}}$. Suppose $a_\varepsilon/\sqrt{\varepsilon} \to a \in [0, \infty]$. The limiting minimal energy is given by*

$$\liminf_{\varepsilon \to 0} \{\mathcal{E}_\varepsilon(y) : y \in \mathcal{A}(a_\varepsilon)\} = \min \left\{ \frac{\alpha l}{\sqrt{3}} a^2, \frac{2\beta}{\sin(\phi + \frac{\pi}{3})} \right\}. \tag{1.23}$$

Although the theory of Section 1.3 applies, we have explicitly formulated the limiting minimal energy for our model problem since we want to give an independent proof as a first approach to the general cleavage law (see Section 3.1). Indeed, the result can be shown in a comparatively elementary way by resorting to slicing methods and convexity estimates in combination with a suitable projection technique. By way of contrast, the analogous result in arbitrary dimensions with general lattices and interaction potentials given in Theorem 1.3.1 is much more involved as it requires (1) new projection estimates for the size of cracks in the specimen, (2) a thorough analysis of all possible crack modes of a possible lattice unit cell and (3) full dimensional analysis of an anisotropic mesoscopic auxiliary problem in various regimes.

In what follows, we specialize to sequences $a_\varepsilon = \sqrt{\varepsilon}a$ for the sake of simplicity. If the assumptions (ii') and (iii') on W hold, we have the following sharp estimate on the discrete minimal energies up to error terms of the order of surface contributions.

Theorem 1.4.2. *For ε small the discrete minimal energy is given by*

$$\inf \mathcal{E}_\varepsilon = \min\left\{\frac{\alpha l}{\sqrt{3}}a^2 + \frac{[6\alpha + 7\alpha' - 2(3\alpha - \alpha')\cos(6\phi)]l}{27\sqrt{3}}\sqrt{\varepsilon}a^3, \frac{2\beta}{\sin(\phi + \frac{\pi}{3})}\right\} + O(\varepsilon).$$

Thus, while the zeroth order contributions in the elastic regime are isotropic, the higher order contributions as well as the fracture energy explicitly depend on the lattice orientation angle ϕ. A proof of this fine energy estimate will be given in Section 4.1. The results presented in this and the following section are content of the work [46]. (Parts of the results published in [46] were already obtained in [45].)

1.5 Limiting minimal configurations

For notational convenience we let $\gamma = \max\{|\mathbf{v}_1 \cdot \mathbf{e}_2|, |\mathbf{v}_2 \cdot \mathbf{e}_2|, |\mathbf{v}_3 \cdot \mathbf{e}_2|\}$ and $\mathbf{v}_\gamma \in \mathcal{V}$ such that $\gamma = |\mathbf{v}_\gamma \cdot \mathbf{e}_2|$. We note that $\gamma = \sin(\phi + \frac{\pi}{3}) = |\mathbf{v}_\gamma \cdot \mathbf{e}_2| = \mathbf{v}_\gamma \cdot \mathbf{e}_2$ takes values in $[\sqrt{3}/2, 1]$ and that $\mathbf{v}_\gamma$ is unique if $\phi \neq 0$, i.e., $\gamma > \sqrt{3}/2$.

Our analysis of the limiting minimal energy so far showed that in terms of the critical boundary displacement

$$a_{\text{crit}} = \sqrt{\frac{2\sqrt{3}\beta}{\alpha\gamma l}}$$

the limit is attained for homogeneously deformed configurations if $a \leq a_{\text{crit}}$ and for configurations cracked along lines parallel to $\mathbb{R}\mathbf{v}_\gamma$, if $a \geq a_{\text{crit}}$. However, it falls short of showing that in fact these configurations are the only possibilities to obtain asymptotically optimal energies. Indeed, in the special case that $\mathbf{v}_\gamma$ is not unique, the limit is also attained if the crack takes a serrated course parallel to $\mathbb{R}(\frac{1}{2}, \frac{\sqrt{3}}{2})^T$ or $\mathbb{R}(-\frac{1}{2}, \frac{\sqrt{3}}{2})^T$.

Our next result shows that energy minimizing configurations converge to a homogeneous continuum deformation for subcritical boundary values, while in the supercritical case they converge to a continuum deformation which is completely cracked and does not store elastic energy. If $\phi \neq 0$ and hence $\mathbf{v}_\gamma$ is unique, the crack path follows the optimal crystallographic line. For $\phi = 0$ such a cleavage behavior fails in general. Nevertheless, we obtain an explicit characterization of all possible limiting crack shapes in this case as well in terms of Lipschitz curves whose tangent vector lies a.e. in the cone generated by $(-\frac{1}{2}, \frac{\sqrt{3}}{2})$ and $(\frac{1}{2}, \frac{\sqrt{3}}{2})$.

The basic idea behind our reasoning will be to 'count' the number of 'broken' springs, i.e. the springs intersected transversally by the crack path. We see that the springs broken by a crack line $(p, 0) + \mathbb{R}\mathbf{v}_\gamma$ do not overlap in the projection onto the x_2-axis and the length of the projection of two adjacent broken springs equals $\varepsilon\gamma$. This leads to a fracture energy of approximately $\frac{2\beta}{\gamma}$. If in the generic case $\phi \neq 0$ we assume that the cleavage is not parallel to $\mathbb{R}\mathbf{v}_\gamma$ we conclude that some springs in $\mathbf{v}_\gamma$ direction must be broken, too. If we consider the adjacent triangles of such a spring and their neighbors we find that the projection onto the x_2-axis of broken springs overlap. A careful analysis of this phenomenon then shows that every broken spring in $\mathbf{v}_\gamma$ direction 'costs' an additional energy of $\approx 2\varepsilon\beta\frac{P(\gamma)}{\gamma}$, where $P(\gamma)$ is the geometrical factor

$$P(\gamma) = \frac{1}{2}\left(1 - \sqrt{3}\frac{\sqrt{1-\gamma^2}}{\gamma}\right). \tag{1.24}$$

(Note that $P(\gamma) = 0 \Leftrightarrow \gamma = \frac{\sqrt{3}}{2} \Leftrightarrow \phi = 0$ in accordance to the above considerations.) For the special case $\phi = 0$ we provide a similar counting argument.

In order to give a precise meaning to the convergence of discrete to continuum deformations, to each discrete deformation $y : \mathcal{L}_\varepsilon \to \mathbb{R}^2$ we assign – as mentioned above – the affine interpolation $\tilde{y}$ on each triangle $\triangle \in \mathcal{C}_\varepsilon$. Accordingly, to the rescaled discrete displacements $u : \mathcal{L}_\varepsilon \to \mathbb{R}^2$ with $y = \mathbf{id} + \sqrt{\varepsilon}u$ ($\mathbf{id}$ denoting the identity mapping $\mathbf{id}(x) = x$) we define $\tilde{u}$ to be its affine interpolation on each triangle $\triangle \in \mathcal{C}_\varepsilon$.

In the cracked regime we may of course only hope for a unique limiting deformation up to translation of the crack path. However, without an additional mild extra assumption on the admissible discrete configurations or their energy even this cannot hold true, as apart from the crack, parts of the specimen could flip their orientation and fold onto other parts on the body at zero energy. In order to avoid such unphysical behavior we add a frame indifferent penalty term $\chi \geq 0$ to $W_\triangle$ with $\chi \geq c_\chi > 0$ in a neighborhood of $O(2) \setminus SO(2)$ and $\chi \equiv 0$ in a neighborhood of $SO(2)$ and ∞, which in particular does not change the energy response in the linear elastic and in the fracture regime:

$$W_{\triangle,\chi}(F) = W_\triangle(F) + \chi(F). \tag{1.25}$$

For instance, an admissible choice for χ is the local orientation preserving condition in the elastic regime

$$\chi(F) = \begin{cases} 0, & \text{if } \det(F) > 0 \text{ or } |F| > R, \\ \infty, & \text{if } \det(F) \le 0 \text{ and } |F| \le R, \end{cases} \tag{1.26}$$

for some threshold $R \gg 1$. (Also $1 \le R = R(\varepsilon) \ll \frac{1}{\sqrt{\varepsilon}}$ would be admissible.) We remark that such an infinitely strong penalization of deformation gradients with non-positive determinant is widely used in the elastic models. Allowing for fracture, however, a penalization of orientation reversion between different cracked parts of the body is no longer physically justifiable, whence we set $\chi = 0$ for very large deformation gradients. We set

$$\mathcal{E}_\varepsilon^\chi(y) = \frac{4}{\sqrt{3}\varepsilon} \int_{\Omega_\varepsilon} W_{\triangle,\chi}(\nabla \tilde{y})\, dx + \varepsilon E_\varepsilon^{\text{boundary}}(y), \tag{1.27}$$

for $y \in \mathcal{A}(a_\varepsilon)$. More generally than a sequence of minimizers we will consider sequences (y_ε) of almost minimizers that satisfy

$$\mathcal{E}_\varepsilon^\chi(y_\varepsilon) = \inf\{\mathcal{E}_\varepsilon^\chi(y) : y \in \mathcal{A}(a_\varepsilon)\} + O(\varepsilon). \tag{1.28}$$

For those deformations we will show in Section 4:

Theorem 1.5.1. *Assume that W satisfies (i), (ii') and (iii'). Let $a_\varepsilon = \sqrt{\varepsilon}a$, $a \neq a_{\text{crit}}$ and suppose (y_ε) satisfies (1.28). Let u_ε such that $y_\varepsilon = \mathbf{id} + \sqrt{\varepsilon}u_\varepsilon$. Then there exist $\bar{u}_\varepsilon : \Omega \to \mathbb{R}^2$ with $|\{x \in \Omega_\varepsilon : \bar{u}_\varepsilon(x) \neq \tilde{u}_\varepsilon(x)\}| = O(\varepsilon)$ such that:*

(i) If $a < a_{\text{crit}}$, then there is a sequence $s_\varepsilon \in \mathbb{R}$ such that

$$\|\bar{u}_\varepsilon - (0, s_\varepsilon) - F^a \cdot \|_{H^1(\Omega)} \to 0,$$

where $F^a = \begin{pmatrix} a & 0 \\ 0 & -\frac{a}{3} \end{pmatrix}$.

(ii) If $a > a_{\text{crit}}$ and $\phi \neq 0$, then there exist sequences $p_\varepsilon \in (0, l)$, $s_\varepsilon, t_\varepsilon \in \mathbb{R}$ such that $(p_\varepsilon, 0) + \mathbb{R}\mathbf{v}_\gamma$ intersects both the segments $(0, l) \times \{0\}$ and $(0, l) \times \{1\}$ and, for the parts to the left and right of $(p_\varepsilon, 0) + \mathbb{R}\mathbf{v}_\gamma$

$$\Omega^{(1)} := \left\{x \in \Omega : 0 < x_1 < p_\varepsilon + \tfrac{\mathbf{v}_\gamma \cdot e_1}{\mathbf{v}_\gamma \cdot e_2} x_2\right\} \quad \textit{and}$$
$$\Omega^{(2)} := \left\{x \in \Omega : p_\varepsilon + \tfrac{\mathbf{v}_\gamma \cdot e_1}{\mathbf{v}_\gamma \cdot e_2} x_2 < x_1 < l\right\},$$

respectively, we have

$$\|\bar{u}_\varepsilon - (0, s_\varepsilon)\|_{H^1(\Omega^{(1)})} + \|\bar{u}_\varepsilon - (al, t_\varepsilon)\|_{H^1(\Omega^{(2)})} \to 0.$$

(iii) If $a > a_{\text{crit}}$ and $\phi = 0$, then there exist sequences of Lipschitz functions $g_\varepsilon : (0,1) \to (0,l)$ satisfying $g_\varepsilon' = \pm\frac{1}{\sqrt{3}}$ a.e. such that for the parts to the left and right of graph(g_ε)

$$\Omega^{(1)}[g_\varepsilon] := \{x \in \Omega : 0 < x_1 < g_\varepsilon(x_2)\} \quad \textit{and}$$
$$\Omega^{(2)}[g_\varepsilon] := \{x \in \Omega : g_\varepsilon(x_2) < x_1 < l\},$$

respectively, we have

$$\|\bar{u}_\varepsilon - (0, s_\varepsilon)\|_{H^1(\Omega^{(1)}[g_\varepsilon])} + \|\bar{u}_\varepsilon - (al, t_\varepsilon)\|_{H^1(\Omega^{(2)}[g_\varepsilon])} \to 0,$$

for suitable sequences $s_\varepsilon, t_\varepsilon \in \mathbb{R}$.

As a consequence, we obtain a complete characterization of limiting continuum displacements, when no mass leaks to infinity.

Corollary 1.5.2. *Under the assumptions and with the notation of Theorem 1.5.1, if $\sup_\varepsilon \|u_\varepsilon\|_\infty < \infty$, up to passing to subsequences, $\tilde{u}_\varepsilon \to u$ in measure where*

(i) if $a < a_{\text{crit}}$, $u(x) = F^a x + (0,s)$ for some constant $s \in \mathbb{R}$,

(ii) if $a > a_{\text{crit}}$ and $\phi \neq 0$, $$u(x) = \begin{cases} (0,s), & \textit{for } x \textit{ to the left of } (p,0) + \mathbb{R}\mathbf{v}_\gamma, \\ (al,t), & \textit{for } x \textit{ to the right of } (p,0) + \mathbb{R}\mathbf{v}_\gamma, \end{cases}$$
for constants $s, t \in \mathbb{R}$ and $p \in (0,l)$ such that $(p,0) + \mathbb{R}\mathbf{v}_\gamma$ intersects both the segments $(0,l) \times \{0\}$ and $(0,l) \times \{1\}$,

(iii) if $a > a_{\text{crit}}$ and $\phi = 0$, $$u(x) = \begin{cases} (0,s), & \textit{if } 0 < x_1 < g(x_2), \\ (al,t), & \textit{if } g(x_2) < x_1 < l, \end{cases}$$
for a Lipschitz function $g : (0,1) \to [0,l]$ with $|g'| \leq \frac{1}{\sqrt{3}}$ a.e. and constants $s, t \in \mathbb{R}$.

Conversely, for every u as given in the cases (i)-(iii) there is a minimizing sequence (y_ε) satisfying (1.28) and $\tilde{u}_\varepsilon \to u$ in measure.

We close this section emphasizing that all the optimal configurations found in Theorem 1.5.1 and Corollary 1.5.2 by minimizing the energy without a priori assumptions show purely elastic behavior in the subcritical case and complete fracture in the supercritical regime. In particular, the elastic minimizer in (i) shows elongation a in $\mathbf{e}_1$-direction and compression $-\frac{a}{3}$ in the perpendicular $\mathbf{e}_2$-direction, a manifestation of the Poisson effect (with Poisson ratio $\frac{1}{3}$), which cannot be derived in scalar valued models. On the other hand, the crack minimizer in (ii) for $\phi \neq 0$ is broken parallel to $\mathbb{R}\mathbf{v}_\gamma$ which proves that cleavage occurs along crystallographic lines, while we see that cleavage in the symmetric case $\phi = 0$ in general fails.

1.6 Limiting variational problem

We finally address the more general question if not only the minimal values or the minimizers but the whole energy functionals (1.19) converge to a continuum energy functional in a variational sense. We will analyze the limiting problem independently of its discrete approximations and will also discuss an application to fractured magnets in an external field. As this analysis not only allows for the derivation of cleavage laws for brittle crystals, we will treat the problem in a slightly more general setting. The results announced here are proved in Section 5 and can be also found in [48].

The macroscopic region $\Omega \subset \mathbb{R}^2$ occupied by the body is now supposed to be a bounded domain with Lipschitz boundary. The deformation are again mappings $y : \mathcal{L}_\varepsilon \cap \Omega \to \mathbb{R}^2$ with the associated energy defined in (1.19). Our convergence result will be formulated in terms of rescaled displacement fields $u = \frac{1}{\sqrt{\varepsilon}}(y - \mathbf{id})$ and we write with a slight abuse of notation

$$\mathcal{E}_\varepsilon(u) := \mathcal{E}_\varepsilon(y) = \varepsilon E_\varepsilon(y) = \varepsilon E_\varepsilon(\mathbf{id} + \sqrt{\varepsilon} u).$$

Likewise we consider the functionals $\mathcal{E}_\varepsilon^\chi$ which arise from $\mathcal{E}_\varepsilon$ by replacing $W_\triangle$ by $W_{\triangle,\chi} = W_\triangle + \chi$. We impose the following more general boundary conditions: Assume that $\tilde{\Omega} \supset \Omega$ is a bounded, open domain in $\mathbb{R}^2$ with Lipschitz boundary defining the Dirichlet boundary $\partial_D \Omega = \partial\Omega \cap \tilde{\Omega}$ of Ω. For (the continuous representative of) $g \in W^{1,\infty}(\tilde{\Omega})$ we define the class of discrete displacements assuming the boundary value g on $\partial_D \Omega$ as

$$\mathcal{A}_g = \{u : \mathcal{L}_\varepsilon \cap \tilde{\Omega} \to \mathbb{R}^2 : u(x) = g(x) \text{ for } x \in \mathcal{L}_\varepsilon \cap \Omega_{D,\varepsilon}\}, \tag{1.29}$$

where $\Omega_{D,\varepsilon} := \{x \in \tilde{\Omega} : \mathrm{dist}(x, \partial_D\Omega) \le \varepsilon\} \cup (\tilde{\Omega} \setminus \Omega)$. Note that in contrast to (1.12) the boundary values are formulated in terms of the displacement field. Moreover, $\mathcal{E}_\varepsilon(u)$, which in fact only depends on the restriction $u|_\Omega$, does not depend on the particular choice of $\tilde{\Omega}$ and on $g|_{\tilde{\Omega}\setminus\Omega}$ as long as the Dirichlet boundary $\partial_D\Omega = \partial\Omega \cap \tilde{\Omega}$ and the values of g on $\{x \in \Omega : \mathrm{dist}(x, \partial_D\Omega) < \varepsilon\}$ remain unchanged. Similarly as before, we let $\tilde{\mathcal{C}}_\varepsilon$ be the set of equilateral triangles $\triangle \subset \tilde{\Omega}$ with vertices in $\mathcal{L}_\varepsilon$ and define $\tilde{\Omega}_\varepsilon = \bigcup_{\triangle \in \tilde{\mathcal{C}}_\varepsilon} \triangle$. By $\tilde{y} : \tilde{\Omega}_\varepsilon \to \mathbb{R}^2$ and $\tilde{u} : \tilde{\Omega}_\varepsilon \to \mathbb{R}^2$ we again denote the piecewise affine interpolation of y and u, respectively.

1.6.1 Convergence of the variational problems

Our convergence analysis applies to discrete deformations which may elongate a number scaling with $\frac{1}{\varepsilon}$ of springs very largely, leading to cracks of finite length in the continuum limit. On triangles not adjacent to such essentially broken springs, the defomations are $\sqrt{\varepsilon}$-close to the identity mapping, so that the accordingly rescaled displacements are of bounded L^2-norm. Note that the first of these assumptions can be inferred from suitable energy bounds. By way of example, however, we see that this cannot be true for the displacement estimates in the bulk: The sequence of functionals $(\mathcal{E}_\varepsilon)_\varepsilon$ is not equicoercive.

Example 1.6.1. Let $B \subset \Omega$ be an arbitrary ball. Assume that the specimen satisfying the boundary conditions is broken into the two parts B and $\Omega \setminus B$, where the inner part is subject to a rotation $R \neq \mathbf{Id}$ so that

$$\nabla \tilde{y}_\varepsilon(x) = R \text{ for } x \in B.$$

In particular, the energy of the configuration is of order 1. But for $x \in B$

$$|\nabla \tilde{u}_\varepsilon(x)| = \left| \frac{1}{\sqrt{\varepsilon}} (R - \mathbf{Id}) \right| \to \infty \text{ for } \varepsilon \to 0.$$

Thus, $\nabla \tilde{u}_\varepsilon$ is not bounded in L^1 and so u_ε does not converge.

Nevertheless, it is interesting to investigate this regime in order to identify a corresponding continuum functional which describes the system in the realm of Griffith models with linearized elasticity. In fact, on the one hand we will discuss two specific problems, our model of crystal cleavage and a model for fractured magnets, where boundary conditions or external fields break the rotational symmetry whence the sequence $(\mathcal{E}^\chi_\varepsilon)_\varepsilon$ satisfies suitable equicoercivity conditions.

On the other hand, in the second part of this thesis we will show that one may establish equicoercivity in a certain sense for the brittle fracture models under consideration if one admits a generalized definition of the displacement field. In Theorem 6.2.1 we show in the setting of continuum fracture mechanics that the example above essentially illustrates the only way that coercitvity may fail: The body breaks apart and in each connected component the deformation is near a different rigid motion. By defining the displacement field on each component appropriately we see that one can establish a compactness result (see (6.12) below). Hereby, the aforementioned assumptions get justified a posteriori as indeed on each connected component the system essentially shows the above properties, in particular we obtain an L^2-bound for the rescaled displacement.

Recall the definition and the main properties of the space $SBV(\Omega;\mathbb{R}^2)$, abbreviated as $SBV(\Omega)$ hereafter, in Section A.1. The sense in which discrete displacements are considered convergent to a limiting displacement in SBV is made precise in the following definition.

Definition 1.6.2. *Suppose $u_\varepsilon : \mathcal{L}_\varepsilon \cap \tilde{\Omega} \to \mathbb{R}^2$ is a sequence of discrete displacements. We say that u_ε converges to some $u \in SBV^2(\tilde{\Omega})$ and write $u_\varepsilon \to u$, if*

(i) $\chi_{\tilde{\Omega}_\varepsilon} \tilde{u}_\varepsilon \to u$ in $L^1(\tilde{\Omega})$

and there exists a sequence $C^*_\varepsilon \subset \tilde{C}_\varepsilon$ with $\# C^*_\varepsilon \le \frac{C}{\varepsilon}$ for a constant C independent of ε such that

(ii) $\|\nabla \tilde{u}_\varepsilon\|_{L^2(\tilde{\Omega}\setminus\cup_{\triangle\in\mathcal{C}^*_\varepsilon}\triangle)} \le C$.

Consider the limiting functional

$$\mathcal{E}(u) = \frac{4}{\sqrt{3}} \int_\Omega \frac{1}{2} Q(e(u))\, dx + \int_{J_u} \sum_{\mathbf{v}\in\mathcal{V}} \frac{2\beta}{\sqrt{3}} |\mathbf{v}\cdot\nu_u|\, d\mathcal{H}^1$$

for $u \in SBV^2(\tilde{\Omega})$, where $e(u) = \frac{1}{2}\left(\nabla u^T + \nabla u\right)$ denotes the symmetric part of the gradient. Q is the linearization of $W_\triangle$ around the identity matrix $\mathbf{Id}$ (see Lemma 2.4.2 for its explicit form). Observe that u is defined on the enlarged set $\tilde{\Omega}$ and therefore also jumps lying in $\tilde{\Omega}\setminus\Omega$ (and thus particularly those lying on $\partial_D\Omega$) contribute to $\mathcal{E}(u)$. For a displacement field u, which is the limit of a sequence $(u_\varepsilon) \subset \mathcal{A}_{g_\varepsilon}$ converging in the sense of Definition 1.6.2, we get $u = g$ on $\tilde{\Omega}\setminus\Omega$, where $g = L^1\text{-}\lim_{\varepsilon\to 0} g_\varepsilon$. Consequently, if $u|_\Omega$ does not attain the boundary condition g on the Dirichlet boundary $\partial_D\Omega$ (in the sense of traces), this will be penalized in the energy $\mathcal{E}(u)$ as then $\mathcal{H}^1(J_u \cap \partial_D\Omega) > 0$.

Moreover, as g by assumption is continuous, for any $u \in \mathcal{A}_g$ the jump set J_u does not intersect $\tilde{\Omega}\setminus\overline{\Omega}$, which shows that $\mathcal{E}(u)$ is in fact independent of the particular choice of $\tilde{\Omega}$ and $g|_{\tilde{\Omega}\setminus\Omega}$ as long as $\partial_D\Omega$ and $g|_{\partial_D\Omega}$ remain unchanged. In Section 5.1 we prove the following Γ-convergence result (see [33] for an exhaustive treatment of Γ-convergence):

Theorem 1.6.3. *(i) Let $(g_\varepsilon)_\varepsilon \subset W^{1,\infty}(\tilde{\Omega})$ with $\sup_\varepsilon \|g_\varepsilon\|_{W^{1,\infty}(\tilde{\Omega})} < +\infty$. If $(u_\varepsilon)_\varepsilon$ is a sequence of discrete displacements with $u_\varepsilon \in \mathcal{A}_{g_\varepsilon}$ and $u_\varepsilon \to u \in SBV^2(\tilde{\Omega})$, then*

$$\liminf_{\varepsilon\to 0} \mathcal{E}_\varepsilon(u_\varepsilon) \ge \mathcal{E}(u).$$

(ii) For every $u \in SBV^2(\tilde{\Omega})$ and $g \in W^{1,\infty}(\tilde{\Omega})$ with $u = g$ on $\tilde{\Omega}\setminus\Omega$ there is a sequence $(u_\varepsilon)_\varepsilon$ of discrete displacements such that $u_\varepsilon \in \mathcal{A}_g$, $u_\varepsilon \to u \in SBV^2(\tilde{\Omega})$ and

$$\lim_{\varepsilon\to 0} \mathcal{E}^\chi_\varepsilon(u_\varepsilon) = \mathcal{E}(u).$$

Note that the recovery sequence is obtained for the energy $\mathcal{E}^\chi_\varepsilon$ which includes the frame indifferent penalty term.

The main idea will be to separate the energy into elastic and crack surface contributions by introducing a threshold such that triangles $\triangle$ with $(y)_\triangle$ beyond that threshold are considered as cracked and $\tilde{y}$ is modified there to a discontinuous function. The treatment of the elastic part draws ideas from [65] and [49]. To derive the crack energy, one could use a slicing technique, see, e.g., [17]. Although also possible in our framework, we follow a different approach here: We carefully construct crack shapes of discrete configurations in an explicit way which allows us to directly appeal to lower semicontinuity results for SBV functions in order to derive the main energy estimates.

In Example 1.6.1 we have seen that $(\mathcal{E}_\varepsilon)$ and $(\mathcal{E}_\varepsilon^\chi)$ are not equicoercive due to the frame indifference of W. We now add a term to $\mathcal{E}_\varepsilon$ such that the sequence becomes equicoercive. Let $\hat{m} : \mathbb{R}^{2\times 2} \to S^1$ be a function satisfying

$$\hat{m}(RF) = R\hat{m}(F) \text{ for all } F \in \mathbb{R}^{2\times 2}, R \in SO(2), \qquad \hat{m}(\mathbf{Id}) = \mathbf{e}_1.$$

Moreover, we assume that $\hat{m}$ is C^2 in a neighborhood of $SO(2)$ and $\mathbb{R}^{2\times 2}_{\mathrm{sym}} \subset \ker(D\hat{m}(\mathbf{Id}))$. Let $\mathcal{F}_\varepsilon(u) = \mathcal{E}_\varepsilon(u) + \frac{1}{\varepsilon}\int_{\Omega_\varepsilon} f_\kappa(\nabla\tilde{y})$ with

$$f_\kappa(F) = \begin{cases} \kappa(1 - \mathbf{e}_1 \cdot \hat{m}(F)), & |F| \le T, \\ 0 & \text{else}, \end{cases} \tag{1.30}$$

for $F \in \mathbb{R}^{2\times 2}$, where $T, \kappa > 0$. Likewise, we define $\mathcal{F}_\varepsilon^\chi$. In Lemma 2.4.6 below we show that $W_{\Delta,\chi}(F) + f_\kappa(F) \ge C|F - \mathbf{Id}|^2$ for all $F \in \mathbb{R}^{2\times 2}$ with $|F| \le T$.

This implies that the sequence $(\mathcal{F}_\varepsilon^\chi)_\varepsilon$ is equicoercive: Given a sequence of displacement fields $(u_\varepsilon)_\varepsilon$ with $\mathcal{F}_\varepsilon^\chi(u_\varepsilon) + \|u_\varepsilon\|_\infty \le C$ we find a subsequence converging in the sense of Definition 1.6.2. Indeed, we get that $\#\mathcal{C}_\varepsilon^* \le \frac{C}{\varepsilon}$, where $\mathcal{C}_\varepsilon^* := \{\Delta \in \tilde{\mathcal{C}}_\varepsilon : |(\mathbf{Id} + \sqrt{\varepsilon}\tilde{u}_\varepsilon)_\Delta| > T\}$. By Lemma 2.4.6 we then get $\|\nabla\tilde{u}_\varepsilon\|_{L^2(\tilde{\Omega}\setminus\bigcup_{\Delta\in\mathcal{C}_\varepsilon^*}\Delta)} \le C$ and therefore condition (ii) in Definition 1.6.2 is satisfied. By an SBV compactness theorem (see Theorem A.1.1) we then find a (not relabeled) subsequence such that $\tilde{u}_\varepsilon\chi_{\tilde{\Omega}_\varepsilon\setminus\bigcup_{\Delta\in\mathcal{C}_\varepsilon^*}\Delta} \to u$ in L^1 for some $u \in SBV^2(\tilde{\Omega})$. This together with $\|u_\varepsilon\|_\infty \le C$ and $|\bigcup_{\Delta\in\mathcal{C}_\varepsilon^*}\Delta| \le C\varepsilon$ implies that also condition (i) in Definition 1.6.2 holds with this function u.

Define $\hat{m}_1 : \mathbb{R}^{2\times 2} \to [-1, 1]$ by $\hat{m}_1 = \mathbf{e}_1 \cdot \hat{m}$ and let $\hat{Q} = D^2\hat{m}_1(\mathbf{Id})$ be the Hessian at the identity. We introduce the limiting functional $\mathcal{F} : SBV^2(\tilde{\Omega}) \to [0, \infty)$ given by

$$\mathcal{F}(u) = \mathcal{E}(u) - \frac{\kappa}{2}\int_\Omega \hat{Q}(\nabla u).$$

We then obtain a Γ-convergence result similar to Theorem 1.6.3.

Theorem 1.6.4. *The assertions of Theorem 1.6.3 remain true when $\mathcal{E}_\varepsilon$, $\mathcal{E}_\varepsilon^\chi$ and $\mathcal{E}$ are replaced by $\mathcal{F}_\varepsilon$, $\mathcal{F}_\varepsilon^\chi$ and $\mathcal{F}$, respectively.*

1.6.2 Analysis of a limiting variational problem

We now analyze the limiting functional $\mathcal{E}$ for a rectangular slab $\Omega = (0, l) \times (0, 1)$ with $l \ge \frac{1}{\sqrt{3}}$ under uniaxial extension in $\mathbf{e}_1$ direction. We determine the minimizers and prove uniqueness up to translation of the specimen and the crack line for the boundary conditions

$$u_1 = 0 \text{ for } x_1 = 0 \quad \text{and} \quad u_1 = al \text{ for } x_1 = l. \tag{1.31}$$

(More precisely: $u \in SBV^2((-\eta, l + \eta) \times (0, 1))$ with $u_1(x) = 0$ for $x \le 0$ and $u_1(x) = al$ for $x \ge l$.) Note that we can investigate the limiting problem without any assumption on the second component of the boundary displacement.

Theorem 1.6.5. *Let $a \neq a_{\text{crit}}$. Then*

$$\min\left\{\mathcal{E}(u) : u \textit{ satisfies } (1.31)\right\} = \min\left\{\frac{\alpha l}{\sqrt{3}}a^2, \frac{2\beta}{\gamma}\right\}.$$

All minimizers of $\mathcal{E}$ subject to (1.31) are of the form given in Corollary 1.5.2(i)-(iii) depending on whether (i) $a < a_{\text{crit}}$, (ii) $a > a_{\text{crit}}$ and $\phi \neq 0$ or (iii) $a > a_{\text{crit}}$ and $\phi = 0$.

This theorem will be addressed in Section 5.2. An analogous result for isotropic, incompressible materials has been obtained recently by Mora-Corral [57]. Theorem 1.6.5 is an extension of this result to anisotropic, compressible brittle materials in the framework of linearized elasticity.

Theorem 1.6.3 in combination with Theorem 1.6.5 gives a new perspective to the results presented in Section 1.4 and Section 1.5. Let $\tilde{\Omega} = (-\eta, l+\eta) \times (0,1)$ and define for $a \geq 0$

$$\begin{aligned}\mathcal{A}(a) = \big\{u = &(u_1, u_2) : \mathcal{L}_\varepsilon \cap \tilde{\Omega} \to \mathbb{R}^2 : \\ &u(x) = g(x) \text{ for } x_1 \leq \varepsilon \text{ and } x_1 \geq l - \varepsilon \text{ for some } g \in \mathcal{G}(a)\big\},\end{aligned}$$

where $\mathcal{G}(a) := \{g \in W^{1,\infty}(\tilde{\Omega}) : g_1(x) = 0 \text{ for } x_1 \leq \varepsilon,\ g_1(x) = al \text{ for } x_1 \geq l - \varepsilon\}$. Note that this is a slight variant of the boundary value problem investigated in Section 1.4.

One implication of Theorem 1.5.1, Corollary 1.5.2 is that, under the tensile boundary conditions $u_\varepsilon \in \mathcal{A}(a)$, the requirement that u_ε be an almost energy minimizer satisfying (1.28) and $\sup_\varepsilon \|u_\varepsilon\|_\infty < \infty$, guarantees the existence of a subsequence converging in the sense of Definition 1.6.2. (The convergence obtained in Theorem 1.5.1 is even stronger.) In particular, the sequence $(\mathcal{E}_\varepsilon^\chi)$ is mildly equicoercive. A fundamental theorem of Γ-convergence (see, e.g., [12, Theorem 1.21]) implies that such low energy sequences converge to limiting configurations given in Theorem 1.6.5 in the sense of Definition 1.6.2. Consequently, in this way we have re-derived the universal cleavage law given in Theorem 1.4.1 and the convergence result Corollary 1.5.2 (in the sense of Definition 1.6.2).

In the second part of this thesis in Section 6.3 we will return to the investigation of a limiting cleavage law. Employing a compactness result presented in Section 6.2 we will see that the above analysis of minimal values and minimizers can be extended to the case of compression.

1.6.3 An application: Fractured magnets in an external field

In the above model we have seen that a mild equicoercivity of the sequence $(\mathcal{E}_\varepsilon^\chi)_\varepsilon$ is guaranteed by investigating a specific boundary value problem. We close this section with an application to fractured magnets, where an external field provides an even stronger equicoercivity condition.

Assume that the material is a permanent magnet and let $\mathbf{e}_1$ be the magnetization direction. We suppose that there is a constitutive relation between $\nabla\tilde{y}(x)$ and the local magnetization direction $\hat{m}(\tilde{y},x) \in S^1$ of the deformed configuration $\tilde{y}$ at some point $x \in \Omega$, which is of the form $\hat{m}(\tilde{y},x) = \hat{m}(\nabla\tilde{y}(x))$ with $\hat{m}$ as defined in Section 1.6.1. Let $H_{\text{ext}} : \mathbb{R}^2 \to \mathbb{R}^2$ be an external magnetic field. The magnetic energy corresponding to the deformation $y = \mathbf{id} + \sqrt{\varepsilon}u$ is then given by

$$\mathcal{E}_\varepsilon^{\text{mag}}(u) = -\frac{1}{\varepsilon}\int_{\Omega_\varepsilon} H_{\text{ext}} \cdot \hat{m}(\nabla\tilde{y}),$$

i.e. alignment of the magnetization direction with the external field is energetically favored. The total energy of the system is given by

$$\mathcal{E}_\varepsilon^{\text{tot}} = \mathcal{E}_\varepsilon^{\chi} + \mathcal{E}_\varepsilon^{\text{mag}}.$$

We now suppose that the external field is homogeneous and satisfies without restriction $H_{\text{ext}} = \kappa\mathbf{e}_1$ for $\kappa > 0$. We then see that

$$\mathcal{F}_\varepsilon = \mathcal{E}_\varepsilon^{\text{tot}} + \frac{\kappa}{\varepsilon}|\Omega_\varepsilon|$$

with f_κ as in (1.30) and corresponding $\mathcal{F}_\varepsilon$. By Theorem 1.6.4 we get that the renormalized functionals $\mathcal{F}_\varepsilon$ Γ-converge to the renormalized total energy functional $\mathcal{E}_{\text{ren}}^{\text{tot}} = \mathcal{F}$. (Obviously, a configuration minimizes $\mathcal{E}_\varepsilon^{\text{tot}}$ if and only if it minimizes $\mathcal{F}_\varepsilon$.)

We consider a boundary value problem $\min_{u\in\mathcal{A}_g} \mathcal{E}_{\text{ren}}^{\text{tot}}(u)$ for $g \in W^{1,\infty}(\tilde{\Omega})$. Since the sequence $(\mathcal{F}_\varepsilon)_\varepsilon$ is equicoercive as discussed in Section 1.6.1, the theory of Γ-convergence implies $\lim_{\varepsilon\to 0}(\frac{\kappa|\Omega_\varepsilon|}{\varepsilon} + \min_{u\in\mathcal{A}_g} \mathcal{E}_\varepsilon^{\text{tot}}(u)) = \lim_{\varepsilon\to 0}\min_{u\in\mathcal{A}_g} \mathcal{F}_\varepsilon(u) = \min_{u\in\mathcal{A}_g} \mathcal{E}_{\text{ren}}^{\text{tot}}(u)$ and also convergence of the corresponding (almost) minimizers of $\mathcal{F}_\varepsilon$, and hence $\mathcal{E}_\varepsilon^{\text{tot}}$, to minimizers of $\mathcal{E}_{\text{ren}}^{\text{tot}}$ in the sense of Definition 1.6.2 is guaranteed. In this context, note that by a truncation argument taking $g \in W^{1,\infty}(\tilde{\Omega})$ into account, we may indeed assume that a low energy sequence satisfies $\sup_\varepsilon \|u_\varepsilon\|_\varepsilon < +\infty$.

Chapter 2

Preliminaries

2.1 Elementary properties of the cell energy

Elastic energy

We first provide a lower bound for the cell energy. For that purpose, let $V_0 = \mathbb{R}^d \otimes (1, \dots 1)$ denote the subspace of infinitesimal translations $(x_1, \dots, x_{2^d}) \to (v, \dots, v)$, $v \in \mathbb{R}^d$.

Lemma 2.1.1. *For every $T > 0$ there is a constant $C > 0$ such that*

$$\operatorname{dist}^2(G, \bar{SO}(d)) \le C W_{\text{cell}}(G)$$

for all $G \in \mathbb{R}^{d \times 2^d}$ with $G \perp V_0$ and $|G| \le T$.

Proof. The proof is essentially contained in [63, Lemma 3.2] and relies on the growth assumptions on W_{cell} near $\bar{SO}(d)$ (see Assumption 1.1.2(iii)). □

We now give a precise characterization of α_A (see (1.16)). We may view a symmetric matrix $F = (f_{ij}) \in \mathbb{R}^{d \times d}_{\text{sym}}$ as a vector $f = (f_1, \dots, f_{\hat{d}}) \in \mathbb{R}^{\hat{d}}$, $\hat{d} = \frac{d(d+1)}{2}$, whose components are the entries f_{ij} with $i \le j$, numbered such that $f_1 = f_{11}$. Then

$$Q_{\text{cell}}(F \cdot Z) = f^T \mathcal{Q} f$$

for some symmetric positive definite $\mathcal{Q} \in \mathbb{R}^{\hat{d} \times \hat{d}}_{\text{sym}}$. For each $r \in \mathbb{R}$ there is consequently a unique $\bar{f} \in \mathbb{R}^{\hat{d}}$ minimizing

$$f^T \mathcal{Q} f \quad \text{subject to} \quad f_1 = r$$

and a corresponding Lagrange multiplier $\lambda \in \mathbb{R}$ such that $\bar{f}^T \mathcal{Q} = \lambda \mathbf{e}_1^T$, i.e.

$$\bar{f} = \lambda \mathcal{Q}^{-1} \mathbf{e}_1,$$

where $\mathbf{e}_1$ denotes the first canonical unit vector in $\mathbb{R}^{\hat{d}}$. Multiplying with $\mathbf{e}_1^T$ and using $\bar{f}_1 = r$, we find that $\lambda = \frac{r}{\mathbf{e}_1^T \mathcal{Q}^{-1} \mathbf{e}_1}$ and thus

$$\bar{f} = \frac{r}{\mathbf{e}_1^T \mathcal{Q}^{-1} \mathbf{e}_1} \mathcal{Q}^{-1} \mathbf{e}_1.$$

For the minimal value we obtain

$$\tilde{Q}(r) = \bar{f}^T \mathcal{Q} \bar{f} = \frac{r^2}{\mathbf{e}_1^T \mathcal{Q}^{-1} \mathbf{e}_1}.$$

With $\hat{\mathcal{Q}}$ denoting the $(\hat{d}-1) \times (\hat{d}-1)$ matrix obtained form $\mathcal{Q}$ by deleting the first row and the first column and using that $\mathcal{Q}^{-1} = \frac{1}{\det \mathcal{Q}} \operatorname{cof} \mathcal{Q}$ and thus $\mathcal{Q}^{-1}\mathbf{e}_1 = \frac{1}{\det \mathcal{Q}} (\operatorname{cof} \mathcal{Q})_{\cdot 1}$ and $\mathbf{e}_1^T \mathcal{Q}^{-1} \mathbf{e}_1 = \frac{(\operatorname{cof} \mathcal{Q})_{11}}{\det \mathcal{Q}} = \frac{\det \hat{\mathcal{Q}}}{\det \mathcal{Q}}$, this can alternatively be written as

$$\bar{f} = \frac{r \det \mathcal{Q}}{\det \hat{\mathcal{Q}}} (\operatorname{cof} \mathcal{Q})_{\cdot 1}, \qquad \tilde{Q}(r) = \frac{r^2 \det \mathcal{Q}}{\det \hat{\mathcal{Q}}}.$$

We summarize these observations in the following lemma:

Lemma 2.1.2. *The reduced energy satisfies*

$$\tilde{Q}(r) = \alpha_A r^2 \quad \textit{with} \quad \alpha_A = \frac{1}{\mathbf{e}_1^T \mathcal{Q}^{-1} \mathbf{e}_1} = \frac{\det \mathcal{Q}}{\det \hat{\mathcal{Q}}}.$$

For each r there exists a unique $\bar{F}(r) \in \mathbb{R}^{d \times d}_{\mathrm{sym}}$ which satisfies $\tilde{Q}(r) = Q_{\mathrm{cell}}(\bar{F}(r) \cdot Z)$ and $f_{11} = r$. $\bar{F}(r)$ depends linearly on r.

Fracture energy

In Theorem 1.3.1 we have seen that the limiting minimal fracture energy has the form (1.14). We now investigate this term in detail and determine the minimizers. We let

$$\Lambda(\varsigma) := \frac{\sum_{\nu \in \mathcal{V}} \beta(\nu) |\nu \cdot \varsigma| - |\mathbf{e}_1 \cdot \varsigma| \beta_A}{|\varsigma|} \tag{2.1}$$

and observe $\Lambda(\varsigma) \geq 0$ for all $\varsigma \in \mathbb{R}^d \backslash \{0\}$. Note that the minimum in the definition (1.14) of β_A and the minumum of Λ in (2.1) are attained on the compact set $S^{d-1} = \{\varsigma \in \mathbb{R}^d : |\varsigma| = 1\}$. Moreover, we note that the minimizers in (1.14) are precisely the minimizers of Λ (with minimal value 0). They are obviously not perpendicular to $\mathbf{e}_1$.

Lemma 2.1.3. *The minimum ($= 0$) of $\Lambda(\xi)$ in S^{d-1} is attained for some $\xi \in \mathcal{P}$.*

Remark 2.1.4. A natural guess would be that in fact every minimizer of Λ lies in $\mathcal{P}$. Surprisingly this turns out to be wrong in general. There are (non-generic) models even leading to a continuum of optimal crack directions, as we will exemplify in Section 3.4 for a basic mass spring model in 2d. As a consequence, for such a model it is not possible to prove that in the fracture regime the body has to break apart along crystallographic hyperplanes.

Proof. For $\delta > 0$ we define

$$\Lambda_\delta(\varsigma) = \Lambda(\varsigma) + \delta \frac{|\varsigma \cdot \mathbf{e}_1|}{|\varsigma|}.$$

Obviously Λ_δ attains its minimum with $0 < \min_{\xi \in \mathbb{R}^d} \Lambda_\delta(\xi) \leq \delta$. We show that for small δ if a $\varphi \in S^{d-1}$ satisfies $\Lambda_\delta(\varphi) = \min_{\xi \in \mathbb{R}^d} \Lambda_\delta(\xi)$ then $\varphi \in \mathcal{P}$.

We define $\mathcal{U} = \{\nu \in \mathcal{V} : \nu \cdot \varphi > 0\} \cup \{\mathbf{e}_1\}$ and $\mathcal{U}_0 = \{\nu \in \mathcal{V} : \nu \cdot \varphi = 0\}$. Note that by (2.1) and (1.14) $\varphi \cdot \mathbf{e}_1 \neq 0$, so without loss of generality we may assume that $\varphi \cdot \mathbf{e}_1 > 0$. For $\nu \in \mathcal{V} \setminus \{\mathbf{e}_1\}$ let $\tilde{\beta}(\nu) = 2\beta(\nu)$. If $\mathbf{e}_1 \in \mathcal{V}$, we set $\tilde{\beta}(\mathbf{e}_1) = 2\beta(\mathbf{e}_1) - \beta_A + \delta$. Otherwise we only set $\tilde{\beta}(\mathbf{e}_1) = -\beta_A + \delta$. If the claim were false, then $\dim \operatorname{span} \mathcal{U}_0 < d - 1$. Therefore, we can choose some $\eta \in \mathbb{R}^d \setminus \{0\}$ such that $\eta \cdot \varphi = 0$ and $\eta \cdot \nu = 0$ for all $\nu \in \mathcal{U}_0$. We now investigate the behavior of Λ_δ at φ in direction η. Using that $\nu \cdot \varphi = \nu \cdot \eta = 0$ for all $\nu \in \mathcal{U}_0$, for $|t|$ sufficiently small we obtain

$$\lambda(t) := \Lambda_\delta(\varphi + t\eta) = \frac{\sum_{\nu \in \mathcal{U}} \tilde{\beta}(\nu)\nu \cdot (\varphi + t\eta)}{|\varphi + t\eta|}.$$

We differentiate and obtain from $\eta \cdot \varphi = 0$

$$\begin{aligned} \lambda'(t) &= \frac{\sum_{\nu \in \mathcal{U}} \tilde{\beta}(\nu)\nu \cdot \eta}{|\varphi + t\eta|} - \frac{\left(\sum_{\nu \in \mathcal{U}} \tilde{\beta}(\nu)\nu \cdot (\varphi + t\eta)\right)(\varphi + t\eta) \cdot \eta}{|\varphi + t\eta|^3} \\ &= \frac{\sum_{\nu \in \mathcal{U}} \tilde{\beta}(\nu)\nu \cdot \eta}{|\varphi + t\eta|} - t|\eta|^2 \frac{\sum_{\nu \in \mathcal{U}} \tilde{\beta}(\nu)\nu \cdot (\varphi + t\eta)}{|\varphi + t\eta|^3}. \end{aligned}$$

If $\lambda'(0) \neq 0$ then φ is not a critical point of Λ_δ which contradicts the above assumption. So we may assume that $\lambda'(0) = \frac{\sum_{\nu \in \mathcal{U}} \tilde{\beta}(\nu)\nu \cdot \eta}{|\varphi|} = 0$ and thus

$$\lambda'(t) = -t|\eta|^2 \frac{\sum_{\nu \in \mathcal{U}} \tilde{\beta}(\nu)\nu \cdot \varphi}{|\varphi + t\eta|^3},$$

leading to the contradiction

$$\lambda''(0) = -|\eta|^2 \Lambda_\delta(\varphi) < 0.$$

Thus, we have shown that Λ_δ attains its minimum for some $\varphi \in \mathcal{P}$ for all $\delta > 0$. As $\min_{\xi\in\mathbb{R}^d} \Lambda_\delta(\xi) \leq \delta$, passing to the limit $\delta \to 0$ we obtain the claim. $\square$

We are now in a position to render more precisely the definition of the minimum length assumed in Section 1.2. Let

$$M_1 = \min_{\xi\in S^{d-1}} \sum_{\nu\in\mathcal{V}} \beta(\nu)|\nu\cdot\xi|, \quad M_2 = \max_{\xi\in S^{d-1}} \sum_{\nu\in\mathcal{V}} \beta(\nu)|\nu\cdot\xi|.$$

It is not hard to see that M_1, M_2 are independent of the particular rotation of the lattice. Then $\beta_A \leq M_2$ and therefore the minimizer $\xi \in S^{d-1}$ of (1.14) satisfies $|\xi\cdot\mathbf{e}_1| \geq \frac{M_1}{M_2}$. Consequently, an elementary argument shows that choosing $C > 0$ large enough independently of $M_1, M_2, l_2, \ldots, l_d$ and setting

$$L = L(\sqrt{A^TA}, W_{\text{cell}}, l_2, \ldots, l_d) = C\max\{l_2, \ldots, l_d\}\,\frac{M_2}{M_1} \tag{2.2}$$

we find that for specimens with $l_1 > L$ it is possible to completely break apart along hyperplanes not passing through the boundary parts B_1 and B_2.

2.2 Interpolation

In the following it will be useful to choose a particular interpolation $\tilde{y}$ of the lattice deformation $y : \mathcal{L}_\varepsilon \cap U \to \mathbb{R}^d$ for $U \subset \Omega$ open. We introduce a threshold value $C_{\text{int}} \geq 1$ to be specified later and first consider some cell $Q_\varepsilon(\bar{x})$, $\bar{x} \in (\mathcal{L}'_\varepsilon(U))^\circ$, where the lattice deformation satisfies $\operatorname{diam}\left\{\bar{\nabla}y(\bar{x}), \mathcal{Z}\right\} \leq C_{\text{int}}$. Here for $G = (g_1, \ldots, g_{2^d}) \in \mathbb{R}^{d\times 2^d}$ and $\tilde{\mathcal{Z}} \subset \mathcal{Z}$ we define $\operatorname{diam}\left\{G, \tilde{\mathcal{Z}}\right\} := \max\left\{|g_i - g_j| : z_i, z_j \in \tilde{\mathcal{Z}}\right\}$, so particularly we have

$$\operatorname{diam}\left\{\bar{\nabla}y(\bar{x}), \tilde{\mathcal{Z}}\right\} = \frac{1}{\varepsilon}\max\left\{|y(\bar{x}+\varepsilon z_i) - y(\bar{x}+\varepsilon z_j)| : z_i, z_j \in \tilde{\mathcal{Z}}\right\}, \tag{2.3}$$

where $\bar{\nabla}y(\bar{x})$ is given in (1.3). We will call cells with this property 'intact cells' and by $\mathcal{C}'_\varepsilon \subset (\mathcal{L}'_\varepsilon(U))^\circ$ we denote the set of their midpoints. The complement $\bar{\mathcal{C}}'_\varepsilon := (\mathcal{L}'_\varepsilon(U))^\circ \setminus \mathcal{C}'_\varepsilon$ labels the centers of cells we consider to be 'broken'.

We first consider $Q_\varepsilon(\bar{x})$ for $\bar{x} \in \mathcal{C}'_\varepsilon$. The interpolation we use was introduced in [65]. We repeat the procedure here for the sake of completeness. Consider the reference cell $Q = A[-\frac{1}{2}, \frac{1}{2})^d$ with deformation $y : A\left\{-\frac{1}{2}, \frac{1}{2}\right\}^d = \mathcal{Z} \to \mathbb{R}^d$. We first interpolate linearly on the one-dimensional faces of $\overline{Q}$, which are given by segments $[z_i, z_j]$, where $z_i - z_j$ is parallel to one of the lattice vectors v_n, $n = 1, \ldots, d$. Subsequently we consider two-dimensional faces and define a triangulation and interpolation as follows: Given a face $\operatorname{co}\{z_{i_1}, z_{i_2}, z_{i_3}, z_{i_4}\}$ with

$$z_{i_2} = z_{i_1} + v_n, \quad z_{i_3} = z_{i_1} + v_n + v_m, \quad z_{i_4} = z_{i_1} + v_m$$

we define

$$\zeta = \frac{1}{4}(z_{i_1} + \ldots + z_{i_4}), \quad y(\zeta) = \frac{1}{4}(y(z_{i_1}) + \ldots + y(z_{i_4}))$$

and interpolate linearly on each of the four triangles $\operatorname{co}\{z_{i_j}, z_{i_{j+1}}, \zeta\}$ for $j = 1, \ldots, 4$ with the convention $i_5 = i_1$. In general, having chosen a simplicical decomposition as well as corresponding linear interpolations on the faces of dimension $n-1$ we decompose and interpolate on an n-dimensional face $F = \operatorname{co}\{z_{i_1}, \ldots, z_{i_{2^n}}\}$ in the following way: Let

$$\zeta = \frac{1}{2^n}\sum_{j=1}^{2^n} z_{i_j}, \quad y(\zeta) = \frac{1}{2^n}\sum_{j=1}^{2^n} y(z_{i_j}).$$

We decompose F by the simplices $\operatorname{co}\{w_1, \ldots, w_n, \zeta\}$, where $\operatorname{co}\{w_1, \ldots, w_n\}$ is a simplex belonging to the decompostion of an $(n-1)$-dimensional face constructed in a previous step. We now interpolate linearly on these simplices.

For cells lying at the boundary B_1^ε, B_2^ε we can repeat the above construction at least for the first component y^1: Let $\bar{x} \in \mathcal{L}'_\varepsilon(\Omega)\backslash(\mathcal{L}'_\varepsilon(\Omega))^\circ$ such that $\bar{x}+\varepsilon z_i \notin \Omega$. This implies $-l_A\varepsilon \le (\bar{x} + \varepsilon z_i)_1 \le 0$ or $l_1 \le (\bar{x} + \varepsilon z_i)_1 \le l_1 + l_A\varepsilon$, respectively. Now let $y^1(\bar{x} + \varepsilon z_i) = (1 + a_\varepsilon)(\bar{x} + \varepsilon z_i)_1$. Thus, the first component of $Y(\bar{x}) = (y_1, \ldots, y_{2^d})$ is well defined and we may proceed as above to construct $\tilde{y}^1$.

We now concern ourselves with 'broken cells' $\bar{x} \in \bar{\mathcal{C}}'_\varepsilon$. Let y be a corresponding lattice deformation defined on $\mathcal{Z}$. Recalling definition (2.3) we first choose a partition $\mathcal{Z}_1\dot{\cup}\ldots\dot{\cup}\mathcal{Z} = \mathcal{Z}$ of the corners with

$$\operatorname{diam}\{\bar{\nabla}y(\bar{x}), \mathcal{Z}_i\} \le \left(\frac{\#\mathcal{Z}_i}{2^d}\right)^2 C_{\text{int}}.$$

We note that this partition can be chosen in a way that $\varepsilon d(\bar{\nabla}y(\bar{x}); \mathcal{Z}_i, \mathcal{Z}_j) > 2^{-2d}\varepsilon C_{\text{int}}$ for all sets $\mathcal{Z}_i, \mathcal{Z}_j$, $i \neq j$. Indeed, if there were $\bar{z}_i \in \mathcal{Z}_i$, $\bar{z}_j \in \mathcal{Z}_j$ such that $|y(\bar{z}_i) - y(\bar{z}_j)| = \varepsilon d(\bar{\nabla}y(\bar{x}); \mathcal{Z}_i, \mathcal{Z}_j) \le 2^{-2d}\varepsilon C_{\text{int}}$ then

$$\begin{aligned}|y(z_i) - y(z_j)| &\le |y(z_i) - y(\bar{z}_i)| + |y(\bar{z}_i) - y(\bar{z}_j)| + |y(z_j) - y(\bar{z}_j)| \\ &\le \frac{(\#\mathcal{Z}_i)^2 + (\#\mathcal{Z}_j)^2 + 1}{2^{2d}}\varepsilon C_{\text{int}} \le \left(\frac{\#(\mathcal{Z}_i \cup \mathcal{Z}_j)}{2^d}\right)^2 \varepsilon C_{\text{int}}\end{aligned}$$

for all $z_i \in \mathcal{Z}_i$, $z_j \in \mathcal{Z}_j$ and we could set $\tilde{\mathcal{Z}} = \mathcal{Z}_i \cup \mathcal{Z}_j$. Clearly, the cardinality of this partition is at least two for every cell $Q_\varepsilon(\bar{x})$, $\bar{x} \in \bar{\mathcal{C}}'_\varepsilon$. Then, by Assumption 1.1.2(ii) and (iv) it is not hard to see that there is a constant $C = C(C_{\text{int}})$ such that

$$W_{\text{cell}}(\bar{\nabla}y(\bar{x})) \ge C \tag{2.4}$$

for $\bar{x} \in \bar{\mathcal{C}}'_\varepsilon$. Note that $C = C(C_{\text{int}})$ can be chosen independently of C_{int} for $C_{\text{int}} \ge 1$.

We now choose the interpolation for each component i separately as follows. If $\operatorname{diam}_i \{\bar{\nabla} y(\bar{x}), \mathcal{Z}\} \leq C_{\text{int}}$ we define $\tilde{y}^i$ as before for 'intact cells' in $\mathcal{C}'_\varepsilon \setminus \bar{\mathcal{C}}'_\varepsilon$. Here for $G = (g_1, \ldots, g_{2^d}) \in \mathbb{R}^{d \times 2^d}$ we define similarly as in (2.3)

$$\operatorname{diam}_i \{G, \mathcal{Z}\} := \max \{|(g_j - g_k) \cdot \mathbf{e}_i| : z_j, z_k \in \mathcal{Z}\}.$$

Otherwise we set

$$\tilde{y}^i(x) = y^i(z_1) + (x - z_1)_i$$

on $Q_\varepsilon(\bar{x})$ and therefore $\nabla \tilde{y}^i = \mathbf{e}_i^T$. Consequently, $|\nabla \tilde{y}| \leq C C_{\text{int}}$ a.e. also on broken cells $Q_\varepsilon(\bar{x})$, $\bar{x} \in \bar{\mathcal{C}}'_\varepsilon$. Finally having constructed the interpolation on all cells $Q_\varepsilon(\bar{x})$ we briefly note that $\tilde{y} \in SBV(\Omega_\varepsilon, \mathbb{R}^d)$, cf. Section A.1.

For every interaction direction $\nu \in \mathcal{V}$ we introduce a further interpolation $\bar{y}_\nu$ as follows. We choose vectors $\{v_{i_1}, \ldots, v_{i_{d-1}}\} \subset \{v_1, \ldots, v_d\}$ such that $\nu, v_{i_1}, \ldots, v_{i_{d-1}}$ are linearly independent and for $D^\nu = (\nu, v_{i_1}, \ldots, v_{i_{d-1}})$ define the lattice $\mathcal{G}^\nu_\varepsilon = \varepsilon D^\nu \mathbb{Z}^d$ partitioning $\mathbb{R}^d$ into cells of the form $Q^\nu_\varepsilon(\lambda) := \varepsilon D^\nu(\lambda + [0,1)^d)$ for $\lambda \in \mathbb{Z}^d$. Note that $\mathcal{L}_\varepsilon = \mathcal{G}^\nu_\varepsilon$.

We describe the interpolation on the reference cell $Q^\nu = D^\nu[0,1)^d$ with deformation $y : D^\nu\{0,1\}^d \to \mathbb{R}^d$. If $|y(\nu) - y(0)| \leq C_{\text{int}}\varepsilon$ we let

$$\bar{y}_\nu(D^\nu x) = (1 - x_1)y(0) + x_1 y(\nu) \tag{2.5}$$

for $x \in [0,1)^d$. If $|y(\nu) - y(0)| > C_{\text{int}}\varepsilon$ we set $\bar{y}_\nu(D^\nu x) = y(0)$. Let $J_{\bar{y}_\nu}$ be the set of discontinuity points of $\bar{y}_\nu$ and denote by $\partial_\nu Q^\nu$ the two faces of ∂Q^ν which are not parallel to ν. Then $\bar{y}_\nu$ is typically discontinuous on $\partial Q^\nu_\varepsilon(\lambda) \setminus \partial_\nu Q^\nu_\varepsilon(\lambda)$, $\lambda \in \mathbb{Z}^d$. This, however, will not affect our analysis. Essentially, we observe that $J_{\bar{y}_\nu} \cap \partial_\nu Q^\nu_\varepsilon(\lambda) \neq \emptyset$ can only occur if there is some cube $Q_\varepsilon(\bar{x})$, $\bar{x} \in \bar{\mathcal{C}}'_\varepsilon$ such that $Q_\varepsilon(\bar{x}) \cap Q^\nu_\varepsilon(\lambda) \neq \emptyset$. This is due to (2.5) and the definition of $\bar{\mathcal{C}}'_\varepsilon$.

For later we compute the $\mathcal{H}^{d-1}$ volume of $\partial_\nu Q^\nu$. We first choose $\hat{\nu} \in \mathbb{R}^d$ such that $|\hat{\nu}| = 1$ and $\hat{\nu} \cdot v_{i_j} = 0$ for $j = 1, \ldots, d-1$, i.e. $\hat{\nu}$ is a unit normal vector to $\partial_\nu Q^\nu$. Then set $\bar{\nu} = |\nu \cdot \hat{\nu}|\hat{\nu}$ and obtain

$$\frac{1}{2}\mathcal{H}^{d-1}(\partial_\nu Q^\nu) = \frac{|\det(\bar{\nu}, v_{i_1}, \ldots, v_{i_{d-1}})|}{|\bar{\nu}|} = \frac{|\det D^\nu|}{|\nu \cdot \hat{\nu}|} = \frac{\det A}{|\nu \cdot \hat{\nu}|}. \tag{2.6}$$

We recall here some important properties of the interpolation $\tilde{y}$ on cells $Q_\varepsilon(\bar{x})$, $\bar{x} \in \mathcal{C}'_\varepsilon$ being proved in [65] for the case $p = 2$. The extension of the results to general p are straightforward.

Lemma 2.2.1. *Let $y : \mathcal{L}_\varepsilon \to \mathbb{R}^d$ a lattice deformation, $\tilde{y}$ the corresponding linear interpolation. Then for every $1 \leq p < \infty$ there are constants $c, C > 0$ such that for every cell $Q = Q_\varepsilon(\bar{x})$, $\bar{x} \in \mathcal{C}'_\varepsilon \setminus \bar{\mathcal{C}}'_\varepsilon$ we have*

(i) $c \operatorname{dist}^p(\bar{\nabla} y|_Q, \bar{SO}(d)) \leq \frac{1}{|Q|}\int_Q \operatorname{dist}^p(\nabla \tilde{y}, SO(d)) \leq C \operatorname{dist}^p(\bar{\nabla} y|_Q, \bar{SO}(d))$

(ii) $c|\bar{\nabla} y|_Q|^p \leq \frac{1}{|Q|}\int_Q |\nabla \tilde{y}|^p \leq C|\bar{\nabla} y|_Q|^p$.

The interpolation $\tilde{y}$ proves useful to show that in the continuum limit the discrete gradient reduces to a classical gradient (again cf. [65]).

Lemma 2.2.2. *Let $U \subset \Omega$, $\varepsilon_k \to 0$ and a sequence $y_k : \mathcal{L}_\varepsilon(U) \to \mathbb{R}^d$ with $\bar{\nabla} y_k \rightharpoonup f$ in L^p, $\tilde{y}_k \rightharpoonup y$ in $W^{1,p}$ for some $f \in L^p(U)$, $y \in W^{1,p}(U)$, $1 \le p < \infty$. Assume that $\bar{C}'_{\varepsilon_k} = \emptyset$ for all k. Then*

$$f = \nabla y \cdot Z.$$

The following lemma shows that passing from $\tilde{y}$ to $\bar{y}_\nu$, $\nu \in \mathcal{V}$, we do not change the limit.

Lemma 2.2.3. *Let $U \subset \Omega$, $\varepsilon_k \to 0$ and $y_k : \mathcal{L}_{\varepsilon_k}(U) \to \mathbb{R}^d$ be a sequence of lattice deformations with $\#\bar{C}'_{\varepsilon_k} \le C\varepsilon_k^{-d+1}$. Let $\tilde{y}_k$, $\bar{y}_{\nu,k}$ be the corresponding interpolations. Then passing to the limit $\varepsilon_k \to 0$ we obtain $\tilde{y}_k - \bar{y}_{\nu,k} \to 0$ in measure for all $\nu \in \mathcal{V}$.*

Proof. Let $\nu \in \mathcal{V}$ and consider the sequences $\tilde{y}_k$ and $\bar{y}_{\nu,k}$. By $\mathcal{D}'_{\varepsilon_k}(U) \subset \mathcal{L}'_{\varepsilon_k}$ we denote the midpoints $\bar{x}$ of all cells being either a broken cell itself or a neighbor of a broken cell, i.e. $\bar{x} \in \bar{C}'_{\varepsilon_k}$ or $\bar{x} + \varepsilon_k \nu \in \bar{C}'_{\varepsilon_k}$ for some $\nu \in \mathcal{V}$. Note that $\#\mathcal{D}'_{\varepsilon_k}(U) \le 3^d \#\bar{C}'_{\varepsilon_k} \le C\varepsilon_k^{-d+1}$.

We first take a cell $Q_{\varepsilon_k}(\bar{x})$ with $\bar{x} \notin \mathcal{D}'_{\varepsilon_k}(U)$ into account. By construction we have $\tilde{y}_k(\bar{x} + \varepsilon z_i) = \bar{y}_{\nu,k}(\bar{x} + \varepsilon z_i)$ for a suitable $z_i \in \mathcal{Z}$. As $|\nabla \tilde{y}_k|, |\nabla \bar{y}_k^\nu| \le CC_{\rm int}$ on $Q_{\varepsilon_k}(\bar{x})$ we deduce that

$$\sup_{x \in Q_{\varepsilon_k}(\bar{x})} |\tilde{y}_k(x) - \bar{y}_{\nu,k}(x)| \le CC_{\rm int}\varepsilon_k$$

for all $\bar{x} \notin \mathcal{D}'_{\varepsilon_k}(U)$. Noting that

$$\sum_{\bar{x} \in \mathcal{D}'_{\varepsilon_k}(U)} |Q_{\varepsilon_k}(\bar{x})| \le C\varepsilon_k^d \#\mathcal{D}'_{\varepsilon_k}(U) \le C\varepsilon_k \to 0$$

as $\varepsilon_k \to 0$, we deduce that $\tilde{y}_k - \bar{y}_{\nu,k} \to 0$ in measure for $\varepsilon_k \to 0$. □

2.3 An estimate on geodesic distances

We now formulate a short lemma about the length of Lipschitz curves in sets $W \subset \mathbb{R}^{d-1}$ of the form (3.7) introduced below: We estimate geodesic distances and the area swept by curves of given length emanating from a common point in terms of the area and surface of W. For this purpose, we define $\mathrm{dist}_W(p,q)$ as the infimum of the length of Lipschitz curves in W connecting the points $p, q \in W$ and let $\mathcal{H}^m$ denote the m-dimensional Hausdorff measure.

Lemma 2.3.1. *There are constants $C, c, c' > 0$ (depending on D and D') such that for all $\tilde{W}, W \subset (0,1)^{d-1}$ of the form (3.7) and ε small enough the following holds:*

(i) $\operatorname{dist}_{\tilde{W}}(p,q) \leq C(1+\varepsilon^{(s-1)(d-3)})$ *for all* $p, q \in W$

(ii) For all $p \in W$, $t \in (0, c\varepsilon^{(s-1)(d-2)})$ *one has for* ε *small enough*

$$\mathcal{H}^{d-1}(\{q \in \tilde{W} : \operatorname{dist}_{\tilde{W}}(p,q) \leq t\}) \geq c' t \varepsilon^{(1-s)(d-2)}.$$

Proof. We cover $\tilde{\Omega} := (0,1)^{d-1}$ up to a set of measure zero with the sets $C_\varepsilon(\bar{x}) = \bar{x} + (0, \frac{1}{l})^{d-1}$, where $l = \lceil \frac{1}{\hat{\varepsilon}} \rceil$, $\hat{\varepsilon} = \varepsilon^{1-s}$ and $\bar{x} \in I_\varepsilon(\tilde{\Omega}) \subset \frac{1}{l}\mathbb{Z}^{d-1}$. Also set $I_\varepsilon(\tilde{W}) = \{\bar{x} \in I_\varepsilon(\tilde{\Omega}) : \overline{C_\varepsilon(\bar{x})} \subset \tilde{W}\}$. We let $\tilde{V}$ be the connected component of

$$\bigcup_{\bar{x} \in I_\varepsilon(\tilde{W})} \overline{C_\varepsilon(\bar{x})} \subset \tilde{W}$$

with largest Lebesgue measure. We note that for D' sufficiently large $W \subset \tilde{V}$ and thus also $\tilde{V}$ satisfies condition (3.7) possibly passing to a larger D. Given $U = \tilde{\Omega}, \tilde{V}$ for two points $p, q \in I_\varepsilon(U)$ we denote the lattice geodesic distance of p and q in U, i.e. the length of the shortest polygonal path $\Gamma_U(p,q) := (x_0 = p, x_1, \ldots, x_n = q)$ with $x_j \in I_\varepsilon(U)$ and $x_{j+1} - x_j = \pm \frac{1}{l}\mathbf{e}_i$ for some $i = 1, \ldots, d-1$ connecting p and q, by $d_U(p,q)$.

Denote the connected components of $\tilde{\Omega} \setminus \tilde{V}$ by $\tilde{V}_1, \ldots, \tilde{V}_n$ and choose $I_\varepsilon(\tilde{V}_i) \subset I_\varepsilon(\tilde{\Omega})$ such that $\overline{\tilde{V}_i} = \bigcup_{\bar{x} \in I_\varepsilon(\tilde{V}_i)} \overline{C_\varepsilon(\bar{x})}$. It is easy to see that for (i) it suffices to show that $d_{\tilde{V}}(p,q) \leq C(1 + \hat{\varepsilon}^{3-d})$ for all $p, q \in I_\varepsilon(\tilde{V})$. Given $p, q \in I_\varepsilon(\tilde{V})$ we first note that $d_{\tilde{\Omega}}(p,q) \leq d-1$. Let $\Gamma_{\tilde{\Omega}}(p,q) = (x_0, \ldots, x_m)$ be a (non unique) shortest lattice path connecting p and q. If $x_j \in I_\varepsilon(\tilde{W})$ for all j we are finished. Otherwise, for the local nature of the arguments we may assume that $\Gamma_{\tilde{\Omega}}(p,q)$ intersects exactly one $\tilde{V}_i$. Let $x_{j_1}, x_{j_2} \in I_\varepsilon(\tilde{V}_i)$ be the first and the last point in $\tilde{V}_i$, i.e. $x_j \notin I_\varepsilon(\tilde{V}_i)$ for $j < j_1$ and $j > j_2$. Then it is elementary to see that

$$d_{\tilde{V}}(x_{j_1-1}, x_{j_2+1}) \leq C l^{d-3} \mathcal{H}^{d-2}(\partial \tilde{V}_i) \leq C \hat{\varepsilon}^{3-d} \mathcal{H}^{d-2}(\partial \tilde{V}_i)$$

for some $C > 0$ not depending on $\tilde{V}$ as the number of cubes at the boundary of $\tilde{V}_i$ can be bounded by $C l^{d-2} \mathcal{H}^{d-2}(\partial \tilde{V}_i)$). Let $\Gamma_{\tilde{V}}(x_{j_1-1}, x_{j_2+1}) = (y_0, \ldots, y_{\tilde{m}})$ be a shortest path. Then

$$(x_0, \ldots, x_{j_1-1}, y_1, \ldots, y_{\tilde{m}-1}, x_{j_2+1}, \ldots, x_m)$$

is a lattice path in $\tilde{V}$ connecting p and q which shows that $d_{\tilde{V}}(p,q) \leq m + \tilde{m} \leq C + C\hat{\varepsilon}^{3-d} \mathcal{H}^{d-2}(\partial \tilde{V}_i) \leq C + CD\varepsilon^{(s-1)(d-3)}$.

To show (ii) we let $p \in W$ and $t \in (0, c\varepsilon^{(s-1)(d-2)})$ for some small $c > 0$. Without restriction we may assume $p \in I_\varepsilon(\tilde{V})$. If $\operatorname{dist}_{\tilde{W}}(p,q) \leq t$ for all $q \in$

$\tilde{V}$ the assertion is clear. Otherwise, there is some $q \in I_\varepsilon(\tilde{V})$ with $d_{\tilde{V}}(p,q) \geq \operatorname{dist}_{\tilde{V}}(p,q) \geq \operatorname{dist}_{\tilde{W}}(p,q) > \frac{t}{2}$ and a corresponding shortest path $\Gamma_{\tilde{V}}(p,q) = (x_0 = p, x_1, \ldots, x_m = q)$ with $x_i \neq x_j$ for $i \neq j$ and $m \geq \bar{m} := \lceil \frac{lt}{2} \rceil$. Now let $U = \bigcup_{j=0}^{\bar{m}} \overline{C_\varepsilon(x_j)}$. Then it is not hard to see that $U \subset \{q \in \tilde{V} : \operatorname{dist}_{\tilde{V}}(p,q) \leq t\}$ for ε small enough and $\mathcal{H}^{d-1}(U) \geq l^{1-d} \cdot c'lt \geq c'\varepsilon^{(1-s)(d-2)}t$, as desired. □

2.4 Cell energy of the triangular lattice

We close the preparatory chapter by collecting some elementary properties of the cell energy $W_\triangle$ of the triangular lattice and the *reduced energy* $\tilde{W}$ defined by

$$\tilde{W}(r) = \inf\{W_\triangle(F) : \mathbf{e}_1^T F \mathbf{e}_1 = r\}. \tag{2.7}$$

Assume that W satisfies the assumptions (i), (ii) and (iii) given in Section 1.4.

Lemma 2.4.1. *$W_\triangle$ is*

(i) frame indifferent: $W_\triangle(QF) = W_\triangle(F)$ for all $F \in \mathbb{R}^{2\times 2}$, $Q \in O(2)$,

(ii) non-negative and satisfies $W_\triangle(F) = 0$ if and only if $F \in O(2)$ and

(iii) $\liminf_{|F|\to\infty} W_\triangle(F) = \liminf_{|F|\to\infty} W_{\triangle,\chi}(F) = \beta$.

Proof. (i) is clear. For (ii) it suffices to note that $\mathbf{v}F^T F\mathbf{v} = 1$ for $\mathbf{v} \in \mathcal{V}$ implies that $F^T F = \mathbf{Id}$. As χ vanishes near ∞, (iii) can be seen by noting that if $|F| \to \infty$, then for at least two vectors $\mathbf{v} \in \mathcal{V}$ one has $|F\mathbf{v}| \to \infty$. Moreover, if $|F| \to \infty$ with $|F\mathbf{v}_1| = 1$, then $W_\triangle(F) \to \beta$. □

We compute the linearization about the identity matrix $\mathbf{Id}$:

Lemma 2.4.2. *Let $F = \mathbf{Id} + G$ for $G \in \mathbb{R}^{2\times 2}$. Then for $|G|$ small*

$$W_\triangle(F) = \frac{1}{2}Q(G) + o(|G|^2),$$

where $Q(G) = \frac{3\alpha}{16}\left(3g_{11}^2 + 3g_{22}^2 + 2g_{11}g_{22} + 4\left(\frac{g_{12}+g_{21}}{2}\right)^2\right)$.

In particular, $Q(G)$ only depends on the symmetric part $\left(G^T + G\right)/2$ of G. Q is positive semidefinite and thus convex on $\mathbb{R}^{2\times 2}$ and positive definite and strictly convex on the subspace $\mathbb{R}^{2\times 2}_{\mathrm{sym}}$ of symmetric matrices.

Proof. Let $\mathbf{v} \in \mathcal{V}$ and $G \in \mathbb{R}^{2\times 2}$ small. We Taylor expand the contributions $W(|F\mathbf{v}|)$ to the energy $W_\triangle$:

$$\begin{aligned} W(|(\mathbf{Id} + G)\mathbf{v}|) &= W\left(\sqrt{\langle \mathbf{v}, (\mathbf{Id} + G^T)(\mathbf{Id} + G)\mathbf{v}\rangle}\right) \\ &= \frac{W''(1)}{2}\left\langle \mathbf{v}, \frac{G^T + G}{2}\mathbf{v}\right\rangle^2 + o(|G|^2). \end{aligned}$$

Now using the elementary identity

$$\begin{aligned}&\langle \mathbf{v}_1, H\mathbf{v}_1\rangle^2 + \langle \mathbf{v}_2, H\mathbf{v}_2\rangle^2 + \langle (\mathbf{v}_2 - \mathbf{v}_1), H(\mathbf{v}_2 - \mathbf{v}_1)\rangle^2 \\ &= \frac{3}{8}\left(2\,\mathrm{trace}(H^2) + (\mathrm{trace}\, H)^2\right)\end{aligned} \tag{2.8}$$

for any symmetric matrix $H \in \mathbb{R}^{2\times 2}$, we obtain by summing over $\mathbf{v} \in \mathcal{V}$

$$\begin{aligned}W_\triangle(F) &= \frac{1}{2}\cdot\frac{\alpha}{2}\cdot\frac{3}{8}\cdot\left(2\,\mathrm{trace}\left(\left(\frac{G^T+G}{2}\right)^2\right) + \left(\mathrm{trace}\,\frac{G^T+G}{2}\right)^2\right) + o(|G|^2) \\ &= \frac{1}{2}Q(G) + o(|G|^2).\end{aligned}$$

As $Q(G) \geq \frac{3\alpha}{16}(2g_{11}^2 + 2g_{22}^2 + (g_{12}+g_{21})^2)$, Q is positive semidefinite on $\mathbb{R}^{2\times 2}$ and positive definite on $\mathbb{R}^{2\times 2}_{\mathrm{sym}}$. □

As a consequence, we have the following properties of the reduced energy $\tilde{W}$.

Lemma 2.4.3. *The reduced energy satisfies*

(i) $\tilde{W}(r) = 0 \iff |r| \leq 1$.

(ii) For $r \geq 1$ *one has*

$$\tilde{W}(r) = W_\triangle\left(\begin{pmatrix} r & 0 \\ 0 & \frac{4-r}{3}\end{pmatrix}\right) + o((r-1)^2) = \frac{\alpha}{4}(r-1)^2 + o((r-1)^2).$$

(iii) $\lim_{|r|\to\infty} \tilde{W}(r) = \beta$.

Proof. (i) If $|r| \leq 1$, then one can choose $Q \in SO(2)$ with $\mathbf{e}_1^T Q \mathbf{e}_1 = r$ and so $0 \leq \tilde{W}(r) \leq W_\triangle(Q) = 0$. If $|r| > 1$, then $\tilde{W}(r) > 0$ for otherwise there would be a sequence $F_k \in \mathbb{R}^{2\times 2}$ with $\mathbf{e}_1^T F_k \mathbf{e}_1 = r$ and $W_\triangle(F_k) \to 0$. But then $\mathrm{dist}(F_k, O(2)) \to 0$ by (ii) and (iii) of Lemma 2.4.1 and thus, up to subsequences, $F_k \to F \in O(2)$ with $\mathbf{e}_1^T F \mathbf{e}_1 = r$, which is impossible.

(ii) This discussion shows that in fact for any $\eta > 0$ there exists $\delta > 0$ such that $W_\triangle(F) \geq \delta$ whenever $\mathrm{dist}(F, O(2)) \geq \eta$. Now since $\tilde{W}(r) \to 0$ as $r \searrow 1$, we obtain that, for sufficiently small $r > 1$ and $\delta > 0$, any F with $W_\triangle(F) < \tilde{W}(r) + \delta$ is contained in a small neighborhood of $O(2)$. If in addition $\mathbf{e}_1^T F \mathbf{e}_1 = r$ holds, then in fact, F must be close to $\mathbf{Id}$ or to $P = \begin{pmatrix} 1 & 0 \\ 0 & -1\end{pmatrix}$. In particular, by continuity of W, the infimum on the right hand side in the definition of $\tilde{W}$ is attained for those r.

We now fix such an $r > 1$ near 1 and choose $F = \mathbf{Id} + G$ such that $\tilde{W}(r) = W_\triangle(F)$ and $\mathbf{e}_1^T F \mathbf{e}_1 = r$. As $W_\triangle$ is invariant under the reflection P, we may without loss of generality assume that G is small. Then Lemma 2.4.2 yields

$$W_\triangle(F) = \frac{3\alpha}{32}\left(3g_{11}^2 + 3g_{22}^2 + 2g_{11}g_{22} + 4\left(\frac{g_{12}+g_{21}}{2}\right)^2\right) + o(|G|^2).$$

Since $g_{11} = r-1$ and $W_\triangle(F) \le W_\triangle\left(\begin{pmatrix} r & 0 \\ 0 & 1\end{pmatrix}\right) = O((r-1)^2)$, by noting that $3g_{11}^2+3g_{22}^2+2g_{11}g_{22} = \frac{8}{3}(r-1)^2+(\frac{1}{\sqrt{3}}g_{11}+\sqrt{3}g_{22})^2$ we decduce from the minimality property of F that $g_{12}+g_{21} = o(r-1)$ and $g_{22} = -\frac{1}{3}g_{11}+o(r-1)$ and F satisfies

$$\frac{F^T+F}{2} = \begin{pmatrix} r & 0 \\ 0 & \frac{4-r}{3}\end{pmatrix} + o(r-1)$$

with energy

$$\begin{aligned} W_\triangle(F) &= W_\triangle\left(\frac{F^T+F}{2}\right) + o((r-1)^2) \\ &= \frac{\alpha}{4}(r-1)^2 + o((r-1)^2). \end{aligned}$$

(iii) $\liminf_{|r|\to\infty} \tilde{W}(r) \ge \beta$ is immediate from Lemma 2.4.1(iii). Considering matrices F with $F\mathbf{e}_1 = r\mathbf{e}_1$ and $F\mathbf{v}_i = \mathbf{v}_i$ for $i=1$ or $i=2$ we see that also $\limsup_{|r|\to\infty} \tilde{W}(r) \le \beta$. □

Under strengthened hypotheses on W we have the following expansion:

Lemma 2.4.4. *If W in addition satisfies the assumptions (ii') and (iii'), then for $r>1$ close to 1 we have*

$$\tilde{W}(r) = \frac{\alpha(r-1)^2}{4} + \frac{1}{108}\Big(5\alpha+7\alpha'-2(3\alpha-\alpha')\cos(6\phi)\Big)(r-1)^3 + O((r-1)^4),$$

where ϕ is such that $R_\mathcal{L} = \begin{pmatrix} \cos\phi & -\sin\phi \\ \sin\phi & \cos\phi\end{pmatrix}$.

Proof. Let $s = r-1$. By definition,

$$\tilde{W}(r) = \min\left\{W_\triangle(F(s,x,y,z)) : x,y,z \in \mathbb{R}\right\},$$

where $F(s,x,y,z) = \begin{pmatrix} 1+s & z+y \\ z-y & 1+x\end{pmatrix}$. Due to the quadratic energy growth near $SO(2)$, we need to minimize only over x,y,z with $|x|,|z|,\sqrt{s}|y| \le Cs$ for a constant C large enough. Indeed, as $W_\triangle(F(s,0,0,0)) = O(s^2)$, for a minimizer one has $\mathrm{dist}(F(s,x,y,z),O(2)) = O(s)$ by the subsequent Lemma 2.4.5(i) and without loss of generality $\mathrm{dist}(F(s,x,y,z),SO(2)) = O(s)$. We first use that the absolute value of the two rows and columns of $F(s,x,y,z)$ is of order $1+O(s)$. Then $\sqrt{(1+s)^2+(z\pm y)^2} = 1+O(s)$, which implies $|z\pm y| = O(\sqrt{s})$ and so $|z|,|y| = O(\sqrt{s})$, and also $\sqrt{(1+x)^2+(z\pm y)^2} = 1+O(s)$, which then implies $\pm(1+x) = 1+O(s)$ and thus without loss of generality $x = O(s)$. Finally using that the scalar product $(1+s)(z+y)+(1+x)(z-y) = 2z+O(s^{3/2})$ of the two columns of $F(s,x,y,z)$ in absolute value is also bounded by $O(s)$, we obtain that $|z| = O(s)$.

Set $x = -\frac{s}{3} + sx_1$, $y = \sqrt{s}y_1$, $z = sz_1$ with $|x_1|, |y_1|, |z_1| \leq C$. Explicit calculation gives

$$W_\triangle(F(s,x,y,z)) = \frac{\alpha}{32}\left(8 + 3x_1^2 + 8y_1^2 + 12z_1^2 + 6(x_1 + y_1^2)^2\right)s^2 + O(s^3).$$

Since $\alpha > 0$, we thus obtain that this expression is minimized in x_1, y_1, z_1 with $x_1^2, y_1^2, z_1^2 = O(s)$ and we may set $x_1 = \sqrt{s}x_2$, $y_1 = \sqrt{s}y_2$ and $z_1 = \sqrt{s}z_2$ with $|x_2|, |y_2|, |z_2| \leq C$ for some $C > 0$. Explicit expansion in powers of s then yields

$$\begin{aligned}
&W_\triangle(F(s,x,y,z)) \\
&= \frac{\alpha s^2}{4} + \frac{1}{864}\Big(48\alpha + 56\alpha' - 16(3\alpha - \alpha')\cos(6\phi) \\
&\qquad\qquad + 3\alpha\left(81x_2^2 + 72y_2^2 + 108z_2^2\right)\Big)s^3 \\
&\quad + \frac{1}{24}\Big(\left(9\alpha y^2 + \alpha' + (3\alpha - \alpha')\cos(6\phi)\right)x_2 \\
&\qquad\qquad + 2(3\alpha - \alpha')\sin(6\phi)z_2)\Big)s^{7/2} + O(s^4) \\
&= \frac{\alpha s^2}{4} + \frac{1}{108}\Big(6\alpha + 7\alpha' - 2(3\alpha - \alpha')\cos(6\phi)\Big)s^3 \\
&\quad + \frac{9\alpha}{32}\Big(x_2^2 + 2A\sqrt{s}x_2\Big)s^3 + \frac{\alpha y_2^2 s^3}{4} + \frac{3\alpha}{8}\Big(z_2^2 + 2B\sqrt{s}z_2\Big)s^3 + O(s^4)
\end{aligned}$$

for A and B bounded uniformly in s and so

$$\begin{aligned}
&W_\triangle(F(s,x,y,z)) \\
&= \frac{\alpha s^2}{4} + \frac{1}{108}\Big(6\alpha + 7\alpha' - 2(3\alpha - \alpha')\cos(6\phi)\Big)s^3 \\
&\quad + \frac{9\alpha}{32}\Big(x_2 + A\sqrt{s}\Big)^2 s^3 + \frac{\alpha y_2^2 s^3}{4} + \frac{3\alpha}{8}\Big(z_2 + B\sqrt{s}\Big)^2 s^3 + O(s^4).
\end{aligned}$$

Minimizing with respect to x_2, y_2 and z_2 we finally obtain that

$$\tilde{W}(1+s) = \frac{\alpha s^2}{4} + \frac{1}{108}\Big(6\alpha + 7\alpha' - 2(3\alpha - \alpha')\cos(6\phi)\Big)s^3 + O(s^4).$$

□

The following lemma provides useful lower bounds for the energy $W_\triangle$ and the reduced energy $\tilde{W}$.

Lemma 2.4.5. *For all $T > 1$ one has:*

(i) There exists some $c > 0$ such that $c\,\mathrm{dist}^2(F, O(2)) \leq W_\triangle(F)$ for all $F \in \mathbb{R}^{2\times 2}$ satisfying $|F| \leq T$.

(ii) *For $\delta > 0$ small enough, there is a convex function $V \geq 0$ with $V(r) \leq \tilde{W}(r)$ for $r \leq T$, $V'(1) = 0$ and such that the second derivative $V''_+(1)$ from the right at 1 exists and satisfies $V''_+(1) = \frac{\alpha}{2} - 2\delta$.*

(iii) *If in addition W satisfies assumptions (ii') and (iii'), then there exists a convex function $V \geq 0$ with $V(r) \leq \tilde{W}(r) \leq V(r) + O((r-1)^4)$ for $r \leq T$.*

(iv) *For $\rho > 0$ there is an increasing, subadditive function $\psi^\rho : [0,\infty) \to (0,\infty)$ which satisfies $\psi^\rho(r) - \rho \leq \tilde{W}(r+1)$ for all $r \geq 0$ and $\psi(r) = \beta$ for all $r \geq c_\rho$ for some constant c_ρ only depending on ρ.*

Proof. (i) Let $F \in \mathbb{R}^{2\times 2}$ satisfying $|F| \leq T$. By polar decomposition we find $R \in O(2)$ and $U = \sqrt{F^T F}$ symmetric and non-negative definite such that $F = RU$. A short computation yields $|U - \mathbf{Id}| = \operatorname{dist}(F, O(2))$. Assume first $|U - \mathbf{Id}| < \eta$ for $\eta > 0$ small enough. Since $W_\triangle(F)$ is invariant under rotation and reflection we obtain applying Lemma 2.4.2:

$$W_\triangle(F) = W_\triangle(R^T RU) \geq \frac{1}{2} Q(U - \mathbf{Id}) + o(|U - \mathbf{Id}|^2).$$

Noting that Q grows quadratically on $\mathbb{R}^{2\times 2}_{\mathrm{sym}}$ (see Lemma 2.4.2) we obtain a constant $c_1 > 0$ such that for $|U - \mathbf{Id}| < \eta$

$$W_\triangle(F) \geq c_1 |U - \mathbf{Id}|^2 = c_1 \operatorname{dist}^2(F, O(2)).$$

Consider the compact set $M := \{F \in \mathbb{R}^{2\times 2}, \operatorname{dist}(F, O(2)) \geq \eta, |F| \leq T\}$. $W_\triangle$ attains its minimum on M, which is strictly positive by Lemma 2.4.1(ii). This provides a second constant $c_2 > 0$ such that for all $F \in M$

$$W_\triangle(F) \geq c_2 |U - \mathbf{Id}|^2 = c_2 \operatorname{dist}^2(F, O(2)).$$

Taking $c = \min\{c_1, c_2\}$ yields the claim.

(ii) We construct such a function directly applying Lemma 2.4.3.

$$V(r) = \begin{cases} 0 & \text{for } r \leq 1, \\ \left(\frac{\alpha}{4} - \delta\right)(r-1)^2 & \text{for } 1 \leq r \leq 1+\eta, \\ \left(\frac{\alpha}{4} - \delta\right)\eta\,(2r - 2 - \eta) & \text{for } r \geq 1+\eta, \end{cases}$$

when $\eta > 0$ is sufficiently small.

(iii) With $f(r) := \frac{\alpha(r-1)^2}{4} + \frac{1}{108}\Big(6\alpha + 7\alpha' - 2(3\alpha - \alpha')\cos(6\phi)\Big)(r-1)^3 - C(r-1)^4$ for sufficiently large C, Lemma 2.4.4 shows that we can choose

$$V(r) = \begin{cases} 0 & \text{for } r \leq 1, \\ f(r) & \text{for } 1 \leq r \leq 1+\eta, \\ f(1+\eta) + f'(1+\eta)(r-1-\eta) & \text{for } r \geq 1+\eta, \end{cases}$$

when $\eta > 0$ is sufficiently small.

(iv) We define

$$\bar{\psi}(r) = \begin{cases} \eta r & \text{for } 0 \le r \le \frac{\beta}{\eta}, \\ \beta & \text{for } r \ge \frac{\beta}{\eta}, \end{cases}$$

for some $\eta > 0$ (depending on ρ) such that $\bar{\psi} - \rho \le \tilde{W}$. Then we set $\psi^\rho(r) = \bar{\psi}(r+1)$. As ψ^ρ is a concave function with $\psi^\rho(0) > 0$, it is subadditive. □

Recall the definition of f_κ in (1.30). Finally, we provide a lower bound for $W_{\Delta,\chi} + f_\kappa$ which implies the equicoercivity of $(\mathcal{F}_\varepsilon^\chi)_\varepsilon$.

Lemma 2.4.6. *Let $T > \sqrt{2}$. Then there are constants $C_1, C_2 > 0$ such that for all $F \in \mathbb{R}^{2\times 2}$ with $|F| \le T$ we obtain*

i) $|\hat{m}(F) - \hat{m}(R(F))| \le C_1 |F - R(F)|^2$, *where $R(F) \in SO(2)$ is a solution of $|F - R(F)| = \min_{R \in SO(2)} |F - R|$,*

(ii) $W_{\Delta,\chi}(F) + f_\kappa(F) \ge C_2 |F - \mathbf{Id}|^2$.

Proof. (i) Without restriction we may assume that $|F - R(F)|$ is small as otherwise the assertion is clear. So in particular, $R(F)$ is uniquely determined. Moreover, it suffices to consider $F \in \mathbb{R}^{2\times 2}_{\text{sym}}$ and $R(F) = \mathbf{Id}$. Indeed, once this is proved, we find $|\hat{m}(F) - \hat{m}(R(F))| = |R(F)\hat{m}(R(F)^T F) - R(F)\hat{m}(\mathbf{Id})| \le C|R(F)^T F - \mathbf{Id}|^2$, as desired.

Let $F \in \mathbb{R}^{2\times 2}_{\text{sym}}$, $R(F) = \mathbf{Id}$ and set $G = F - \mathbf{Id}$ with $G \in \mathbb{R}^{2\times 2}_{\text{sym}}$ small. As $\hat{m}$ is C^2 in a neighborhood of $SO(2)$ we derive $|\hat{m}(F) - \hat{m}(\mathbf{Id})| \le |D\hat{m}(\mathbf{Id})\, G| + C|G|^2 = C|G|^2$ as $\mathbb{R}^{2\times 2}_{\text{sym}} \subset \ker(D\hat{m}(\mathbf{Id}))$.

(ii) By Lemma 2.4.5(i) the assertion is clear for all $|F| \le T$ with $c_0 \le \operatorname{dist}(F, O(2))$ for $c_0 > 0$ and $C_2 = C_2(c_0, T)$ sufficiently small. Otherwise, we again apply Lemma 2.4.5(i) to obtain for c_0 small enough

$$W_{\Delta,\chi}(F) \ge C \operatorname{dist}^2(F, O(2)) + \chi(F) \ge C \operatorname{dist}^2(F, SO(2)) = C|F - R(F)|^2.$$

For convenience we write $r_{ij} = \mathbf{e}_i^T R(F) \mathbf{e}_j$ for $i, j = 1, 2$. As $r_{12}^2 = r_{21}^2 = 1 - r_{11}^2$ we find $1 - r_{11} = 1 - r_{11}^2 + r_{11}(r_{11} - 1) = r_{12}^2 + (1 - r_{11})^2 - (1 - r_{11})$. Thus, recalling $\hat{m}(R) = R\mathbf{e}_1$ for all $R \in SO(2)$ and applying (i) we get for $0 < c \le \kappa$ small enough

$$\begin{aligned} W_{\Delta,\chi}(F) &+ f_\kappa(F) \\ &\ge C|F - R(F)|^2 + c(1 - \mathbf{e}_1 \cdot \hat{m}(R(F))) + c\mathbf{e}_1 \cdot (\hat{m}(R(F)) - \hat{m}(F)) \\ &\ge C|F - R(F)|^2 + c(1 - \mathbf{e}_1^T R(F)\mathbf{e}_1) - cC_1|F - R(F)|^2 \\ &\ge \frac{C}{2}|F - R(F)|^2 + \frac{c}{2}(1 - r_{11})^2 + \frac{c}{2} r_{12}^2 \ge C_2|F - \mathbf{Id}|^2, \end{aligned} \tag{2.9}$$

as desired. □

Chapter 3

Limiting minimal energy and cleavage laws

This chapter is devoted to the derivation of the limiting minimal energy. First, in Section 3.1 we give an independent proof for the triangular lattice. Section 3.2 contains the preliminary results concerning a mesoscopic localization technique and the proof of Theorem 1.3.1 is addressed in Section 3.3. Finally, in Section 3.4 we give some examples to which our cleavage law applies.

3.1 Warm up: Proof for the triangular lattice

As a first approach to the proof of the cleavage law we present an elementary proof for the planar model where the atoms in the reference configuration are given by a portion of a triangular lattice. We will first establish a lower bound for the limiting minimal energy by considering slices of the form $(0, l) \times \{x_2\}$ for $x_2 \in (0, 1)$ and using the reduced energy defined in (2.7). In a second step we show that this bound is attained by either elastic deformations or by configurations with cleavage along a specific crystallographic line depending on the boundary displacements.

We can classify (or 'color') all triangles in $\mathcal{C}_\varepsilon$ into two types, say 'type one' and 'type two', such that all triangles of the same type are translates of each other. Then only triangles of different type can share a common side. Denote the sets by $\mathcal{C}_\varepsilon^{(1)}$ and $\mathcal{C}_\varepsilon^{(2)}$, respectively.

Proof of Theorem 1.4.1. We first show that the expression on the right hand side is a lower bound for the limiting minimal energy. For every deformation $y \in \mathcal{A}(a_\varepsilon)$ we have by (1.20) and (1.21)

$$\mathcal{E}_\varepsilon(y) \geq \frac{4}{\sqrt{3}\varepsilon} \int_{\Omega_\varepsilon \cap (0,l) \times (\varepsilon, 1-\varepsilon)} W_\triangle(\nabla \tilde{y})\, dx.$$

Let $0 < \delta < \frac{\alpha}{4}$ and choose R so large that $W(r) > \beta - \delta$ if $r \geq R$. Define $\bar{\mathcal{C}}_\varepsilon^{(1)}$ to be the set of those triangles $\triangle$ of type one for which at least one side in the deformed configuration $y(\triangle)$ is larger than $2R\varepsilon$. By $I \subset (\varepsilon, 1-\varepsilon)$ we denote the set of those points x_2 for which there exists $x_1 \in (0, l)$ such that (x_1, x_2) lies in one of these triangles.

We can then estimate the energy integral by splitting the x_2-integration into a first part where $x_2 \notin I$ and a second part with $x_2 \in I$.

1. If $x_2 \notin I$, then all sidelengths of $y(\triangle)$ for a triangle $\triangle$ whose interior intersects the segment $(0, l) \times \{x_2\}$ are less or equal to $4R\varepsilon$. This is clear for triangles of type one by construction. For triangles of type two it follows from the fact that the two sides of $\triangle$ intersecting $(0, l) \times \{x_2\}$ are also sides of triangles of type one and therefore bounded by $2R\varepsilon$. The third side is thus less than $4R\varepsilon$, too.

It is elementary to see that for $F \in \mathbb{R}^{2\times 2}$

$$|\mathbf{e}_1^T F \mathbf{e}_1| \leq 8R, \quad \text{if } |\mathbf{v}^T F \mathbf{v}| \leq 4R \text{ for all } \mathbf{v} \in \mathcal{V}. \tag{3.1}$$

Indeed, if λ_1, λ_2 are the eigenvalues of $\frac{1}{2}(F^T + F)$, then by (2.8) one has $\frac{3}{4}(\lambda_1^2 + \lambda_2^2) = \frac{3}{4} \operatorname{trace} \left(\frac{1}{2}(F^T + F)\right)^2 \leq 3 \cdot (4R)^2$ and thus $|\mathbf{e}_1^T F \mathbf{e}_1| \leq \max\{|\lambda_1|, |\lambda_2|\} \leq 8R$. Consequently, for almost every $x_2 \notin I$ we have $\mathbf{e}_1^T \nabla \tilde{y}(x_1, x_2) \mathbf{e}_1 \leq 8R$ for all $x_1 \in (0, l)$.

By Lemma 2.4.5(ii) choose a convex function with $V(r) \leq \tilde{W}(r)$ for $r \leq 8R$ and $V_+''(1) = \frac{\alpha}{2} - 2\delta$. For $x_2 \in (\varepsilon, 1-\varepsilon)$ define $\Omega_\varepsilon^{x_2} \subset (0, l)$ such that $\Omega_\varepsilon^{x_2} \times \{x_2\} = \Omega_\varepsilon \cap (0, l) \times \{x_2\}$. Then for the first part one obtains, if $a < \infty$, by convexity of V

$$\begin{aligned}
\frac{4}{\sqrt{3}\varepsilon} \int_{(\varepsilon,1-\varepsilon)\setminus I} \int_{\Omega_\varepsilon^{x_2}} W_\triangle(\nabla \tilde{y})\, dx_1\, dx_2 &\geq \frac{4}{\sqrt{3}\varepsilon} \int_{(\varepsilon,1-\varepsilon)\setminus I} \int_{\Omega_\varepsilon^{x_2}} V\left(\mathbf{e}_1^T \nabla \tilde{y}\, \mathbf{e}_1\right) dx_1\, dx_2 \\
&\geq \frac{4}{\sqrt{3}\varepsilon} \int_{(\varepsilon,1-\varepsilon)\setminus I} |\Omega_\varepsilon^{x_2}| V(1 + a_\varepsilon)\, dx_2 \\
&\geq \frac{2}{\sqrt{3}\varepsilon} (1 - 2\varepsilon - |I|)(l - 2\varepsilon)(V_+''(1) a_\varepsilon^2 + o(\varepsilon)) \\
&\to \frac{2}{\sqrt{3}} (1 - |I|) l V_+''(1) a^2
\end{aligned} \tag{3.2}$$

as $\varepsilon \to 0$. It is not hard to see that this asymptotic estimate remains true also for $a = \infty$.

2. On the other hand, the energy of the second part can be estimated by the energy of all springs lying on the side of a triangle in $\bar{\mathcal{C}}_\varepsilon^{(1)}$, which yields

$$\frac{4}{\sqrt{3}\varepsilon} \int_I \int_{\Omega_\varepsilon^{x_2}} W_\triangle(\nabla \tilde{y})\, dx_1\, dx_2 \geq 2(\beta - \delta)\varepsilon \# \bar{\mathcal{C}}_\varepsilon^{(1)}, \tag{3.3}$$

as the length of at least two springs in each of these triangles is larger than $R\varepsilon$ in the deformed configuration. Now the projection of any triangle onto the x_2-axis is an interval of length $\varepsilon \sin(\phi + \frac{\pi}{3}) = \varepsilon\gamma$, and so $\varepsilon\gamma \# \bar{\mathcal{C}}_\varepsilon^{(1)} \geq |I|$, i.e.,

$$\frac{4}{\sqrt{3}\varepsilon} \int_I \int_{\Omega_\varepsilon^{x_2}} W_\triangle(\nabla \tilde{y}) \, dx_1 \, dx_2 \geq 2(\beta - \delta)\gamma^{-1}|I|. \tag{3.4}$$

Summarizing (3.2) and (3.4) we find

$$\begin{aligned}
&\liminf_{\varepsilon \to \infty} \inf\{\mathcal{E}_\varepsilon(y) : y \in \mathcal{A}(a_\varepsilon)\} \\
&\geq \min\left\{\frac{2}{\sqrt{3}}\left(\frac{\alpha}{2} - 2\delta\right) la^2(1 - |I|) + 2(\beta - \delta)\gamma^{-1}|I| : |I| \in [0, 1]\right\} \\
&= \min\left\{\frac{2}{\sqrt{3}}\left(\frac{\alpha}{2} - 2\delta\right) la^2, \frac{2(\beta - \delta)}{\gamma}\right\}.
\end{aligned}$$

Now $\delta \to 0$ shows

$$\liminf_{\varepsilon \to \infty} \inf\{\mathcal{E}_\varepsilon(y) : y \in \mathcal{A}(a_\varepsilon)\} \geq \min\left\{\frac{\alpha l}{\sqrt{3}} a^2, \frac{2\beta}{\gamma}\right\}.$$

This establishes the lower bound.

It remains to prove that the right hand side in Theorem 1.4.1 is attained for some sequence of deformations. In order to do so, we consider two specific sequences of deformations. First, for $a < \infty$ let

$$y_\varepsilon^{\mathrm{el}}(x) = (\mathbf{Id} + F^{a_\varepsilon})x = \begin{pmatrix} 1 + a_\varepsilon & 0 \\ 0 & 1 - \frac{a_\varepsilon}{3} \end{pmatrix} x. \tag{3.5}$$

By Lemma 2.4.2 we have that $W_\triangle(F) = \frac{\alpha}{4} a_\varepsilon^2 + o(\varepsilon)$ and so

$$\lim_{\varepsilon \to 0} \mathcal{E}_\varepsilon(y_\varepsilon^{\mathrm{el}}) = \frac{\alpha l}{\sqrt{3}} a^2$$

by (1.20).

To define y^{cr} we choose any line $(s, 0) + \mathbb{R}\mathbf{v}_\gamma$ intersecting both the segments $(0, l) \times \{0\}$ and $(0, l) \times \{1\}$ (as in Corollary 1.5.2). This is possible since $l > \frac{1}{\sqrt{3}}$. Let $a > 0$ and set

$$y_\varepsilon^{\mathrm{cr}}(x) = \begin{cases} x & \text{for } x \text{ to the left of } (s, 0) + \mathbb{R}\mathbf{v}_\gamma, \\ x + a_\varepsilon l \mathbf{e}_1 & \text{for } x \text{ to the right of } (s, 0) + \mathbb{R}\mathbf{v}_\gamma \end{cases} \tag{3.6}$$

for atoms x with $\varepsilon < x_1 < l - \varepsilon$. Except for a negligible contribution from the boundary layers, the energy of this configuration can be estimated as in Step 2 of the proof of the lower bound: It is given by the energy of springs intersecting $(s, 0) + \mathbb{R}\mathbf{v}_\gamma$, i.e., by the two springs lying on the boundary of the triangles of type one which are intersected by $(s, 0) + \mathbb{R}\mathbf{v}_\gamma$. These springs are elongated by a factor scaling with $a_\varepsilon/\varepsilon$, thus yielding a contribution β in the limit $\varepsilon \to 0$. □

3.2 Estimates on a mesoscopic cell

3.2.1 Mesoscopic localization

The goal of this section is the derivation of a lower comparison potential on 'large cells' defined on a mesoscopic scale ε^s with

$$s = \frac{3d-1}{3d}.$$

We define the domain $U_\varepsilon = (0, \varepsilon^s\lambda) \times (0, \varepsilon^s)^{d-1}$ for $\lambda \in [\lambda_0, 2\lambda_0]$, where $\lambda_0 \geq L = L(\sqrt{A^T A}, W_{\text{cell}}, 1, \ldots, 1)$. We consider the Bravais-lattice defined in Section 1.1 with a possible translation. For $\rho \in A[0,1)^d$ set $\mathcal{L}_{\varepsilon,\rho} = \varepsilon\rho + \mathcal{L}_\varepsilon$. Let $D, D' > 0$ and suppose that $\tilde{W} \subset (0,1)^{d-1}$ is connected with

$$\frac{1}{2} < \mathcal{H}^{d-1}(\tilde{W}) \leq 1, \qquad \mathcal{H}^{d-2}(\partial\tilde{W}) < D, \tag{3.7}$$

such that the connected component W of $\{x \in \tilde{W} : \text{dist}(x, \partial\tilde{W}) > D'\varepsilon^{1-s}\}$ with largest Lebesgue measure also satisfies (3.7) for ε small enough.

We define

$$\partial_W(\mathcal{L}'_{\varepsilon,\rho}(U_\varepsilon)) = \{\bar{x} \in \mathcal{L}'_{\varepsilon,\rho} : \overline{Q}_\varepsilon(\bar{x}) \cap (\{0, \varepsilon^s\lambda\} \times \varepsilon^s W) \neq \emptyset\}. \tag{3.8}$$

Let $y : \mathcal{L}_{\varepsilon,\rho}(U_\varepsilon) \to \mathbb{R}^d$ be the lattice deformations on U_ε with corresponding energy

$$E(U_\varepsilon, y) = \varepsilon^{d-1} \sum_{\bar{x} \in (\mathcal{L}'_{\varepsilon,\rho}(U_\varepsilon))^\circ} W_{\text{cell}}(\bar{\nabla} y(\bar{x})) + \frac{1}{2}\varepsilon^{d-1} \sum_{\bar{x} \in \partial_W(\mathcal{L}'_{\varepsilon,\rho}(U_\varepsilon))} W_{\text{cell}}(\bar{\nabla} y(\bar{x})).$$

The factor $\frac{1}{2}$ takes account of the fact that in the proof of Theorem 1.3.1 half of the energy of the boundary cells will be assigned to each of the two adjacent mesoscopic cells. Let $\tilde{y}$ denote the interpolation for y defined in Section 2.2. For $r \in \mathbb{R}$ we will investigate the minimization problem of finding $\inf E(U_\varepsilon, y)$ under certain boundary conditions given as follows. We define the averaged boundary conditions

$$\frac{1}{\varepsilon^{s(d-1)}|W|} \int_{\varepsilon^s W} \Big(\tilde{y}^1(\lambda\varepsilon^s, x') - \tilde{y}^1(0, x')\Big)\, dx' = \lambda\varepsilon^s(1+r). \tag{3.9}$$

Here, $x' = (x_2, \ldots, x_d)$, $|W| = \mathcal{H}^{d-1}(W)$ and $\tilde{y}^1$ denotes the first component of $\tilde{y}$. Moreover, we introduce the condition

$$\bigcup_{\bar{x} \in \bar{\mathcal{B}}'_\varepsilon} Q_\varepsilon(\bar{x}) \cap \big(\{0, \varepsilon^s\lambda\} \times \varepsilon^s\tilde{W}\big) = \emptyset, \tag{3.10}$$

where $\tilde{W}$ is of the form (3.7) and similar as in Section 2.2, $\bar{\mathcal{B}}'_\varepsilon \subset \mathcal{L}'_{\varepsilon,\rho}(U_\varepsilon)$ denotes the cells, where $\operatorname{diam}_1 \left\{\bar{\nabla} y(\bar{x}), \mathcal{Z}\right\} > C^*_{\text{int}}$ for a fixed $C^*_{\text{int}} > 0$ (which may differ from C_{int} to be chosen later). Note that in contrast to the definition of $\bar{\mathcal{C}}'_\varepsilon$ we consider $\operatorname{diam}_1\{\cdot, \mathcal{Z}\}$ only instead of $\operatorname{diam}\{\cdot, \mathcal{Z}\}$.

We now concern ourselves with the minimization problem

$$M(U_\varepsilon, r) := \inf\left\{E(U_\varepsilon, y) : y \text{ satisfies (3.9) and (3.10)}\right\}.$$

Before we state the main theorem of this section we briefly note that $M(U_\varepsilon, r) = 0$ for $-2 \leq r \leq 0$ as the averaged boundary conditions may be satisfied by a suitable rotation of the specimen, i.e. $y(x) = Rx$ for $R \in SO(d)$.

Theorem 3.2.1. *Let $\lambda_0 \geq L$ and $C_2 > C_1 > 0$, C_2 sufficiently large. Let $\delta > 0$ small. Then for all $W \subset (0,1)^{d-1}$ as in (3.7) and $\rho \in A[0,1)^d$ there is a function*

$$f : \mathbb{R} \times [\varepsilon^s \lambda_0, 2\varepsilon^s \lambda_0] \to \mathbb{R}, \qquad (r, \lambda) \mapsto f(r, \lambda),$$

which for $r \leq \max\{\varepsilon^{(s-1)(d-3)}, 1\}C_2$ is convex in r and linear in λ and for ε small enough, independently of ρ and W, satisfies

- *for $r \in \mathbb{R}$, $\lambda \in [\varepsilon^s \lambda_0, 2\varepsilon^s \lambda_0]$: $f(r, \lambda) \leq M(U_\varepsilon, r)$,*
- *for $r \in [0, C_1\sqrt{\varepsilon}]$, $\lambda \in [\varepsilon^s \lambda_0, 2\varepsilon^s \lambda_0]$:*

$$\begin{aligned} f(r, \lambda) &= \varepsilon^{s(d-1)} \lambda \omega(|W|) \left(\frac{\alpha_A}{2 \det A} \frac{r^2}{\varepsilon} - \delta\right) \\ &\leq M(U_\varepsilon, r) \leq \frac{1}{\omega(|W|)} f(r, \lambda) + 4\varepsilon^{sd} \lambda_0 \delta, \end{aligned} \tag{3.11}$$

- *for $r \geq \max\{\varepsilon^{(s-1)(d-3)}, 1\}C_2$, $\lambda \in [\varepsilon^s \lambda_0, 2\varepsilon^s \lambda_0]$:*

$$\begin{aligned} f(r, \lambda) &= \varepsilon^{s(d-1)} \omega(|W|) \left(\frac{\beta_A}{\det A} - \delta\right) \\ &\leq M(U_\varepsilon, r) \leq \frac{1}{\omega(|W|)} f(r, \lambda) + 2\varepsilon^{s(d-1)} \delta \end{aligned} \tag{3.12}$$

for a continuous function $\omega : [0,1] \to \mathbb{R}$ with $\omega(1) = 1$.

The theorem shows that $f(r, \lambda)$ is a lower bound for $M(U_\varepsilon, r)$ which becomes sharp in the regimes $r \in [0, C_1\sqrt{\varepsilon}]$ and $r \in [\max\{\varepsilon^{(s-1)(d-3)}, 1\}C_2, \infty)$ provided that $|W| \approx 1$.

The proof of Theorem 3.2.1 is essentially divided into three steps each of which dealing with one particular regime: The elastic regime (Lemma 3.2.2), the fracture regime (Lemma 3.2.4) and the one in between (Lemma 3.2.3). In addition, in the case $d \geq 4$ we need a short additional argument in the intermediate regime (Lemma 3.2.5). It will be convenient to rescale the system in order to obtain a problem on a macroscopic domain not depending on ε. Therefore, we let $\hat{\varepsilon} = \varepsilon^{1-s} = \varepsilon^{1/3d}$, $\hat{U} = (0, \lambda) \times (0,1)^{d-1}$, $\hat{y} : \mathcal{L}_{\hat{\varepsilon},\rho}(\hat{U}) \to \mathbb{R}^d$ and

$$\begin{aligned}\hat{E}(\hat{U}, \hat{y}) &= \hat{\varepsilon}^d \varepsilon^{-1} \sum_{\bar{x} \in (\mathcal{L}'_{\hat{\varepsilon},\rho}(\hat{U}))^\circ} W_{\mathrm{cell}}(\bar{\nabla}\hat{y}(\bar{x})) + \frac{1}{2}\hat{\varepsilon}^d \varepsilon^{-1} \sum_{\bar{x} \in \partial_W (\mathcal{L}'_{\hat{\varepsilon},\rho}(\hat{U}))} W_{\mathrm{cell}}(\bar{\nabla}\hat{y}(\bar{x})) \\ &= \varepsilon^{-\frac{2}{3}} \sum_{\bar{x} \in (\mathcal{L}'_{\hat{\varepsilon},\rho}(\hat{U}))^\circ} W_{\mathrm{cell}}(\bar{\nabla}\hat{y}(\bar{x})) + \frac{1}{2}\varepsilon^{-\frac{2}{3}} \sum_{\bar{x} \in \partial_W (\mathcal{L}'_{\hat{\varepsilon},\rho}(\hat{U}))} W_{\mathrm{cell}}(\bar{\nabla}\hat{y}(\bar{x})),\end{aligned} \tag{3.13}$$

where $\bar{\nabla}\hat{y}$ is defined as in (1.3) replacing ε by $\hat{\varepsilon}$ and $\partial_W(\mathcal{L}'_{\hat{\varepsilon},\rho}(\hat{U}))$ as in (3.8) replacing ε^s by 1. The averaged boundary conditions now become

$$\fint_W \left(\tilde{\hat{y}}^1(\lambda, x') - \tilde{\hat{y}}^1(0, x')\right) dx' = \lambda(1 + r), \tag{3.14}$$

where $\fint$ denotes the averaged integral and the interpolation $\tilde{\hat{y}}$ is defined as in Section 2.2. The condition (3.10) on the boundary cells reads as

$$\bigcup_{\bar{x} \in \mathcal{B}'_{\hat{\varepsilon}}} Q_{\hat{\varepsilon}}(\bar{x}) \cap \left(\{0, \lambda\} \times \tilde{W}\right) = \emptyset \tag{3.15}$$

and the minimum problem in the rescaled variant becomes

$$\hat{M}(\hat{U}, r) := \inf \left\{ \hat{E}(\hat{U}, \hat{y}) : \hat{y} \text{ satisfies } (3.14), (3.15) \right\}. \tag{3.16}$$

It is not hard to see that

$$M(U_\varepsilon, r) = \varepsilon^{sd} \hat{M}(\hat{U}, r) = \varepsilon^{\frac{3d-1}{3}} \hat{M}(\hat{U}, r). \tag{3.17}$$

3.2.2 Estimates in the elastic regime

We first determine $\hat{M}(\hat{U}, r)$ for r near zero.

Lemma 3.2.2. *For $0 \leq r \leq C_{\mathrm{el}}\sqrt{\varepsilon}$ the minimizing problem (3.16) satisfies*

$$|W| \frac{\lambda \alpha_A}{2 \det A} \frac{r^2}{\varepsilon} + o(1) \leq \hat{M}(\hat{U}, r) \leq \frac{\lambda \alpha_A}{2 \det A} \frac{r^2}{\varepsilon} + o(1) \tag{3.18}$$

for $\varepsilon \to 0$ with α_A as in (1.16). Here $o(1)$ is independent of $\rho \in A[0,1)^d$, $W \subset (0,1)^{d-1}$ and $\lambda \in [\lambda_0, 2\lambda_0]$ and depends only on C_{el}.

Proof. In the following we drop the superscript $\hat{\cdot}$ if no confusion arises. We first show that

$$M(U,r) \geq |W| \frac{\lambda \alpha_A}{2 \det A} \frac{r^2}{\varepsilon} + o(1) \tag{3.19}$$

for $\varepsilon \to 0$, where $o(1)$ only depends on C_{el}. We argue by contradiction. If the claim were false, there would exist a $\delta > 0$, sequences $\varepsilon_k \to 0$, $C_{\mathrm{el}}\sqrt{\varepsilon_k} \geq r_k \to 0$, $\lambda_k \in [\lambda_0, 2\lambda_0]$, $\rho_k \in A[0,1)^d$, $W_k \subset (0,1)^{d-1}$ satisfying (3.7) as well as a sequence $y_k : \mathcal{L}_{\hat{\varepsilon}_k,\rho_k}(U_k) \to \mathbb{R}^d$ satisfying (3.14) with respect to r_k, (3.15) and $E(U_k, y_k) \leq M(U_k, r_k) + \frac{1}{k}$ such that

$$E(U_k, y_k) \leq |W_k| \frac{\lambda_k \alpha_A}{2 \det A} \frac{r_k^2}{\varepsilon_k} - \delta. \tag{3.20}$$

Passing to a subsequence we may assume that $\rho_k \to \rho \in A[0,1)^d$, $\lambda_k \to \lambda \in [\lambda_0, 2\lambda_0]$ and $\varepsilon_k^{-\frac{1}{2}} r_k \to r \geq 0$. Moreover, as $\mathcal{H}^{d-2}(\partial W_k)$ is uniformly bounded in k, we may assume that $\chi_{W_k} \to \chi_W$ in measure for some $W \subset (0,1)^{d-1}$ with $\frac{1}{2} \leq |W| \leq 1$ by Theorem A.1.2. As discussed in Section 1.2, there is an obvious choice for an elastic deformation, namely $y_k^*(x) = (1 + r_k)x$ for all $x \in \mathcal{L}_{\hat{\varepsilon}_k,\rho_k}(U_k)$. We note that $\tilde{y}_k^*$ satisfies (3.14) as this interpolation by construction is equal to the linear map y_k^*. It is elementary to see that

$$\begin{aligned} E(U_k, y_k^*) = \frac{(1 + O(\hat{\varepsilon}_k))\, \lambda_k}{\det A} \frac{1}{\varepsilon_k} W_{\mathrm{cell}}((1 + r_k)Z) &\leq \frac{C}{\varepsilon_k} \left(Q_{\mathrm{cell}}(r_k Z) + o(r_k^2) \right) \\ &\leq \frac{C (C_{\mathrm{el}} \sqrt{\varepsilon_k})^2}{\varepsilon_k} \leq C. \end{aligned} \tag{3.21}$$

As usual we follow the convention of denoting different constants with the same letter. As by (2.4) and (3.13) a broken cell contributes an energy of order $\varepsilon_k^{-\frac{2}{3}}$ the comparison with (3.21) yields

$$\bar{\mathcal{C}}'_{\hat{\varepsilon}_k} = \emptyset \tag{3.22}$$

for $k \in \mathbb{N}$ sufficiently large. This together with (3.15) shows that $\tilde{y}_k$ is a continuous, piecewise linear interpolation on the set V_k, where

$$V_k^\circ = \bigcup_{\bar{x} \in (\mathcal{L}'_{\hat{\varepsilon}_k,\rho_k}(U_k))^\circ} Q_{\hat{\varepsilon}_k}(\bar{x}), \quad V_k = V_k^\circ \cup \bigcup_{\bar{x} \in \partial_{W_k}(\mathcal{L}'_{\hat{\varepsilon}_k,\rho_k}(U_k))} Q_{\hat{\varepsilon}_k}(\bar{x}). \tag{3.23}$$

Applying Lemma 2.2.1(i) and Lemma 2.1.1 we obtain

$$\begin{aligned} \int_{V_k} \mathrm{dist}^2(\nabla \tilde{y}_k, SO(d)) &\leq C \int_{V_k} \mathrm{dist}^2(\bar{\nabla} y_k(\bar{x}), \bar{SO}(d)) \leq C\varepsilon_k \, E(U_k, y_k) \\ &\leq C\varepsilon_k \left(M(U_k, r_k) + k^{-1} \right) \leq C\varepsilon_k \left(E(U_k, y_k^*) + k^{-1} \right) \leq C\varepsilon_k. \end{aligned}$$

In order to estimate $\int_{V_k} \mathrm{dist}^2(\nabla \tilde{y}_k, SO(d))$ we use the geometric rigidity result in Theorem B.1. We find a constant $C = C(\lambda_0)$ and rotations $R_k \in SO(d)$ such that

$$\|\nabla \tilde{y}_k - R_k\|^2_{L^2(V_k)} \leq C\varepsilon_k. \tag{3.24}$$

In fact, C depends only on λ_0 as all shapes V_k are related to $(0,\lambda_0) \times (0,1)^{d-1}$ through bi-Lipschitzian homeomorphisms with Lipschitz constants of both the homeomorphism itself and its inverse uniformly bounded in k, see Section B. Up to a not relabeled subsequence we may assume that $R_k \to R$ for some $R \in SO(d)$. By Poincaré's inequality we obtain

$$\|\tilde{y}_k - (R_k\, x + c_k)\|^2_{L^2(V_k)} \leq C\varepsilon_k \tag{3.25}$$

for suitable constants $c_k \in \mathbb{R}^d$. We let $u_k = \frac{1}{\sqrt{\varepsilon_k}}(y_k - (R_k \;\cdot + c_k))$ and obtain by Lemma 2.2.1(ii)

$$\|\bar{\nabla} u_k\|^2_{L^2(V_k)} \leq \frac{C}{\varepsilon_k}\|\nabla \tilde{y}_k - R_k\|^2_{L^2(V_k)} \leq C. \tag{3.26}$$

From (3.24) and (3.25) we deduce that for a suitable subsequence (not relabeled) $\chi_{V_k}\tilde{u}_k \rightharpoonup \chi_U u$ and $\chi_{V_k}\nabla\tilde{u}_k \rightharpoonup \chi_U \nabla u$ in L^2 for some $u \in H^1(U,\mathbb{R}^d)$, where $U = (0,\lambda) \times (0,1)^{d-1}$. Then by (3.26) and possibly passing to a further subsequence we find some $f \in L^2(U, \mathbb{R}^{d\times 2^d})$ such that $\chi_{V_k}\bar{\nabla} u_k \rightharpoonup \chi_U f$ in L^2. By (3.22) and Lemma 2.2.2 we obtain $f = \nabla u \cdot Z$, i.e. in particular $\chi_{V_k}\bar{\nabla} u_k \rightharpoonup \chi_U \nabla u \cdot Z$ in L^2.

We now concern ourselves with the averaged boundary condition (3.14) for the displacement fields u_k. We obtain

$$\begin{aligned}\lambda_k(1 + r_k) &= \fint_{W_k} (\tilde{y}_k^1(\lambda_k, x') - \tilde{y}_k^1(0, x'))\, dx' \\ &= (R_k)_{11}\lambda_k + \sqrt{\varepsilon_k}\fint_{W_k} (\tilde{u}_k^1(\lambda_k, x') - \tilde{u}_k^1(0, x'))\, dx'\end{aligned}$$

and therefore

$$\fint_{W_k} (\tilde{u}_k^1(\lambda_k, x') - \tilde{u}_k^1(0, x'))\, dx' = \lambda_k \frac{r_k}{\sqrt{\varepsilon_k}} + \frac{\lambda_k(1 - (R_k)_{11})}{\sqrt{\varepsilon_k}}. \tag{3.27}$$

We extend $\tilde{u}_k \in H^1(V_k, \mathbb{R}^d)$ to $\hat{u}_k \in H^1(U_k \cup V_k, \mathbb{R}^d)$ such that $\|\hat{u}_k\|_{H^1(U_k \cup V_k)} \leq C\|\tilde{u}_k\|_{H^1(V_k)}$, where C may be chosen independently of ε for the same reasoning as in (3.24). In particular, we note that $\hat{u}_k^1 = \tilde{u}_k^1$ on $\{0, \lambda_k\} \times W_k$. Defining $\bar{u}_k \in H^1(U, \mathbb{R}^d)$ by $\bar{u}_k(x_1, x') := \hat{u}_k(\frac{\lambda_k}{\lambda}x_1, x')$ and possibly passing to a further subsequence we may assume $\bar{u}_k \rightharpoonup u$ weakly in $H^1(U)$. Now choosing $r^* \in \mathbb{R}$ such that the trace of u satisfies

$$\fint_{W} (u^1(\lambda, x') - u^1(0, x'))\, dx = \lambda r^*, \tag{3.28}$$

by the weak continuity of the trace operator, (3.27) yields

$$r^* = \lim_{k\to\infty} \varepsilon_k^{-\frac{1}{2}}(r_k + 1 - (R_k)_{11}) \in [0,\infty). \tag{3.29}$$

For later use we remark that, since $r_k \geq 0$, the existence of this limit also implies that $R_{11} = 1$.

We now derive a lower bound for the limiting energy. To this end, note that by Assumption 1.1.2 for $G \in \mathbb{R}^{d\times 2^d}$ small we can write $W_{\text{cell}}(Z+G) = \frac{1}{2}Q_{\text{cell}}(G) + \eta(G)$ with $\sup\left\{\frac{\eta(G)}{|G|^2} : |G| \leq \rho\right\} \to 0$ as $\rho \to 0$. Furthermore, let $\chi_k(x) := \chi_{[0,\varepsilon_k^{-1/4}]}(|\bar{\nabla} u_k(x)|)$ and estimate

$$\begin{aligned}
E(U_k, y_k) &\geq \hat{\varepsilon}_k^d \varepsilon_k^{-1} \sum_{\bar{x}\in(\mathcal{L}'_{\hat{\varepsilon}_k,\rho_k}(U))^\circ} W_{\text{cell}}(R_k \cdot Z + \sqrt{\varepsilon_k}\,\bar{\nabla} u_k(\bar{x})) \\
&= \hat{\varepsilon}_k^d \varepsilon_k^{-1} \sum_{\bar{x}\in(\mathcal{L}'_{\hat{\varepsilon}_k,\rho_k}(U))^\circ} W_{\text{cell}}(Z + \sqrt{\varepsilon_k}\,R_k^{-1}\,\bar{\nabla} u_k(\bar{x})) \\
&\geq \frac{1}{\varepsilon_k \det A}\int_{V_k^\circ} \chi_k(x)\, W_{\text{cell}}(Z + \sqrt{\varepsilon_k} R_k^{-1}\,\bar{\nabla} u_k(x))\,dx \\
&\geq \frac{1}{2\det A}\int_{V_k^\circ} \chi_k(x)\Big(Q_{\text{cell}}(R_k^{-1}\,\bar{\nabla} u_k(x)) + \varepsilon_k^{-1}\eta(\sqrt{\varepsilon_k} R_k^{-1}\,\bar{\nabla} u_k(x))\Big)\,dx.
\end{aligned}$$

The second term may be bounded by

$$\chi_k |\bar{\nabla} u_k|^2 \frac{\eta(\sqrt{\varepsilon_k}\,\bar{\nabla} u_k)}{|\sqrt{\varepsilon_k}\,\bar{\nabla} u_k|^2}.$$

Since $\chi_{V_k^\circ}\chi_k|\bar{\nabla} u_k|^2$ is bounded in L^1 and $\chi_k \frac{\eta(\sqrt{\varepsilon_k}\,\bar{\nabla} u_k)}{|\sqrt{\varepsilon_k}\,\bar{\nabla} u_k|^2} \to 0$ uniformly, we deduce that $\chi_{V_k^\circ}\chi_k|\bar{\nabla} u_k|^2 \frac{\eta(\sqrt{\varepsilon_k}\,\bar{\nabla} u_k)}{|\sqrt{\varepsilon_k}\,\bar{\nabla} u_k|^2}$ converges to zero in L^1 as $k \to \infty$. Consequently,

$$\liminf_{k\to\infty} E(U_k, y_k) \geq \liminf_{k\to\infty} \frac{1}{2\det A}\int_{V_k^\circ} Q_{\text{cell}}(\chi_k(x) R_k^{-1}\,\bar{\nabla} u_k(x))\,dx.$$

As $R_k \to R$ and $\chi_k \to 1$ boundedly in measure, we obtain $\chi_{V_k^\circ}\chi_k R_k^{-1}\bar{\nabla} u_k \rightharpoonup \chi_U R^{-1}\nabla u \cdot Z$ weakly in L^2 and thus

$$\liminf_{k\to\infty} E(U_k, y_k) \geq \frac{1}{2\det A}\int_U Q_{\text{cell}}(R^{-1}\,\nabla u(x)\cdot Z)\,dx =: E_{\text{lim}}(U,u). \tag{3.30}$$

We now derive a lower bound for $E_{\text{lim}}(U,u)$ which will give a contradiction to (3.20) and thus (3.19) is proved. Since $R_{11} = 1$, we deduce that $R_{1i} = R_{i1} = 0$ for $i = 2,\dots,d$ and therefore $(R^{-1}\,\nabla u(x))_{11} = (\nabla u(x))_{11}$. Applying (1.15), Lemma

2.1.2 and (3.28) we obtain

$$
\begin{aligned}
E_{\text{lim}}(U,u) &\geq \frac{1}{2\det A}\int_U \tilde{Q}((\nabla u(x))_{11})\,dx \\
&\geq \frac{|W|}{2\det A}\fint_W \int_0^\lambda \alpha_A(\partial_1 u^1(x_1,x'))^2\,dx_1\,dx' \\
&\geq \frac{|W|\alpha_A}{2\det A}\lambda\left(\lambda^{-1}\fint_W\int_0^\lambda \partial_1 u^1(x_1,x')\,dx_1\,dx'\right)^2 \\
&= \frac{|W|\lambda\alpha_A}{2\det A}(r^*)^2,
\end{aligned}
\tag{3.31}
$$

where we have used Jensen's inequality. Recalling (3.29) we then obtain $r^* \geq \lim_{k\to\infty}\frac{r_k}{\sqrt{\varepsilon_k}} = r$ and thus by (3.20), (3.30) and (3.31)

$$
\infty > \frac{|W|\lambda\alpha_A}{2\det A}r^2 - \delta \geq \liminf_{k\to\infty} E(U_k,y_k) \geq E_{\text{lim}}(U,u) \geq \frac{|W|\lambda\alpha_A}{2\det A}r^2,
$$

giving the desired contradiction.

To see the upper bound in (3.18), for given $0 \leq r \leq C_{\text{el}}\sqrt{\varepsilon}$, $\rho \in A[0,1)^d$, $\lambda \in [\lambda_0, 2\lambda_0]$ we consider the deformation $y : \mathcal{L}_{\hat{\varepsilon},\rho} \to \mathbb{R}^d$,

$$
y(x) = x + \bar{F}(r)\,x \tag{3.32}
$$

with $\bar{F}(r)$ as in Lemma 2.1.2. As $\bar{F}(r)$ depends linearly on r we obtain

$$
\begin{aligned}
E(U,y) &= \frac{\hat{\varepsilon}^d}{\varepsilon}\sum_{\bar{x}\in(\mathcal{L}'_{\varepsilon,\rho}(\hat{U}))^\circ} W_{\text{cell}}(Z+\bar{F}(r)\cdot Z) + \frac{\hat{\varepsilon}^d}{2\varepsilon}\sum_{\bar{x}\in\partial_W(\mathcal{L}'_{\varepsilon,\rho}(\hat{U}))} W_{\text{cell}}(Z+\bar{F}(r)\cdot Z) \\
&= \frac{\lambda(1+O(\hat{\varepsilon}))}{\det A}\frac{1}{\varepsilon}W_{\text{cell}}(Z+\sqrt{\varepsilon}\,\bar{F}(r/\sqrt{\varepsilon})\cdot Z) \\
&= \frac{\lambda(1+O(\hat{\varepsilon}))}{2\det A}Q_{\text{cell}}(\bar{F}(r/\sqrt{\varepsilon})\cdot Z) + \frac{\lambda(1+O(\hat{\varepsilon}))}{2\det A}\frac{\eta(\sqrt{\varepsilon}\,\bar{F}(r/\sqrt{\varepsilon}))}{\varepsilon},
\end{aligned}
$$

where η is as before. Note that the second term converges uniformly to zero as $\varepsilon \to 0$ since $\sqrt{\varepsilon}\,\bar{F}(r/\sqrt{\varepsilon}) \leq Cr \leq \sqrt{\varepsilon}\,CC_{\text{el}}$. Thus, we obtain

$$
E(U,y) = \frac{\lambda\alpha_A}{2\det A}\frac{r^2}{\varepsilon} + o(1).
$$

Since $\tilde{y} = y$ satisfies (3.14) and (3.15), this shows that $M(U,r) \leq \frac{\lambda\alpha_A}{2\det A}\frac{r^2}{\varepsilon} + o(1)$. □

3.2.3 Estimates in the intermediate regime

We now determine $\hat{M}(\hat{U},r)$ in an intermediate regime.

Lemma 3.2.3. *Let $C_{\mathrm{med},2} > 0$, $C_{\mathrm{med},1} > 0$ sufficiently large, $1 < p < \frac{4}{3}$. Then there is a constant $C > 0$ such that the minimizing problem (3.16) satisfies*

$$\hat{M}(\hat{U}, r) \geq C|W|\lambda\varepsilon^{-\frac{p}{2}} r^p \tag{3.33}$$

for $\sqrt{\varepsilon}\, C_{\mathrm{med},1} \leq r \leq C_{\mathrm{med},2}$ as $\varepsilon \to 0$. The constant C is independent of $\rho \in A[0,1)^d$, $W \subset (0,1)^{d-1}$ and $\lambda \in [\lambda_0, 2\lambda_0]$.

Note that we only provide a lower bound which might not be sharp.

Proof. We follow the previous proof and only indicate the necessary changes. We again drop the superscript $\hat{\cdot}$ if no confusion arises. By Lemma 2.1.1 for a suitable constant $c > 0$ and some $1 < p < \frac{4}{3}$ the cell energy W_{cell} may be bounded from below by a function of the form

$$V_\varepsilon(G) = \begin{cases} \varepsilon^{1-\frac{p}{2}}\, c\, \chi_{\{\mathrm{dist}(G, \bar{SO}(d)) \geq \sqrt{\varepsilon}\}}\, \mathrm{dist}^p(G, \bar{SO}(d)) & G \perp V_0,\ |G| \leq C_{\mathrm{int}}, \\ W_{\mathrm{cell}}(G) & \text{else.} \end{cases}$$

Then $\varepsilon^{\frac{p}{2}} r^{-p} E(U, y) \geq \mathcal{E}(U, y; r)$, where

$$\mathcal{E}(U, y; r) := \hat{\varepsilon}^d \varepsilon^{\frac{p}{2}-1} r^{-p} \Big(\sum_{\bar{x} \in (\mathcal{L}'_{\hat{\varepsilon},\rho}(U))^\circ} V_\varepsilon(\bar{\nabla} y(\bar{x})) + \frac{1}{2} \sum_{\bar{x} \in \partial_W(\mathcal{L}'_{\hat{\varepsilon},\rho}(U))} V_\varepsilon(\bar{\nabla} y(\bar{x})) \Big). \tag{3.34}$$

We also note that

$$V_\varepsilon(G) \geq \varepsilon^{1-\frac{p}{2}}\, c \left(\mathrm{dist}^p(G, \bar{SO}(d)) - \varepsilon^{\frac{p}{2}} \right) \tag{3.35}$$

for $G \in \mathbb{R}^{d \times 2^d}$ with $G \perp V_0$ and $|G| \leq C_{\mathrm{int}}$. We show that for sufficiently small ε

$$\mathcal{M}(U, r) := \inf \{\mathcal{E}(U, y; r) : y \text{ satisfies } (3.14), (3.15)\} \geq C|W|\lambda \tag{3.36}$$

for some $C > 0$ and argue again by contradiction. If (3.36) were false, there would exist sequences $\varepsilon_k \to 0$, $C_{\mathrm{med},2} \geq r_k \geq C_{\mathrm{med},1}\sqrt{\varepsilon_k}$, $\lambda_k \in [\lambda_0, 2\lambda_0]$, $\rho_k \in A[0,1)^d$, $W_k \subset (0,1)^{d-1}$ satisfying (3.7) as well as a sequence $y_k : \mathcal{L}_{\hat{\varepsilon}_k,\rho_k}(U_k) \to \mathbb{R}^d$ satisfying (3.14) with respect to r_k, (3.15) and $\mathcal{E}(U_k, y_k; r_k) \leq \mathcal{M}(U_k, r_k) + \frac{1}{k}$ such that

$$\mathcal{E}(U_k, y_k; r_k) \leq \frac{|W_k|\lambda_k}{k}. \tag{3.37}$$

As above we assume that $\rho_k \to \rho$, $\lambda_k \to \lambda$, $\chi_{W_k} \to \chi_W$ in measure and $r_k \to r$ up to subsequences. Plugging in the obvious choice for an elastic deformation $y_k^*(x) = (1 + r_k)x$, $x \in \mathcal{L}_{\hat{\varepsilon}_k,\rho_k}(U_k)$ we see that

$$\mathcal{E}(U_k, y_k^*; r_k) \leq C\hat{\varepsilon}_k^d \varepsilon_k^{\frac{p}{2}-1} r_k^{-p}\, \hat{\varepsilon}_k^{-d}\, c\, \varepsilon_k^{1-\frac{p}{2}} r_k^p = C, \tag{3.38}$$

and thus as in (3.22) we deduce that $\bar{\mathcal{C}}'_{\hat{\varepsilon}_k} = \emptyset$ for all k sufficiently large since by (2.4) and (3.34) a broken cell contributes an energy of order $r_k^{-p}\varepsilon_k^{\frac{p}{2}-\frac{2}{3}} \geq C_{\mathrm{med},2}^{-p}\varepsilon_k^{\frac{p}{2}-\frac{2}{3}} \gg 1$. Similarly as in the previous proof we obtain by using Lemma 2.2.1(i) and (3.35)

$$\begin{aligned}\int_{V_k} \operatorname{dist}^p(\nabla\tilde{y}_k, SO(d)) &\leq C\int_{V_k} \operatorname{dist}^p(\bar{\nabla}y_k(\bar{x}), \bar{SO}(d)) \\ &\leq C\int_{V_k} \varepsilon_k^{\frac{p}{2}-1}V_{\varepsilon_k}(\bar{\nabla}y_k(\bar{x})) + C\,c\,\varepsilon_k^{\frac{p}{2}}|V_k| \\ &\leq Cr_k^p\,\mathcal{E}(U_k, y_k^*; r_k) + \frac{Cr_k^p}{k} + C\varepsilon_k^{\frac{p}{2}}|V_k|\end{aligned}$$

and thus together with (3.38) and $r_k \geq C_{\mathrm{med},1}\sqrt{\varepsilon_k}$

$$\int_{V_k} \operatorname{dist}^p(\nabla\tilde{y}_k, SO(d)) \leq Cr_k^p$$

if C is sufficiently large. By geometric rigidity (see Theorem B.1) and Poincaré's inequality there are rotations $R_k \in SO(d)$ and constants $c_k \in \mathbb{R}^d$ such that

$$\|\nabla\tilde{y}_k - R_k\|^p_{L^p(V_k)} \leq C\int_{V_k} \operatorname{dist}^p(\nabla\tilde{y}_k, SO(d)) \leq Cr_k^p \tag{3.39}$$

and

$$\|\tilde{y}_k - (R_k\,\cdot + c_k)\|^p_{L^p(V_k)} \leq Cr_k^p.$$

Letting $u_k = \frac{1}{r_k}(y_k - (R_k\,\cdot + c_k))$ we obtain by Lemma 2.2.1(ii)

$$\left\|\bar{\nabla}u_k\right\|^p_{L^p(V_k)} \leq \frac{C}{r_k^p}\|\nabla\tilde{y}_k - R_k\|^p_{L^p(V_k)} \leq C \tag{3.40}$$

and deduce that for a suitable subsequence (not relabeled) $R_k \to R$ and $\chi_{V_k}\tilde{u}_k \rightharpoonup \chi_U u$ for some $R \in SO(d)$ and $u \in H^1(U, \mathbb{R}^d)$, where $U = (0,\lambda) \times (0,1)^{d-1}$. Then as in the previous proof we derive $\chi_{V_k}\bar{\nabla}u_k \rightharpoonup \chi_U\nabla u \cdot Z$ in L^p possibly after extracting a further subsequence.

As before, in particular applying the argument in (3.27) and (3.28), we obtain that the limit function satisfies the constraint

$$\fint_W (u^1(\lambda, x') - u^1(0, x'))\,dx' = \lambda r^* \in [0, \infty),$$

where $r^* = \lim_{k\to\infty}\frac{1}{r_k}(r_k + 1 - (R_k)_{11}) = 1 + \lim_{k\to\infty}\frac{1-(R_k)_{11}}{r_k} \geq 1$. Applying

(3.35), Lemma 2.2.1(i), (3.39) and (3.40) we compute

$$
\begin{aligned}
&\liminf_{k\to\infty} \mathcal{E}(U_k, y_k; r_k) + \limsup_{k\to\infty} \frac{|V_k| c \varepsilon_k^{\frac{p}{2}}}{r_k^p \det A} \\
&\geq \liminf_{k\to\infty} \frac{c}{r_k^p \det A} \int_{V_k} \operatorname{dist}^p(\bar{\nabla} y_k(x), \bar{SO}(d))\, dx \\
&\geq \liminf_{k\to\infty} \frac{C}{r_k^p} \int_{V_k} \operatorname{dist}^p(\nabla \tilde{y}_k(x), SO(d))\, dx \geq \bar{C} \int_U |\nabla u(x)|^p\, dx
\end{aligned}
$$

for some $\bar{C} > 0$. We define

$$
\mathcal{E}_{\lim}(U, u; r^*) = \bar{C} \int_U |\nabla u(x)|^p\, dx
$$

and then the arguments in (3.31), in particular a slicing argument and Jensen's inequality, yield

$$
\mathcal{E}_{\lim}(U, u; r^*) \geq \bar{C}|W|\lambda(r^*)^p \geq \bar{C}|W|\lambda
$$

since $r^* \geq 1$. Consequently, for $C_{\mathrm{med},1}$ sufficiently large (independent of ε_k) we derive

$$
\liminf_{k\to\infty} \mathcal{E}(U_k, y_k; r_k) \geq \frac{\bar{C}}{2}|W|\lambda.
$$

In view of (3.37) this gives the desired contradiction. Thus, (3.36) holds and then by (3.34) the claim (3.33) follows. □

3.2.4 Estimates in the fracture regime

We now determine $\hat{M}(\hat{U}, r)$ for large r.

Lemma 3.2.4. *Let $\lambda_0 \geq L$. For $\max\{\hat{\varepsilon}^{3-d}, 1\}C_{\mathrm{cr}} \leq r$, C_{cr} sufficiently large, the minimizing problem* (3.16) *satisfies*

$$
\varepsilon^{-s}(|W| - C(1-|W|))\Big(\frac{\beta_A}{\det A} + o(1)\Big) \leq \hat{M}(\hat{U}, r) \leq \varepsilon^{-s}\Big(\frac{\beta_A}{\det A} + o(1)\Big) \quad (3.41)
$$

for $\varepsilon \to 0$. Here $C > 0$ and $o(1)$ are independent of $\rho \in A[0,1)^d$, $W \subset (0,1)^{d-1}$ and $\lambda \in [\lambda_0, 2\lambda_0]$.

Proof. We again drop the superscript $\hat{\cdot}$ if no confusion arises. We first show that

$$
\varepsilon^s M(\bar{U}, r) \geq \frac{|W|\beta_A}{\det A} - \hat{C}(1-|W|) + o(1) \quad (3.42)
$$

for $\varepsilon \to 0$ and some fixed $\hat{C}$ large enough. We again argue by contradiction. If the claim were false, there would exist a $\delta > 0$, sequences $\varepsilon_k \to 0$, $\max\{\hat{\varepsilon}_k^{3-d}, 1\}C_{\mathrm{cr}} \leq r_k$, $\lambda_k \in [\lambda_0, 2\lambda_0]$, $\rho_k \in A[0,1)^d$, $W_k \subset (0,1)^{d-1}$ satisfying (3.7) as well as a sequence $y_k : \mathcal{L}_{\hat{\varepsilon}_k,\rho_k}(U_k) \to \mathbb{R}^d$ satisfying (3.14) with respect to r_k, (3.15) and $E(U_k, y_k) \le M(U_k, r_k) + \frac{1}{k}$ such that

$$\varepsilon_k^s E(U_k, y_k) \le \frac{|W_k|\beta_A}{\det A} - \hat{C}(1 - |W_k|) - 2\delta. \tag{3.43}$$

Up to choosing subsequences we may assume that $\rho_k \to \rho \in A[0,1)^d$, $\lambda_k \to \lambda \in [\lambda_0, 2\lambda_0]$ and $W_k \to W \subset (0,1)^{d-1}$.

We again derive a first upper bound of the minimal energy, now by testing with $y_k^*(x) = x\,\chi_{\{x_1 \le \lambda_k/2\}} + (x + \lambda_k r_k \mathbf{e}_1)\,\chi_{\{x_1 > \lambda_k/2\}}$. It is easy to see that only cells intersecting the set $\{\frac{\lambda_k}{2}\} \times (0,1)^{d-1}$ give an energy contribution. As the quantity of these cells scales like $\hat{\varepsilon}_k^{-d+1}$, by (1.9) we obtain

$$E(U_k, y_k^*) \le C\hat{\varepsilon}_k^{-d+1}\hat{\varepsilon}_k^d\varepsilon_k^{-1}C_2 = C\varepsilon_k^{-s}C_2$$

and thus $\varepsilon_k^s E(U_k, y_k^*) \le C$ for all $k \in \mathbb{N}$ and some C large enough. Then, by (2.4) it is not hard to see that there is some $\tilde{C}$ such that

$$\#\bar{\mathcal{C}}'_{\hat{\varepsilon}_k} \le \tilde{C}\hat{\varepsilon}_k^{-d+1}, \tag{3.44}$$

where $\tilde{C}$ can be chosen independently of $C_{\rm int} \ge 1$. We now choose $C_{\rm int}$ large enough (depending on δ and possibly larger than the fixed $C^*_{\rm int}$) such that for every partition $\mathcal{Z} = \mathcal{Z}_1 \cup \ldots \cup \mathcal{Z}_n$ and $G \in \mathbb{R}^{d\times 2^d}$ with $\operatorname{diam}\{G, \mathcal{Z}_i\} \le C_{\rm int}$ and $\min_{i,j} d(G; \mathcal{Z}_i, \mathcal{Z}_j) \ge 2^{-2d}C_{\rm int}$ (see Section 2.2) we obtain (cf. Assumption 1.1.2(iv))

$$W_{\rm cell}(G) - \sum_{i=1}^n W^{\mathcal{Z}_i}(G[\mathcal{Z}_i]) \ge \frac{1}{2}\sum_{\substack{1\le i,j\le n\\ i\neq j}}\sum_{z_s\in\mathcal{Z}_i}\sum_{z_t\in\mathcal{Z}_j}\beta(z_s,z_t) - \frac{\delta}{\tilde{C}}. \tag{3.45}$$

We obtain from Lemma 2.1.1 and Lemma 2.2.1(i),

$$\begin{aligned}\varepsilon_k^s E(U_k, y_k) &\ge \frac{1}{2}\hat{\varepsilon}_k^{d-1}\sum_{\bar{x}\in\mathcal{F}'_{\hat{\varepsilon}_k}\setminus\bar{\mathcal{C}}'_{\hat{\varepsilon}_k}} W_{\rm cell}(\bar{\nabla}y_k(\bar{x})) + \hat{\varepsilon}_k^{d-1}\sum_{\bar{x}\in\bar{\mathcal{C}}'_{\hat{\varepsilon}_k}} W_{\rm cell}(\bar{\nabla}y_k(\bar{x}))\\ &\ge C\hat{\varepsilon}_k^{-1}\sum_{\bar{x}\in\mathcal{F}'_{\hat{\varepsilon}_k}\setminus\bar{\mathcal{C}}'_{\hat{\varepsilon}_k}}\int_{Q_{\hat{\varepsilon}_k}(\bar{x})}\operatorname{dist}^2(\nabla\tilde{y}_k, SO(d)),\end{aligned}$$

where $\mathcal{F}'_{\hat{\varepsilon}_k} = \mathcal{C}'_{\hat{\varepsilon}_k} \cup \partial_{W_k}(\mathcal{L}'_{\hat{\varepsilon}_k,\rho_k}(U_k))$. Note that $\operatorname{dist}(F, B_{\sqrt{d}}(0)) \le \operatorname{dist}(F, SO(d))$ for all $F \in \mathbb{R}^{d\times d}$, where $B_{\sqrt{d}}(0) \subset \mathbb{R}^{d\times d}$ denotes the ball centered at 0 with radius $\sqrt{d}$. Therefore, with V_k as in (3.23) and recalling the construction of $\tilde{y}$ in Section 2.2 with uniformly bounded $\nabla\tilde{y}$ we derive

$$\begin{aligned}\int_{V_k}\operatorname{dist}^2(\nabla\tilde{y}_k, B_{\sqrt{d}}(0)) &\le \sum_{\bar{x}\in\mathcal{F}'_{\hat{\varepsilon}_k}\setminus\bar{\mathcal{C}}'_{\hat{\varepsilon}_k}}\int_{Q_{\hat{\varepsilon}_k}(\bar{x})}\operatorname{dist}^2(\nabla\tilde{y}_k, SO(d)) + CC_{\rm int}^2\hat{\varepsilon}_k^d\#\bar{\mathcal{C}}'_{\hat{\varepsilon}_k}\\ &\le C\hat{\varepsilon}_k\varepsilon_k^s E(U_k, y_k) + C\hat{\varepsilon}_k \le C\hat{\varepsilon}_k \to 0\end{aligned} \tag{3.46}$$

for $\varepsilon_k \to 0$.

In the following we only consider the first component $\tilde{w}_k := \tilde{y}_k^1$ of the deformations. For $\eta > 0$ we enlarge the set $(0, \lambda_k) \times W_k$ and define $W_k^\eta = ((-\eta, \lambda_k + \eta) \times W_k) \cup U_k$. We extend $\tilde{w}_k$ to W_k^η by $\tilde{w}_k(x_1, x') = \tilde{w}_k(0, x') + x_1 \mathbf{e}_1$ for $-\eta < x_1 \leq 0$ and $\tilde{w}_k(x_1, x') = \tilde{w}_k(\lambda_k, x') + (x_1 - \lambda_k)\mathbf{e}_1$ for $\lambda_k \leq x_1 \leq \lambda_k + \eta$. We note that $J_{\tilde{w}_k} \cap W_k^\eta \subset U_k$ by (3.15). Due to the boundary condition (3.14) there are points $q_k^1 \in \{0\} \times W_k$ and $q_k^2 \in \{\lambda_k\} \times W_k$ such that

$$|\tilde{w}_k(q_k^1) - \tilde{w}_k(q_k^2)| \geq \lambda_k(1 + \max\{\hat{\varepsilon}^{3-d}, 1\} C_{\mathrm{cr}}). \tag{3.47}$$

Due to Lemma 2.3.1(i) and condition (3.15) there is a constant $C = C(D)$ such that

$$\sup\{|\tilde{w}_k(j, s) - \tilde{w}_k(j, t)| : s, t \in W_k\} \leq C(D)(1 + \hat{\varepsilon}^{3-d}) C^*_{\mathrm{int}}$$

for $j = 0, \lambda_k$. Choosing C_{cr} sufficiently large this together with (3.47) shows

$$\inf\{|\tilde{w}_k(p) - \tilde{w}_k(q)| : p, q \in W_k^\eta, p \cdot \mathbf{e}_1 = 0, q \cdot \mathbf{e}_1 = \lambda_k\} \geq \frac{\lambda_k \max\{\hat{\varepsilon}^{3-d}, 1\} C_{\mathrm{cr}}}{2}.$$

Let $U_k^\eta = (-\eta, \lambda_k + \eta) \times (0,1)^{d-1}$. For $M = \frac{\lambda_0 C_{\mathrm{cr}}}{2}$ we now introduce the truncated function $\tilde{u}_k : U_k^\eta \to \mathbb{R}$ defined by

$$\tilde{u}_k(x) := \max\left\{ \ \min\{(\tilde{w}_k(x) - \tilde{w}_k(0, x')), M\}, -M \ \right\}$$

on W_k^η and zero elsewhere, where $x' = (x_2, \ldots, x_d)$. Then it is not hard to see that $\tilde{u}_k(0, x') = 0$, $|\tilde{u}_k(\lambda_k, x')| = M$ for $x' \in W_k$ a.e. and thus

$$|\tilde{u}_k(x_1, x')| \leq \eta, \ |\tilde{u}_k(\lambda_k - x_1, x')| \geq M - \eta \text{ for } x_1 \in (-\eta, 0), x' \in W_k \text{ a.e.} \tag{3.48}$$

Moreover, $|\nabla \tilde{u}_k| \leq |\nabla \tilde{w}_k| + |\nabla_{x'} \ \tilde{w}_k(0, x')| \leq |\nabla \tilde{w}_k| + C C^*_{\mathrm{int}}$ a.e. on W_k^η. The lattice deformation corresponding to $\tilde{u}_k$ is denoted by u_k, i.e. $u_k = \tilde{u}_k|_{\mathcal{L}_{\hat{\varepsilon}_k, \rho_k}(U_k)}$. Keeping in mind that W_k^η is open, it is elementary to see that by truncation of the function no further discontinuity points arise, i.e. $J_{\tilde{u}_k} \cap W_k^\eta \subset J_{\tilde{w}_k} \cap U_k$. Moreover, by (3.7) we obtain

$$\mathcal{H}^{d-1}(J_{\tilde{u}_k} \setminus W_k^\eta) \leq 2(D\eta + 1 - |W_k|). \tag{3.49}$$

We now show that we can find a weakly converging subsequence of $(\tilde{u}_k)_k$. To see this, we first note that by (3.44) and (3.49) there is some $C > 0$ such that $\mathcal{H}^{d-1}(J_{\tilde{u}_k}) \leq \mathcal{H}^{d-1}(J_{\tilde{w}_k}) + \mathcal{H}^{d-1}(J_{\tilde{u}_k} \setminus W_k^\eta) \leq C$ for all $k \in \mathbb{N}$. Moreover $\|\nabla \tilde{u}_k\|_\infty \leq C C^*_{\mathrm{int}} + \|\nabla \tilde{w}_k\|_\infty \leq C(C_{\mathrm{int}} + C^*_{\mathrm{int}})$ and $\|\tilde{u}_k\|_\infty \leq M$ for all $k \in \mathbb{N}$ by construction. Now applying Theorem A.1.1 we deduce that there is some $u \in SBV(U^\eta)$ such that up to a subsequence (not relabeled) $\tilde{u}_k \to u$ in the sense of (A.4) and a.e., where $U^\eta = (-\eta, \lambda + \eta) \times (0,1)^{d-1}$. By (3.48) for M large enough with respect to η the limit function satisfies

$$\operatorname{ess\,inf}\{|u(p) - u(q)| : p \in (-\eta, 0) \times W, \ q \in (\lambda, \lambda + \eta) \times W\} \geq \frac{M}{2}. \tag{3.50}$$

Note that the above compactness theorem implies $\|\nabla u\|_\infty \leq C(C_{\text{int}} + C^*_{\text{int}})$. We now improve this bound by showing that $\|\nabla u\|_\infty \leq T$ for some $T > 0$ large enough independent of δ (recall that C_{int} may depend on δ). Let $R = \sqrt{d} + CC^*_{\text{int}}$ for some $C > 0$ sufficiently large. Then (3.46) yields

$$\int_{V_k} \text{dist}^2(\nabla \tilde{u}_k, B_R(0)) \leq \int_{V_k} \text{dist}^2(\nabla \tilde{y}_k, B_{\sqrt{d}}(0)) \leq C\hat{\varepsilon}_k \to 0,$$

for $\varepsilon_k \to 0$. Consequently, we get

$$\int_U \text{dist}^2(\nabla u, B_R(0)) \leq \liminf_{k\to\infty} \int_{V_k} \text{dist}^2(\nabla \tilde{u}_k, B_R(0)) = 0,$$

where we used the convexity of $\text{dist}^2(\cdot, B_R(0))$. Therefore, $|\nabla u| \leq R$ a.e. in U and by the extension of $\tilde{y}_k$ to W_k^η we get $|\nabla u| \leq CC^*_{\text{int}}$ a.e. on $U^\eta \setminus U$. Consequently, choosing $T > 0$ sufficiently large we obtain

$$|\nabla u| \leq T \text{ a.e. in } U^\eta. \tag{3.51}$$

We now concern ourselves with the energy contribution of the broken cells $\bar{\mathcal{C}}'_{\hat{\varepsilon}_k}$. For all $\nu \in \mathcal{V}$ we let $\bar{y}_{\nu,k} : U_k \to \mathbb{R}^d$, $\bar{w}_{\nu,k} : U_k \to \mathbb{R}$ and $\bar{u}_{\nu,k} : U_k^\eta \to \mathbb{R}$ be the interpolations introduced in Section 2.2. Recalling the construction of the interpolations we obtain by (1.10) and (3.45)

$$\begin{aligned}\varepsilon_k^s E(U_k, y_k) &\geq \hat{\varepsilon}_k^{d-1} \sum_{\bar{x}\in\bar{\mathcal{C}}'_{\hat{\varepsilon}_k}} W_{\text{cell}}(\bar{\nabla} y_k(\bar{x})) \\ &\geq \sum_{\nu\in\mathcal{V}} \int_{J_{\bar{y}_{\nu,k}}\cap\Gamma(\nu)} \frac{\hat{\varepsilon}_k^{d-1}}{\frac{1}{2}\mathcal{H}^{d-1}(\partial_\nu Q^\nu_{\hat{\varepsilon}_k})} \beta(\nu)\, d\mathcal{H}^{d-1} - \hat{\varepsilon}_k^{d-1} \#\bar{\mathcal{C}}'_{\hat{\varepsilon}_k} \frac{\delta}{\tilde{C}},\end{aligned}$$

where $\Gamma(\nu) = \bigcup_{\lambda\in\mathbb{Z}^d} \partial_\nu Q^\nu_{\hat{\varepsilon}_k}(\lambda)$. Then by (2.6) and (3.44) we get

$$\varepsilon_k^s E(U_k, y_k) \geq \sum_{\nu\in\mathcal{V}} \int_{J_{\bar{y}_{\nu,k}}} \frac{\beta(\nu)}{\det A} |\nu \cdot \xi_{\bar{y}_{\nu,k}}|\, d\mathcal{H}^{d-1} - \delta.$$

Applying (3.49) it is not hard to see that there is some $\Gamma_{k,\nu}$ with $\mathcal{H}^{d-1}(\Gamma_{k,\nu}) \leq C(D\eta + 1 - |W_k|)$ such that $J_{\bar{u}_{\nu,k}} \subset J_{\bar{w}_{\nu,k}} \cup \Gamma_{k,\nu} \subset J_{\bar{y}_{\nu,k}} \cup \Gamma_{k,\nu}$. Furthermore, the normals $\xi_{\bar{y}_{\nu,k}}$ and $\xi_{\bar{u}_{\nu,k}}$ coincide on $J_{\bar{u}_{\nu,k}} \cap J_{\bar{y}_{\nu,k}}$. Therefore, we derive

$$\begin{aligned}\delta + C(D\eta + 1 - |W_k|) &+ \varepsilon_k^s E(U_k, y_k) \\ &\geq \sum_{\nu\in\mathcal{V}} \int_{J_{\bar{u}_{\nu,k}}} \frac{\beta(\nu)}{\det A} |\nu \cdot \xi_{\bar{u}_{\nu,k}}|\, d\mathcal{H}^{d-1} =: E_S(U_k, u_k).\end{aligned} \tag{3.52}$$

With the notation introduced in Section A.1 we get using Theorem A.1.5

$$E_S(U_k, u_k) \geq \frac{1}{\det A} \sum_{\nu\in\mathcal{V}} \int_{\Pi^\nu} \#J_{\bar{u}^{\nu,s}_{\nu,k}} \beta(\nu)\, d\mathcal{H}^{d-1}(s).$$

Then by the equiboundedness of $E_S(U_k, u_k)$ and Fatou's lemma we deduce that $\liminf_{k\to\infty} \#J_{\bar{u}^{\nu,s}_{\nu,k}} < +\infty$ for a.e. $s \in \Pi^\nu$ and all $\nu \in \mathcal{V}$. As $\bar{u}_{\nu,k}$ and $\nabla \bar{u}_{\nu,k}$ are uniformly bounded, by Theorem A.1.1 and Lemma 2.2.3 $\bar{u}^{\nu,s}_{\nu,k}$ converges (up to a subsequence) to $u^{\nu,s}$ in the sense of (A.4) for a.e. $s \in \Pi^\nu$. In particular, we get

$$\liminf_{k\to\infty} \#J_{u^{\nu,s}_{\nu,k}} \geq \#J_{u^{\nu,s}_\nu}.$$

Applying Fatou's lemma and the slicing theorem once more we then derive

$$\begin{aligned}\liminf_{k\to\infty} E_S(U_k, u_k) &\geq \frac{1}{\det A} \sum_{\nu\in\mathcal{V}} \int_{\Pi^\nu} \#J_{u^{\nu,s}} \beta(\nu) \, d\mathcal{H}^{d-1}(s) \\ &= \frac{1}{\det A} \int_{J_u} \sum_{\nu\in\mathcal{V}} \beta(\nu) |\nu \cdot \xi_u| \, d\mathcal{H}^{d-1} =: E_{S,\lim}(U, u).\end{aligned} \tag{3.53}$$

By (1.14) and slicing in $\mathbf{e}_1$-direction we get

$$\begin{aligned}E_{S,\lim}(U, u) &\geq \frac{1}{\det A} \int_{J_u} \beta_A |\mathbf{e}_1 \cdot \xi_u| \, d\mathcal{H}^{d-1} \\ &= \frac{1}{\det A} \int_{(0,1)^{d-1}} \beta_A \, \#J_{u^{\mathbf{e}_1,s}} \, d\mathcal{H}^{d-1}(s).\end{aligned}$$

We now choose $M = \frac{\lambda_0 C_{c:}}{2}$ sufficiently large (independently of δ) such that $M \geq 4T\lambda$. Then due to (3.50) and (3.51) it is not hard to see that $\#J_{u^{\mathbf{e}_1,s}} \geq 1$ for a.e. $s \in W$ and therefore $E_{S,\lim}(U, u) \geq \frac{|W|\beta_A}{\det A}$. Letting $\eta \to 0$ and choosing $\hat{C}$ sufficiently large we now conclude by (3.43), (3.52) and (3.53):

$$\begin{aligned}\infty > \frac{|W|\beta_A}{\det A} - 2\delta &\geq \liminf_{k\to\infty} \varepsilon_k^s E(U_k, y_k) + \hat{C}(1 - |W|) \geq \liminf_{k\to\infty} E_S(U_k, y_k) - \delta \\ &\geq E_{S,\lim}(U, u) - \delta \geq \frac{|W|\beta_A}{\det A} - \delta.\end{aligned}$$

This gives the desired contradiction.

To see the upper bound in (3.41) we choose $\xi \in S^{d-1}$ such that (1.14) is minimized and define the hyperplane $\Pi = \{x \in \mathbb{R}^d : x \cdot \xi = c\}$ for a suitable $c \in \mathbb{R}$ such that $\Pi \cap U \subset \{\delta \leq x_1 \leq \lambda - \delta\}$ for some $\delta > 0$. We set

$$y(x) = \begin{cases} x, & x \cdot \xi \leq c, \\ x + r\lambda \mathbf{e}_1, & x \cdot \xi > c. \end{cases} \tag{3.54}$$

The energy corresponding to the deformation y is given by the bonds intersecting Π. These springs, associated to the lattice directions $\nu \in \mathcal{V}$, are elongated by a factor scaling with $r/\hat{\varepsilon}$ and yield a contribution $\beta(\nu)$ in the limit $\varepsilon \to 0$ by (1.10).

As the projection in ν-direction onto the hyperplane $\{x \cdot \xi = c\}$ of the face $\partial_\nu Q^\nu$ has $\mathcal{H}^{d-1}$-volume

$$\frac{1}{2}\mathcal{H}^{d-1}(\partial_\nu Q^\nu)\left|\frac{\nu}{|\nu|}\cdot\hat{\nu}\right|\frac{1}{\left|\frac{\nu}{|\nu|}\cdot\xi\right|} = \frac{\hat{\varepsilon}^{d-1}\det A}{|\nu\cdot\hat{\nu}|}\frac{|\nu\cdot\hat{\nu}|}{|\nu\cdot\xi|} = \frac{\hat{\varepsilon}^{d-1}\det A}{|\nu\cdot\xi|}$$

(see (2.6)) it is not hard so see that

$$\frac{|\nu\cdot\xi|}{\hat{\varepsilon}^{d-1}\det A|e_1\cdot\xi|} + O\Big(\frac{1}{\hat{\varepsilon}^{d-2}}\Big) \tag{3.55}$$

springs in ν-direction are broken. This yields the energy

$$\varepsilon^{-s}\Big(\frac{\beta_A}{\det A} + o(1)\Big) + O(\hat{\varepsilon}^2\varepsilon^{-1}),$$

for $\varepsilon \to 0$ as desired. □

3.2.5 Estimates in a second intermediate regime

We provide an additional lemma needed in the case $d \geq 4$.

Lemma 3.2.5. *Let $d \geq 4$, $C^*_{\mathrm{med},2} > 0$ and $C^*_{\mathrm{med},1} > 0$ sufficiently large. Then there is a constant $C > 0$ such that the minimization problem* (3.16) *satisfies*

$$\hat{M}(\hat{U}, r) \geq C\varepsilon^{d-2+s(1-d)}r$$

for $C^*_{\mathrm{med},1} \leq r \leq C^*_{\mathrm{med},2}\hat{\varepsilon}^{3-d}$ as $\varepsilon \to 0$. The constant C is independent of $\rho \in A[0,1)^d$, $\tilde{W} \subset (0,1)^{d-1}$ and $\lambda \in [\lambda_0, 2\lambda_0]$.

Proof. The superscript $\hat{\cdot}$ is again dropped where no confusion arises. Let $C^*_{\mathrm{med},2} > 0$. Let $\rho, \tilde{W}, \lambda$ and r with $C^*_{\mathrm{med},1} \leq r \leq C^*_{\mathrm{med},2}\hat{\varepsilon}^{3-d}$ be given and consider a deformation $y : \mathcal{L}_{\hat{\varepsilon},\rho}(U) \to \mathbb{R}^d$ satisfying (3.14) with respect to r, (3.15) and $E(U,y) \leq 2M(U,r)$. Due to (3.14) there is a $q \in W$ such that $|\tilde{y}^1(\lambda,q) - \tilde{y}^1(0,q)| \geq \lambda(1+r)$. Applying Lemma 2.3.1(ii) for $t = \frac{\lambda(1+r)}{4CC^*_{\mathrm{int}}}$ and (3.15) we find a set $V \subset \tilde{W}$ with $\mathcal{H}^{d-1}(V) \geq c'\lambda_0 r\varepsilon^{(1-s)(d-2)}$ such that

$$|\tilde{y}(\lambda,q) - \tilde{y}(0,q)| \geq \frac{\lambda(1+r)}{2} \geq \frac{\lambda_0 C^*_{\mathrm{med},1}}{2}$$

for all $q \in V$. Fix C_{int} as defined in Section 2.2 and recall $\|\nabla\tilde{y}\|_\infty \leq CC_{\mathrm{int}}$. Choose $C^*_{\mathrm{med},1}$ large enough such that $((0,\lambda)\times\{q\}) \cap \bigcup_{\bar{x}\in\bar{\mathcal{C}}'_{\hat{\varepsilon}}} Q_{\hat{\varepsilon}}(\bar{x}) \neq \emptyset$ for all $q \in V$. As the orthogonal projection of a cell onto $\{0\}\times\mathbb{R}^{d-1}$ has $\mathcal{H}^{d-1}$-measure smaller than $C\hat{\varepsilon}^{d-1}$ we deduce

$$\#\bar{\mathcal{C}}'_{\hat{\varepsilon}} \geq C\varepsilon^{s-1}r$$

for some $C > 0$. As every broken cell provides at least the energy $C\varepsilon^{-\frac{2}{3}} = C\varepsilon^{d(1-s)-1}$ by (2.4) we derive

$$M(U,r) \geq C\varepsilon^{d(1-s)-1} \ \#\bar{\mathcal{C}}'_{\hat{\varepsilon}} \geq C\varepsilon^{d-2+s(1-d)}r.$$

□

Proof of Theorem 3.2.1. We begin to construct such a function $\hat{f} : \mathbb{R}\times[\lambda_0, 2\lambda_0] \to \mathbb{R}$ for the rescaled problem $\hat{M}(\hat{U}, r)$. Let $\omega(|W|) = |W|(1 - C(1 - |W|))$ with the constant C of Lemma 3.2.4. For $\delta > 0$ small we set $\hat{f}(r, \lambda) = -\delta\lambda\omega(|W|)$ for $r \leq 0$ and for $C_1 > 0$ sufficiently large we define

$$\hat{f}(r,\lambda) = \omega(|W|)\lambda\Big(\frac{\alpha_A}{2\det A}\frac{r^2}{\varepsilon} - \delta\Big)$$

for $r \in [0, C_1\sqrt{\varepsilon}]$, $\lambda \in [\lambda_0, 2\lambda_0]$. Choose the affine function $g : \mathbb{R} \to \mathbb{R}$ satisfying $\lambda\omega(|W|)g(C_1\sqrt{\varepsilon}) = \hat{f}(C_1\sqrt{\varepsilon}, \lambda)$ and $\lambda\omega(|W|)g' = \partial_r\hat{f}(C_1\sqrt{\varepsilon}, \lambda)$. For $t > C_1C$ sufficiently large we let $h(r) = C\varepsilon^{-\frac{p}{2}}r^p - t$ for $r \geq C_{\mathrm{med},1}$ with $C_{\mathrm{med},1}$ as in Lemma 3.2.3, so that there is a (unique) intersection point of the graphs of g and h, $(\bar{r}_t, g(\bar{r}_t)) = (\bar{r}_t, h(\bar{r}_t))$, for which $h'(\bar{r}_t) \geq g'$. Note that $\bar{r}_t \sim \sqrt{\varepsilon}$. Then we set $\hat{f}(r, \lambda) = \lambda\omega(|W|)g(r)$ for $r \in [C_1\sqrt{\varepsilon}, \bar{r}_t]$ and $\hat{f}(r, \lambda) = \lambda\omega(|W|)h(r)$ for $r \in [\bar{r}_t, \max\{\hat{\varepsilon}^{3-d}, 1\}C_2]$ for $C_2 > 0$ large enough. Finally, we let

$$\hat{f}(r,\lambda) = \varepsilon^{-s}\omega(|W|)\Big(\frac{\beta_A}{\det A} - \delta\Big)$$

for $r \geq \max\{\hat{\varepsilon}^{3-d}, 1\}C_2$. In the case $d \geq 4$ we observe that for $C^*_{\mathrm{med},1} \leq r \leq C_2\hat{\varepsilon}^{3-d}$ we have

$$C\omega(|W|)\lambda\varepsilon^{-\frac{p}{2}}r^p \leq C\varepsilon^{d-2+s(1-d)}r \leq C\varepsilon^{-s}\hat{\varepsilon} \leq C\varepsilon^{-s}$$

for some $p > 1$ small enough. Consequently, by Lemmas 3.2.2, 3.2.3, 3.2.4, 3.2.5 it is not hard to see that $\hat{f}$ is convex for $r \leq \max\{\hat{\varepsilon}^{3-d}, 1\}C_2$ and satisfies $\hat{f} \leq \hat{M}(\hat{U}, \cdot)$ for ε small enough independently of $\rho \in A[0,1)^d$ and W. Moreover, we obtain

$$\hat{f}(r,\lambda) \leq \hat{M}(\hat{U},r) \leq \frac{1}{\omega(|W|)}\hat{f}(r,\lambda) + 4\lambda_0\delta$$

for $r \in [0, C_1\sqrt{\varepsilon}]$ and

$$\hat{f}(r,\lambda) \leq \hat{M}(\hat{U},r) \leq \frac{1}{\omega(|W|)}\hat{f}(r,\lambda) + 2\varepsilon^{-s}\delta$$

for $r \geq \max\{\hat{\varepsilon}^{3-d}, 1\}C_2$. To finish the proof it suffices to recall $M(U_\varepsilon, r) = \varepsilon^{sd}\hat{M}(\hat{U}, r)$ by (3.17) and to set $f(r,\lambda) = \varepsilon^{sd}\hat{f}(r, \varepsilon^{-s}\lambda)$ for all $r \in \mathbb{R}$ and $\lambda \in \varepsilon^s[\lambda_0, 2\lambda_0]$. □

3.3 Proof of the cleavage law

We are now in a position to prove the main theorem about the limiting minimal energy.

Proof of Theorem 1.3.1. Let $y \in \mathcal{A}(a_\varepsilon)$. We partition $(0, l_2) \times \ldots \times (0, l_d)$ up to a set of size $O(\varepsilon^s)$ with sets V_i, $i \in I$, which are translates of the cube $\varepsilon^s(0,1)^{d-1}$. Furthermore, we set $V_i^d = (0, l_1) \times V_i$ for all $i \in I$. For $C_{\text{int}}^* > 0$ we denote the set of broken cells by $\bar{\mathcal{B}}'_\varepsilon$ as defined at the beginning of Section 3.2. We let

$$\bar{I} := \left\{ i \in I : \#\{\bar{x} \in \bar{\mathcal{B}}'_\varepsilon : Q_\varepsilon(\bar{x}) \subset V_i^d\} > \frac{2\beta_A}{\varepsilon^{(1-s)(d-1)}\bar{C} \det A} \right\}$$

with $\bar{C} = \bar{C}(C_{\text{int}}^*)$ as in (2.4). Then for $i \in \bar{I}$ we estimate

$$\varepsilon^{d-1} \sum_{\bar{x} \in (\mathcal{L}'_\varepsilon(V_i^d))^\circ} W_{\text{cell}}(\bar{\nabla} y(\bar{x})) \geq \varepsilon^{d-1} \frac{2\beta_A}{\varepsilon^{(1-s)(d-1)}\bar{C} \det A} \bar{C} = \frac{\varepsilon^{s(d-1)} 2\beta_A}{\det A}. \tag{3.56}$$

Now consider some V_i for $i \in I \backslash \bar{I}$. For $\lambda_0 \geq L(\sqrt{A^T A}, W_{\text{cell}}, 1, \ldots, 1)$ we partition V_i^d into sets of the form $U_1 = (u_0, u_1) \times V_i, \ldots, U_n = (u_{n-1}, u_n) \times V_i$, where $u_0 = l_A \varepsilon$, $u_n = l_1 - l_A \varepsilon$ and $u_j - u_{j-1} \in \varepsilon^s[\lambda_0, 2\lambda_0]$ for all $j = 1, \ldots, n$. This can and will be done so that

$$N(u_j) := \#\mathcal{T}(u_j) = \min_{\bar{u} \in J(u_j)} \#\mathcal{T}(\bar{u}), \tag{3.57}$$

for all $j = 1, \ldots, n-1$, where

$$\mathcal{T}(s) = \{\bar{x} \in \bar{\mathcal{B}}'_\varepsilon : Q_\varepsilon(\bar{x}) \subset V_i^d \text{ and } Q_\varepsilon(\bar{x}) \cap (\{s\} \times V_i) \neq \emptyset\}$$

and $J(u_j) = [u_j - \frac{\varepsilon^s \lambda_0}{2}, u_j]$ or $[u_j, u_j + \frac{\varepsilon^s \lambda_0}{2}]$. Moreover, we have $N(u_0) = N(u_n) = 0$ due to the boundary conditions (1.12). We now show that

$$\sum_{j=0}^{n} N(u_j) \leq \frac{C\varepsilon^{(s-1)(d-2)}}{\lambda_0}. \tag{3.58}$$

We cover $J(u_j) \times V_i$ with translates of $(0, \varepsilon l_A) \times (0, \varepsilon^s)^{d-1}$, where l_A is as defined in (1.11). As every cell is contained in at most two of these translates we derive

$$\#\{\bar{x} \in \bar{\mathcal{B}}'_\varepsilon : Q_\varepsilon(\bar{x}) \subset V_i^d \text{ and } Q_\varepsilon(\bar{x}) \cap (J(u_j) \times V_i) \neq \emptyset\} \geq \left\lfloor \frac{\lambda_0 \varepsilon^s}{4 l_A \varepsilon} \right\rfloor N(u_j)$$

for $j = 1, \ldots, n-1$ due to the construction (3.57). Summing over j, we find

$$\sum_{j=0}^{n} N(u_j) \leq \frac{C\varepsilon}{\varepsilon^s \lambda_0} \#\{\bar{x} \in \bar{\mathcal{B}}'_\varepsilon : Q_\varepsilon(\bar{x}) \subset V_i^d\} \leq \frac{C\varepsilon^{1-s}\varepsilon^{(1-s)(1-d)}}{\lambda_0} = \frac{C\varepsilon^{(s-1)(d-2)}}{\lambda_0}$$

since $i \in I \setminus \bar{I}$. Note that the estimate (3.58) relies only on the fact that $i \in I \setminus \bar{I}$ but is independent of the particular set V_i, the deformation y and ε.

Let $T_i = \bigcup_{j=1}^{n-1} \bigcup_{\bar{x} \in \mathcal{T}(u_j)} \overline{Q_\varepsilon(\bar{x})}$ and $S_i = \pi_1 T_i$, where $\pi_1 T_i \subset \mathbb{R}^{d-1}$ denotes the set which arises from T_i by orthogonal projection onto $\{0\} \times V_i$ and cancellation of the first component. Using (3.58) we find

$$\mathcal{H}^{d-2}(\partial S_i) \leq \sum_{j=1}^{n-1} N(u_j) \mathcal{H}^{d-2}(\partial \pi_1 Q_\varepsilon) \leq C\varepsilon^{d-2} \sum_{j=1}^{n-1} N(u_j) \leq C\lambda_0^{-1} \varepsilon^{s(d-2)}.$$

Choose λ_0 so large that $\mathcal{H}^{d-2}(\partial S_i) \leq \delta \varepsilon^{s(d-2)}$. Let $V_{i,\varepsilon} = \{x \in V_i : \operatorname{dist}(x, \partial V_i) \geq C\varepsilon\}$ with C so big that $\pi_1 Q_\varepsilon(\bar{x}) \cap V_{i,\varepsilon} = \emptyset$ whenever $Q_\varepsilon(\bar{x}) \not\subset V_i^d$. By the isoperimetric inequality we deduce that there is a unique connected component $\tilde{W}_i$ of $V_{i,\varepsilon} \setminus S_i$ satisfying $|\tilde{W}_i| := \mathcal{H}^{d-1}(\tilde{W}_i) \geq (1 - C\varepsilon^{1-s} - C\delta^{\frac{d-1}{d-2}})\varepsilon^{s(d-1)}$, where C is a constant only depending on the dimension. Moreover, we have $\mathcal{H}^{d-2}(\partial \tilde{W}_i) \leq C\varepsilon^{s(d-2)}$ and so we see that for δ small enough $\tilde{W}_i$ is of the form (3.7) (after rescaling by ε^{-s}). Furthermore, by a similar argument (e.g. by enlarging the cubes which form T_i) we find that

$$\mathcal{H}^{d-2}\big(\{x \in \tilde{W}_i : \operatorname{dist}(x, \partial \tilde{W}_i) = D'\varepsilon \text{ and } \operatorname{dist}(x, \partial V_{i,\varepsilon}) \neq D'\varepsilon\}\big) \leq C\delta\varepsilon^{s(d-2)}.$$

Consequently, we define W_i corresponding to $\tilde{W}_i$ as described in (3.7) (replacing ε^{1-s} by ε due to the different scaling) and obtain $|W_i| := \mathcal{H}^{d-1}(W_i) \geq (1 - C\varepsilon^{1-s} - C\delta^{\frac{d-1}{d-2}})\varepsilon^{s(d-1)}$ and $\mathcal{H}^{d-2}(\partial W_i) \leq C\varepsilon^{s(d-2)}$ for some possibly larger constant C. Clearly, W_i is of the form (3.7). The sets U_j defined above correspond to U_ε considered in Section 3.2 up to a translation. In particular, the sets $\tilde{W}_i$ satisfy condition (3.10) due to the construction of S_i.

We define

$$r_j := -1 + \frac{1}{u_j - u_{j-1}} \fint_{W_i} \left(\tilde{y}^1(u_j, x') - \tilde{y}^1(u_{j-1}, x')\right) dx'$$

for $j = 1, \dots, n$. Note that this definition is meaningful as $\tilde{y}^1$ is defined on all of $(0, l_1) \times W_i$ (see Section 2.2). As $y \in \mathcal{A}(a_\varepsilon)$ it is not hard to see that

$$\begin{aligned}\sum_{j=1}^{n} (u_j - u_{j-1}) r_j &= -(l_1 - 2l_A\varepsilon) + \fint_{W_i} \left(\tilde{y}^1(l_1 - l_A\varepsilon, x') - \tilde{y}^1(l_A\varepsilon, x')\right) dx' \\ &= -(l_1 - 2l_A\varepsilon) + l_1 - 2l_A\varepsilon + (l_1 - 2l_A\varepsilon)a_\varepsilon = (l_1 - 2l_A\varepsilon)a_\varepsilon.\end{aligned}$$

We define $W_i^d = (0, l_1) \times W_i$ and the energy

$$\mathcal{E}_\varepsilon^i(y) := \varepsilon^{d-1} \sum_{\bar{x} \in (\mathcal{L}'_\varepsilon(V_i^d))^\circ} W_{\text{cell}}(\bar{\nabla} y(\bar{x})).$$

For $C_1 \geq 2a_{\text{crit}}\sqrt{\varepsilon}$, $C_2 > 0$ sufficiently large and for $\delta > 0$ as before choose f as in Theorem 3.2.1. Then for ε small enough

$$\begin{aligned}
\mathcal{E}^i_\varepsilon(y) &\geq \varepsilon^{d-1} \sum_{j=1}^n \Big(\sum_{\bar{x} \in (\mathcal{L}'_\varepsilon(U_j))^\circ} W_{\text{cell}}(\bar{\nabla} y(\bar{x})) + \frac{1}{2} \sum_{\bar{x} \in \partial_{W_i}(\mathcal{L}'_\varepsilon(U_j))} W_{\text{cell}}(\bar{\nabla} y(\bar{x})) \Big) \\
&\geq \sum_{j=1}^n M(U_j, r_j) \geq \sum_{j=1}^n f(r_j, u_j - u_{j-1}).
\end{aligned}$$

Here we observe that due to the construction of the sets W_i we have $\partial_{W_i}(\mathcal{L}'_\varepsilon(U_j)) \subset (\mathcal{L}'_\varepsilon(V_i^d))^\circ$ for all $j = 1, \ldots, n$, $i \in I$. If there is some j such that $r_j \geq C_2 \max\{1, \varepsilon^{(s-1)(3-d)}\}$ then $\mathcal{E}^i_\varepsilon(y) \geq \tilde{\omega}(|W_i|)\big(\frac{\beta_A}{\det A} - \delta\big)$ by (3.12), where $\tilde{\omega}(\cdot) = \varepsilon^{s(d-1)}\omega(\varepsilon^{-s(d-1)}\cdot)$ (note that now $|W_i| \sim \varepsilon^{s(d-1)}$). Otherwise all r_j lie in the regime, where f is convex in r and linear in λ. We then compute using Jensen's inequality

$$\begin{aligned}
\mathcal{E}^i_\varepsilon(y) &\geq \sum_{j=1}^n f(r_j, u_j - u_{j-1}) = \sum_{j=1}^n \frac{u_j - u_{j-1}}{\lambda_0 \varepsilon^s} f(r_j, \lambda_0 \varepsilon^s) \\
&\geq \frac{\sum_{j=1}^n u_j - u_{j-1}}{\lambda_0 \varepsilon^s} \, f\Big(\frac{\sum_{j=1}^n (u_j - u_{j-1})\, r_j}{\sum_{j=1}^n u_j - u_{j-1}}, \lambda_0 \varepsilon^s \Big) \\
&= \frac{l_1 - 2 l_A \varepsilon}{\lambda_0 \varepsilon^s} \, f\Big(a_\varepsilon, \lambda_0 \varepsilon^s \Big),
\end{aligned}$$

whence for $a_\varepsilon \geq 2a_{\text{crit}}\sqrt{\varepsilon}$, due to the monotonicity of f, also

$$\begin{aligned}
\mathcal{E}^i_\varepsilon(y) &\geq \frac{l_1 - 2 l_A \varepsilon}{\lambda_0 \varepsilon^s} \, f\Big(2a_{\text{crit}}\sqrt{\varepsilon}, \lambda_0 \varepsilon^s \Big) \\
&= \tilde{\omega}(|W_i|)\Big(\frac{(l_1 - 2 l_A \varepsilon)\alpha_A 4 a_{\text{crit}}^2}{2 \det A} - (l_1 - 2 l_A \varepsilon)\delta \Big) \geq \tilde{\omega}(|W_i|)\Big(\frac{\beta_A}{\det A} - \delta \Big)
\end{aligned}$$

by (3.11), where the last inequality holds for δ small enough. Repeating the calculation for $a_\varepsilon \leq 2a_{\text{crit}}\sqrt{\varepsilon}$ and using (3.11) yields

$$\mathcal{E}^i_\varepsilon(y) \geq \tilde{\omega}(|W_i|) m_\varepsilon := \tilde{\omega}(|W_i|) \min\Big\{ \frac{(l_1 - 2 l_A \varepsilon)\alpha_A a_\varepsilon^2}{2\varepsilon \det A} - (l_1 - 2 l_A \varepsilon)\delta, \frac{\beta_A}{\det A} - \delta \Big\}.$$

Using that $\mathcal{E}_\varepsilon(y) \geq \sum_{i \in I} \mathcal{E}^i_\varepsilon(y)$ and $\tilde{\omega}(|W_i|) \geq \sigma(\delta)\varepsilon^{s(d-1)}$, where $\sigma(\delta) = \min\{\omega(s) : 1 - C\delta^{\frac{d-1}{d-2}} \leq s \leq 1\} \leq 1$ for all $i \in I$ and recalling (3.56) we get for δ small enough

$$\begin{aligned}
\liminf_{\varepsilon \to 0} \inf\{\mathcal{E}_\varepsilon(y) : y \in \mathcal{A}(a_\varepsilon)\} &\geq \liminf_{\varepsilon \to 0} \Big(\#\bar{I} \; \frac{\varepsilon^{s(d-1)} 2\beta_A}{\det A} + \#(I \setminus \bar{I})\, \sigma(\delta)\varepsilon^{s(d-1)} \; m_\varepsilon \Big) \\
&\geq \liminf_{\varepsilon \to 0} \#I \;\; \sigma(\delta)\varepsilon^{s(d-1)} \; m_\varepsilon \\
&\geq \sigma(\delta) \prod_{j=2}^d l_j \min\Big\{ \frac{l_1 \alpha_A a^2}{2 \det A} - l_1 \delta, \frac{\beta_A}{\det A} - \delta \Big\},
\end{aligned}$$

as $a_\varepsilon/\sqrt{\varepsilon} \to a$. Letting $\delta \to 0$ shows

$$\liminf_{\varepsilon\to 0} \inf\{\mathcal{E}_\varepsilon(y) : y \in \mathcal{A}(a_\varepsilon)\} \geq \frac{\prod_{j=2}^d l_j}{\det A} \min\Big\{\frac{1}{2} l_1 \alpha_A a^2, \beta_A\Big\}.$$

Here we used that $\lim_{\delta\to 1}\sigma(\delta) = 1$. It remains to prove that the right hand side in Theorem 1.3.1 is attained for some sequence of deformations. This essentially follows from the sharpness of the estimates (3.11) and (3.12). In particular, as in (3.32) for $a < \infty$ we consider

$$y^{\mathrm{el}}_\varepsilon(x) = x + \bar{F}(a_\varepsilon)\,x, \quad x \in \mathcal{L}_\varepsilon \cap \Omega, \tag{3.59}$$

and as in the proof of Lemma 3.2.2 it is not hard to see that

$$\lim_{\varepsilon\to 0} \mathcal{E}_\varepsilon(y^{\mathrm{el}}_\varepsilon) = \prod_{j=1}^d l_j \frac{\alpha_A}{2\det A} \lim_{\varepsilon\to 0}\Big(\frac{a_\varepsilon}{\varepsilon}\Big)^2 = \prod_{j=1}^d l_j \frac{\alpha_A}{2\det A} a^2.$$

For $y^{\mathrm{cr}}_\varepsilon$ we proceed as in (3.54): We choose ξ such that (1.14) is satisfied. Due to the assumption $l_1 \geq L$ it is possible to define a hyperplane $\Pi = \{x \in \mathbb{R}^d : x\cdot\xi = c\}$ such that $\Pi \cap \overline{\Omega} \subset \overline{\Omega} \setminus (B^\varepsilon_1 \cup B^\varepsilon_2)$. We let

$$y^{\mathrm{cr}}_\varepsilon(x) = \begin{cases} x, & x\cdot\xi < c, \\ x + l_1 a_\varepsilon \mathbf{e}_1, & x\cdot\xi > c, \end{cases} \quad x \in \mathcal{L}_\varepsilon \cap \Omega. \tag{3.60}$$

Again counting the quantity of broken springs as in (3.55) we derive $\lim_{\varepsilon\to 0}\mathcal{E}_\varepsilon(y^{\mathrm{cr}}_\varepsilon) = \prod\limits_{j=2}^d l_j \frac{\beta_A}{\det A}$. □

3.4 Examples: mass-spring models

In the following we examine several mass-spring models to which the above results apply. We calculate the constants α_A, β_A explicitly and thus we can provide the limiting energy of Theorem 1.3.1 as well as the critical value of boundary displacements $a_{\mathrm{crit.}}$ Moreover, we specify minimizing configurations and discuss their behavior depending on the properties of the cell energy.

Note that the cell energies under consideration which consist of pair interaction energies are typically minimized on $\bar{O}(d)$. Similarly as in Section 1.5 we have to introduce a frame indifferent penalty term to avoid unphysical behavior and to satisfy Assumption 1.1.2. Choose some $\chi \geq 0$ which vanishes in a neighborhood of $\bar{SO}(d)$ and ∞ and satisfies $\chi \geq c_\chi > 0$ in a neighborhood of $\bar{O}(d) \setminus \bar{SO}(d)$. For example, as in (1.26) we may set

$$\chi(\bar{\nabla} y(\bar{x})) = \begin{cases} 0, & \text{if } \det(\nabla\tilde{y}) > 0 \text{ a.e. on } Q(\bar{x}) \text{ or } |\bar{\nabla} y(\bar{x})| \geq R \\ \infty & \text{otherwise} \end{cases}$$

for some $R \gg 1$. The penalty term does not change the energy in the elastic and fracture regime.

3.4.1 Triangular lattices with NN interaction

We first concern ourselves with the triangular lattice. Although the cleavage law was presented in detail in Section 1.4, we include this model here for the sake of completeness and briefly indicate that it fits into the framework of Section 1.1. Recall $\Omega = (0,l) \times (0,1)$ and $\mathcal{L} = A\mathbb{Z}^2 = T_\phi \begin{pmatrix} 1 & \frac{1}{2} \\ 0 & \frac{\sqrt{3}}{2} \end{pmatrix} \mathbb{Z}^2$ for $\phi \in [0, \frac{\pi}{3})$, where $T_\phi = \begin{pmatrix} \cos\phi & -\sin\phi \\ \sin\phi & \cos\phi \end{pmatrix}$. For a deformation $y : \mathcal{L}_\varepsilon \cap \Omega \to \mathbb{R}^2$ let

$$\mathcal{E}_\varepsilon(y) = \frac{\varepsilon}{2} \sum_{\substack{x,x' \in \mathcal{L}_\varepsilon \\ |x-x'|=\varepsilon}} W\Big(\frac{|y(x) - y(x')|}{\varepsilon}\Big) + \varepsilon \sum_{\bar{x} \in (\mathcal{L}'_\varepsilon(\Omega))^\circ} \chi(\bar{\nabla} y(\bar{x})),$$

where $W : [0,\infty) \to [0,\infty)$ satisfies the assumptions (i), (ii) and (iii) for $\alpha := W''(1) > 0$ and $\lim_{r\to\infty} W(r) = \beta > 0$. Denoting the i-th column of G by G_i and letting $Z = A\frac{1}{2}\begin{pmatrix} -1 & 1 & 1 & -1 \\ -1 & -1 & 1 & 1 \end{pmatrix}$ the cell energy can be written as

$$\begin{aligned} W_{\text{cell}}(G) = \frac{1}{2}\big(&W(|G_2 - G_1|) + W(|G_3 - G_2|) + W(|G_4 - G_3|) \\ &+ W(|G_1 - G_4|) + 2W(|G_4 - G_2|) + \chi(G)\big). \end{aligned}$$

Note that in Lemma 2.4.1 and Lemma 2.4.2 we have shown that this is an admissible cell energy in the sense of Assumption 1.1.2. We compute

$$\mathcal{Q} = \frac{3\alpha}{8} \begin{pmatrix} 3 & 1 & 0 \\ 1 & 3 & 0 \\ 0 & 0 & 2 \end{pmatrix}$$

and therefore $\alpha_A = \alpha$. With $\nu_1^\phi = T_\phi(1,0)^T$, $\nu_2^\phi = T_\phi(\frac{1}{2}, \frac{\sqrt{3}}{2})^T$ and $\nu_3^\phi = T_\phi(-\frac{1}{2}, \frac{\sqrt{3}}{2})^T$ we get

$$\beta_A = \min_{\varsigma \in S^1} \beta \frac{\sum_{i=1}^3 |\nu_i^\phi \cdot \varsigma|}{|\mathbf{e}_1 \cdot \varsigma|} = \frac{\sqrt{3}\beta}{\sin(\phi + \frac{\pi}{3})}$$

and then we re-derive

$$\mathcal{E}_{\text{lim}}(a) = \frac{2}{\sqrt{3}} \min\Big\{\frac{1}{2} l\alpha a^2, \frac{\sqrt{3}\beta}{\sin(\phi + \frac{\pi}{3})}\Big\}.$$

3.4.2 Square lattices with NN and NNN interaction

The behavior in the elastic regime of the following two dimensional model comprising nearest and next to nearest neighbor atomic interactions was treated by

Friesecke and Theil in [50]. We let $\Omega = (0, l_1) \times (0, l_2)$, set $\mathcal{L} = A\mathbb{Z}^2 = T_\phi \mathbb{Z}^2$ for $\phi \in [0, \frac{\pi}{2})$ and

$$\mathcal{E}_\varepsilon(y) = \frac{\varepsilon}{2} \sum_{\substack{x, x' \in \mathcal{L}_\varepsilon \\ |x - x'| = \varepsilon}} W_1\Big(\frac{|y(x) - y(x')|}{\varepsilon}\Big)$$
$$+ \frac{\varepsilon}{2} \sum_{\substack{x, x' \in \mathcal{L}_\varepsilon \\ |x - x'| = \sqrt{2}\varepsilon}} W_2\Big(\frac{|y(x) - y(x')|}{\sqrt{2}\varepsilon}\Big) + \varepsilon \sum_{\bar{x} \in (\mathcal{L}'_\varepsilon(\Omega))^\circ} \chi(\bar{\nabla} y(\bar{x}))$$

for deformations $y : \mathcal{L}_\varepsilon \cap \Omega \to \mathbb{R}^2$ and potentials W_1, W_2 as above with α_1, β_1 and α_2, β_2, respectively. The associated cell energy is given by

$$W_{\mathrm{cell}}(G) = \frac{1}{4} \sum_{|z_i - z_j| = 1} W_1(|G_i - G_j|) + \frac{1}{2} \sum_{|z_i - z_j| = \sqrt{2}} W_2\Big(\frac{|G_i - G_j|}{\sqrt{2}}\Big) + \chi(G)$$

In [50] it is shown that the cell energy is admissible. We calculate

$$\mathcal{Q} = \frac{1}{2} T_\phi^{*T} \begin{pmatrix} 2\alpha_1 + \alpha_2 & \alpha_2 & 0 \\ \alpha_2 & 2\alpha_1 + \alpha_2 & 0 \\ 0 & 0 & 2\alpha_2 \end{pmatrix} T_\phi^*,$$

with

$$T_\phi^* = \begin{pmatrix} c^2 & s^2 & -\sqrt{2}cs \\ s^2 & c^2 & \sqrt{2}cs \\ \sqrt{2}cs & -\sqrt{2}cs & c^2 - s^2 \end{pmatrix}, \quad c := \cos\phi, \ s := \sin\phi.$$

An elementary computation then shows

$$\alpha_A = \frac{2\alpha_1\alpha_2(\alpha_1 + \alpha_2)}{2\alpha_1\alpha_2 + 4\alpha_1^2 c^2 s^2 + \alpha_2^2(c^2 - s^2)^2}.$$

Letting $\nu_1^\phi = T_\phi \mathbf{e}_1$, $\nu_2^\phi = T_\phi \mathbf{e}_2$, $\nu_3^\phi = T_\phi(\mathbf{e}_1 + \mathbf{e}_2)$ and $\nu_4^\phi = T_\phi(\mathbf{e}_1 - \mathbf{e}_2)$ and $\gamma_1 = \max\{c, s\}$, $\gamma_2 = c + s$ we obtain for the fracture constant

$$\begin{aligned} \beta_A &= \min_{\varsigma \in S^1} \frac{\bar{\beta}_1(|\nu_1^\phi \cdot \varsigma| + |\nu_2^\phi \cdot \varsigma|) + \bar{\beta}_2(|\nu_3^\phi \cdot \varsigma| + |\nu_4^\phi \cdot \varsigma|)}{|\mathbf{e}_1 \cdot \varsigma|} \\ &= \min\Big\{\frac{\beta_1 + 2\beta_2}{\gamma_1}, \frac{2\beta_1 + 2\beta_2}{\gamma_2}\Big\}. \end{aligned}$$

Here we used that it suffices to minimize over the set $\mathcal{P} = \{\frac{\varsigma}{|\varsigma|} : \varsigma = \nu_i^\phi, i = 1, \dots, 4\} \subset S^1$. Below the critical value a_{crit} energetically optimal configurations are given by functions of the form (3.59) with

$$\bar{F}(a_\varepsilon) = \begin{pmatrix} a_\varepsilon & 0 \\ \frac{(\alpha_2^2 - \alpha_1^2)cs(c^2 - s^2)}{2\alpha_1\alpha_2 + \alpha_2^2(c^2 - s^2)^2 + 4\alpha_1^2 c^2 s^2} a_\varepsilon & \frac{-\alpha_2^2(c^2 - s^2)^2 - 4\alpha_1^2 c^2 s^2}{2\alpha_1\alpha_2 + \alpha_2^2(c^2 - s^2)^2 + 4\alpha_1^2 c^2 s^2} a_\varepsilon \end{pmatrix}.$$

In particular, the configurations show the Poisson-effect and in the case that $\alpha_1 \neq \alpha_2$ and $\phi \in (0, \frac{\pi}{2}) \setminus \{\frac{\pi}{4}\}$ also shear effects occur. Limiting minimal configurations beyond critical loading are given by deformations of the form (3.60), where the normal ξ to the hyperplane Π is an element of $\{\nu_i^\phi : i = 1, \dots 4\}$. While in the previous example the cleavage direction was determined only by the geometry of the problem (i.e. by ϕ), it now depends also on the ratio of β_1, β_2.

We note that here for every $\phi \in (0, \frac{\pi}{2}) \setminus \{\frac{\pi}{4}\}$ by choosing the specific values $\beta_1 = 1$ and $\beta_2 = \frac{1}{2}\max\{\cot\phi, \tan\phi\} - \frac{1}{2}$ the minimum in the expression for β_A is attained at ν_2^ϕ and ν_3^ϕ, respectively, at ν_1^ϕ and ν_4^ϕ. As a consequence, unlike for the triangular lattice in the previous example, also for general lattice orientations there may be deformations with almost optimal energy whose rescaled displacements in the continuum limit have a serrated jump set.

3.4.3 Cubic lattices with NN and NNN interaction

We consider the following three dimensional model with nearest and next nearest interactions in the domain $\Omega = (0, l_1) \times (0, l_2) \times (0, l_3)$. We let $\mathcal{L} = A\mathbb{Z}^3 = T_{\phi,\psi}\mathbb{Z}^3$, where

$$T_{\phi,\psi} = \begin{pmatrix} \cos\psi & -\sin\psi & 0 \\ \sin\psi & \cos\psi & 0 \\ 0 & 0 & 1 \end{pmatrix} \begin{pmatrix} 1 & 0 & 0 \\ 0 & \cos\phi & -\sin\phi \\ 0 & \sin\phi & \cos\phi \end{pmatrix} \text{ for } \phi, \psi \in [0, \frac{\pi}{2}).$$

We let

$$\mathcal{E}_\varepsilon(y) = \frac{\varepsilon^2}{2} \sum_{\substack{x, x' \in \mathcal{L}_\varepsilon \\ |x - x'| = \varepsilon}} W_1\Big(\frac{|y(x) - y(x')|}{\varepsilon}\Big)$$

$$+ \frac{\varepsilon^2}{2} \sum_{\substack{x, x' \in \mathcal{L}_\varepsilon \\ |x - x'| = \sqrt{2}\varepsilon}} W_2\Big(\frac{|y(x) - y(x')|}{\sqrt{2}\varepsilon}\Big) + \varepsilon^2 \sum_{\bar{x} \in (\mathcal{L}'_\varepsilon(\Omega))^\circ} \chi(\bar{\nabla} y(\bar{x}))$$

for deformations $y : \mathcal{L}_\varepsilon \cap \Omega \to \mathbb{R}^2$ and potentials W_1, W_2 as above with α_1, β_1 and α_2, β_2, respectively. The associated cell energy is given by

$$W_{\text{cell}}(G) = \frac{1}{8} \sum_{|z_i - z_j| = 1} W_1(|G_i - G_j|) + \frac{1}{4} \sum_{|z_i - z_j| = \sqrt{2}} W_2\Big(\frac{|G_i - G_j|}{\sqrt{2}}\Big) + \chi(G).$$

In [63] it has been shown that the cell energy is admissible. As before, an elementary computation shows

$$\mathcal{Q} = T_\psi^{*T} T_\phi^{*T} \frac{1}{2} \begin{pmatrix} 2\alpha_1 + 2\alpha_2 & \alpha_2 & \alpha_2 & 0 & 0 & 0 \\ \alpha_2 & 2\alpha_1 + 2\alpha_2 & \alpha_2 & 0 & 0 & 0 \\ \alpha_2 & \alpha_2 & 2\alpha_1 + 2\alpha_2 & 0 & 0 & 0 \\ 0 & 0 & 0 & 2\alpha_2 & 0 & 0 \\ 0 & 0 & 0 & 0 & 2\alpha_2 & 0 \\ 0 & 0 & 0 & 0 & 0 & 2\alpha_2 \end{pmatrix} T_\phi^* T_\psi^*,$$

where

$$T_\phi^* = \begin{pmatrix} 1 & 0 & 0 & 0 & 0 & 0 \\ 0 & c_1^2 & s_1^2 & 0 & 0 & -2c_1s_1 \\ 0 & s_1^2 & c_1^2 & 0 & 0 & 2c_1s_1 \\ 0 & 0 & 0 & c_1 & -s_1 & 0 \\ 0 & 0 & 0 & s_1 & c_1 & 0 \\ 0 & c_1s_1 & -c_1s_1 & 0 & 0 & c_1^2 - s_1^2 \end{pmatrix}, T_\psi^* = \begin{pmatrix} c_2^2 & s_2^2 & 0 & -2c_2s_2 & 0 & 0 \\ s_2^2 & c_2^2 & 0 & 2c_2s_2 & 0 & 0 \\ 0 & 0 & 1 & 0 & 0 & 0 \\ c_2s_2 & -c_2s_2 & 0 & c_2^2 - s_2^2 & 0 & 0 \\ 0 & 0 & 0 & 0 & c_2 & -s_2 \\ 0 & 0 & 0 & 0 & s_2 & c_2 \end{pmatrix},$$

with the abbreviations $c_1 = \cos\phi$, $c_2 = \cos\psi$, $s_1 = \sin\phi$ and $s_2 = \sin\psi$. Applying Lemma 2.1.2 we then obtain

$$\alpha_A = \frac{\alpha_2(2\alpha_1 + \alpha_2)^2(\alpha_1 + 2\alpha_2)}{8\alpha_1^3c_2^2s_2^2 + 2\alpha_1\alpha_2^2(4 - c_2^2s_2^2) + 4\alpha_1^2\alpha_2(4c_2^2s_2^2 - 1) + \alpha_2^3(3 - 4c_2^2s_2^2)}.$$

In particular, α_A is independent of c_1 and s_1. We let $\mathcal{V}_1^{\phi,\psi} = T_{\phi,\psi}\{\mathbf{e}_1, \mathbf{e}_2, \mathbf{e}_3\}$, $\mathcal{V}_2^{\phi,\psi} = T_{\phi,\psi}\{\mathbf{e}_1 + \mathbf{e}_2, \mathbf{e}_1 - \mathbf{e}_2, \mathbf{e}_1 + \mathbf{e}_3, \mathbf{e}_1 - \mathbf{e}_3, \mathbf{e}_2 + \mathbf{e}_3, \mathbf{e}_2 - \mathbf{e}_3\}$ as well as $\gamma_1 = \max\{|c_2|, |c_1s_2|, |s_1s_2|\}$, $\gamma_2 = \max\{|c_2 \pm c_1s_2|, |c_2 \pm s_1s_2|, |c_1s_2 \pm s_1s_2|\}$, $\gamma_3 = \max\{|c_2 \pm c_1s_2 \pm s_1s_2|\}$, $\gamma_4 = \max\{|2c_2 \pm c_1s_2 \pm s_1s_2|, |c_2 \pm 2c_1s_2 \pm s_1s_2|, |c_2 \pm c_1s_2 \pm 2s_1s_2|\}$. One can show that

$$\begin{aligned} \mathcal{P} &= \{\varsigma/|\varsigma| : \varsigma = T_{\phi,\psi}\mathbf{e}_1, i = 1,2,3\} \cup \{\varsigma/|\varsigma| : \varsigma = T_{\phi,\psi}(\mathbf{e}_1 \pm \mathbf{e}_2 \pm \mathbf{e}_3)\} \\ &\quad \cup \{\varsigma/|\varsigma| : \varsigma = T_{\phi,\psi}(\mathbf{e}_1 \pm \mathbf{e}_2 \pm \mathbf{e}_3 \pm \mathbf{e}_i), i = 1,2,3\}. \end{aligned}$$

Then

$$\begin{aligned} \beta_A &= \min_{\varsigma \in S^1} \frac{\sum_{\nu \in \mathcal{V}_1^{\phi,\psi}} \beta_1 |\nu \cdot \varsigma| + \sum_{\nu \in \mathcal{V}_2^{\phi,\psi}} \beta_2 |\nu \cdot \varsigma|}{|\mathbf{e}_1 \cdot \varsigma|} \\ &= \min\left\{\frac{\beta_1 + 4\beta_2}{\gamma_1}, \frac{2\beta_1 + 6\beta_2}{\gamma_2}, \frac{3\beta_1 + 6\beta_2}{\gamma_3}, \frac{4\beta_1 + 10\beta_2}{\gamma_4}\right\}. \end{aligned}$$

Chapter 4

Limiting minimal energy configurations

This chapter is devoted to the proofs of Theorem 1.5.1 and Corollary 1.5.2. First, in Section 4.1 we derive a fine estimate on the limiting minimal energy (see Theorem 1.4.2). This will be an essential ingredient to analyze the number and position of broken triangles in more detail (see Section 4.2). Finally, the main convergence results for almost minimizers are addressed in Section 4.3.

4.1 Fine estimates on the limiting minimal energy

We first prove Theorem 1.4.2 about the sharp estimate on the discrete minimal energies. Assume that W in addition satisfies assumptions (ii'), (iii') and recall the definition of $\mathcal{C}_\varepsilon^{(1)}$ and $\mathcal{C}_\varepsilon^{(2)}$ in Section 3.1. In order to investigate a deformation y again we let $\bar{\mathcal{C}}_\varepsilon$ and $\bar{\mathcal{C}}_\varepsilon^{(1)}$ denote the set of triangles $\triangle$ (of type one respectively) for which at least one side in $y(\triangle)$ is larger than $2R\varepsilon$, where now the threshold value $R > 1$ is chosen in such a way that $c_R := \inf\{W(r) : r \geq R\} \geq \frac{\beta}{2}$. According to Lemma 2.4.5(iii) we may choose a convex function V such that

$$0 \leq V(r) \leq \tilde{W}(r) \leq V(r) + O((r-1)^4) \text{ for } r \leq 8R. \tag{4.1}$$

As in (3.1) we observe that $|\mathbf{e}_1^T(y)_\triangle \mathbf{e}_1|$ is bounded by $8R$ on triangles with bond length not exceeding $4R\varepsilon$ and thus lies in the convex regime of V. Moreover, we find that every triangle in $\bar{\mathcal{C}}_\varepsilon$ provides at least the energy $\frac{4}{\sqrt{3}\varepsilon}\int_\triangle W_\triangle(\nabla\tilde{y}) \geq c_R\varepsilon$.

For given $0 < \eta < a$ we also define $R_{\varepsilon,\eta} = \frac{a-\eta}{\sqrt{\varepsilon}}$ as a threshold for triangles we consider 'essentially broken':

$$\bar{\mathcal{C}}_{\varepsilon,\eta} = \left\{\triangle \in \bar{\mathcal{C}}_\varepsilon, |\nabla y_\varepsilon \mathbf{v}| > R_{\varepsilon,\eta} \text{ for at least two } \mathbf{v} \in \mathcal{V}\right\}. \tag{4.2}$$

The minimal energy contribution of all the springs on such a triangle in $\bar{\mathcal{C}}_{\varepsilon,\eta}$ is given by

$$2\beta^\eta\varepsilon := 2\inf\left\{W(r) : r \geq \frac{a-\eta}{\sqrt{\varepsilon}}\right\}\varepsilon = (2\beta + O(\varepsilon))\varepsilon$$

by the assumption (iii') on W. By $I \subset (\varepsilon, 1-\varepsilon)$ we denote the set of points x_2 for which the segment $(0,l) \times \{x_2\}$ intersects a broken triangle (of type one) in $\bar{\mathcal{C}}_\varepsilon^{(1)}$. In addition, we say $x_2 \in I^\eta \subset I$ if one of the intersected triangles lies in $\bar{\mathcal{C}}_{\varepsilon,\eta} \cap \bar{\mathcal{C}}_\varepsilon^{(1)}$.

With these preparations we can now proceed to prove Theorem 1.4.2:

Proof of Theorem 1.4.2. Let $\mathcal{E}_\varepsilon(y) = \inf \mathcal{E}_\varepsilon + O(\varepsilon)$. Inspired by (3.2) and (3.3) we establish a lower bound for the energies additionally taking the set $I \setminus I^\eta$ into account. Since the sidelength of any triangle whose interior intersects $(0,l) \times (I \setminus I^\eta)$ is bounded by $4R_{\varepsilon,\eta}$, we find

$$|\mathbf{e}_1^T \nabla \tilde{y}(x_1, x_2)\, \mathbf{e}_1| \le 8R_{\varepsilon,\eta}$$

for all $(x_1, x_2) \in (0,l) \times (I \setminus I^\eta)$ as in (3.1). Let $k = k(x_2)$ count the number of triangles in $\bar{\mathcal{C}}_\varepsilon$ on the slice $(0,l) \times \{x_2\}$, $x_2 \in I \setminus I^\eta$, and define $\bar{\mathcal{C}}_\varepsilon^{x_2} \subset (0,l)$ such that $((0,l) \times \{x_2\}) \cap \bigcup_{\triangle \in \bar{\mathcal{C}}_\varepsilon} \triangle = \bar{\mathcal{C}}_\varepsilon^{x_2} \times \{x_2\}$. Then

$$\int_{\bar{\mathcal{C}}_\varepsilon^{x_2}} \mathbf{e}_1^T \nabla \tilde{y}(x_1, x_2)\, \mathbf{e}_1 \, dx_1 \le 8k\varepsilon R_{\varepsilon,\eta}. \tag{4.3}$$

and so

$$\begin{aligned}\int_{\Omega_\varepsilon^{x_2} \setminus \bar{\mathcal{C}}_\varepsilon^{x_2}} \mathbf{e}_1^T \nabla \tilde{y}(x_1, x_2)\, \mathbf{e}_1 &\ge (1 + \sqrt{\varepsilon} a)(l + O(\varepsilon)) - 8k\varepsilon R_{\varepsilon,\eta} \\ &= \left(1 + \sqrt{\varepsilon}\left(a - \frac{8k(a-\eta)}{l} + O(\sqrt{\varepsilon})\right)\right) l.\end{aligned}$$

Since $\#(\bar{\mathcal{C}}_\varepsilon \setminus \bar{\mathcal{C}}_{\varepsilon,\eta}) \ge \frac{1}{\varepsilon\gamma} \int_{I \setminus I^\eta} k(x_2)\, dx_2$, a convexity argument as in the proof of Theorem 1.4.1 on slices $(0,l) \times \{x_2\}$ with $x_2 \in (\varepsilon, 1-\varepsilon) \setminus I$ and on the unbroken part $(\Omega_\varepsilon^{x_2} \setminus \bar{\mathcal{C}}_\varepsilon^{x_2}) \times \{x_2\}$ of slices with x_2 in $I \setminus I^\eta$ then shows that

$$\mathcal{E}_\varepsilon(y) \ge \frac{4(l-2\varepsilon)}{\sqrt{3}\varepsilon} V(1+\sqrt{\varepsilon}a)(1-2\varepsilon-|I|) + G_{\eta,\varepsilon}|I \setminus I^\eta| + \frac{2\beta^\eta}{\gamma}|I^\eta| + O(\varepsilon), \tag{4.4}$$

where

$$G_{\eta,\varepsilon} = \min_{k \in \mathbb{N}} \left(\frac{4l}{\sqrt{3}\varepsilon} V \left(1 + \sqrt{\varepsilon}\left(a - \frac{8k(a-\eta)}{l} + O(\sqrt{\varepsilon})\right)\right) + \frac{kc_R}{\gamma}\right).$$

We note that this minimum exists and can be taken over $1 \le k \le K_0$ for some $K_0 \in \mathbb{N}$ large enough and independent of η as $\frac{kc_R}{\gamma} \to \infty$ for $k \to \infty$. We choose $0 < \eta < a$ large enough such that

$$\frac{l\alpha}{\sqrt{3}} a^2 < \min_{1 \le k \le K_0} \left(\frac{\alpha l}{\sqrt{3}} \left(a - \frac{8k}{l}(a-\eta)\right)^2 + \frac{kc_R}{\gamma}\right).$$

Recalling that, by (4.1) and Lemma 2.4.3, $\frac{4l}{\sqrt{3}\varepsilon}V(1+\sqrt{\varepsilon}r) = \frac{4l}{\sqrt{3}\varepsilon}\tilde{W}(1+\sqrt{\varepsilon}r) + O(\varepsilon) \to \frac{l\alpha}{\sqrt{3}}r^2$ uniformly in r on bounded sets in $\mathbb{R}$, we see that thus $G_{\eta,\varepsilon}$ exceeds the elastic term $\frac{4l}{\sqrt{3}\varepsilon}V(1+\sqrt{\varepsilon}a)$ for ε sufficiently small. So from (4.4) we obtain

$$\mathcal{E}_\varepsilon(y) \geq \frac{4l}{\sqrt{3}\varepsilon}V(1+\sqrt{\varepsilon}a)(1-2\varepsilon-|I^\eta|) + \frac{2\beta^\eta}{\gamma}|I^\eta| + O(\varepsilon). \tag{4.5}$$

As $\frac{4l}{\sqrt{3}\varepsilon}V(1+\sqrt{\varepsilon}a) \to \frac{l\alpha}{\sqrt{3}}a^2$ and $\beta^\eta \to \beta$ for all $\eta > 0$, for ε small enough we thus obtain by minimizing over $|I^\eta| \in [0,1]$ that $\inf \mathcal{E}_\varepsilon \geq \frac{4}{\sqrt{3}}\frac{l}{\varepsilon}V(1+\sqrt{\varepsilon}a)(1-2\varepsilon) + O(\varepsilon) = \frac{4}{\sqrt{3}}\frac{l}{\varepsilon}\tilde{W}(1+\sqrt{\varepsilon}a) + O(\varepsilon)$ or $\inf \mathcal{E}_\varepsilon \geq \frac{2\beta+O(\varepsilon)}{\gamma}(1-2\varepsilon) = \frac{2\beta}{\gamma} + O(\varepsilon)$, respectively, depending on a.

Applying (3.5) and (3.6) we then get indeed

$$\inf \mathcal{E}_\varepsilon = \frac{4}{\sqrt{3}}\frac{l}{\varepsilon}\tilde{W}(1+\sqrt{\varepsilon}a) + O(\varepsilon) \quad \text{or} \quad \inf \mathcal{E}_\varepsilon = \frac{2\beta}{\gamma} + O(\varepsilon), \tag{4.6}$$

respectively. The claim now follows from Lemma 2.4.4. □

Remark 4.1.1. From the proof of Theorem 1.4.2, especially taking (3.5) and (3.6) into account, it follows that Theorem 1.4.2 still holds if $\mathcal{E}_\varepsilon$ is replaced by $\mathcal{E}_\varepsilon^\chi$.

4.2 Sharp estimates on the number of the broken triangles

Throughout this section we will assume that $a_\varepsilon = \sqrt{\varepsilon}a$, y_ε is a sequence of deformations satisfying (1.28) and $u_\varepsilon = \frac{1}{\sqrt{\varepsilon}}(y_\varepsilon - \mathbf{id})$ are the corresponding rescaled displacements. Moreover, we suppose that the threshold value R is chosen as above Equation (4.1) implying $c_R \geq \frac{\beta}{2}$ and that $\bar{\mathcal{C}}_\varepsilon$ is defined accordingly.

For a rescaled displacement $\tilde{u}$ we denote by $D^\mu \subset (\varepsilon, 1-\varepsilon)$ for $\mu > 0$ the set of x_2 such that there is precisely one triangle $\triangle_{x_2} \in \bar{\mathcal{C}}_\varepsilon^{(1)}$ with $\mathrm{int}(\triangle_{x_2}) \cap ((0,l)\times\{x_2\}) \neq \emptyset$ and

$$\int_{\Omega_\varepsilon^{x_2}\setminus\bar{\mathcal{C}}_\varepsilon^{x_2}} \mathbf{e}_1^T \nabla\tilde{u}(x_1,x_2)\mathbf{e}_1\,dx_1 \leq l\mu. \tag{4.7}$$

(for the definition of $\bar{\mathcal{C}}_\varepsilon^{x_2}$ recall (4.3).) Note that $D^\mu \subset I^\eta$ for μ small enough: For $x_2 \in D^\mu$ we have

$$\int_{\bar{\mathcal{C}}_\varepsilon^{x_2}} \mathbf{e}_1^T \nabla\tilde{y}(x_1,x_2)\mathbf{e}_1\,dx_1 \geq \sqrt{\varepsilon}l(a-\mu) + O(\varepsilon)$$

and using the arguments in (3.1) we see that for given η (not too small) we can choose μ small enough such that $\triangle_{x_2} \in \bar{\mathcal{C}}_{\varepsilon,\eta}$ and thus $x_2 \in I^\eta$. We also define

$\bar{\mathcal{C}}^{\mu}_{\varepsilon,\eta} \subset \bar{\mathcal{C}}_{\varepsilon,\eta}$ as the set of those essentially broken triangles $\triangle$ for which there exists some $x_2 \in D^{\mu}$ such that $\operatorname{int}(\triangle) \cap ((0,l) \times \{x_2\}) \neq \emptyset$. The projection of a triangle $\triangle$ onto the linear subspace spanned by $\mathbf{v}_{\gamma}^{\perp}$ is an interval of length $\frac{\sqrt{3}}{2}\varepsilon$. We denote the center of this interval by $m_{\triangle}$.

The following lemmas give sharp estimates on the number of broken triangles and their position.

Lemma 4.2.1. *Let $a < a_{\text{crit}}$ and suppose $\tilde{u}_{\varepsilon}$ is a minimizing sequence satisfying*

$$\mathcal{E}_{\varepsilon}(\mathbf{id} + \sqrt{\varepsilon} u_{\varepsilon}) = \inf \mathcal{E}_{\varepsilon} + O(\varepsilon).$$

Then $\varepsilon \# \bar{\mathcal{C}}_{\varepsilon} = O(\varepsilon)$.

Proof. Using (4.4), (4.6) and Lemma 2.4.5(iii) we find

$$\begin{aligned}
\mathcal{E}_{\varepsilon}(y_{\varepsilon}) &= \frac{4l}{\sqrt{3}\varepsilon} \tilde{W}(1 + \sqrt{\varepsilon} a) + O(\varepsilon) \\
&\geq \frac{4(l - 2\varepsilon)}{\sqrt{3}\varepsilon} \tilde{W}(1 + \sqrt{\varepsilon} a)(1 - 2\varepsilon - |I|) + \min\left\{G_{\eta,\varepsilon}, \frac{2\beta^{\eta}}{\gamma}\right\} |I| + O(\varepsilon) \\
&= \frac{4l}{\sqrt{3}\varepsilon} \tilde{W}(1 + \sqrt{\varepsilon} a)(1 - |I|) + \min\left\{G_{\eta,\varepsilon}, \frac{2\beta^{\eta}}{\gamma}\right\} |I| + O(\varepsilon).
\end{aligned}$$

An elementary computation yields, whenever ε is small enough,

$$\begin{aligned}
|I| &\leq \left(\min\left\{G_{\eta,\varepsilon}, \frac{2\beta^{\eta}}{\gamma}\right\} - \frac{4l}{\sqrt{3}\varepsilon} \tilde{W}(1 + \sqrt{\varepsilon} a)\right)^{-1} \cdot O(\varepsilon) \\
&= \left(\min\left\{G_{\eta,\varepsilon}, \frac{2\beta}{\gamma}\right\} - \frac{\alpha l}{\sqrt{3}} a^2 + o(1)\right)^{-1} \cdot O(\varepsilon) = O(\varepsilon).
\end{aligned}$$

(The argument leading to (4.5) together with $a < a_{\text{crit}}$ shows that the term in parentheses is bounded from below by a positive constant independent of ε). Then the elastic energy is $\frac{4l}{\sqrt{3}\varepsilon} \tilde{W}(1 + \sqrt{\varepsilon} a) + O(\varepsilon)$ and consequently, the crack energy coming from triangles in $\bar{\mathcal{C}}_{\varepsilon}$ is of order $O(\varepsilon)$. As every broken triangle in $\bar{\mathcal{C}}_{\varepsilon}$ provides at least energy εc_R we conclude $\varepsilon \# \bar{\mathcal{C}}^{(1)}_{\varepsilon} = O(\varepsilon)$. But then, possibly after replacing R by $2R$, also $\varepsilon \# \bar{\mathcal{C}}^{(2)}_{\varepsilon} = O(\varepsilon)$ as those triangles are neighbors of broken triangles of type 1. □

Lemma 4.2.2. *Let $a > a_{\text{crit}}$, $\phi \neq 0$ and suppose $\tilde{u}_{\varepsilon}$ is a minimizing sequence satisfying*

$$\mathcal{E}_{\varepsilon}(\mathbf{id} + \sqrt{\varepsilon} u_{\varepsilon}) = \inf \mathcal{E}_{\varepsilon} + O(\varepsilon).$$

Then $|I^{\eta}| = 1 - O(\varepsilon)$ for $0 < \eta < a$. Furthermore, for μ sufficiently small, $\varepsilon \# \left(\bar{\mathcal{C}}_{\varepsilon} \setminus \bar{\mathcal{C}}^{\mu}_{\varepsilon,\eta}\right) = O(\varepsilon)$ and

$$\sup\left\{|m_{\triangle_1} - m_{\triangle_2}| : \triangle_1, \triangle_2 \in \bar{\mathcal{C}}^{\mu}_{\varepsilon,\eta}\right\} = O(\varepsilon).$$

Proof. Without loss of generality we choose η sufficiently large such that by (4.5) and (4.6) we get

$$\mathcal{E}_\varepsilon(y_\varepsilon) = \frac{2\beta}{\gamma} + O(\varepsilon) \geq \frac{4l}{\sqrt{3}\varepsilon}\tilde{W}(1+\sqrt{\varepsilon}a)(1-2\varepsilon-|I^\eta|) + \frac{2\beta^\eta}{\gamma}|I^\eta|.$$

So for ε small enough we obtain

$$1 - |I^\eta| \leq \left(\frac{\alpha l}{\sqrt{3}}a^2 + o(1) - \frac{2\beta}{\gamma}\right)^{-1} \cdot O(\varepsilon) = O(\varepsilon)$$

since $a > a_{\text{crit}}$. Consequently, the crack energy from triangles in $\bar{\mathcal{C}}_{\varepsilon,\eta}$ is given by $\frac{2\beta}{\gamma} + O(\varepsilon)$ and thus the energy contribution from $\bar{\mathcal{C}}_\varepsilon \setminus \bar{\mathcal{C}}_{\varepsilon,\eta}$ is of order $O(\varepsilon)$. As in the proof of Lemma 4.2.1 we find $\varepsilon\#\left(\bar{\mathcal{C}}_\varepsilon \setminus \bar{\mathcal{C}}_{\varepsilon,\eta}\right) = O(\varepsilon)$. Let $k_\eta(x_2)$ and $k^C_\eta(x_2)$ count the number of triangles in $\bar{\mathcal{C}}_{\varepsilon,\eta} \cap \bar{\mathcal{C}}^{(1)}_\varepsilon$ and $(\bar{\mathcal{C}}_\varepsilon \setminus \bar{\mathcal{C}}_{\varepsilon,\eta}) \cap \bar{\mathcal{C}}^{(1)}_\varepsilon$ intersected by $(0,l) \times \{x_2\}$, respectively. We dissect $I^\eta \setminus D^\mu$ into two disjoint sets: By $D_1 \subset I^\eta \setminus D^\mu$ we denote the set where we find more than one triangle $\triangle_{x_2} \in \bar{\mathcal{C}}^{(1)}_\varepsilon$ with $\operatorname{int}(\triangle_{x_2}) \cap ((0,l) \times \{x_2\}) \neq \emptyset$. The complement D_2 is the set where (4.7) does not hold. Using a convexity argument for $x_2 \in D_2$ we obtain

$$\begin{aligned}
\frac{2\beta}{\gamma} + O(\varepsilon) &\geq 2\beta\varepsilon(\#\bar{\mathcal{C}}_{\varepsilon,\eta} \cap \bar{\mathcal{C}}^{(1)}_\varepsilon) + 2c_R\varepsilon\#\left((\bar{\mathcal{C}}_\varepsilon \setminus \bar{\mathcal{C}}_{\varepsilon,\eta}) \cap \bar{\mathcal{C}}^{(1)}_\varepsilon\right) \\
&\quad + \frac{4}{\sqrt{3}\varepsilon}\int_\varepsilon^{1-\varepsilon}\int_{\Omega^{x_2}_\varepsilon \setminus \bar{\mathcal{C}}^{x_2}_\varepsilon} W_\triangle(\nabla\tilde{y})\,dx_1\,dx_2 \\
&\geq \frac{2\beta}{\gamma}\int_\varepsilon^{1-\varepsilon} k_\eta(x_2)\,dx_2 + \frac{2c_R}{\gamma}\int_{D_1} k^C_\eta(x_2)\,dx_2 + \left(\frac{\alpha l}{\sqrt{3}}\mu^2 + o(1)\right)|D_2| \\
&\geq \frac{2\beta}{\gamma}|I^\eta| + \frac{2c_R}{\gamma}|D_1| + \left(\frac{\alpha l}{\sqrt{3}}\mu^2 + o(1)\right)|D_2| \\
&\geq \frac{2\beta}{\gamma}|I^\eta| + \min\left\{\frac{2c_R}{\gamma}, \frac{\alpha l}{\sqrt{3}}\mu^2 + o(1)\right\}|I^\eta \setminus D^\mu|.
\end{aligned}$$

It follows $|I^\eta \setminus D^\mu| = O(\varepsilon)$ and $|D^\mu| = 1 - O(\varepsilon)$, whence the crack energy from triangles in $\bar{\mathcal{C}}^\mu_{\varepsilon,\eta}$ is given by $\frac{2\beta}{\gamma} + O(\varepsilon)$ and then also $\varepsilon\#\left(\bar{\mathcal{C}}_\varepsilon \setminus \bar{\mathcal{C}}^\mu_{\varepsilon,\eta}\right) = O(\varepsilon)$.

Finally, we concern ourselves with the projected distance of triangles in $\bar{\mathcal{C}}^\mu_{\varepsilon,\eta}$. We first note that it suffices to show

$$\sup\left\{|m_{\triangle_1} - m_{\triangle_2}| : \triangle_1, \triangle_2 \in \bar{\mathcal{C}}^\mu_{\varepsilon,\eta} \cap \bar{\mathcal{C}}^{(1)}_\varepsilon\right\} = O(\varepsilon)$$

since for a suitable $\tilde{\eta} \geq \eta$ for any $\triangle \in \bar{\mathcal{C}}^\mu_{\varepsilon,\eta} \cap \bar{\mathcal{C}}^{(2)}_\varepsilon$ there is a $\tilde{\triangle} \in \bar{\mathcal{C}}^\mu_{\varepsilon,\tilde{\eta}} \cap \bar{\mathcal{C}}^{(1)}_\varepsilon$ with $|m_\triangle - m_{\tilde{\triangle}}| \leq \varepsilon$. Let $x_2, z_2 \in D^\mu$, $x_2 < z_2$ with $z_2 - x_2 \leq C\varepsilon$ and $|m_{\triangle_1} - m_{\triangle_2}| > 0$ for the corresponding broken triangles $\triangle_1, \triangle_2 \in \bar{\mathcal{C}}^{(1)}_\varepsilon$. We may assume if a triangle intersects $(0,l) \times \{z_2\}$ or $(0,l) \times \{x_2\}$ then its interior does so, too. Denote by $\bar{d} = \gamma^{-1}|m_{\triangle_1} - m_{\triangle_2}|$ the distances of the centers in $\mathbf{v}_\gamma$-projection onto the x_1-axis.

Let $x_1, z_1 \in (0, l)$ be the points on the slices $(0, l) \times \{x_2\}$ and $(0, l) \times \{z_2\}$ satisfying $\pi_{\mathbf{v}_\gamma^\perp}(x_1, x_2) = m_{\triangle_1}$ and $\pi_{\mathbf{v}_\gamma^\perp}(z_1, z_2) = m_{\triangle_2}$, respectively, where $\pi_{\mathbf{v}_\gamma^\perp}$ denotes the orthogonal projection onto the linear subspace spanned by $\mathbf{v}_\gamma^\perp$. Let $w = \mathbf{e}_1 \cdot \mathbf{v}_\gamma |x_2 - z_2|/\gamma$. Then the $\mathbf{v}_\gamma$-projection of $z = (z_1, z_2)$ onto the x_2-slice is given by $(\tilde{z}_1, x_2)$ with $\tilde{z}_1 = z_1 - w$. Then $\bar{d} = |x_1 - \tilde{z}_1|$ and without restriction we may assume $x_1 > \tilde{z}_1$.

Let $s_\varepsilon = \frac{\sqrt{3}\varepsilon}{4\gamma}$. We now consider the area bounded by the parallelogram with corners $(\tilde{z}_1 + s_\varepsilon, x_2)$, $(x_1 - s_\varepsilon, x_2)$, $(z_1 + \bar{d} - s_\varepsilon, z_2)$, $(z_1 + s_\varepsilon, z_2)$. It is covered by $\frac{2\gamma\bar{d}}{\sqrt{3}\varepsilon} - 1$ stripes of width $\frac{\sqrt{3}}{2}\varepsilon$ in $\mathbf{v}_\gamma$-direction consisting of lattice triangles intersecting the parallelogram, the first of these stripes touching $\triangle_1$, the last one touching $\triangle_2$ (note that if $\gamma\bar{d} = \frac{\sqrt{3}}{2}\varepsilon$ the parallelogram is degenerated to a segment). For the intermediate stripes (4.7) shows that

$$\begin{aligned} y_1(t, x_2) &\le t + \sqrt{\varepsilon} l \mu \quad \forall t < x_1 - s_\varepsilon \qquad \text{and} \\ y_1(t, z_2) &\ge t + \sqrt{\varepsilon} l (a - \mu) \quad \forall t > z_1 + s_\varepsilon. \end{aligned}$$

This shows that if (t, x_2) and $(t + w, z_2)$, $x_1 - \bar{d} + s_\varepsilon < t < x_1 - s_\varepsilon$ lie in the bottom and top triangles of some intermediate stripe, respectively, which are unbroken by construction of D^μ, then

$$|y(t + w, z_2) - y(t, x_2)| \ge y_1(t + w, z_2) - y_1(t, x_2) \ge w + \sqrt{\varepsilon} l (a - 2\mu) \sim \sqrt{\varepsilon}.$$

Consider the $\frac{2\gamma\bar{d}}{\sqrt{3}\varepsilon}$ atomic chains in $\mathbf{v}_\gamma$ direction that lie on the boundary of these stripes. They are of length $\gamma^{-1}(z_2 - x_2) + O(\varepsilon) \le C\varepsilon \ll \sqrt{\varepsilon}$. So there is a constant $c > 0$ such that each of these chains contains at least one spring elongated by a factor of more than $\frac{c}{\sqrt{\varepsilon}}$. By passing, if necessary, to a lower threshold $\tilde{\eta} \ge \eta$, we obtain that the triangles sharing such a spring are broken and additionally one neighbor of each. As broken triangles for such springs on neighboring chains might overlap, we only consider every second atom chain and denote the set of type one triangles adjacent to such a spring on atom chains of odd numbers by $\bar{\mathcal{C}}^{(1)}_{\mathbf{v}_\gamma}(\triangle_1, \triangle_2)$. We note that

$$\gamma\bar{d} \le \sqrt{3}\varepsilon \# \bar{\mathcal{C}}^{(1)}_{\mathbf{v}_\gamma}(\triangle_1, \triangle_2). \tag{4.8}$$

The projection onto the x_2-axis of the spring in $\mathbf{v}_\gamma$-direction is an interval J of length $\gamma\varepsilon$. Counting broken springs, it is elementary to see that the energy contribution $\frac{4}{\sqrt{3}\varepsilon} \int_{(\varepsilon, l-\varepsilon) \times J} W_\triangle(\nabla \tilde{y}_\varepsilon)$ of the part of these broken triangles that lies in the stripe $(0, l) \times J$ is bounded from below by

$$2\varepsilon(1 + P(\gamma))\beta^{\tilde{\eta}}, \tag{4.9}$$

where $P(\gamma)$ is the projection coefficient from (1.24) satisfying $P(1) = \frac{1}{2}$ and in particular $P(\gamma) = 0$ if and only if $\gamma = \frac{\sqrt{3}}{2}$. On the other hand, the energy within stripes $(0,l) \times J'$ when J' is the projection of an arbitrary broken triangle is still bounded from below by $2\varepsilon\beta^{\tilde{\eta}}$.

Now let $\triangle_i$, $i = 1, \ldots, M_\varepsilon$, denote all triangles $\triangle$ in $\bar{\mathcal{C}}^\mu_{\varepsilon,\tilde{\eta}} \cap \bar{\mathcal{C}}^{(1)}_\varepsilon$ such that there exists $x_2^{(i)} \in D^\mu$ with $(0,l) \times \{x_2^{(i)}\}$ intersecting with the interior of $\triangle$. The numbering shall be chosen so as to satisfy $x_2^{(1)} < \ldots < x_2^{(M_\varepsilon)}$. As $1 - |D^\mu| = O(\varepsilon)$, there exists a constant $C > 0$ such that $x_2^{(i+1)} - x_2^{(i)} < C\varepsilon$, $i = 1, \ldots, M_\varepsilon - 1$. We define the subset $\{x_2^{(i_j)}\}_{j=1,\ldots N_\varepsilon}$ of $\{x_2^{(i)}\}_{i=1,\ldots,M_\varepsilon}$ such that $x_2^{(i)} = x_2^{(i_j)}$ for a $j = 1, \ldots N_\varepsilon$ if and only if $|m_{\triangle_i} - m_{\triangle_{i+1}}| > 0$. According to our previous considerations, if $I^{\tilde{\eta}}_{\mathbf{v}_\gamma}$ is the projection of $\bar{\mathcal{C}}^{(1)}_{\mathbf{v}_\gamma} := \bigcup_{j=1}^{N_\varepsilon} \bar{\mathcal{C}}^{(1)}_{\mathbf{v}_\gamma}(\triangle_{i_j}, \triangle_{i_j+1})$ onto the x_2-axis, then

$$|I^{\tilde{\eta}}_{\mathbf{v}_\gamma}| \le \gamma\varepsilon \# \bar{\mathcal{C}}^{(1)}_{\mathbf{v}_\gamma}. \tag{4.10}$$

As before using (4.9) and (4.10) we see that the total energy is greater or equal to

$$\begin{aligned}
&\#\bar{\mathcal{C}}^{(1)}_{\mathbf{v}_\gamma} 2\varepsilon(1 + P(\gamma))\beta^{\tilde{\eta}} + |I^{\tilde{\eta}} \setminus I^{\tilde{\eta}}_{\mathbf{v}_\gamma}| \frac{2\beta^{\tilde{\eta}}}{\gamma} + O(\varepsilon) \\
&= |I^{\tilde{\eta}}| \frac{2\beta^{\tilde{\eta}}}{\gamma} + 2\#\bar{\mathcal{C}}^{(1)}_{\mathbf{v}_\gamma} \varepsilon P(\gamma)\beta^{\tilde{\eta}} + 2\#\bar{\mathcal{C}}^{(1)}_{\mathbf{v}_\gamma} \varepsilon\beta^{\tilde{\eta}} - |I^{\tilde{\eta}}_{\mathbf{v}_\gamma}| \frac{2\beta^{\tilde{\eta}}}{\gamma} + O(\varepsilon) \\
&\ge \frac{2\beta}{\gamma} + 2\#\bar{\mathcal{C}}^{(1)}_{\mathbf{v}_\gamma} \varepsilon P(\gamma)\beta^{\tilde{\eta}} + O(\varepsilon),
\end{aligned}$$

and so $\#\bar{\mathcal{C}}^{(1)}_{\mathbf{v}_\gamma} = O(1)$. As every $\triangle \in \bar{\mathcal{C}}^{(1)}_{\mathbf{v}_\gamma}$ is in at most two different $\bar{\mathcal{C}}^{(1)}_{\mathbf{v}_\gamma}(\triangle_{i_j}, \triangle_{i_j+1})$, this also yields $\sum_{j=1}^{N_\varepsilon} \#\bar{\mathcal{C}}^{(1)}_{\mathbf{v}_\gamma}(\triangle_{i_j}, \triangle_{i_j+1}) = O(1)$.

Applying (4.8) we find that

$$O(1) = \sum_{j=1}^{N_\varepsilon} \#\bar{\mathcal{C}}^{(1)}_{\mathbf{v}_\gamma}(\triangle_{i_j}, \triangle_{i_j+1}) \ge \sum_{j=1}^{N_\varepsilon} \frac{\gamma \bar{d}_{i_j}}{\sqrt{3}\varepsilon} \ge \frac{c}{\varepsilon} \sum_{j=1}^{N_\varepsilon} |m_{\triangle_{i_j}} - m_{\triangle_{i_j+1}}|$$

for a constant $c > 0$, when $\bar{d}_i = \gamma^{-1}|m_{\triangle_i} - m_{\triangle_{i+1}}|$. This concludes the proof. □

The above Lemmas 4.2.1 and 4.2.2 show that for a sequence of almost minimizers $(\tilde{y}_\varepsilon)$ satisfying (1.28), the number $\#\bar{\mathcal{C}}_\varepsilon$ of largely deformed triangles is bounded independently of ε for $a < a_{\text{crit}}$, while in the supercritical case for $\phi \neq 0$ there are two subsets

$$\begin{aligned}
\Omega^{(1)}_\varepsilon &:= \{x \in \Omega_\varepsilon : 0 \le x_1 \le p_\varepsilon - c\varepsilon + (\mathbf{v}_\gamma \cdot \mathbf{e}_1)x_2\}, \\
\Omega^{(2)}_\varepsilon &:= \{x \in \Omega_\varepsilon : p_\varepsilon + c\varepsilon + (\mathbf{v}_\gamma \cdot \mathbf{e}_1)x_2 \le x_1 \le l\},
\end{aligned} \tag{4.11}$$

$c > 0$ independent of ε and p_ε to be chosen appropriately, such that the number of triangles in $\bar{\mathcal{C}}_\varepsilon$ intersecting $\Omega^{(1)}_\varepsilon \cup \Omega^{(2)}_\varepsilon$ is bounded uniformly in ε. We recall that the last claim in Lemma 4.2.2 does not hold if $\mathbf{v}_\gamma$ is not unique ($\gamma = \frac{\sqrt{3}}{2}$). Indeed, if $P(\gamma)$ vanishes, we cannot conlude that $\#\bar{\mathcal{C}}^{(1)}_{\mathbf{v}_\gamma} = O(1)$ in the above proof. In this case we do not expect that the essential part of the broken triangles lies in in a small stripe parallel to $\mathbb{R}(\frac{1}{2}, \frac{\sqrt{3}}{2})^T$ or $\mathbb{R}(\frac{1}{2}, \frac{\sqrt{3}}{2})^T$ as we have already seen that the crack can take a serrated course. Nevertheless, if $\gamma = \frac{\sqrt{3}}{2}$ (or equivalently $\phi = 0$) one can show that up to a number being uniformly bounded in ε the broken triangles $\bar{\mathcal{C}}_\varepsilon$ lie in a stripe around the graph of a Lipschitz function. Recall that $\psi(\varepsilon)$ is the width of the lateral boundaries (see (1.22)).

Lemma 4.2.3. *Let $\tilde{u}_\varepsilon$ be a minimizing sequence satisfying*

$$\mathcal{E}_\varepsilon(\mathbf{id} + \sqrt{\varepsilon}u_\varepsilon) = \inf \mathcal{E}_\varepsilon + O(\varepsilon).$$

Let $a > a_{\text{crit}}$ and $\phi = 0$. Then there exist Lipschitz functions $g_\varepsilon : (0,1) \to (\psi(\varepsilon), l - \psi(\varepsilon))$ with $|g_\varepsilon'| = \frac{1}{\sqrt{3}}$ a.e. such that for μ sufficiently small $\varepsilon\#(\bar{\mathcal{C}}_\varepsilon \setminus \bar{\mathcal{C}}^\mu_{\varepsilon,\eta}) = O(\varepsilon)$ and

$$\cup_{\triangle \in \bar{\mathcal{C}}^\mu_{\varepsilon,\eta}} \triangle \subset \{(x,y) \in \Omega : g_\varepsilon(y) - C\varepsilon \le x \le g_\varepsilon(y) + C\varepsilon\}, \qquad (4.12)$$

for some $C > 0$ independent of g_ε and ε.

Proof. We have $\mathbf{v}_2 = (\frac{1}{2}, \frac{\sqrt{3}}{2})^T$, $\mathbf{v}_3 := \mathbf{v}_2 - \mathbf{v}_1 = (-\frac{1}{2}, \frac{\sqrt{3}}{2})^T$ and $\mathbf{v}_2^\perp = (-\frac{\sqrt{3}}{2}, \frac{1}{2})^T$, $\mathbf{v}_3^\perp = -(\frac{\sqrt{3}}{2}, \frac{1}{2})^T$. By Lemma 4.2.2 we immediately get $|I^\eta| = 1 - O(\varepsilon)$ and $\varepsilon\#(\bar{\mathcal{C}}_\varepsilon \setminus \bar{\mathcal{C}}^\mu_{\varepsilon,\eta}) = O(\varepsilon)$ for μ sufficiently small recalling that these properties were derived independently of the choice of γ. Similarly as before we note that after passing to a suitable $\tilde{\eta} \ge \eta$ it suffices to show the claim for $\tilde{\triangle} \in \bar{\mathcal{C}}^\mu_{\varepsilon,\tilde{\eta}} \cap \bar{\mathcal{C}}^{(1)}_\varepsilon$. We estimate the difference of broken triangles $\bar{\mathcal{C}}^\mu_{\varepsilon,\eta} \cap \bar{\mathcal{C}}^{(1)}_\varepsilon$ projected onto the linear subspaces spanned by $\mathbf{v}_2^\perp$ and $\mathbf{v}_3^\perp$. We recall that the projection of some triangle $\triangle$ on these subspaces are intervals of length $\frac{\sqrt{3}}{2}\varepsilon$ and denote the centers of the intervals by $m^{(2)}_\triangle$ and $m^{(3)}_\triangle$, respectively.

Let $x_2, z_2 \in D^\mu$, $x_2 < z_2$ with $z_2 - x_2 \le C\varepsilon$ and

$$(m^{(2)}_{\triangle_1} - m^{(2)}_{\triangle_2}) \cdot \mathbf{v}_2^\perp > 0 \qquad (4.13)$$

or

$$(m^{(3)}_{\triangle_1} - m^{(3)}_{\triangle_2}) \cdot \mathbf{v}_3^\perp < 0 \qquad (4.14)$$

for the corresponding broken triangles $\triangle_1, \triangle_2 \in \bar{\mathcal{C}}^{(1)}_\varepsilon$. Without restriction we treat the case (4.13). As in the proof of Lemma 4.2.2 we may assume if a triangle intersects $(0,l) \times \{z_2\}$ or $(0,l) \times \{x_2\}$ then its interior does so, too. Denote by $\bar{d}^{(i)} = \frac{2}{\sqrt{3}}|m^{(i)}_{\triangle_1} - m^{(i)}_{\triangle_2}|$, $i = 2,3$, the distances of the centers in $\mathbf{v}_i$-projection onto the x_1-axis.

Let $x_1^{(i)}, z_1^{(i)} \in (0,l)$ such that $\pi_{\mathbf{v}_i^\perp}(x_1^{(i)}, x_2) = m^{(i)}_{\triangle_1}$ and $\pi_{\mathbf{v}_i^\perp}(z_1^{(i)}, z_2) = m^{(i)}_{\triangle_2}$, respectively, where $\pi_{\mathbf{v}_i^\perp}$ denotes the orthogonal projection onto the linear subspace spanned by $\mathbf{v}_i^\perp$, $i = 2,3$. Let $w^{(2)} = \frac{1}{\sqrt{3}}|x_2 - z_2|$ and $w^{(3)} = -\frac{1}{\sqrt{3}}|x_2 - z_2|$. Then the $\mathbf{v}_i$-projection of $z^{(i)} = (z_1^{(i)}, z_2)$ onto the x_2-slice is given by $(\tilde{z}_1^{(i)}, x_2)$ with $\tilde{z}_1^{(i)} = z_1^{(i)} - w^{(i)}$ for $i = 2, 3$. We note that $\bar{d}^{(i)} = |x_1^{(i)} - \tilde{z}_1^{(i)}|$. Taking (4.13) into account we obtain $x_1^{(2)} < \tilde{z}_1^{(2)} < \tilde{z}_1^{(3)}$.

Let $s_\varepsilon = \frac{\varepsilon}{2}$. As in the previous proof we consider areas bounded by parallelograms. For $i = 2, 3$, let $P^{(i)}$ be the parallelogram with corners $(x_1^{(i)} + s_\varepsilon, x_2)$, $(\tilde{z}_1^{(i)} - s_\varepsilon, x_2)$, $(z_1^{(i)} - s_\varepsilon, z_2)$, $(z_1^{(i)} - \bar{d}^{(i)} + s_\varepsilon, z_2)$. They are covered by $\frac{\bar{d}^{(i)}}{\varepsilon} - 1$ stripes of width $\frac{\sqrt{3}}{2}\varepsilon$ in $\mathbf{v}_i$-direction, respectively (note that $P^{(2)}$ can again be degenerated to a segment if $\bar{d}^{(2)} = \varepsilon$). It is not hard to see that both parallelograms cover $\left\lceil \frac{2|z_2 - x_2|}{\sqrt{3}\varepsilon} \right\rceil$ or $\left\lceil \frac{2|z_2 - x_2|}{\sqrt{3}\varepsilon} \right\rceil + 1$ stripes of width $\frac{\sqrt{3}}{2}\varepsilon$ in $\mathbf{e}_1$-direction, where the stripes at the top and at the bottom are only partially covered (the exact number depends of the precise location of the slices $(0, l) \times \{x_2\}$ and $(0, l) \times \{z_2\}$). We denote the number of these covered stripes by $N(\triangle_1, \triangle_2)$ and the orthogonal projection onto the x_2-axis by $I(\triangle_1, \triangle_2)$. Setting

$$n^{(i)}(\triangle_1, \triangle_2) = \frac{2}{\sqrt{3}\varepsilon} |m_{\triangle_1}^{(i)} - m_{\triangle_2}^{(i)}| \tag{4.15}$$

it is elementary to see that $\bar{d}^{(2)} = n^{(2)}(\triangle_1, \triangle_2)\varepsilon$ and $\bar{d}^{(3)} = (n^{(2)}(\triangle_1, \triangle_2) + N(\triangle_1, \triangle_2) - 1)\varepsilon$.

Following the lines of the previous proof we see that each of the $\frac{\bar{d}^i}{\varepsilon}$ atomic chains in $\mathbf{v}_i$ direction lying on the boundary of the stripes which cover $P^{(i)}$, contains at least one spring elongated by a factor of more than $\frac{c}{\sqrt{\varepsilon}}$. Consequently, on the $N(\triangle_1, \triangle_2)$ stripes in $\mathbf{e}_1$-direction we have at least $2n^{(2)}(\triangle_1, \triangle_2) + N(\triangle_1, \triangle_2) - 1 > N(\triangle_1, \triangle_2)$ broken springs orientated in $\mathbf{v}_2$ or $\mathbf{v}_3$ direction. Let J be an interval of length $\frac{\sqrt{3}}{2}\varepsilon$ such that the stripe $(0, l) \times J$ consists of lattice triangles. It is elementary to see that if two broken springs in $\mathbf{v}_2$ and $\mathbf{v}_3$ lie in the stripe at least three triangles are broken, i.e. lie in the set $\bar{\mathcal{C}}_{\varepsilon,\tilde{\eta}}$. Thus, the energy contribution $\frac{4}{\sqrt{3}\varepsilon} \int_{(\varepsilon, l-\varepsilon) \times J} W_\triangle(\nabla \tilde{y}_\varepsilon)$ of the stripe can be bounded from below by $3\varepsilon\beta^{\tilde{\eta}}$. More generally, if on a stripe there are $k \in \mathbb{N}$ broken springs in $\mathbf{v}_2$ and $\mathbf{v}_3$ the energy contribution is at least $(k+1)\varepsilon\beta^{\tilde{\eta}}$.

On the other hand, we recall that on an arbitrary stripe $(0, l) \times J'$ consisting of lattice triangles the energy is always bounded from below by $2\varepsilon\beta^{\tilde{\eta}}$. Consequently, we derive that in the above situation the energy contribution of the $N(\triangle_1, \triangle_2)$ stripes $\frac{4}{\sqrt{3}\varepsilon} \int_{(\varepsilon, l-\varepsilon) \times I(\triangle_1, \triangle_2)} W_\triangle(\nabla \tilde{y}_\varepsilon)$ is bounded from below by

$$N(\triangle_1, \triangle_2)\varepsilon\beta^{\tilde{\eta}} + (2n^{(2)}(\triangle_1, \triangle_2) + N(\triangle_1, \triangle_2) - 1)\varepsilon\beta^{\tilde{\eta}} \tag{4.16}$$

and note that the energy contribution of $N(\triangle_1, \triangle_2)$ stripes is always bounded from below by $2N(\triangle_1, \triangle_2)\varepsilon\beta^{\tilde{\eta}}$.

Now let $\triangle_i$, $i = 1, \ldots, M_\varepsilon$, denote all triangles $\triangle$ in $\bar{\mathcal{C}}_{\varepsilon,\tilde{\eta}}^\mu \cap \bar{\mathcal{C}}_\varepsilon^{(1)}$ such that there exists $x_2^{(i)} \in D^\mu$ with $(0, l) \times \{x_2^{(i)}\}$ intersecting with the interior of $\triangle$. The numbering shall be chosen so as to satisfy $x_2^{(1)} < \ldots < x_2^{(M_\varepsilon)}$. As $1 - |D^\mu| = O(\varepsilon)$, there exists a constant $C > 0$ such that $x_2^{(i+1)} - x_2^{(i)} < C\varepsilon$, $i = 1, \dots, M_\varepsilon - 1$. We define the subset $\{x_2^{(i_j)}\}_{j=1,\dots N_\varepsilon}$ of $\{x_2^{(i)}\}_{i=1,\dots,M_\varepsilon}$ such that $x_2^{(i)} = x_2^{(i_j)}$ for a $j = 1, \dots N_\varepsilon$ if and only if $m_{\triangle_i}$ and $m_{\triangle_{i+1}}$ satisfy (4.13) or (4.14). We let $p(j) = 2$ or $p(j) = 3$ if (4.13) or (4.14) holds, respectively. Let $I_\varepsilon = \cup_{j=1}^{N_\varepsilon} I(\triangle_{i_j}, \triangle_{i_j+1})$. Taking (4.16) into account the energy contribution $\frac{4}{\sqrt{3}\varepsilon}\int_{(\varepsilon,l-\varepsilon)\times I_\varepsilon} W_\triangle(\nabla\tilde{y}_\varepsilon)$ can be bounded from below by

$$\frac{4}{\sqrt{3}\varepsilon}|I_\varepsilon|\varepsilon\beta^{\tilde{\eta}} + \frac{1}{2}\sum_{i=1}^{N_\varepsilon}(2n^{(p(j))}(\triangle_{i_j}, \triangle_{i_j+1}) - 1)\varepsilon\beta^{\tilde{\eta}} + O(\varepsilon).$$

The factor $\frac{1}{2}$ accounts for the possibility that two adjacent intervals $I(\triangle_{i_j}, \triangle_{i_j+1})$, $I(\triangle_{i_{j+1}}, \triangle_{i_{j+1}+1})$ may overlap. We thus see that the total energy is greater or equal to

$$\frac{4}{\sqrt{3}}\beta^{\tilde{\eta}} + \frac{1}{2}\sum_{i=1}^{N_\varepsilon}(2n^{(p(j))}(\triangle_{i_j}, \triangle_{i_j+1}) - 1)\varepsilon\beta^{\tilde{\eta}} + O(\varepsilon)$$

and so $\sum_{j=1}^{N_\varepsilon} n^{(p(j))}(\triangle_{i_j}, \triangle_{i_j+1}) = O(1)$. We now construct the function $g_\varepsilon : (0,1) \to (0,l)$. For $i \in M_\varepsilon$ let $M_i^{(1)}$ and $M_i^{(2)}$ be the orthogonal projections of the center of $\triangle_i$ onto the x_1 and x_2-axis, respectively. If $i \notin N_\varepsilon$ set

$$\tilde{\tilde{g}}_\varepsilon = \frac{M_{i+1}^{(1)} - M_i^{(1)}}{M_{i+1}^{(2)} - M_i^{(2)}}$$

on the interval $[M_i^{(2)}, M_{i+1}^{(2)}]$. Now let $\tilde{g}$ be the Lipschitz function satisfying $\tilde{g}_\varepsilon(M_1^{(2)}) = M_1^{(1)}$ and $\tilde{g}_\varepsilon' = \tilde{\tilde{g}}_\varepsilon$. By construction it is easy to see that $|\tilde{g}_\varepsilon'| \leq \frac{1}{\sqrt{3}}$ on $[M_1^{(2)}, M_{N_\varepsilon}^{(2)}]$. We extend $\tilde{g}_\varepsilon$ arbitrarily to $(0,1)$ such that $\|\tilde{g}_\varepsilon'\|_\infty \leq \frac{1}{\sqrt{3}}$. By (4.15) we have

$$\sum_{j=1}^{N_\varepsilon} |m_{\triangle_{i_j}}^{(p(j))} - m_{\triangle_{i_j+1}}^{(p(j))}| = O(\varepsilon),$$

and then is not hard to see that there is some $C > 0$ independent of $\tilde{g}_\varepsilon$ and ε such that (4.12) holds. Recalling (1.22) it remains to choose $g_\varepsilon : (0,1) \to (\psi(\varepsilon), l - \psi(\varepsilon))$ with $g_\varepsilon' = \pm\frac{1}{\sqrt{3}}$ a.e. and $\|g_\varepsilon - \tilde{g}_\varepsilon\|_\infty \leq C\varepsilon$. □

We conclude that for $\phi = 0$ in the supercritical case there are two subsets

$$\begin{aligned} \Omega_{g_\varepsilon}^{(1)} &:= \{x \in \Omega_\varepsilon : 0 \leq x_1 \leq g_\varepsilon(x_2) - c\varepsilon\}, \\ \Omega_{g_\varepsilon}^{(2)} &:= \{x \in \Omega_\varepsilon : c\varepsilon + g_\varepsilon(x_2) \leq x_1 \leq l\}, \end{aligned} \tag{4.17}$$

where g_ε is chosen appropriately as in Lemma 4.2.3 and $c > 0$ independent of ε, such that the number of triangles in $\bar{\mathcal{C}}_\varepsilon$ intersecting $\Omega_{g_\varepsilon}^{(1)} \cup \Omega_{g_\varepsilon}^{(2)}$ is bounded uniformly in ε. Note that with $\varepsilon \ll \bar{\psi}(\varepsilon) = \psi(\varepsilon) - c\varepsilon \ll 1$ one has

$$\begin{aligned} \big((0, \bar{\psi}(\varepsilon)) \times (0,1)\big) \cap \Omega_\varepsilon &\subset \Omega_{g_\varepsilon}^{(1)}, \\ \big((l - \bar{\psi}(\varepsilon), l) \times (0,1)\big) \cap \Omega_\varepsilon &\subset \Omega_{g_\varepsilon}^{(2)}, \end{aligned} \tag{4.18}$$

so that, in particular, $\Omega_{g_\varepsilon}^{(1)}$ and $\Omega_{g_\varepsilon}^{(2)}$ are connected.

4.3 Convergence of almost minimizers

As a further preparation we shows that broken triangles can be 'healed'. In order to treat the different cases simultaneously in the following we will call these sets the 'good set'

$$\Omega_{\text{good}} = \begin{cases} \Omega_\varepsilon & \text{for } a < a_{\text{crit}}, \\ \Omega_\varepsilon^{(1)} \cup \Omega_\varepsilon^{(2)} & \text{for } a > a_{\text{crit}},\ \phi \neq 0 \text{ and} \\ \Omega_{g_\varepsilon}^{(1)} \cup \Omega_{g_\varepsilon}^{(2)} & \text{for } a > a_{\text{crit}},\ \phi = 0, \end{cases}$$

with $\Omega_\varepsilon^{(i)}$ and $\Omega_{g_\varepsilon}^{(i)}$, $i = 1, 2$, as defined in (4.11) and (4.17).

Lemma 4.3.1. *Suppose $\tilde{y}_\varepsilon$ is a minimizing sequence satisfying $\mathcal{E}_\varepsilon(y_\varepsilon) = \inf \mathcal{E}_\varepsilon + O(\varepsilon)$. There exists $\bar{y}_\varepsilon \in W^{1,\infty}(\Omega_{\text{good}}; \mathbb{R}^2)$ with $\nabla \bar{y}_\varepsilon$ bounded in $L^\infty(\Omega_{\text{good}})$ uniformly in ε such that*

$$|\{x \in \Omega_{\text{good}} : \bar{y}_\varepsilon(x) \neq \tilde{y}_\varepsilon(x)\}| = O(\varepsilon^2)$$

and

$$\int_{\Omega_{\text{good}}} \operatorname{dist}^2(\nabla \bar{y}_\varepsilon(x), SO(2))\, dx \leq C \int_{\Omega_{\text{good}} \setminus \bigcup_{\triangle \in \bar{\mathcal{C}}_\varepsilon} \triangle} \operatorname{dist}^2(\nabla \tilde{y}_\varepsilon, SO(2))\, dx.$$

Proof. For notational convenience we drop the subscript ε in the following proof. By Lemmas 4.2.1, 4.2.2 and 4.2.3 we can partition the area covered by the (closed) triangles in $\bar{\mathcal{C}}$ intersecting Ω_{good} into connected components $C_1, \ldots, C_N$ such that

$$\bigcup_{\triangle \in \bar{\mathcal{C}} : \triangle \cap \Omega_{\text{good}} \neq \emptyset} \triangle = C_1 \dot{\cup} \ldots \dot{\cup} C_N,$$

where N is bounded uniformly in ε. Then the maximal diameter of each set C_i is bounded by a term $O(\varepsilon)$. For each i, the largest connected component D_i of the complement $\Omega_{\text{good}} \setminus C_i$ lying in the same component of Ω_{good} is unique (with area of the order 1 while all the other components of the complement are of size $O(\varepsilon^2)$). Let V_i be the union of triangles whose interior is contained in D_i that touch the boundary of C_i.

We now proceed to define $\bar{y}$ by modifying $\tilde{y}$ on all the triangles not contained in $\overline{D}_i$, successively for $i = 1, \ldots, N$. For each i this modification is done iteratively on triangles $\triangle$ which share at least one side with a triangle that has been modified previously or with a triangle lying in V_i in such a way that $\bar{y}$ is continuous along such sides and $\bar{y}|_\triangle$ is affine and minimizes $\operatorname{dist}((\bar{y})_\triangle, SO(2))$.

In order to estimate $\operatorname{dist}(\nabla\bar{y}, SO(2))$ we will use the geometric rigidity result in Theorem B.1 and recall that the constant is invariant under rescaling of the domain. For later use we mention that if $\operatorname{dist}^2(\nabla f(x), SO(2))$ is equiintegrable, then R can be chosen in such a way that also $|\nabla f(x) - R|^2$ is equiintegrable, cf. [28].

Consider a single step in the modification process, when $\tilde{y}$ is modified to $\bar{y}$ on $\triangle$, and let U be the union of triangles that have been modified previously or lie in V_i. By Theorem B.1, there is a rotation $R \in SO(2)$ such that

$$\int_U |\nabla\bar{y}(x) - R|^2\,dx \le C \int_U \operatorname{dist}^2(\nabla\bar{y}(x), SO(2))\,dx$$

holds. Since $\nabla\bar{y}$ is piecewise constant, this means

$$\sum_{\triangle'\subset U} |(\bar{y})_{\triangle'} - R|^2 \le C \sum_{\triangle'\subset U} \operatorname{dist}^2((\bar{y})_{\triangle'}, SO(2)).$$

It is not hard to see that there exists an extension w of $\bar{y}$ from U to $U \cup \triangle$ such that

$$|(w)_\triangle - R|^2 \le C \sum_{\triangle'\subset U} |(\bar{y})_{\triangle'} - R|^2.$$

(If there is only one side of $\triangle$ on the boundary of U, say adjacent to $\triangle' \subset U$, then one can take w with $(w)_\triangle = (\bar{y})_{\triangle'}$. If at least two sides, say in $\mathbf{v}_1$ and $\mathbf{v}_2$ direction, are shared by triangles $\triangle_1, \triangle_2 \subset U$, respectively, then these sides have a common corner and the unique extension w satisfies $(w)_\triangle \mathbf{v}_i = (\bar{y})_{\triangle_i}\mathbf{v}_i = R\mathbf{v}_i + ((\bar{y})_{\triangle_i} - R)\mathbf{v}_i$, $i = 1, 2$.) Now by construction of $\bar{y}$ on $\triangle$ we see that

$$\operatorname{dist}^2((\bar{y})_\triangle, SO(2)) \le C \sum_{\triangle'\subset U} |(\bar{y})_{\triangle'} - R|^2$$

and so

$$\int_{U\cup\triangle} \operatorname{dist}^2(\nabla\bar{y}(x), SO(2))\,dx \le C \int_U \operatorname{dist}^2(\nabla\bar{y}(x), SO(2))\,dx.$$

Iterating this estimate we finally arrive at

$$\int_{\Omega_{\text{good}}} \operatorname{dist}^2(\nabla\bar{y}(x), SO(2))\,dx \le C \int_{\Omega_{\text{good}}\setminus\bigcup_i C_i} \operatorname{dist}^2(\nabla\tilde{y}, SO(2))\,dx.$$

Here the constant C can be chosen independently of ε. This is due to the facts that the number of modification steps is bounded uniformly in ε and – after rescaling the shapes U with $\frac{1}{\varepsilon}$ – there is also only a uniformly bounded number of shapes U involved in the previous rigidity estimates. Moreover, each triangle is covered by no more than three of the sets V_i.

The uniform boundedness of the number of modification steps also shows that $|\{x \in \Omega_{\text{good}} : \bar{y}(x) \neq \tilde{y}(x)\}| = O(\varepsilon^2)$ and, by definition of $\bar{\mathcal{C}}$ and construction of $\bar{y}$, that $\|\nabla \bar{y}\|_{L^\infty(\Omega_{\text{good}})} = O(1)$. □

Note that up to a set of small size $\bar{y}_\varepsilon$ satisfies the same boundary conditions as $\tilde{y}_\varepsilon$ on the lateral boundary. More precisely, there are $\Gamma_\varepsilon^{(i)} \subset (0,1)$, $|\Gamma_\varepsilon^{(i)}| = O(\varepsilon)$, $i = 1, 2$, such that $\bar{y}_\varepsilon$ and $\tilde{y}_\varepsilon$ coincide on $(0, \varepsilon) \times ((0,1) \setminus \Gamma_\varepsilon^{(1)})$ and $(l - \varepsilon, l) \times ((0,1) \setminus \Gamma_\varepsilon^{(2)})$. With these boundary conditions and the geometric rigidity estimate in Theorem B.1 we can now derive strong convergence results for $\bar{y}_\varepsilon$ and even the corresponding rescaled displacement $\bar{u}_\varepsilon = \frac{1}{\sqrt{\varepsilon}}(\bar{y}_\varepsilon - \mathbf{id})$ on Ω_{good}. We first consider the supercritical case and treat the cases $\phi \neq 0$ and $\phi = 0$ separately.

4.3.1 The supercritical case

Lemma 4.3.2. *If $a > a_{\text{crit}}$ and $\phi \neq 0$, then there exist sequences $s_\varepsilon, t_\varepsilon \in \mathbb{R}$ such that*

$$\|\bar{u}_\varepsilon - (0, s_\varepsilon)\|_{H^1(\Omega_\varepsilon^{(1)})} + \|\bar{u}_\varepsilon - (al, t_\varepsilon)\|_{H^1(\Omega_\varepsilon^{(2)})} \to 0.$$

Proof. We again drop the subscript ε. By applying the geometric rigidity estimate in Theorem B.1 to $\Omega^{(1)}$ and to $\Omega^{(2)}$, we obtain rotations $R^{(1)}, R^{(2)} \in SO(2)$ such that

$$\|\nabla \bar{y} - R^{(i)}\|_{L^2(\Omega_\varepsilon^{(i)})} \leq C \|\operatorname{dist}(\nabla \bar{y}, SO(2))\|_{L^2(\Omega_\varepsilon^{(i)})}, \quad i = 1, 2. \tag{4.19}$$

Here C can be chosen independently of ε as all the possible shapes of $\Omega^{(i)}$ are related through bi-Lipschitzian homeomorphisms with Lipschitz constants of both the homeomorphism itself and its inverse bounded uniformly in ε, see Section B. Now using that $\nabla \bar{y}$ is uniformly bounded in L^∞, we obtain from Lemmas 4.3.1 and 2.4.5(i)

$$\begin{aligned}
\sum_{i=1}^{2} \|\nabla \bar{y} - R^{(i)}\|^2_{L^2(\Omega_\varepsilon^{(i)})} &\leq C \int_{\Omega_{\text{good}} \setminus \bigcup_{\triangle \in \bar{\mathcal{C}}_\varepsilon} \triangle} \operatorname{dist}^2(\nabla \tilde{y}, SO(2))\, dx \\
&\leq C \int_{\Omega_{\text{good}} \setminus \bigcup_{\triangle \in \bar{\mathcal{C}}_\varepsilon} \triangle} \operatorname{dist}^2(\nabla \tilde{y}, O(2)) + \chi(\nabla \tilde{y})\, dx \\
&\leq C \int_{\Omega_{\text{good}} \setminus \bigcup_{\triangle \in \bar{\mathcal{C}}_\varepsilon} \triangle} W_{\triangle,\chi}(\nabla \tilde{y})\, dx.
\end{aligned}$$

But, as seen before,

$$\begin{aligned}
\frac{4}{\sqrt{3}\varepsilon} \int_{\Omega_{\text{good}} \setminus \bigcup_{\triangle \in \bar{\mathcal{C}}_\varepsilon} \triangle} W_{\triangle,\chi}(\nabla \tilde{y})\, dx &\leq \mathcal{E}^\chi(y) - \frac{4}{\sqrt{3}\varepsilon} \int_{\bigcup_{\triangle \in \bar{\mathcal{C}}_\varepsilon} \triangle} W_{\triangle,\chi}(\nabla \tilde{y})\, dx \\
&\leq \inf \mathcal{E}^\chi + O(\varepsilon) - \frac{2\beta^\eta}{\gamma} |I^\eta| = O(\varepsilon),
\end{aligned}$$

where the last step followed from Lemma 4.2.2 and (4.6), and so

$$\sum_{i=1}^{2} \|\nabla \bar{y} - R^{(i)}\|^2_{L^2(\Omega_\varepsilon^{(i)})} = O(\varepsilon^2).$$

By Poincaré's inequality we then deduce that there are $\zeta^{(i)} \in \mathbb{R}^2$ such that

$$\sum_{i=1}^{2} \|\bar{y} - R^{(i)} \cdot -\zeta^{(i)}\|_{H^1(\Omega_\varepsilon^{(i)})} = O(\varepsilon). \tag{4.20}$$

We extend $\bar{y}$ as an H^1-function from $\Omega_\varepsilon^{(i)}$ to $\Omega^{(i)}$ (as defined in Theorem 1.5.1), $i = 1, 2$, such that (4.20) still holds and $\bar{y}_1(0, x_2) = 0$ for $x_2 \in (0,1) \setminus \Gamma_\varepsilon^{(1)}$, $\bar{y}_1(l, x_2) = l(1 + a_\varepsilon)$ for $x_2 \in (0,1) \setminus \Gamma_\varepsilon^{(2)}$. The trace theorem for Sobolev functions with $x_1 = 0$ or $x_1 = l$ according to $i = 1$ and $i = 2$, respectively, gives

$$\sum_{i=1}^{2} \|\bar{y}(x_1, \cdot) - R^{(i)}(x_1, \cdot) - \zeta^{(i)}\|_{L^2(0,1)} = O(\varepsilon).$$

In particular, setting $\tilde{\zeta}^{(1)} = \zeta^{(1)}$ and $\tilde{\zeta}^{(2)} = \zeta^{(2)} - la_\varepsilon \mathbf{e}_1$, the first components satisfy

$$\sum_{i=1}^{2} \|x_1 - R_{11}^{(i)} x_1 - R_{12}^{(i)} \cdot -\tilde{\zeta}_1^{(i)}\|_{L^2((0,1)\setminus\Gamma_\varepsilon^{(i)})} = O(\varepsilon). \tag{4.21}$$

But then also the constant function

$$\frac{1}{2} R_{12}^{(i)} = \left(x_1 - R_{11}^{(i)} x_1 - R_{12}^{(i)} \left(\cdot - \frac{1}{2} \right) - \tilde{\zeta}_1^{(i)} \right) - \left(x_1 - R_{11}^{(i)} x_1 - R_{12}^{(i)} \cdot -\tilde{\zeta}_1^{(i)} \right)$$

is of order ε in $L^2((\frac{1}{2}, 1) \setminus (\Gamma_\varepsilon^{(i)} \cup (\Gamma_\varepsilon^{(i)} + \frac{1}{2})))$ and thus $|R_{12}^{(i)}| \leq C\varepsilon$. An elementary argument now yields

$$|R^{(i)} - \mathbf{Id}| = O(\varepsilon) \qquad \text{or} \qquad |R^{(i)} + \mathbf{Id}| = O(\varepsilon).$$

It is not hard to see that $|R^{(i)} - \mathbf{Id}| = O(\varepsilon)$ as otherwise, e.g. for $i = 1$, on the set $T = \{\triangle \in \mathcal{C}_\varepsilon : \triangle \subset (0, \varepsilon) \times (0,1)\}$ we get, due to the boundary conditions,

$$O(\varepsilon^2) = \int_T |\nabla \bar{y} - R^{(1)}|^2 \geq \int_T |1 + a_\varepsilon + 1|^2 + O(\varepsilon^2) \geq C\varepsilon,$$

which is clearly impossible. Returning to (4.21) and (4.20), it now follows that $|\tilde{\zeta}_1^{(i)}| = O(\varepsilon)$ and then

$$\|\bar{u} - (0, s_\varepsilon)\|_{H^1(\Omega_\varepsilon^{(1)})} + \|\bar{u} - (al, t_\varepsilon)\|_{H^1(\Omega_\varepsilon^{(2)})} = O(\sqrt{\varepsilon}),$$

where $s_\varepsilon = \frac{1}{\sqrt{\varepsilon}} \zeta_2^{(1)}$ and $t_\varepsilon = \frac{1}{\sqrt{\varepsilon}} \zeta_2^{(2)}$. □

Lemma 4.3.3. *If $a > a_{\mathrm{crit}}$ and $\phi = 0$, then there exist sequences $s_\varepsilon, t_\varepsilon \in \mathbb{R}$ and Lipschitz functions g_ε as in Lemma 4.2.3 such that*

$$\|\bar{u}_\varepsilon - (0, s_\varepsilon)\|_{H^1(\Omega^{(1)}_{g_\varepsilon})} + \|\bar{u}_\varepsilon - (al, t_\varepsilon)\|_{H^1(\Omega^{(2)}_{g_\varepsilon})} \to 0.$$

Proof. Without restriction we only estimate $\bar{y}$ (again dropping subscripts ε) on $\Omega^{(1)}_{g_\varepsilon}$. We first note that we may not simply proceed as in (4.19) and (4.20) as, due to a possibly complicated shape of the set $\Omega^{(1)}_{g_\varepsilon}$, the corresponding constants cannot be controlled. As a remedy, we claim that we can find a partition $(T_j)_j$, $j = 1, \ldots, M_\varepsilon$ of $\Omega^{(1)}_{g_\varepsilon}$ of the form $T_j = \left\{x \in \Omega^{(1)}_{g_\varepsilon} : t_{j-1} \le x_2 \le t_j\right\}$ for suitable $t_j \in [0, 1]$, $j = 0, \ldots, M_\varepsilon$ with $t_0 = 0$ and $t_{M_\varepsilon} = 1$ such that the T_j are related through bi-Lipschitzian homeomorphism with uniformly bounded Lipschitz constants to cubes of sidelength $d_j = t_j - t_{j-1} \ge \psi(\varepsilon) \gg \varepsilon$. We will show this at the end of the proof. Recalling that the constant in (B.1) is invariant under rescaling of the domain and repeating the above arguments in (4.20) we now obtain $R^{(j)} \in SO(2)$ and $\xi^{(j)} \in \mathbb{R}^2$, $j = 1, \ldots, M_\varepsilon$, such that

$$\sum_{j=1}^{M_\varepsilon} \left\|\nabla \bar{y} - R^{(j)}\right\|^2_{L^2(T_j)} = O(\varepsilon^2) \quad \text{and} \quad \sum_{j=1}^{M_\varepsilon} d_j^{-2} \left\|\bar{y} - R^{(j)} \cdot - \xi^j\right\|^2_{L^2(T_j)} = O(\varepsilon^2).$$

Let $\tilde{T}_j = (t_{j-1}, t_j)$ for $j = 1, \ldots, M_\varepsilon$ and $T^* = \bigcup_{j=1}^{M_\varepsilon} (t_{j-1} + \frac{d_j}{2}, t_j)$. A standard rescaling argument and the trace theorem yield

$$\sum_{j=1}^{M_\varepsilon} d_j^{-1} \left\|\bar{y}(0, \cdot) - R^{(j)}(0, \cdot) - \xi^j\right\|^2_{L^2(\tilde{T}_j)} = O(\varepsilon^2). \tag{4.22}$$

Similarly as above we calculate the norm in $L^2(T^* \setminus (\Gamma^{(1)}_\varepsilon \cup (\Gamma^{(1)}_\varepsilon + \frac{d_j}{2})))$ on the trace $\{x_1 = 0\}$ of the piecewise constant function

$$\frac{d_j}{2} R^{(j)}_{12} = \left(x_1 - R^{(j)}_{11} x_1 - R^{(j)}_{12}\left(\cdot - \frac{d_j}{2}\right) - \xi^{(j)}_1\right) - \left(x_1 - R^{(j)}_{11} x_1 - R^{(j)}_{12} \cdot - \xi^{(j)}_1\right)$$

and now find that $\sum_{j=1}^{M_\varepsilon} d_j^2 |R^{(i)}_{12}|^2 = O(\varepsilon^2)$. Consequently, noting that $d_j \ge \psi(\varepsilon) \gg \varepsilon$ for all $j = 1, \ldots, M_\varepsilon$ and proceeding as before, we obtain

$$\sum_{j=1}^{M_\varepsilon} d_j^2 |R^{(j)} - \mathbf{Id}|^2 = O(\varepsilon^2) \tag{4.23}$$

so that

$$\sum_{j=1}^{M_\varepsilon} \left\|\bar{y} - \mathbf{id} - \xi^j\right\|^2_{H^1(T_j)} = O(\varepsilon^2).$$

Due to the boundary conditions, (4.22) and (4.23) yield $\sum_{j=1}^{M_\varepsilon} d_j^2 |\xi_1^j|^2 = O(\varepsilon^2)$ and therefore

$$\sum_{j=1}^{M_\varepsilon} \left\| \bar{y} - \mathbf{id} - (0, \xi_2^j) \right\|_{H^1(T_j)}^2 = O(\varepsilon^2). \tag{4.24}$$

We define the stripe $S = (0, \bar{\psi}(\varepsilon)) \times (0,1)$ and note that $S \cap \Omega_\varepsilon \subset \Omega_{g_\varepsilon}^{(1)}$ by (4.18). From Poincaré's inequality we obtain a $\zeta \in \mathbb{R}^2$ such that

$$\|\bar{y} - \mathbf{id} - \zeta\|_{H^1(S)}^2 \leq C \|\nabla \bar{y} - \mathbf{Id}\|_{L^2(S)}^2 = O(\varepsilon^2). \tag{4.25}$$

Note that the constant C can be chosen independently of the length of S, i.e. independently of ε. Applying (4.24) we may suppose $\zeta = (0, \zeta_2)$.

Moreover, by (4.24) and (4.25) there is some $\rho_\varepsilon \in (0, \bar{\psi}(\varepsilon))$ such that the trace on the slice $\Gamma = \{\rho_\varepsilon\} \times (0,1)$ satisfies

$$\int_\Gamma |\bar{y} - \mathbf{id} - (0,\zeta_2)|^2 = \frac{O(\varepsilon^2)}{\bar{\psi}(\varepsilon)} \quad \text{and} \quad \sum_{j=1}^{M_\varepsilon} \left\| \bar{y} - \mathbf{id} - (0, \xi_2^j) \right\|_{L^2(\Gamma \cap \overline{T_j})}^2 = \frac{O(\varepsilon^2)}{\bar{\psi}(\varepsilon)}.$$

We compare the trace on Γ and deduce from $d_j = \mathcal{H}^1(\Gamma \cap \overline{T_j})$

$$\begin{aligned}\sum_{j=1}^{M_\varepsilon} d_j |\zeta - \xi^j|^2 &\leq C \sum_{j=1}^{M_\varepsilon} \left(\|\bar{y} - \mathbf{id} - \xi^j\|_{L^2(\Gamma\cap\overline{T_j})}^2 + \|\bar{y} - \mathbf{id} - \zeta\|_{L^2(\Gamma\cap\overline{T_j})}^2 \right) \\ &= \frac{O(\varepsilon^2)}{\bar{\psi}(\varepsilon)} = \frac{O(\varepsilon^2)}{\psi(\varepsilon)}.\end{aligned}$$

Thus, returning to (4.24) we conclude

$$\begin{aligned}\|\bar{y} - \mathbf{id} - (0,\zeta_2)\|_{H^1(\Omega_{g_\varepsilon}^{(1)})}^2 &\leq C \sum_{j=1}^{M_\varepsilon} \|\bar{y} - \mathbf{id} - \xi^j\|_{H^1(T_j)}^2 + C \sum_{j=1}^{M_\varepsilon} d_j^2 |\xi_j - \zeta|^2 \\ &\leq O(\varepsilon^2) + C \sum_{j=1}^{M_\varepsilon} d_j |\xi_j - \zeta|^2 = \frac{O(\varepsilon^2)}{\psi(\varepsilon)}\end{aligned}$$

and finally

$$\|\bar{u} - (0, s_\varepsilon)\|_{H^1(\Omega_{g_\varepsilon}^{(1)})}^2 = \frac{O(\varepsilon)}{\psi(\varepsilon)} \to 0$$

for $\varepsilon \to 0$, where $s_\varepsilon = \frac{1}{\sqrt{\varepsilon}} \zeta_2$. For $\Omega_{g_\varepsilon}^{(2)}$ we proceed likewise.

To finish the proof it suffices to show the existence of a partition $(T_j)_j$ with the above properties. Recall that $\Omega_{g_\varepsilon}^{(1)} = \{x \in \Omega : 0 < x_1 < g(x_2) - c\varepsilon\}$ and $\|g'\|_\infty =$

$\frac{1}{\sqrt{3}}$, $g \geq \psi(\varepsilon)$. Let $r_0 = 0$ and define $r_1, \ldots, r_{M_\varepsilon} \in (0,1)$ inductively by setting $r_{j+1} = r_j + g(r_j)$, so that $r_{M_\varepsilon} + g(r_{M_\varepsilon}) \geq 1$. Now setting

$$T_j = \begin{cases} \{x \in \Omega_{g_\varepsilon}^{(1)} : r_{j-1} \leq x_2 \leq r_j\} & \text{for } 1 \leq j \leq M_\varepsilon - 1, \\ \{x \in \Omega_{g_\varepsilon}^{(1)} : r_{M_\varepsilon - 1} \leq x_2 \leq 1 & \text{for } j = M_\varepsilon, \end{cases}$$

it is not hard so see that every T_j is related to $\lambda(0,1)^2$ for a suitable λ through some bi-Lipschitzian homeomorphism with uniformly bounded Lipschitz constants. By construction, $t_j - t_{j-1} \geq g(t_j) \geq \psi(\varepsilon) \gg \varepsilon$ for $j = 1, \ldots, M_\varepsilon$. □

4.3.2 The subcritical case

Strong convergence in the subcritical case can be shown along the lines of the proofs of the main linearization results in [64] and [65]. We include a simplified proof adapted to the present situation here for the sake of completeness.

Lemma 4.3.4. *If $a < a_{\text{crit}}$, then there is a sequence $s_\varepsilon \in \mathbb{R}$ such that*

$$\|\bar{u}_\varepsilon - (0, s_\varepsilon) - F^a \cdot\|_{H^1(\Omega_{\text{good}})} \to 0.$$

where $F^a = \begin{pmatrix} a & 0 \\ 0 & -\frac{a}{3} \end{pmatrix}$.

Proof. We again drop subscripts ε if no confusion arises. With the help of the geometric rigidity estimate (Theorem B.1) we find by arguing as in the proof of Lemma 4.3.2 that

$$\|\nabla \bar{y} - R\|_{L^2(\Omega_\varepsilon)}^2 \leq C \int_{\Omega_\varepsilon \setminus \bigcup_{\triangle \in \bar{\mathcal{C}}_\varepsilon} \triangle} W_{\triangle,\chi}(\nabla \tilde{y})\, dx = O(\varepsilon)$$

for a suitable rotation $R \in SO(2)$ with

$$|R \pm \mathbf{Id}| = O(\sqrt{\varepsilon}) \tag{4.26}$$

and

$$\|\bar{y} \pm \mathbf{id} - \zeta\|_{H^1(\Omega_\varepsilon)} = O(\sqrt{\varepsilon})$$

for some $\zeta \in \mathbb{R}^2$ with $\zeta_1 = O(\sqrt{\varepsilon})$ and thus, due to the boundary conditions,

$$\|\bar{u} - (0, \zeta_2)\|_{H^1(\Omega_\varepsilon)} = O(1).$$

In particular, $\bar{u}_\varepsilon - (\zeta_\varepsilon)_2 \mathbf{e}_2$ converges – up to passing to a subsequence – weakly. It now suffices to prove that $\|e(\bar{u}_\varepsilon) - F^a\|_{L^2(\Omega_\varepsilon)} \to 0$, where $e(u) = \frac{(\nabla u)^T + \nabla u}{2}$ denotes the symmetrized gradient, for then the assertion follows from Korn's inequality.

To this end, we let $V_\varepsilon(F) = \frac{1}{\varepsilon} W_\triangle(\mathbf{Id} + \sqrt{\varepsilon}F)$ and $V_{\varepsilon,\chi}(F) = V_\varepsilon(F) + \frac{1}{\varepsilon}\chi(\mathbf{Id} + \sqrt{\varepsilon}F)$, so that $V_{\varepsilon,\chi}(F) \to \frac{1}{2}D^2 W_\triangle(\mathbf{Id})[F,F] = \frac{1}{2}Q(F)$ uniformly on compact subsets of $\mathbb{R}^{2\times 2}$. Then by frame indifference (see Lemma 2.4.1)

$$\begin{aligned} W_{\triangle,\chi}(\mathbf{Id} + \sqrt{\varepsilon}F) &= W_{\triangle,\chi}\left(\sqrt{(\mathbf{Id} + \sqrt{\varepsilon}F)^T(\mathbf{Id} + \sqrt{\varepsilon}F)}\right) \\ &= \varepsilon V_{\varepsilon,\chi}\left(\frac{F^T + F}{2} + \frac{1}{\sqrt{\varepsilon}} f(\sqrt{\varepsilon}F)\right) \end{aligned} \tag{4.27}$$

with $f(F) = \sqrt{(\mathbf{Id} + F)^T(\mathbf{Id} + F)} - \mathbf{Id} - \frac{F^T+F}{2}$, so that $|f(F)| \le C\min\{|F|, |F|^2\}$. Then by Lemma 2.4.5(i) and (4.27), $V_{\varepsilon,\chi}$ satisfies

$$\begin{aligned} V_{\varepsilon,\chi}\left(\frac{F^T + F}{2} + \frac{1}{\sqrt{\varepsilon}} f(\sqrt{\varepsilon}F)\right) &\ge \frac{c}{\varepsilon}\operatorname{dist}^2(\mathbf{Id} + \sqrt{\varepsilon}F, O(2)) + \frac{1}{\varepsilon}\chi(\mathbf{Id} + \sqrt{\varepsilon}F) \\ &\ge \frac{c}{\varepsilon}\operatorname{dist}^2(\mathbf{Id} + \sqrt{\varepsilon}F, SO(2)) \\ &\ge \frac{c}{\varepsilon}\left|\sqrt{(\mathbf{Id} + \sqrt{\varepsilon}F)^T(\mathbf{Id} + \sqrt{\varepsilon}F)} - \mathbf{Id}\right|^2 \\ &= c\left|\frac{F^T + F}{2} + \frac{1}{\sqrt{\varepsilon}} f(\sqrt{\varepsilon}F)\right|^2. \end{aligned} \tag{4.28}$$

In the sequel we set $A_\varepsilon(F) = \frac{F^T+F}{2} + \frac{1}{\sqrt{\varepsilon}} f(\sqrt{\varepsilon}F)$. Choose convex functions $\psi_k : \mathbb{R}^{2\times 2} \to \mathbb{R}$ with linear growth at infinity such that $\psi_1 \le \psi_2 \le \ldots$ and $\psi_k(F) \to \frac{1}{2}Q(F)$ uniformly on compact subsets of $\mathbb{R}^{2\times 2}$. The previous quadratic estimate on $V_{\varepsilon,\chi}(A_\varepsilon(F))$ from below and the fact that $V_{\varepsilon,\chi} \to \frac{1}{2}Q$ uniformly on compacts then shows that we can also choose $\delta > 0$ and a sequence $r_k \to \infty$ such that

$$V_{\varepsilon,\chi}\left(A_\varepsilon(F)\right) - \delta\chi_{\{|A_\varepsilon(F)| \ge r_k\}}|A_\varepsilon(F)|^2 \ge \psi_k\left(A_\varepsilon(F)\right) - \frac{1}{k},$$

whenever ε (depending on k) is sufficiently small.

With (4.27) we now obtain that

$$\begin{aligned} \frac{1}{\varepsilon}\int_{\Omega_\varepsilon} W_{\triangle,\chi}(\bar{y})\,dx &= \int_{\Omega_\varepsilon} V_{\varepsilon,\chi}\left(A_\varepsilon(\nabla\bar{u})\right)\,dx \\ &\ge \int_{\Omega_\varepsilon} \psi_k\left(A_\varepsilon(\nabla\bar{u})\right)\,dx + \delta\int_{\Omega_\varepsilon} \chi_{\{|A_\varepsilon(\nabla\bar{u})| \ge r_k\}}|A_\varepsilon(\nabla\bar{u})|^2\,dx - \frac{1}{k}. \end{aligned}$$

As ψ_k has linear growth at infinity and $\frac{1}{\sqrt{\varepsilon}} f(\sqrt{\varepsilon}\nabla\bar{u}_\varepsilon) \le C\min\{|\nabla\bar{u}_\varepsilon|, \sqrt{\varepsilon}|\nabla\bar{u}_\varepsilon|^2\}$, $\nabla\bar{u}_\varepsilon$ bounded in L^2, by splitting the integration into two parts according to $|\nabla\bar{u}_\varepsilon| \le M$ or $|\nabla\bar{u}_\varepsilon| > M$ and eventually sending M to infinity, we find

$$\liminf_{\varepsilon\to 0}\int_{\Omega_\varepsilon} \psi_k\left(A_\varepsilon(\nabla\bar{u}_\varepsilon)\right)\,dx = \liminf_{\varepsilon\to 0}\int_{\Omega_\varepsilon} \psi_k\left(e(\bar{u}_\varepsilon)\right)\,dx.$$

When $\bar{u}_\varepsilon - (\zeta_\varepsilon)_2 \mathbf{e}_2 \rightharpoonup u$ in H^1, by Theorem 1.4.1 it then follows that

$$\begin{aligned}\frac{\alpha l a^2}{\sqrt{3}} &= \lim_{\varepsilon\to 0} \frac{4}{\sqrt{3}} \int_{\Omega_\varepsilon} V_{\varepsilon,\chi}\left(A_\varepsilon(\nabla \bar{u}_\varepsilon)\right) dx \\ &\geq \liminf_{\varepsilon\to 0} \frac{4}{\sqrt{3}} \int_{\Omega} \chi_{\{\mathrm{dist}(x,\partial\Omega)\geq k^{-1}\}} \psi_k\left(e(\bar{u}_\varepsilon)\right) dx \\ &\quad + \limsup_{\varepsilon\to 0} \frac{4\delta}{\sqrt{3}} \int_{\Omega_\varepsilon} \chi_{\{|A_\varepsilon(\nabla\bar{u}_\varepsilon)|\geq r_k\}} |A_\varepsilon(\nabla \bar{u}_\varepsilon)|^2 \, dx - \frac{4}{\sqrt{3}k}.\end{aligned}$$

Using that by convexity of ψ_k the first term on the right hand side is lower semicontinuous in $\nabla \bar{u}_\varepsilon$ and that $\chi_{\{\mathrm{dist}(\cdot,\partial\Omega)\geq k^{-1}\}}\psi_k \to \frac{1}{2}Q$ monotonically, we finally find by letting $k \to \infty$

$$\begin{aligned}\frac{\alpha l a^2}{\sqrt{3}} &\geq \frac{2}{\sqrt{3}} \int_\Omega Q\left(e(u)\right) \\ &\quad + \lim_{k\to\infty} \limsup_{\varepsilon\to 0} \frac{4\delta}{\sqrt{3}} \int_{\Omega_\varepsilon} \chi_{\{|A_\varepsilon(\nabla\bar{u}_\varepsilon)|\geq r_k\}} |A_\varepsilon(\nabla \bar{u}_\varepsilon)|^2 \, dx.\end{aligned} \tag{4.29}$$

A slicing and convexity argument similar to (3.2) now shows that $\frac{2}{\sqrt{3}} \int_\Omega Q(e(w)) \geq \frac{\alpha l a^2}{\sqrt{3}}$ for all $w \in H^1$ subject to $w_1(0, x_2) = 0$ and $w_1(l, x_2) = al$ and thus

$$\lim_{k\to\infty} \limsup_{\varepsilon\to 0} \frac{4\delta}{\sqrt{3}} \int_{\Omega_\varepsilon} \chi_{\{|A_\varepsilon(\nabla\bar{u}_\varepsilon)|\geq r_k\}} |A_\varepsilon(\nabla \bar{u}_\varepsilon)|^2 \, dx = 0,$$

or, in other words, $|A_\varepsilon(\nabla \bar{u}_\varepsilon)|^2$ is equiintegrable. By the estimate $|V_{\varepsilon,\chi}(F)| = |\frac{1}{\varepsilon} W_{\triangle,\chi}(\mathbf{Id} + \sqrt{\varepsilon}F)| \leq C(1 + |F|^2)$, (4.28) shows that also

$$\frac{c}{\varepsilon} \operatorname{dist}^2(\nabla \bar{y}_\varepsilon, SO(2)) \leq V_{\varepsilon,\chi}(A_\varepsilon(\nabla \bar{u}_\varepsilon))$$

is equiintegrable, so that by the discussion in the proof of Lemma 4.3.1 in fact we may assume that $\frac{1}{\varepsilon}\|\nabla \bar{y}_\varepsilon - R\|^2_{L^2(\Omega_\varepsilon)}$ is equiintegrable, too, and $|R - \mathbf{Id}| = O(\sqrt{\varepsilon})$ by (4.26). But then also $|\nabla \bar{u}_\varepsilon|^2$ is equiintegrable and this together with (4.29) yields

$$\lim_{\varepsilon\to 0} \frac{2}{\sqrt{3}} \int_{\Omega_\varepsilon} Q(e(\bar{u}_\varepsilon)) = \frac{2}{\sqrt{3}} \int_\Omega Q(e(u)) = \frac{\alpha l a^2}{\sqrt{3}}.$$

For some $\delta > 0$ small enough we finally obtain that

$$\begin{aligned}
\frac{\alpha la^2}{\sqrt{3}} &= \frac{2}{\sqrt{3}} \int_\Omega Q(F^a)\,dx \\
&= \inf\left\{ \frac{2}{\sqrt{3}} \int_\Omega Q(e(w)) - \delta|e(w) - F^a|^2\,dx : \right. \\
&\qquad\qquad \left. w \in H^1(\Omega), w(0,x_2) = 0, w(l,x_2) = al \right\} \\
&\le \liminf_{\varepsilon\to 0} \frac{2}{\sqrt{3}} \int_{\Omega_\varepsilon} Q(e(\bar{u}_\varepsilon)) - \delta|e(\bar{u}_\varepsilon) - F^a|^2\,dx \\
&= \frac{\alpha la^2}{\sqrt{3}} - \delta \limsup_{\varepsilon\to 0} \|e(\bar{u}_\varepsilon) - F^a\|^2_{L^2(\Omega_\varepsilon)}
\end{aligned}$$

and therefore $\lim_{\varepsilon\to 0} \|e(\bar{u}_\varepsilon) - F^a\|^2_{L^2(\Omega_\varepsilon)} = 0$ indeed. □

4.3.3 Proof of the main limiting result

After all these preparatory lemmas, the proof of our main limiting result Theorem 1.5.1 is now straightforward.

Proof of Theorem 1.5.1. Choose s_ε as in Lemmas 4.3.4 if $a < a_{\text{crit}}$, p_ε, s_ε and t_ε as in (4.11) and Lemma 4.3.2 if $a > a_{\text{crit}}$ and $\phi \neq 0$ and finally g_ε and s_ε and t_ε as in Lemma 4.3.3 if $a > a_{\text{crit}}$ and $\phi = 0$. By Lemmas 4.3.4, 4.3.2 and 4.3.3, $\bar{u}_\varepsilon$ can be extended as an H^1-function from Ω_ε to Ω, $\Omega_\varepsilon^{(i)}$ to $\Omega^{(i)}$, $i = 1, 2$, or $\Omega_{g_\varepsilon}^{(i)}$ to $\Omega^{(i)}[g_\varepsilon]$, $i = 1, 2$, respectively, such that still, respectively,

$$\|\bar{u}_\varepsilon - (0, s_\varepsilon) - F^a \cdot\|_{H^1(\Omega)} \to 0, \tag{4.30}$$

$$\|\bar{u}_\varepsilon - (0, s_\varepsilon)\|_{H^1(\Omega^{(1)})} + \|\bar{u}_\varepsilon - (al, t_\varepsilon)\|_{H^1(\Omega^{(2)})} \to 0, \tag{4.31}$$

$$\|\bar{u}_\varepsilon - (0, s_\varepsilon)\|_{H^1(\Omega^{(1)}[g_\varepsilon])} + \|\bar{u}_\varepsilon - (al, t_\varepsilon)\|_{H^1(\Omega^{(2)}[g_\varepsilon])} \to 0. \tag{4.32}$$

This completes the proof as by Lemma 4.3.1 we also still have $|\{x \in \Omega_\varepsilon : \bar{u}_\varepsilon(x) \neq \tilde{u}_\varepsilon(x)\}| = O(\varepsilon)$. □

Finally, we give the proof of Corollary 1.5.2.

Proof of Corollary 1.5.2 . First, let (y_ε) be a minimizing sequence satisfying (1.28). Then by Theorem 1.5.1 we obtain (4.30), (4.31) or (4.32), respectively. Taking the condition $\sup_\varepsilon \|u_\varepsilon\|_\infty < \infty$ into account, in the cases (i) and (ii) we get $\sup_\varepsilon |s_\varepsilon| < \infty$ and $\sup_\varepsilon |s_\varepsilon|, \sup_\varepsilon |t_\varepsilon| < \infty$ such that, passing to subsequences, we obtain $s_\varepsilon \to s$ and $s_\varepsilon \to s, t_\varepsilon \to t, p_\varepsilon \to p$, respectively, for suitable constants $s, t \in \mathbb{R}$, $p \in (0, l)$. In (iii) we first note that up to subsequences g_ε converges uniformly to some Lipschitz function $g : (0, 1) \to [0, l]$ satisfying $|g'| \le \frac{1}{\sqrt{3}}$ a.e. Then using again the uniform bound $\sup_\varepsilon \|u_\varepsilon\|_\infty < \infty$ we get constants s, t such that $s_\varepsilon \to s$ and $t_\varepsilon \to t$ up to subsequences. It follows that $\tilde{u}_\varepsilon \to u$ as given in (i), (ii) and (iii), respectively.

Conversely, we assume that u is given as in Corollary 1.5.2 and show that there is a minimizing sequence (y_ε) satisfying (1.28) with $\tilde{u}_\varepsilon \to u$ in measure. For (i) and (ii) this is obvious by the proof of Theorem 1.4.1 taking the configurations in (3.5) and (3.6) up to suitable translations. For given u in (iii) with corresponding function g and constants s, t we approximate $g : (0,1) \to [0,l]$ uniformly by Lipschitz functions $g_\varepsilon : (0,1) \to (0,l)$ being affine on intervals of length $\frac{\sqrt{3}\varepsilon}{2}$ with $g'_\varepsilon = \pm\frac{1}{\sqrt{3}}$ a.e. We set

$$y_\varepsilon(x) = \begin{cases} x + (0, \sqrt{\varepsilon} s), & \text{if } 0 < x_1 < g_\varepsilon(x_2), \\ x + (a_\varepsilon l, \sqrt{\varepsilon} t), & \text{if } g_\varepsilon(x_2) < x_1 < l, \end{cases}$$

so that $\tilde{u}_\varepsilon = \frac{y_\varepsilon - \mathbf{id}}{\sqrt{\varepsilon}} \to u$ in measure. As in the proof of Theorem 1.4.1, except for negligible contributions of the boundary layers, $\mathcal{E}^{\chi}_{\varepsilon}(y_\varepsilon)$ is given by the energy of the springs intersected transversally by $\mathrm{graph}(g_\varepsilon)$. These springs are elongated by a factor scaling with $\frac{1}{\sqrt{\varepsilon}}$ yielding a contribution $\varepsilon\beta$ in the limit. It is elementary to see that on every stripe in $\mathbf{e}_1$ direction of length $\frac{\sqrt{3}\varepsilon}{2}$ the graph intersects two springs, and consequently $\mathcal{E}^{\chi}_{\varepsilon}(y_\varepsilon) \to \frac{4\beta}{\sqrt{3}}$. □

Chapter 5

The limiting variational problem

The first part of this chapter is devoted to the the Γ-convergence result for $\mathcal{E}_\varepsilon$. Afterwards we will investigate the limiting variational problem.

5.1 Convergence of the variational problems

Recall the definition of the sets $\mathcal{C}_\varepsilon$ and $\tilde{\mathcal{C}}_\varepsilon$ in Section 1.6. As a further preparation we modify the interpolation $\tilde{y}$ on triangles with large deformation: We fix a threshold explicitly as $R = 7$ and let $\bar{\mathcal{C}}_\varepsilon \subset \tilde{\mathcal{C}}_\varepsilon$ be the set of those triangles where $|(\tilde{y})_\triangle| > R$. By definition of the boundary values in (1.29) we find $\bar{\mathcal{C}}_\varepsilon \subset \mathcal{C}_\varepsilon$ for ε small enough. We introduce another interpolation y' which leaves $\tilde{y}$ unchanged on $\triangle \in \tilde{\mathcal{C}}_\varepsilon \setminus \bar{\mathcal{C}}_\varepsilon$ and replaces $\tilde{y}$ on $\triangle \in \bar{\mathcal{C}}_\varepsilon$ by a discontinuous function with constant derivative satisfying $|(y')_\triangle| \le R$. In fact, by introducing jumps we achieve a release of the elastic energy. Note that $y' \in SBV(\tilde{\Omega}_\varepsilon)$.

More precisely, observe that on $\triangle \in \bar{\mathcal{C}}_\varepsilon$ we have $|(\tilde{y})_\triangle \, \mathbf{v}| \ge 2$ for at least two springs $\mathbf{v} \in \mathcal{V}$. Indeed, using the elementary identity (2.8) we find that $|F| > 7$ implies

$$\sum_{\mathbf{v}\in\mathcal{V}} |F\mathbf{v}|^4 = \sum_{\mathbf{v}\in\mathcal{V}} \langle \mathbf{v}, F^T F\mathbf{v}\rangle^2 \ge \frac{3}{8}(\mathrm{trace}(F^T F))^2 = \frac{3}{8}|F|^4$$

and so $\max_{\mathbf{v}\in\mathcal{V}} |F\mathbf{v}|^4 > \frac{7^4}{8} > 4^4$. Hence, $|F\mathbf{v}| > 4$ for at least one $\mathbf{v} \in \mathcal{V}$ and at least two springs are elongated by a factor larger than 2. For $m = 2, 3$ let $\bar{\mathcal{C}}_{\varepsilon,m} \subset \bar{\mathcal{C}}_\varepsilon$ be the set of triangles where $|(\tilde{y})_\triangle \, \mathbf{v}| \ge 2$ holds for exactly m springs $\mathbf{v} \in \mathcal{V}$. For $i, j, k = 1, 2, 3$ pairwise distinct let h_i denote the segment between the centers of the sides in $\mathbf{v}_j$ and $\mathbf{v}_k$ direction and define the set $V_i = h_j \cup h_k$.

We now construct $y' \in SBV^2(\tilde{\Omega}_\varepsilon)$. On $\triangle \in \tilde{\mathcal{C}}_\varepsilon \setminus \bar{\mathcal{C}}_\varepsilon$ we simply set $y' = \tilde{y}$. On $\triangle \in \bar{\mathcal{C}}_{\varepsilon,2}$, assuming $|(\tilde{y})_\triangle \, \mathbf{v}_i| \le 2$, we choose y' such that $\nabla y'$ assumes the constant value $(y')_\triangle$ on $\triangle$ with $(y')_\triangle \, \mathbf{v}_i = (\tilde{y})_\triangle \, \mathbf{v}_i$ and $|(y')_\triangle \, \mathbf{v}| = 1$ for $\mathbf{v} \in \mathcal{V} \setminus \{\mathbf{v}_i\}$. Moreover, we ask that $y' = \tilde{y}$ at the three vertices and on the side orientated in $\mathbf{v}_i$ direction. This can and will be done in such a way that y' is continuous on $\mathrm{int}(\triangle) \setminus h_i$. We note that the definition of $(y')_\triangle$ is unique up to a reflection, unless $(\tilde{y})_\triangle \mathbf{v}_i = 0$. We may and will assume that

$$\mathrm{dist}\left((y')_\triangle, SO(2)\right) \le \mathrm{dist}\left((y')_\triangle, O(2) \setminus SO(2)\right). \tag{5.1}$$

For $\triangle \in \bar{\mathcal{C}}_{\varepsilon,3}$ we set $(y')_\triangle = \mathbf{Id}$ and $y' = \tilde{y}$ at the three vertices such that y' is continuous on $\mathrm{int}(\triangle) \setminus V_i$ for some $i \in \{1,2,3\}$. Here, the index i can be taken arbitrarily at first. However, in what follows it will also be necessary to use the following unambiguously defined 'variants' of y': If on every $\triangle \in \bar{\mathcal{C}}_{\varepsilon,3}$ the set V_i is chosen as the jump set of y' we denote this interpolation explicitly as y'_{V_i}.

We define the interpolation u' for the rescaled displacement field by $u' = \frac{1}{\sqrt{\varepsilon}}(y' - \mathbf{id})$. We note that by construction also on an edge $[p,q] \subset \partial\triangle$ for $\triangle \in \bar{\mathcal{C}}_\varepsilon$ jumps may occur. There, however, the jump height $|[u'_\varepsilon]|$ can be bounded by

$$|[u'_\varepsilon](x)| \le \varepsilon \left\|\nabla u'_\varepsilon\right\|_\infty \le \varepsilon \cdot c\varepsilon^{-\frac{1}{2}} = c\sqrt{\varepsilon} \tag{5.2}$$

for a constant $c > 0$ independent of ε and $x \in [p,q]$. This holds since the interpolations are continuous at the vertices.

The following lemma shows that we may pass from $\tilde{u}_\varepsilon$ to u'_ε without changing the limit.

Lemma 5.1.1. *If $u_\varepsilon \to u$ in the sense of Definition 1.6.2 and $\mathcal{E}_\varepsilon(u_\varepsilon)$ is uniformly bounded, then $\chi_{\tilde{\Omega}_\varepsilon} u'_\varepsilon \to u$ in $L^1(\tilde{\Omega})$, $\chi_{\tilde{\Omega}_\varepsilon} \nabla u'_\varepsilon \rightharpoonup \nabla u$ in $L^2(\tilde{\Omega})$ and $\mathcal{H}^1(J_{u'_\varepsilon})$ is uniformly bounded.*

Proof. We first note that there is some $M > 0$ such that

$$\#\bar{\mathcal{C}}_\varepsilon \le \frac{M}{\varepsilon} \tag{5.3}$$

for all $\varepsilon > 0$. To see this, we just recall that every triangle $\triangle \in \bar{\mathcal{C}}_\varepsilon$ provides at least the energy $\varepsilon \inf\{W(r) : r \ge 2\}$. In fact we may assume that $\mathcal{C}^*_\varepsilon = \bar{\mathcal{C}}_\varepsilon$ in Definition 1.6.2 as for $\Delta \in \mathcal{C}^*_\varepsilon \setminus \bar{\mathcal{C}}_\varepsilon$ we have $|(\tilde{u}_\varepsilon)_\triangle| \le \frac{C}{\sqrt{\varepsilon}}|(\tilde{y}_\varepsilon)_\triangle - \mathbf{Id}| \le \frac{C}{\sqrt{\varepsilon}}$ and so

$$\begin{aligned}
\|\nabla \tilde{u}_\varepsilon\|_{L^2(\tilde{\Omega}_\varepsilon \setminus \cup_{\triangle \in \bar{\mathcal{C}}_\varepsilon} \triangle)} &\le \|\nabla \tilde{u}_\varepsilon\|_{L^2(\tilde{\Omega}_\varepsilon \setminus \cup_{\triangle \in \mathcal{C}^*_\varepsilon} \triangle)} + \|\nabla \tilde{u}_\varepsilon\|_{L^2(\cup_{\triangle \in \mathcal{C}^*_\varepsilon \setminus \bar{\mathcal{C}}_\varepsilon} \triangle)} \\
&\le C + \left(\#(\mathcal{C}^*_\varepsilon \setminus \bar{\mathcal{C}}_\varepsilon)\frac{\sqrt{3}\varepsilon^2}{4} \cdot \frac{C}{\varepsilon}\right)^{\frac{1}{2}} \le C.
\end{aligned}$$

It follows that $\chi_{\tilde{\Omega}_\varepsilon} \nabla u'_\varepsilon$ is bounded uniformly in L^2 and, in particular, equiintegrable. Finally, the jump lengths $\mathcal{H}^1(J_{u'_\varepsilon})$ are readily seen to be bounded by $C\varepsilon \#\bar{\mathcal{C}}_\varepsilon \le C$. But then Ambrosio's compactness Theorem for GSBV (see Theorem A.1.3 or [4, Theorem 2.2]) shows that indeed $\chi_{\tilde{\Omega}_\varepsilon} \nabla u'_\varepsilon \rightharpoonup \nabla u$ in $L^2(\tilde{\Omega})$. □

5.1.1 The Γ-$\liminf$-inequality

With the above preparations at hand, we may now prove the Γ-$\liminf$-inequality in Theorem 1.6.3.

Proof of Theorem 1.6.3(i). Let $(g_\varepsilon)_\varepsilon \in W^{1,\infty}(\tilde{\Omega})$ with $\sup_\varepsilon \|g_\varepsilon\|_{W^{1,\infty}(\tilde{\Omega})} < +\infty$ be given. Let $u \in SBV^2(\tilde{\Omega})$ and consider a sequence $u_\varepsilon \subset SBV^2(\tilde{\Omega}_\varepsilon)$ with $u_\varepsilon \in \mathcal{A}_{g_\varepsilon}$ converging to u in SBV^2 in the sense of Definition 1.6.2. We split up the energy into bulk and crack parts neglecting the contribution $\varepsilon E_\varepsilon^{\text{boundary}}$ from the boundary layers:

$$\begin{aligned}\mathcal{E}_\varepsilon(u_\varepsilon) &\geq \varepsilon \sum_{\triangle \in \mathcal{C}_\varepsilon \setminus \bar{\mathcal{C}}_\varepsilon} W_\triangle((\tilde{y}_\varepsilon)_\triangle) + \varepsilon \sum_{\triangle \in \bar{\mathcal{C}}_\varepsilon} W_\triangle((\tilde{y}_\varepsilon)_\triangle) \\ &= \frac{4}{\sqrt{3}\varepsilon} \int_{\Omega_\varepsilon} W_\triangle\left(\mathbf{Id} + \sqrt{\varepsilon}\nabla u_\varepsilon'\right) + \varepsilon \sum_{\triangle \in \bar{\mathcal{C}}_\varepsilon} \sum_{\substack{\mathbf{v} \in \mathcal{V}, \\ |(\tilde{y}_\varepsilon)_\triangle \mathbf{v}| > 2}} \frac{1}{2} W\left(|(\tilde{y}_\varepsilon)_\triangle \mathbf{v}|\right) \qquad (5.4) \\ &=: \mathcal{E}_\varepsilon^{\text{elastic}}(u_\varepsilon) + \mathcal{E}_\varepsilon^{\text{crack}}(u_\varepsilon).\end{aligned}$$

We note that by contruction of the interpolation u_ε' we may take the integral over Ω_ε. As both parts separate completely in the limit, we discuss them individually.

Elastic energy. We first concern ourselves with the elastic part of the energy. We recall $W_\triangle(\mathbf{Id} + G) = \frac{1}{2}Q(G) + \omega(G)$ with $\sup\left\{\frac{\omega(F)}{|F|^2} : |F| \leq \rho\right\} \to 0$ as $\rho \to 0$. Let $\chi_\varepsilon(x) := \chi_{[0,\varepsilon^{-1/4})}(|\nabla u_\varepsilon'(x)|)$. Note that for $F \in \mathbb{R}^{2\times 2}$, $r > 0$ one has $Q(rF) = r^2 Q(F)$. We compute

$$\mathcal{E}_\varepsilon^{\text{elastic}}(u_\varepsilon) \geq \frac{4}{\sqrt{3}} \int_{\Omega_\varepsilon} \chi_\varepsilon(x) \left(\frac{1}{2}Q(\nabla u_\varepsilon') + \frac{1}{\varepsilon}\omega\left(\sqrt{\varepsilon}\nabla u_\varepsilon'(x)\right)\right) dx.$$

The second term of the integral can be bounded by

$$\chi_\varepsilon |\nabla u_\varepsilon'|^2 \frac{\omega\left(\sqrt{\varepsilon}\nabla u_\varepsilon'\right)}{|\sqrt{\varepsilon}\nabla u_\varepsilon'|^2}.$$

Since $\nabla u_\varepsilon'$ is bounded in L^2 and $\chi_\varepsilon \frac{\omega\left(\sqrt{\varepsilon}\nabla u_\varepsilon'\right)}{|\sqrt{\varepsilon}\nabla u_\varepsilon'|^2}$ converges uniformly to 0 as $\varepsilon \to 0$ it follows that

$$\begin{aligned}\liminf_{\varepsilon \to 0} \mathcal{E}_\varepsilon^{\text{elastic}}(u_\varepsilon) &\geq \liminf_{\varepsilon \to 0} \frac{4}{\sqrt{3}} \int_{\Omega_\varepsilon} \chi_\varepsilon(x) \frac{1}{2} Q(\nabla u_\varepsilon'(x))\, dx \\ &\geq \liminf_{\varepsilon \to 0} \frac{4}{\sqrt{3}} \int_{\Omega} \frac{1}{2} Q(\chi_{\Omega_\varepsilon}\chi_\varepsilon(x)\nabla u_\varepsilon'(x))\, dx.\end{aligned}$$

By assumption $\chi_{\Omega_\varepsilon}\nabla u_\varepsilon' \rightharpoonup \nabla u$ weakly in L^2. As $\chi_\varepsilon \to 1$ boundedly in measure on Ω, it follows $\chi_{\Omega_\varepsilon}\chi_\varepsilon \nabla u_\varepsilon' \rightharpoonup u$ weakly in $L^2(\Omega)$. By lower semicontinuity (Q is convex by Lemma 2.4.2) we conclude recalling that Q only depends on the symmetric part of the gradient:

$$\liminf_{\varepsilon\to 0} \mathcal{E}_\varepsilon^{\text{elastic}}(u_\varepsilon) \geq \frac{4}{\sqrt{3}} \int_\Omega \frac{1}{2} Q(e(u(x)))\, dx.$$

Crack energy. By construction the functions u'_ε have jumps on destroyed triangles $\triangle \in \bar{\mathcal{C}}_\varepsilon$. We now write the energy of such a triangle in terms of the jump height $[u] = u^+ - u^-$. We first concern ourselves with a triangle $\triangle \in \bar{\mathcal{C}}_{\varepsilon,3}$. For the variant u'_{ε,V_i}, $i = 1,2,3$ we consider the springs in $\mathbf{v}_j, \mathbf{v}_k$ direction for $j,k \neq i$. Thus, we compute

$$\varepsilon(\tilde{y}_\varepsilon)_\triangle \, \mathbf{v}_j = \varepsilon(y'_\varepsilon)_\triangle \, \mathbf{v}_j + [y'_{\varepsilon,V_i}]_{h_k} = \varepsilon \mathbf{v}_j + \sqrt{\varepsilon}[u'_{\varepsilon,V_i}]_{h_k}, \tag{5.5}$$

where $[u'_{\varepsilon,V_i}]_{h_k}$ denotes the jump height on the set h_k. Here and in the following equations, the same holds true if we interchange the roles of j and k. We claim that

$$|(\tilde{y}_\varepsilon)_\triangle \, \mathbf{v}_j| \geq \varepsilon^{\frac{1}{4}} \left| \frac{1}{\sqrt{\varepsilon}} [u'_{\varepsilon,V_i}]_{h_k} \right| + 1. \tag{5.6}$$

Indeed, for $|\frac{1}{\sqrt{\varepsilon}}[u'_{\varepsilon,V_i}]_{h_k}| \leq \varepsilon^{-\frac{1}{4}}$ this is clear since $|(\tilde{y}_\varepsilon)_\triangle \, \mathbf{v}_j| \geq 2$. Otherwise, applying (5.5) we compute for ε small enough:

$$\begin{aligned} |(\tilde{y}_\varepsilon)_\triangle \, \mathbf{v}_j| &= \left| \frac{1}{\sqrt{\varepsilon}} [u'_{\varepsilon,V_i}]_{h_k} + \mathbf{v}_j \right| \geq \left| \frac{1}{\sqrt{\varepsilon}} [u'_{\varepsilon,V_i}]_{h_k} \right| - 1 \\ &\geq \varepsilon^{\frac{1}{4}} \left| \frac{1}{\sqrt{\varepsilon}} [u'_{\varepsilon,V_i}]_{h_k} \right| + \left(1 - \varepsilon^{\frac{1}{4}}\right) \varepsilon^{-\frac{1}{4}} - 1 \\ &= \varepsilon^{\frac{1}{4}} \left| \frac{1}{\sqrt{\varepsilon}} [u'_{\varepsilon,V_i}]_{h_k} \right| - 2 + \varepsilon^{-\frac{1}{4}} \geq \varepsilon^{\frac{1}{4}} \left| \frac{1}{\sqrt{\varepsilon}} [u'_{\varepsilon,V_i}]_{h_k} \right| + 1. \end{aligned}$$

Let $\rho > 0$ sufficiently small. Applying Lemma 2.4.5(iv) there is an increasing subadditive function ψ^ρ with $\psi^\rho(r-1) - \rho \leq W(r)$ for $r \geq 1$. We define $\tilde{\psi}^\rho = \psi^\rho - \rho$. The monotonicity of ψ^ρ and (5.6) yield

$$W(|(\tilde{y}_\varepsilon)_\triangle \, \mathbf{v}_j|) \geq \tilde{\psi}^\rho(|(\tilde{y}_\varepsilon)_\triangle \, \mathbf{v}_j| - 1) \geq \tilde{\psi}^\rho\left(\left| \varepsilon^{-\frac{1}{4}} [u'_{\varepsilon,V_i}]_{h_k} \right| \right). \tag{5.7}$$

Now for $\triangle \in \bar{\mathcal{C}}_{\varepsilon,3}$ we may estimate the energy as follows:

$$\begin{aligned} W_\triangle\left((\tilde{y}_\varepsilon)_\triangle\right) &= \frac{1}{2} \sum_{l=1}^{3} W(|(\tilde{y}_\varepsilon)_\triangle \, \mathbf{v}_l|) \\ &\geq \frac{1}{4} \sum_{i=1}^{3} \left\{ \tilde{\psi}^\rho\left(\varepsilon^{-\frac{1}{4}} |[u'_{\varepsilon,V_i}]_{h_k}| \right) + \tilde{\psi}^\rho\left(\varepsilon^{-\frac{1}{4}} |[u'_{\varepsilon,V_i}]_{h_j}| \right) \right\} =: W_{\triangle,3}\left((\tilde{y}_\varepsilon)_\triangle\right), \end{aligned}$$

where $i, j, k = 1, 2, 3$ are pairwise distinct. With $\nu_u^{(i)} = \nu_{u'_{\varepsilon,V_i}}$ we can also write

$$W_{\triangle,3}\left((\tilde{y}_\varepsilon)_\triangle\right) = \frac{1}{4} \cdot \frac{2}{\varepsilon} \cdot \frac{2}{\sqrt{3}} \sum_{i=1}^{3} \int_{h_j \cup h_k} \tilde{\psi}^\rho \left(\varepsilon^{-\frac{1}{4}} |[u'_{\varepsilon,V_i}]|\right) \left(|\mathbf{v}_j \cdot \nu_u^{(i)}| + |\mathbf{v}_k \cdot \nu_u^{(i)}|\right) d\mathcal{H}^1.$$

The factors in front occur since $\mathcal{H}^1(h_j) = \frac{\varepsilon}{2}$ and, letting ν_j be a normal of h_j, one has $|\nu_j \cdot \mathbf{v}_j| = 0$ and $|\nu_j \cdot \mathbf{v}_k| = \frac{\sqrt{3}}{2}$. Consequently, defining $\phi_i^\rho(r,\nu) = \psi^\rho(r)\left(|\mathbf{v}_j \cdot \nu| + |\mathbf{v}_k \cdot \nu|\right)$ and $\tilde{\phi}_i^\rho(r,\nu) = \tilde{\psi}^\rho(r)\left(|\mathbf{v}_j \cdot \nu| + |\mathbf{v}_k \cdot \nu|\right)$, respectively, we get

$$W_{\triangle,3}\left((\tilde{y}_\varepsilon)_\triangle\right) = \frac{1}{\sqrt{3}\varepsilon} \sum_{i=1}^{3} \int_{J_{u'_{\varepsilon,V_i}} \cap \operatorname{int}(\triangle)} \tilde{\phi}_i^\rho(\varepsilon^{-\frac{1}{4}} |[u'_{\varepsilon,V_i}]|, \nu_u^{(i)}) \, d\mathcal{H}^1$$

on every $\triangle \in \bar{\mathcal{C}}_{\varepsilon,3}$. For $\triangle \in \bar{\mathcal{C}}_{\varepsilon,2}$ we proceed analogously. Assuming $|(\tilde{y}_\varepsilon)_\triangle \mathbf{v}_i| \le 2$ we compute for the springs in $\mathbf{v}_j, \mathbf{v}_k$ direction (abbreviated by $\mathbf{v}_{j,k}$) as in (5.5)

$$\varepsilon(\tilde{y}_\varepsilon)_\triangle \mathbf{v}_{j,k} = \varepsilon(y'_\varepsilon)_\triangle \mathbf{v}_{j,k} + \sqrt{\varepsilon}[u'_\varepsilon]_{h_i}. \tag{5.8}$$

Note that in this case we do not have to take a special variant of u'_ε into account. Repeating the steps (5.6) and (5.7) we find

$$\frac{1}{2}\left(W(|(\tilde{y}_\varepsilon)_\triangle \mathbf{v}_j|) + W(|(\tilde{y}_\varepsilon)_\triangle \mathbf{v}_k|)\right) \ge \tilde{\psi}^\rho \left(\varepsilon^{-\frac{1}{4}} |[u'_\varepsilon]_{h_i}|\right) =: W_{\triangle,2}\left((\tilde{y}_\varepsilon)_\triangle\right).$$

Noting that $|\mathbf{v}_j \cdot \nu_i| = |\mathbf{v}_k \cdot \nu_i| = \frac{\sqrt{3}}{2}$, $|\mathbf{v}_i \cdot \nu_i| = 0$ and that every of these terms occurs twice in the sum of the right hand side of the following formula, it is not hard to see that this energy satisfies the same integral representation formula as $W_{\triangle,3}$:

$$W_{\triangle,2}\left((\tilde{y}_\varepsilon)_\triangle\right) = \frac{1}{\sqrt{3}\varepsilon} \sum_{i=1}^{3} \int_{J_{u'_{\varepsilon,V_i}} \cap \operatorname{int}(\triangle)} \tilde{\phi}_i^\rho(\varepsilon^{-\frac{1}{4}} |[u'_{\varepsilon,V_i}]|, \nu_u^{(i)}) \, d\mathcal{H}^1.$$

(Recall that the interpolation variant u'_{ε,V_i} and its crack normal $\nu_u^{(i)}$ do not depend on i on $\triangle \in \bar{\mathcal{C}}_{\varepsilon,2}$.) Let $\sigma > 0$. Note that $\bar{\mathcal{C}}_\varepsilon \subset \mathcal{C}_\varepsilon$ for ε sufficiently small as $\sup_\varepsilon \|g_\varepsilon\|_{W^{1,\infty}(\tilde{\Omega})} < +\infty$. Thus, the crack energy can be estimated by

$$\begin{aligned}
\mathcal{E}_\varepsilon^{\text{crack}}(u_\varepsilon) &\ge \frac{1}{\sqrt{3}} \sum_i \int_{J_{u'_{\varepsilon,V_i}} \cap \tilde{\Omega}_\varepsilon} \tilde{\phi}_i^\rho(\varepsilon^{-\frac{1}{4}} |[u'_{\varepsilon,V_i}]|, \nu_u^{(i)}) \, d\mathcal{H}^1 - E^\rho_{\varepsilon,\cup\partial\triangle}(\tilde{y}_\varepsilon) \\
&\ge \frac{1}{\sqrt{3}} \sum_i \int_{J_{u'_{\varepsilon,V_i}} \cap \tilde{\Omega}_\varepsilon} \left(\phi_i^\rho(\sigma^{-1} |[u'_{\varepsilon,V_i}]|, \nu_u^{(i)}) - 2\rho\right) d\mathcal{H}^1 - E^\rho_{\varepsilon,\cup\partial\triangle}(\tilde{y}_\varepsilon),
\end{aligned}$$

where $E^\rho_{\varepsilon,\cup\partial\triangle}(\tilde{y}_\varepsilon)$ compensates for the extra contribution provided by jumps lying on the boundary of some $\triangle \in \bar{\mathcal{C}}_\varepsilon$. We will show that this term vanishes in the limit.

Now by construction the $\phi_i^\rho(r,\nu)$, $i = 1,2,3$, are products of a positive, increasing and concave function in r and a norm in ν. Moreover, u'_ε and its variants converge to u in L^1 with $\nabla u'_\varepsilon$ bounded in L^2 and thus equiintegrable. By Ambrosio's lower semicontinuity Theorem [4, Theorem 3.7] we obtain

$$\liminf_{\varepsilon\to 0} \mathcal{E}_\varepsilon^{\mathrm{crack}}(u_\varepsilon) \geq \frac{1}{\sqrt{3}} \int_{J_u} \sum_i \phi_i^\rho(\sigma^{-1}|[u]|, \nu_u)\, d\mathcal{H}^1 - CM\rho - \limsup_{\varepsilon\to 0} E^\rho_{\varepsilon,\cup\partial\triangle}(\tilde{y}_\varepsilon),$$

where we used that $\sup_\varepsilon \mathcal{H}^1(J_{u'_\varepsilon}) \leq CM$ for a constant $C > 0$ by (5.3). We recall that $\psi^\rho(r) \to \beta$ for $r \to \infty$. In the limit $\sigma \to 0$ this yields

$$\liminf_{\varepsilon\to 0} \mathcal{E}_\varepsilon^{\mathrm{crack}}(u_\varepsilon) \geq \frac{1}{\sqrt{3}} \int_{J_u} 2\beta \sum_{\mathbf{v}\in\mathcal{V}} |\mathbf{v}\cdot\nu_u|\, d\mathcal{H}^1 - CM\rho - \limsup_{\varepsilon\to 0} E^\rho_{\varepsilon,\cup\partial\triangle}(\tilde{y}_\varepsilon). \quad (5.9)$$

Taking (5.2) and (5.3) into account we compute

$$\begin{aligned}\limsup_{\varepsilon\to 0} \sum_{\triangle\in\bar{\mathcal{C}}_\varepsilon} \int_{\partial\triangle} |\tilde{\psi}^\rho\left(\varepsilon^{-\frac{1}{4}}|[u'_\varepsilon]|\right)| &\leq \lim_{\varepsilon\to 0} CM \sup\left\{|\psi^\rho(r) - \rho| : r \leq \varepsilon^{-\frac{1}{4}}\cdot c\varepsilon^{\frac{1}{2}}\right\} \\ &= CM\rho.\end{aligned}$$

This proves $\limsup_\varepsilon |E^\rho_{\varepsilon,\cup\partial\triangle}(\tilde{y}_\varepsilon)| \leq \tilde{C}M\rho$ for some $\tilde{C} > 0$. We finally let $\rho \to 0$ in (5.9). This finishes the proof of (i). □

We now prove the Γ-lim inf-inequality in Theorem 1.6.4.

Proof of Theorem 1.6.4, first part. Following the proof of Theorem 1.6.3(i) it suffices to show

$$\liminf_{\varepsilon\to 0} \frac{1}{\varepsilon} \int_{\Omega_\varepsilon} \chi_\varepsilon f_\kappa(\nabla y'_\varepsilon) \geq -\frac{\kappa}{2} \int_\Omega \hat{Q}(\nabla u),$$

where $\hat{Q} = D^2\hat{m}_1(\mathbf{Id})$. Let $u'_\varepsilon = \frac{1}{\sqrt{\varepsilon}}(y'_\varepsilon - \mathbf{id})$. With a slight abuse of notation we set $e(F) = \frac{1}{2}(F^T + F)$ and $a(F) = F - e(F)$ for matrices $F \in \mathbb{R}^{2\times 2}$. Let $F = \mathbf{Id} + \sqrt{\varepsilon}G$ for $G \in \mathbb{R}^{2\times 2}$. Linearization around the identity matrix yields $\mathrm{dist}(F, SO(2)) = \sqrt{\varepsilon}|e(G)| + \varepsilon O(|G|^2)$. It is not hard to see that this implies

$$R(F) = \mathbf{Id} + \sqrt{\varepsilon}a(G) + \varepsilon O(|G|^2), \quad (5.10)$$

where $R(F) \in SO(2)$ is defined as in Lemma 2.4.6. As $\hat{m}(\mathbf{Id}) = \mathbf{e}_1$ and $e(G) \in \ker(D\hat{m}(\mathbf{Id}))$, we find by expanding $\hat{m}_1$

$$\hat{m}_1(F) = 1 + \sqrt{\varepsilon}D\hat{m}_1(\mathbf{Id})a(G) + \frac{\varepsilon}{2}\hat{Q}(G) + \omega(\sqrt{\varepsilon}G) \quad (5.11)$$

with $\sup\left\{\frac{\omega(H)}{|H|^2} : |H| \leq \rho\right\} \to 0$ as $\rho \to 0$.

We concern ourselves with the term $D\hat{m}_1(\mathbf{Id})a(G)$. Recall that $|\hat{m}(R(F)) - \hat{m}(F)| \leq C|R(F) - F|^2$ by Lemma 2.4.6(i). For $F = \mathbf{Id} + \sqrt{\varepsilon}G$ this implies by (5.10)

$$\begin{aligned} D\hat{m}_1(\mathbf{Id})a(G) &= \mathbf{e}_1 \cdot D\hat{m}(\mathbf{Id})G = \lim_{\varepsilon\to 0} \mathbf{e}_1 \cdot \frac{\hat{m}(F) - \hat{m}(\mathbf{Id})}{\sqrt{\varepsilon}} \\ &= \lim_{\varepsilon\to 0} \mathbf{e}_1 \cdot \frac{\hat{m}(R(F)) - \mathbf{e}_1}{\sqrt{\varepsilon}} + O(\sqrt{\varepsilon}) = \lim_{\varepsilon\to 0} \mathbf{e}_1 \cdot a(G)\mathbf{e}_1 + O(\sqrt{\varepsilon}) = 0. \end{aligned}$$

In particular, (5.11) then implies $0 \leq \frac{1}{\varepsilon} f_\kappa(F) = -\frac{\kappa}{2}\hat{Q}(G) - \frac{1}{\varepsilon}\omega(\sqrt{\varepsilon}G)$ and thus $-\hat{Q}$ is positive semidefinite. We proceed exactly as in the proof of Theorem 1.6.3(i) and conclude

$$\begin{aligned} \liminf_{\varepsilon\to 0} \frac{1}{\varepsilon} \int_{\Omega_\varepsilon} \chi_\varepsilon f_\kappa(\nabla y'_\varepsilon) &\geq \liminf_{\varepsilon\to 0} -\int_{\Omega_\varepsilon} \chi_\varepsilon \Big(\frac{\kappa}{2}\hat{Q}(\nabla u'_\varepsilon) + \frac{\kappa}{\varepsilon}\omega(\sqrt{\varepsilon}\nabla u'_\varepsilon)\Big) \\ &\geq -\frac{\kappa}{2}\int_\Omega \hat{Q}(\nabla u). \end{aligned}$$

□

5.1.2 Recovery sequences

It remains to construct recovery sequences in order to complete the proof of Theorem 1.6.3.

Proof of Theorem 1.6.3(ii).

The basic tool for the proof of the Γ-limsup-inequality is the density result given in Theorem A.1.7. Let $u \in SBV^2(\tilde{\Omega}, \mathbb{R}^2)$ with $u = g$ on $\tilde{\Omega} \setminus \Omega$. Without restriction we can assume $u \in L^\infty(\tilde{\Omega}, \mathbb{R}^2)$ as this hypothesis may be dropped by applying a truncation argument and taking $Q(F) \leq C|F|^2$ into account. In fact, it suffices to provide a recovery sequence for an approximation u_n defined in Theorem A.1.7. Although our notion of convergence in Definition 1.6.2 is not given in terms of a specific metric, similarly to a general density result in the theory of Γ-convergence this can be seen by a diagonal sequence argument. The crucial point is that due to (5.12) below we may assume that for ε sufficiently small (depending on n)

$$\#\mathcal{C}^*_\varepsilon = \#\mathcal{D}_\varepsilon \leq \frac{C\mathcal{H}^1(J_{u_n})}{\varepsilon} \leq \frac{C\mathcal{H}^1(J_u)}{\varepsilon},$$

where C is independent of n and ε. If $(u_{n,\varepsilon})_\varepsilon$ is a recovery for u_n, one may therefore pass to a diagonal sequence which is a recovery sequence for u, in particular converging to u the sense of Definition 1.6.2. For simplicity write u instead of u_n in what follows.

Let $\delta > 0$ and define $J_u^\delta = \{x \in J_u, |[u](x)| \geq \delta\}$. Since $|[u]|$ is Lipschitz continuous on J_u, it cannot oscillate infinitely often between values $\leq \delta$ and values $\geq 2\delta$ on a single segment. Consequently, there is a finite number N_u^δ of disjoint subsegments $S_1, \ldots, S_{N_u^\delta}$ in J_u such that $|[u]| < 2\delta$ on every S_j and $|[u]| > \delta$ on $J_u \setminus (S_1 \cup \ldots \cup S_{N_u^\delta})$. Note that $\mathcal{H}^1(\bigcup_{i=1}^{N_u^\delta} S_i) \leq \mathcal{H}^1(J_u \setminus J_u^{2\delta}) =: \rho(\delta) \to 0$ for $\delta \to 0$. We cover $S_1, \ldots, S_{N_u^\delta}$ by pairwise disjoint rectangles $Q_1, \ldots Q_{N_u^\delta}$ which satisfy $\sum_j \mathcal{H}^1(\partial Q_i) + |Q_i| \leq C\rho(\delta)$. It is not hard to see that $|u(x) - u(y)| \leq C\mathcal{H}^1(\partial Q_i) + 2\delta$ for $x, y \in Q_j$ as $\nabla u \in L^\infty(\tilde{\Omega})$.

We modify u on the rectangles Q_i: Let $u_\delta = u$ on $\tilde{\Omega} \setminus \bigcup_{i=1}^{N_u^\delta} Q_j$ and define $u_\delta = c_j$ on Q_j for $c_j \in \mathbb{R}^2$ in such a way that $J_{u_\delta} = J_{u_\delta}^\delta$ up to an $\mathcal{H}^1$-negligible set. As $u \in L^\infty(\tilde{\Omega})$, $\nabla u \in L^\infty(\tilde{\Omega})$ we find $u_\delta \to u$ in $L^1(\tilde{\Omega})$ and $\nabla u_\delta \to \nabla u$ in $L^2(\tilde{\Omega})$. Moreover, we have $\mathcal{H}^1(J_u \Delta J_{u_\delta}) \leq C\rho(\delta) \to 0$ for $\delta \to 0$.

Consequently, it suffices to establish a recovery sequence for a function $u \in \mathcal{W}(\tilde{\Omega})$ with $u = g$ in a neighborhood of $\tilde{\Omega} \setminus \Omega$ and $J_u = J_u^\delta$ for some $\delta > 0$. Note after the above modification the segments of J_u might not be pairwise disjoint.

We define $u_\varepsilon(x) = u(x)$ for $x \in \mathcal{L}_\varepsilon \cap \tilde{\Omega}$ and let $y_\varepsilon(x) = \mathbf{id} + \sqrt{\varepsilon} u_\varepsilon(x)$. Clearly we have $u_\varepsilon \in \mathcal{A}_{g_\varepsilon}$ for all ε. By $\tilde{u}_\varepsilon, u'_\varepsilon$ we again denote the interpolations on $\tilde{\Omega}_\varepsilon$. Up to considering a translation of u of order ε, we may assume that $J_u \cap \mathcal{L}_\varepsilon = \emptyset$. Let $\mathcal{D}_\varepsilon$ be the sets of triangles where J_u crosses at least one side of the triangle. Then

$$\#\mathcal{D}_\varepsilon \leq \frac{C\mathcal{H}^1(J_u)}{\varepsilon} + CN_u \tag{5.12}$$

for a constant $C > 0$ independent of $u \in \mathcal{W}(\tilde{\Omega}, \mathbb{R}^2)$ and ε, where N_u denotes the (smallest) number of segments whose union gives J_u. From now on for the local nature of the arguments we may assume that J_u consists of one segment only. Indeed, if J_u consists of segments $S_1, \ldots, S_{N_u}$, which are possibly not disjoint, the number of triangles $\triangle \in \bar{\mathcal{C}}_\varepsilon$ with $\triangle \cap S_{i_1} \cap S_{i_2} \neq \emptyset$ for $1 \leq i_i < i_2 \leq N_u$ scales like N_u and therefore their energy contribution is negligible in the limit. We show

$$\bar{\mathcal{C}}_\varepsilon = \mathcal{D}_\varepsilon$$

for ε small enough. Let $\triangle \in \mathcal{D}_\varepsilon$. We see that, if $J_u = J_u^\delta$ crosses a spring $\mathbf{v}$ at point x_*, say, then a computation similar as in (5.8) together with $\nabla u \in L^\infty$ shows

$$|(\tilde{y}_\varepsilon)_\triangle \mathbf{v}| = \left| \frac{1}{\sqrt{\varepsilon}}[u(x_*)] + O(1) \right| \geq \frac{\delta}{\sqrt{\varepsilon}} + O(1). \tag{5.13}$$

Thus, $\triangle \in \bar{\mathcal{C}}_\varepsilon$ for ε small enough. On the other hand, if we assume $\triangle \notin \mathcal{D}_\varepsilon$, then for at least two springs $\mathbf{v} \in \mathcal{V}$ we have $|(\tilde{y}_\varepsilon)_\triangle \mathbf{v}| \leq 1 + \sqrt{\varepsilon}\, \|\nabla u\|_\infty < 2$ for ε small enough leading to $\triangle \notin \bar{\mathcal{C}}_\varepsilon$.

We claim that

$$\|\nabla u'_\varepsilon\|_{L^\infty(\tilde{\Omega})} \leq C. \tag{5.14}$$

This is clear for $\triangle \notin \mathcal{D}_\varepsilon = \bar{\mathcal{C}}_\varepsilon$ as $\nabla u \in L^\infty$. For $\triangle \in \bar{\mathcal{C}}_{\varepsilon,3}$ it follows by construction. For $\triangle \in \bar{\mathcal{C}}_{\varepsilon,2}$ there is a $\mathbf{v} \in \mathcal{V}$ such that $(y'_\varepsilon)_\triangle \mathbf{v} = (\tilde{y}_\varepsilon)_\triangle \mathbf{v} = \mathbf{v} + O(\sqrt{\varepsilon})$. By Lemma 2.4.5(i) and (5.1) we get a rotation $R_\varepsilon \in SO(2)$ such that

$$|R_\varepsilon - (y'_\varepsilon)_\triangle|^2 = \operatorname{dist}^2((y'_\varepsilon)_\triangle, SO(2)) = \operatorname{dist}^2((y'_\varepsilon)_\triangle, O(2)) \le CW_\triangle((y'_\varepsilon)_\triangle) = O(\varepsilon).$$

This yields $|(y'_\varepsilon)_\triangle - \mathbf{Id}| = O(\sqrt{\varepsilon})$ and thus $|(u'_\varepsilon)_\triangle| = O(1)$.

We note that $\chi_{\tilde{\Omega}_\varepsilon} \tilde{u}_\varepsilon \to u$ in L^1 as u and thus every $\tilde{u}_\varepsilon$ is bounded uniformly in L^∞ and, u being Lipschitz away from J_u, $\tilde{u}_\varepsilon \to u$ uniformly on $\tilde{\Omega}_\varepsilon \setminus \bigcup_{\triangle \in \mathcal{D}_\varepsilon} \triangle$, where $|\bigcup_{\triangle \in \mathcal{D}_\varepsilon} \triangle| \le C\varepsilon$. Letting $\mathcal{C}^*_\varepsilon = \mathcal{D}_\varepsilon$ this shows that $u_\varepsilon \to u$ in the sense of Definition 1.6.2 recalling (5.12) and the fact that $|(\tilde{u}_\varepsilon)_\triangle| = O(1)$ for $\triangle \notin \mathcal{D}_\varepsilon$. We next establish an even stronger convergence of the derivatives. Consider $\nabla \tilde{u}_\varepsilon$ on triangles in $\mathcal{C}_\varepsilon \setminus \mathcal{D}_\varepsilon$. As ∇u is Lipschitz there, the oscillation on such a triangle, $\operatorname{osc}^\triangle_\varepsilon(\nabla u) := \sup\{|\nabla u(x) - \nabla u(x')|, x, x' \in \triangle\}$, tends to zero uniformly (i.e., not depending on the choice of the triangle). We thus obtain

$$\int_{\tilde{\Omega}_\varepsilon \setminus \bigcup_{\triangle \in \mathcal{D}_\varepsilon} \triangle} \|\nabla \tilde{u}_\varepsilon - \nabla u\|^2_\infty \le \int_{\tilde{\Omega}_\varepsilon \setminus \bigcup_{\triangle \in \mathcal{D}_\varepsilon} \triangle} (\operatorname{osc}^\triangle_\varepsilon(\nabla u))^2 \to 0$$

for $\varepsilon \to 0$, so that even $\chi_{\tilde{\Omega}_\varepsilon \setminus \bigcup_{\triangle \in \mathcal{D}_\varepsilon} \triangle} \nabla \tilde{u}_\varepsilon \to \nabla u$ strongly in $L^2(\tilde{\Omega})$. Note that in fact $\chi_{\tilde{\Omega}_\varepsilon} \nabla u'_\varepsilon \to \nabla u$ in $L^2(\tilde{\Omega})$. Indeed, recall $\#\mathcal{D}_\varepsilon \le C\varepsilon^{-1}$ by (5.12). Using (5.14) on the set of broken triangles we then get

$$\int_{\bigcup_{\triangle \in \mathcal{D}_\varepsilon} \triangle} |\nabla u'_\varepsilon - \nabla u|^2 \le C\#\bar{\mathcal{D}}_\varepsilon \varepsilon^2 \to 0$$

for $\varepsilon \to 0$. We now split up the energy in bulk and surface parts

$$\mathcal{E}^\chi_\varepsilon(u_\varepsilon) = \mathcal{E}^{\text{elastic}}_\varepsilon(u_\varepsilon) + \mathcal{E}^{\text{crack}}_\varepsilon(u_\varepsilon) + O(\varepsilon) + \frac{1}{\varepsilon} \int_{\Omega_\varepsilon} \chi(\nabla \tilde{y}_\varepsilon) \tag{5.15}$$

as defined in (5.4). Note that indeed the contribution $\varepsilon E^{\text{boundary}}_\varepsilon$ is of order $O(\varepsilon)$ as $\nabla u \in L^\infty(\tilde{\Omega})$ and $J_u \subset \Omega$ since $u = g$ in a neighborhood of $\tilde{\Omega} \setminus \Omega$. We first observe that $\frac{1}{\varepsilon} \int_{\Omega_\varepsilon} \chi(\nabla \tilde{y}_\varepsilon) = 0$ for ε small enough. Indeed, for $\Delta \in \bar{\mathcal{C}}_\varepsilon$ this follows from (5.13). For $\Delta \notin \mathcal{D}_\varepsilon$ it suffices to recall $|(\tilde{u}_\varepsilon)_\triangle| = O(1)$ which implies that $(\tilde{u}_\varepsilon)_\triangle$ is near $SO(2)$. Repeating the steps in the elastic energy estimate in (i), applying $\chi_{\Omega_\varepsilon} \nabla u'_\varepsilon \to \nabla u$ strongly in $L^2(\Omega)$, (5.14) and $Q(F) \le C|F|^2$ for a constant $C > 0$ we conclude that

$$\limsup_{\varepsilon \to 0} \mathcal{E}^{\text{elastic}}_\varepsilon(u_\varepsilon) = \frac{4}{\sqrt{3}} \int_\Omega \frac{1}{2} Q(e(u(x)))\, dx. \tag{5.16}$$

It is elementary to see that J_u crosses

$$\mathcal{H}^1(J_u) \frac{2|\nu_u \cdot \mathbf{v}|}{\sqrt{3}\varepsilon} + O(1) \tag{5.17}$$

springs in $\mathbf{v}$-direction for $\mathbf{v} \in \mathcal{V}$, where ν_u is a normal to the segment J_u. Recalling (5.13), the crack energy may be estimated by

$$\begin{aligned}
&\limsup_{\varepsilon\to 0} \mathcal{E}_\varepsilon^{\mathrm{crack}}(u_\varepsilon) \\
&\leq \limsup_{\varepsilon\to 0} \mathcal{H}^1(J_u) \ \sup\left\{W(r) : r \geq \delta\varepsilon^{-\frac{1}{2}} + O(1)\right\} \frac{2}{\sqrt{3}} \sum_{\mathbf{v}\in\mathcal{V}} |\nu_u \cdot \mathbf{v}| + O(\varepsilon) \\
&= \mathcal{H}^1(J_u) \ \beta \ \frac{2}{\sqrt{3}} \sum_{\mathbf{v}\in\mathcal{V}} |\nu_u \cdot \mathbf{v}|.
\end{aligned}$$

This together with (5.15) and (5.16) shows that u_ε is a recovery sequence for u. □

Finally, we construct recovery sequences for the functionals $\mathcal{F}_\varepsilon^\chi$ to conclude the proof of Theorem 1.6.4.

Proof of Theorem 1.6.4, second part. Following the proof of Theorem 1.6.3(ii) it suffices to show

$$\lim_{\varepsilon\to 0} \frac{1}{\varepsilon} \int_{\Omega_\varepsilon} f_\kappa(\nabla\tilde{y}_\varepsilon) = -\frac{\kappa}{2} \int_\Omega \hat{Q}(\nabla u).$$

First, by (5.13) and the definition of f_κ we get $\int_{\bigcup_{\Delta\in\mathcal{D}_\varepsilon}\Delta} f_\kappa(\nabla\tilde{y}_\varepsilon) = 0$ for ε small enough. For $\Delta \notin \mathcal{D}_\varepsilon$ we have $(\nabla\tilde{y}_\varepsilon)_\Delta = (\nabla y'_\varepsilon)_\Delta$ and thus we find $f_\kappa((\nabla\tilde{y}_\varepsilon)_\Delta) = -\varepsilon\frac{\kappa}{2}\hat{Q}((\nabla u'_\varepsilon)_\Delta) - \kappa\omega(\sqrt{\varepsilon}\nabla(u'_\varepsilon)_\Delta)$ by (5.11). We obtain

$$\begin{aligned}
\frac{1}{\varepsilon} \int_{\Omega_\varepsilon} f_\kappa(\nabla\tilde{y}_\varepsilon) &= \frac{1}{\varepsilon} \int_{\Omega_\varepsilon \setminus \bigcup_{\Delta\in\mathcal{D}_\varepsilon}\Delta} f_\kappa(\nabla y'_\varepsilon) \\
&\leq -\frac{\kappa}{2} \int_{\Omega_\varepsilon \setminus \bigcup_{\Delta\in\mathcal{D}_\varepsilon}\Delta} \hat{Q}(\nabla u'_\varepsilon) + \frac{C}{\varepsilon} \int_{\Omega_\varepsilon} \omega(\sqrt{\varepsilon}\nabla u'_\varepsilon).
\end{aligned}$$

Using (5.14) and the definition of ω we observe $\frac{1}{\varepsilon}\|\omega(\sqrt{\varepsilon}\nabla u'_\varepsilon)\|_\infty \to 0$ for $\varepsilon \to 0$. This together with strong convergence $\chi_{\Omega_\varepsilon}\nabla u'_\varepsilon \to \nabla u$ in $L^2(\Omega)$ shows

$$\limsup_{\varepsilon\to 0} \frac{1}{\varepsilon} \int_{\Omega_\varepsilon} f_\kappa(\nabla\tilde{y}_\varepsilon) \leq -\frac{\kappa}{2} \int_\Omega \hat{Q}(\nabla u).$$

□

5.2 Analysis of the limiting variational problem

We finally give the proof of Theorem 1.6.5 determining the minimizers of the limiting functional $\mathcal{E}$. An analogous result for isotropic energy functionals has been obtained in [57]. We thus do not repeat all the steps of the proof provided in [57] but rather concentrate on the additional arguments necessary to handle anisotropic surface contributions.

Proof of Theorem 1.6.5. We first establish a lower bound for the energy $\mathcal{E}$. To this end, we begin to estimate $\sum_{\mathbf{v}\in\mathcal{V}} |\mathbf{v}\cdot\nu|$ for $\nu \in S^1$. We recall that $\gamma \in [\frac{\sqrt{3}}{2}, 1]$ and define $P : [\frac{\sqrt{3}}{2}, 1] \times S^1 \to [0,\infty)$ by

$$P(\gamma,\nu) = \begin{cases} \left(1 - \sqrt{3}\frac{\sqrt{1-\gamma^2}}{\gamma}\right)|\mathbf{v}_\gamma\cdot\nu|, & \gamma > \frac{\sqrt{3}}{2}, \\ \max\left\{\sqrt{3}|\mathbf{e}_2\cdot\nu| - |\mathbf{e}_1\cdot\nu|, 0\right\}, & \gamma = \frac{\sqrt{3}}{2}. \end{cases}$$

As $\mathbf{v}_\gamma$ is unique for $\gamma > \frac{\sqrt{3}}{2}$, the function P is well defined. In the generic case, i.e. for $\gamma > \frac{\sqrt{3}}{2}$, an elementary computation yields

$$\begin{aligned}\sum_{\mathbf{v}\in\mathcal{V}} |\mathbf{v}\cdot\nu| &\geq |\mathbf{v}_\gamma\cdot\nu| + \sqrt{3}|\mathbf{v}_\gamma^\perp\cdot\nu| = |\mathbf{v}_\gamma\cdot\nu| + \sqrt{3}\left|\pm\frac{1}{\gamma}\mathbf{e}_1\cdot\nu \pm \frac{\sqrt{1-\gamma^2}}{\gamma}\mathbf{v}_\gamma\cdot\nu\right| \\ &\geq \frac{\sqrt{3}}{\gamma}|\mathbf{e}_1\cdot\nu| + P(\gamma,\nu)\end{aligned}$$

for $\nu \in S^1$. In the first step we used that $\sum_{\mathbf{v}\in\mathcal{V}\setminus\{\mathbf{v}_\gamma\}} \mathbf{v} = \pm\sqrt{3}\mathbf{v}_\gamma^\perp$. In the special case $\phi = 0 \Leftrightarrow \gamma = \frac{\sqrt{3}}{2}$, i.e. $\mathbf{v}_1 = \mathbf{e}_1$, $\mathbf{v}_{2,3} = \pm\frac{1}{2}\mathbf{e}_1 + \frac{\sqrt{3}}{2}\mathbf{e}_2$ we obtain $\sum_{\mathbf{v}\in\mathcal{V}} |\mathbf{v}\cdot\nu| = |\mathbf{e}_1\cdot\nu| + \sqrt{3}|\mathbf{e}_2\cdot\nu|$ for $|\nu_2| > \frac{1}{2}$ and $\sum_{\mathbf{v}\in\mathcal{V}} |\mathbf{v}\cdot\nu| = 2|\mathbf{e}_1\cdot\nu|$ for $|\nu_2| \leq \frac{1}{2}$, $\nu \in S^1$. Consequently, it is not hard to see that

$$\sum_{\mathbf{v}\in\mathcal{V}} |\mathbf{v}\cdot\nu| \geq \frac{\sqrt{3}}{\gamma}|\mathbf{e}_1\cdot\nu| + P(\gamma,\nu)$$

also holds for $\gamma = \frac{\sqrt{3}}{2}$. Thus, we get

$$\mathcal{E}(u) \geq \frac{4}{\sqrt{3}}\int_\Omega \frac{1}{2}Q(e(u(x)))\,dx + \int_{J_u} \frac{2\beta}{\gamma}|\mathbf{e}_1\cdot\nu_u| + \frac{2\beta}{\sqrt{3}}P(\gamma,\nu_u)\,d\mathcal{H}^1.$$

By Lemma 2.4.2 we obtain $\min\{Q(F) : \mathbf{e}_1^T F\mathbf{e}_1 = r\} = \frac{\alpha}{2}r^2$. Then using the slicing method (see Theorem A.1.5) we get

$$\mathcal{E}(u) \geq \int_0^1 \left(\int_0^l \frac{\alpha}{\sqrt{3}}\left(\mathbf{e}_1^T\nabla u(x_1,x_2)\mathbf{e}_1\right)^2\,dx_1 + \frac{2\beta}{\gamma}\#S^{x_2}(u)\right)\,dx_2 + \mathcal{E}^\gamma(u), \quad (5.18)$$

where $\#S^{x_2}$ denotes the number of jumps on a slice $(0,l)\times\{x_2\}$ and

$$\mathcal{E}^\gamma(u) = \int_{J_u} \frac{2\beta}{\sqrt{3}P(\gamma,\nu_u)}\,d\mathcal{H}^1.$$

In case $\#S^{x_2}(u) \geq 1$, the inner integral in (5.18) is obviously bounded from below by $\frac{2\beta}{\gamma}$. If $\#S^{x_2}(u) = 0$, by applyig Jensen's inequality we find that this term is bounded from below by $\alpha l a^2$ due to the boundary conditions. We thus obtain $\inf \mathcal{E} \geq \min\{\frac{\alpha l a^2}{\sqrt{3}}, \frac{2\beta}{\gamma}\}$. On the other hand, it is straighforward to check that $\mathcal{E}(u^{\mathrm{el}}) = \alpha l a^2$ and $\mathcal{E}(u^{\mathrm{cr}}) = \frac{2\beta}{\gamma}$, which shows that u^{el} is a minimizer for $a < a_{\mathrm{crit}}$ and u^{cr} is a minimizer for $a > a_{\mathrm{crit}}$. It remains to prove uniqueness:

(i) Let $a < a_{\mathrm{crit}}$ and u be a minimizer of $\mathcal{E}$. Since $\mathcal{E}(u) = \mathcal{E}(u^{\mathrm{el}})$ we infer from (5.18) that u has no jump on a.e. slice $(0,l) \times \{x_2\}$ and satisfies $\mathbf{e}_1^T \nabla u\, \mathbf{e}_1 = a$ a.e. by the imposed boundary values and strict convexity of the mapping $t \mapsto t^2$ on $[0,\infty)$. Thus, if $J_u \neq \emptyset$, a crack normal must satisfy $\nu_u = \pm \mathbf{e}_2$ $\mathcal{H}^1$-a.e. Taking $\mathcal{E}^\gamma(u)$ and the fact that $P(\gamma, \mathbf{e}_2) > 0$ for $\gamma \in [\frac{\sqrt{3}}{2}, 1]$ into account, we then may assume $J_u = \emptyset$ up to an $\mathcal{H}^1$ negligible set, i.e., $u \in H^1(\Omega)$. We find $u_1(x_1,x_2) = ax_1 + f(x_2)$ a.e. for a suitable function f, and the boundary condition $u_1(0,x_2) = 0$ yields $f = 0$ a.e. In particular, $\mathbf{e}_1^T \nabla u\, \mathbf{e}_2 = 0$ a.e. Applying strict convexity of Q on symmetric matrices (Lemma 2.4.2) we now observe $\mathbf{e}_2^T \nabla u\, \mathbf{e}_2 = -\frac{a}{3}$ and $\mathbf{e}_1^T \nabla u\, \mathbf{e}_2 + \mathbf{e}_2^T \nabla u\, \mathbf{e}_1 = 0$ a.e. So the derivative has the form

$$\nabla u(x) = \begin{pmatrix} a & 0 \\ 0 & -\frac{a}{3} \end{pmatrix} \quad \text{for a.e. } x.$$

Since Ω is connected, we conclude $u(x) = (0,s) + F^a x = u^{\mathrm{el}}(x)$ a.e.

(ii) Let $a > a_{\mathrm{crit}}$, $\phi \neq 0$ and u be a minimizer of $\mathcal{E}$. We again consider the lower bound (5.18) for the energy $\mathcal{E}$ and now obtain that on a.e. slice $(0,l) \times \{x_2\}$ a minimizer u has precisely one jump and that $\mathbf{e}_1^T \nabla u\, \mathbf{e}_1 = 0$ a.e. Now Lemma 2.4.2 shows that ∇u is antisymmetric a.e. As a consequence, the linearized rigidity estimate for SBD functions of Chambolle, Giacomini and Ponsiglione [25] yields that there is a Caccioppoli partition (E_i) of Ω such that

$$u(x) = \sum_i (A_i x + b_i)\chi_{E_i} \quad \text{and} \quad J_u = \bigcup_i \partial^* E_i \cap \Omega,$$

where $A_i^T = -A_i \in \mathbb{R}^{2\times 2}$ and $b_i \in \mathbb{R}^2$. (See Section A.2 for the definition and basic properties of Caccioppoli partitions.) As $\mathcal{E}^\gamma(u) = 0$, we also note that $\nu_u \perp \mathbf{v}_\gamma$ a.e. on J_u. Following the arguments in [57], in particular using regularity results for boundary curves of sets of finite perimeter and exhausting the sets $\partial^* E_i$ with Jordan curves, we find that

$$J_u = \bigcup_i \partial^* E_i \cap \Omega \subset (p,0) + \mathbb{R}\mathbf{v}_\gamma$$

for some p such that $(p,0) + \mathbb{R}\mathbf{v}_\gamma$ intersects both segments $(0,l) \times \{0\}$ and $(0,l) \times \{1\}$. We thus obtain that (E_i) consists of only two sets: E_1 to the left and E_2 to the right of $(p,0) + \mathbb{R}\mathbf{v}_\gamma$, say. Due to the boundary conditions we conclude that $A_1 = A_2 = 0$ and $b_1 = (0,s)$, $b_2 = (al, t)$ for suitable $s, t \in \mathbb{R}$.

(iii) Let $a > a_{\mathrm{crit}}$, $\phi = 0$ and u be a minimizer of $\mathcal{E}$. We follow the lines of the proof in (ii). The only difference is that $\mathcal{E}^\gamma(u) = 0$ now implies that $|\nu_u \cdot \mathbf{e}_1| \geq \frac{\sqrt{3}}{2}$

a.e. and then arguing similarly as before we obtain

$$J_u \subset g((0,1))$$

up to an $\mathcal{H}^1$-negligible set, where $g : (0,1) \to [0,l]$ is a Lipschitz function with $|g'| \leq \frac{1}{\sqrt{3}}$ a.e. We now conclude as in (ii). □

Part II

A quantitative geometric rigidity result in SBD and the derivation of linearized models from nonlinear Griffith energies

Chapter 6

The model and main results

The aim in Section 1.6 was the investigation of the convergence of energies in brittle fracture of the form (1.19). We have already seen that the analysis involves a simultaneous passage from discrete-to-continuum and from nonlinear to linearized elastic energies. It turned out that in the derivation of a small strain limit one has to face major difficulties concerning coercivity of the functionals due to the frame indifference of the energy density. The main goal of this part is the analysis of the passage from nonlinear to linearized Griffith models in a general framework. For the sake of simplicity and to avoid further complicacies of technical nature we treat the problem in a continuum setting in two dimensions.

Let $\Omega \subset \mathbb{R}^2$ open, bounded with Lipschitz boundary. Recall the properties of the space $SBV(\Omega, \mathbb{R}^2)$, frequently abbreviated as $SBV(\Omega)$ hereafter, in Section A.1. For $M > 0$ we define

$$SBV_M(\Omega) = \Big\{ y \in SBV(\Omega, \mathbb{R}^2) : \|y\|_\infty + \|\nabla y\|_\infty \leq M, \ \mathcal{H}^1(J_y) < +\infty \Big\}. \quad (6.1)$$

Here M may be chosen arbitrarily large (but fixed) and therefore the constraint $\|y\|_\infty + \|\nabla y\|_\infty \leq M$ is not a real restriction as we are interested in the small displacement regime in the regions of the domain where elastic behavior occurs. The uniform bound on the absolute continuous part of the gradient is natural when dealing with discrete energies where the corresponding deformations are piecewise affine on cells of microscopic size (see e.g. [17] or the construction of the interpolation in Section 5.1). Moreover, the uniform bound on the function is assumed only to simplify the exposition and may be dropped.

Let $W : \mathbb{R}^{2\times 2} \to [0, \infty)$ be a frame-indifferent stored energy density with $W(F) = 0$ iff $F \in SO(2)$. Assume that W is continuous, C^3 in a neighborhood of $SO(2)$ and scales quadratically at $SO(2)$ in the direction perpendicular to infinitesimal rotations. In other words, we have $W(F) \geq c\,\mathrm{dist}^2(F, SO(2))$ for all $F \in \mathbb{R}^{2\times 2}$ and a positive constant c. We briefly note that we can also treat inhomogeneous materials where the energy density has the form $W : \Omega \times \mathbb{R}^{2\times 2} \to [0, \infty)$. Moreover, it suffices to assume $W \in C^{2,\alpha}$, where $C^{2,\alpha}$ is the Hölder space with exponent $\alpha > 0$. For $\varepsilon > 0$ define the Griffith-energy $E_\varepsilon : SBV_M(\Omega) \to [0, \infty)$ by

$$E_\varepsilon(y) = \frac{1}{\varepsilon} \int_\Omega W(\nabla y(x))\, dx + \mathcal{H}^1(J_y). \tag{6.2}$$

We denote the small parameter occurring in the energy by ε to remind of functionals of the form (1.20). Having the application to discrete systems in mind, we will sometimes refer to ε as the 'atomic length scale'. For later we also introduce a relaxed energy functional. For $\rho > 0$, $\varepsilon > 0$ and $U \subset \Omega$ define $f_\varepsilon^\rho(x) = \min\{\frac{x}{\sqrt{\varepsilon\rho}}, 1\}$ and

$$E_\varepsilon^\rho(y, U) = \frac{1}{\varepsilon} \int_U W(\nabla y(x))\, dx + \int_{J_y \cap U} f_\varepsilon^\rho(|[y](x)|)\, d\mathcal{H}^1(x). \tag{6.3}$$

Clearly, we have $E_\varepsilon^\rho(y, U) \le E_\varepsilon(y)$ for all $y \in SBV_M(\Omega)$ and $U \subset \Omega$.

6.1 Rigidity estimates

We first concern ourselves with the question if the functionals E_ε can be related to a limiting functional for $\varepsilon \to 0$. We observe that for configurations with uniform bounded energy $E_\varepsilon(y_\varepsilon)$ the absolute continuous part of the gradient satisfies $\nabla y_\varepsilon \approx SO(2)$ as the stored energy density is frame-indifferent and minimized on $SO(2)$. Assuming that $y_\varepsilon \to y$ in L^1, one can show that $\nabla y \in SO(2)$ a.e. applying lower semicontinuity results for SBV functions (see [56]) and the fact that the quasiconvex envelope of W is minimized exactly on $SO(2)$ (see [67]).

A classical result due to Liouville states that a smooth function y satisfying the constraint $\nabla y \in SO(2)$ is a rigid motion. In the theory of fracture mechanics global rigidity can fail if the crack disconnects the body. More precisely, Chambolle, Giacomini and Ponsiglione have proven that for configurations which do not store elastic energy (i.e. $\nabla y \in SO(2)$ a.e.) and have finite Griffith energy (i.e. $\mathcal{H}^1(J_y) < +\infty$) the only way that rigidity may fail is that the body is divided into at most countably many parts each of which subject to a different rigid motion (see [25]).

Consequently, the limit of the sequence E_ε (in the sense of Γ-convergence) is given by the functional which is finite for piecewise rigid motions and measures the *segmentation energy* which is necessary to disconnect the body. The exact statement is formulated in Corollary 6.3.2 as a direct consequence of our main Γ-convergence result in Theorem 6.3.1.

To obtain a better understanding of the problem it is interesting to pass to rescaled configurations and to derive a limiting linearized energy as it was performed in [36] in the framework of nonlinear elasticity theory. The main ingredient in that analysis is a quantitative rigidity result due to Friesecke, James and Müller (see Theorem B.1). Extending the classical Liouville results it states that, loosely speaking, if the deformation gradient is close to $SO(2)$ (in L^2) then it is in fact close to one single rotation $R \in SO(2)$ (in L^2).

The first goal of this part of the thesis is to 'combine' the rigidity results of the pure elastic and pure brittle regime in order to derive a rigidity estimate for general Griffith functionals (6.2) where both energy forms are coexistent. Recall the notion of a *Caccioppoli partition* in Section A.2 and the definition of the *perimeter* $P(E, \Omega)$ of a set $E \subset \mathbb{R}^2$ in Ω (see (A.7)). Let $\Omega_\rho = \{x \in \Omega : \operatorname{dist}(x, \partial\Omega) > C\rho\}$ for $\rho > 0$ and for some sufficiently large constant C.

Theorem 6.1.1. *Let $\Omega \subset \mathbb{R}^2$ open, bounded with Lipschitz boundary. Let $M > 0$ and $0 < \eta, \rho \ll 1$. Then there is a constant $C = C(\Omega, M, \eta)$ and a universal $c > 0$ such that the following holds for $\varepsilon > 0$ small enough:*
For each $y \in SBV_M(\Omega, \mathbb{R}^2)$ with $\mathcal{H}^1(J_y) \le M$ and $\int_\Omega \operatorname{dist}^2(\nabla y, SO(2)) \le M\varepsilon$, there is an open set Ω_y with $|\Omega \setminus \Omega_y| \le C\rho$, a modification $\hat{y} \in SBV_{cM}(\Omega, \mathbb{R}^2)$ with $\|\hat{y} - y\|^2_{L^2(\Omega_y)} + \|\nabla\hat{y} - \nabla y\|^2_{L^2(\Omega_y)} \le C\varepsilon\rho$ and

$$E^\rho_\varepsilon(\hat{y}, \Omega_\rho) \le E_\varepsilon(y) + C\rho \tag{6.4}$$

with the following properties: We find a Caccioppoli partition $\mathcal{P} = (P_j)_j$ of Ω_ρ with $\sum_j P(P_j, \Omega_\rho) \le C$ and for each P_j a corresponding rigid motion $R_j\,x + c_j$, $R_j \in SO(2)$ and $c_j \in \mathbb{R}^2$, such that the function $u : \Omega \to \mathbb{R}^2$ defined by

$$u(x) := \begin{cases} \hat{y}(x) - (R_j\,x + c_j) & \text{for } x \in P_j \\ 0 & \text{for } x \in \Omega \setminus \Omega_\rho \end{cases} \tag{6.5}$$

satisfies the estimates

$$\begin{aligned} &(i)\ \ \mathcal{H}^1(J_u) \le C, && (ii)\ \ \|u\|^2_{L^2(\Omega_\rho)} \le \hat{C}\varepsilon, \\ &(iii)\ \ \sum\nolimits_j \|e(R_j^T \nabla u)\|^2_{L^2(P_j)} \le \hat{C}\varepsilon, && (iv)\ \ \|\nabla u\|^2_{L^2(\Omega_\rho)} \le \hat{C}\varepsilon^{1-\eta} \end{aligned} \tag{6.6}$$

for some constant $\hat{C} = \hat{C}(\rho)$, where $e(G) = \frac{G+G^T}{2}$ for all $G \in \mathbb{R}^{2\times 2}$.

This result will be addressed in Section 9. We remark that estimate (6.6) might be wrong without allowing for a small modification of the deformation as we show by way of example in Section 7.1. Moreover, we get a sufficiently strong bound only for the symmetric part of the gradient (see (iii)) which is not surprising due to the fact that there is no analogue of Korn's inequality in SBV. However, there is at least a weaker bound on the total absolutely continuous part of the gradient (see (iv)) which will essentially be needed to derive the Γ-convergence result. We emphasize that also (ii) is highly nontrivial as Poincaré's inequality cannot be applied due to the presence of discontinuity points.

In Section 10 we show that the qualitative piecewise rigidity result in two dimensions can be obtained as a corollary of Theorem 6.1.1.

Corollary 6.1.2. [Chambolle, Giacomini, Ponsiglione] *Let $y \in SBV(\Omega, \mathbb{R}^2)$ such that $\mathcal{H}^1(J_y) < +\infty$ and $\nabla y \in SO(2)$ a.e. Then y is a collection of an at most countable family of rigid deformations, i.e., there exists a Caccioppoli partition $\mathcal{P} = (P_j)_j$ subordinated to J_y such that*

$$y(x) = \sum\nolimits_j (R_j\, x + c_j)\chi_{P_j}(x),$$

where $R_j \in SO(2)$ and $c_j \in \mathbb{R}^2$.

There is also a linearized version of Theorem 6.1.1 which can be interpreted as a 'piecewise Korn-Poincaré-inequality in SBD'. Let $\mathbb{R}^{2\times 2}_{\rm skew} = \{A \in \mathbb{R}^{2\times 2} : A^T = -A\}$ be the set of skew symmetric matrices. Set

$$F^\rho_\varepsilon(y, U) = \frac{1}{\varepsilon}\int_U V(e(\nabla u)(x))\, dx + \int_{J_u \cap U} f^\rho_\varepsilon(|[u]|)\, d\mathcal{H}^1 \tag{6.7}$$

for a coercive quadratic form V, i.e. $V(G) \geq c|G|^2$ for $c > 0$ and $G \in \mathbb{R}^{2\times 2}_{\rm sym}$. Furthermore, define $F_\varepsilon = F^0_\varepsilon(\cdot, \Omega)$, where $f^0_\varepsilon \equiv 1$. For the definition of the space SBD we refer to Section A.1.

Theorem 6.1.3. *Let $\Omega \subset \mathbb{R}^2$ open, bounded with Lipschitz boundary. Let $M > 0$, and $0 < \rho \ll 1$. Then there is a constant $C = C(\Omega, M)$ such that for $\varepsilon > 0$ small enough the following holds:*
For each $u \in SBD^2(\Omega, \mathbb{R}^2) \cap L^2(\Omega, \mathbb{R}^2)$ with $\mathcal{H}^1(J_u) \leq M$ and

$$\int_\Omega |e(\nabla u)(x)|^2\, dx \leq M\varepsilon,$$

there is an open set Ω_u with $|\Omega \setminus \Omega_u| \leq C\rho$, a modification $\hat{u} : \Omega \to \mathbb{R}^2$ with $\|\hat{u} - u\|^2_{L^2(\Omega_u)} + \|e(\nabla \hat{u} - \nabla u)\|^2_{L^2(\Omega_u)} \leq C\rho\varepsilon$ and

$$F^\rho_\varepsilon(\hat{u}, \Omega_\rho) \leq F_\varepsilon(u) + C\rho$$

with the following properties: We find a Caccioppoli partition $\mathcal{P} = (P_j)_j$ of Ω_ρ with $\sum_j P(P_j, \Omega_\rho) \leq C$ and for each P_j a corresponding infinitesimal rigid motion $A_j\, x + c_j$, $A_j \in \mathbb{R}^{2\times 2}_{\rm skew}$ and $c_j \in \mathbb{R}^2$, such that $\mathcal{H}^1(J_{\hat{u}}) \leq C$ and

$$(i)\ \|e(\nabla \hat{u})\|^2_{L^2(\Omega_\rho)} \leq C\varepsilon, \quad (ii)\ \sum\nolimits_j \|\hat{u} - (A_j \,\cdot - c_j)\|^2_{L^2(P_j)} \leq \hat{C}\varepsilon. \tag{6.8}$$

for some constant $\hat{C} = \hat{C}(\rho)$.

The main technical results to prove the linearized version are addressed in Section 8 where we establish a local Korn-Poincaré-type inequality. The estimates in the linearized regime are crucial for the derivation of the nonlinear rigidity result in Theorem 6.1.1 which can be compared with the fact that in the proof of the geometric rigidity result in nonlinear elasticity (Theorem B.1) the usage of Korn's inequality is essential.

6.2 Compactness

For a given (ordered) Caccioppoli partition $\mathcal{P} = (P_j)_j$ of Ω let

$$\mathcal{R}(\mathcal{P}) = \Big\{T : \Omega \to \mathbb{R}^2 : T(x) = \sum\nolimits_j \chi_{P_j}(R_j\,x + c_j), R_j \in SO(2), c_j \in \mathbb{R}^2\Big\} \quad (6.9)$$

be the set of corresponding piecewise rigid motions. Likewise we define the set of piecewise infinitesimal rigid motions, denoted by $\mathcal{A}(\mathcal{P})$, replacing $R_j \in SO(2)$ by $A_j \in \mathbb{R}^{2\times 2}_{\text{skew}}$. Moreover, we define the triples

$$\begin{aligned}
\mathcal{D} &:= \big\{(u, \mathcal{P}, T) : \; u \in SBV(\Omega), \; \mathcal{P} \text{ C.-partition of } \Omega, \; T \in \mathcal{R}(\mathcal{P})\big\},\\
\mathcal{D}_\infty &:= \big\{(u, \mathcal{P}, T) : \; \mathcal{P} \text{ C.-partition of } \Omega, \; T \in \mathcal{R}(\mathcal{P}), \; (\nabla T)^T u \in GSBD^2(\Omega, \mathbb{R}^2)\big\}.
\end{aligned}$$

The space $GSBD^2(\Omega, \mathbb{R}^2)$, abbreviated by $GSBD^2(\Omega)$ hereafter, generalizes the definition of the space $SBD(\Omega, \mathbb{R}^2)$ based on certain slicing properties, see Section A.1. We now formulate the main compactness theorem.

Theorem 6.2.1. *Let $\Omega \subset \mathbb{R}^2$ open, bounded with Lipschitz boundary. Let $M > 0$ and $\varepsilon_k \to 0$. If $E_{\varepsilon_k}(y_k) \le C$ for a sequence $y_k \in SBV_M(\Omega)$, then there exists a subsequence (not relabeled) such that the following holds:*
There are triples $(u_k, \mathcal{P}^k, T_k) \in \mathcal{D}$, where $\mathcal{P}^k = (P_j^k)_j$ and

$$\begin{aligned}
&(i) \;\; u_k(x) - \varepsilon_k^{-1/2}(y_k(x) - T_k(x)) \to 0 \; a.e.,\\
&(ii) \;\; \frac{1}{\varepsilon_k}\int_\Omega W(\mathbf{Id} + \sqrt{\varepsilon_k}\nabla T_k^T \nabla u_k) \le \frac{1}{\varepsilon_k}\int_\Omega W(\nabla y_k) + o(1)
\end{aligned} \quad (6.10)$$

for $\varepsilon_k \to 0$, such that we find a limiting triple $(u, \mathcal{P}, T) \in \mathcal{D}_\infty$ with

$$\begin{aligned}
&(i) \;\; \chi_{P_j^k} \to \chi_{P_j} \quad \text{in measure for all } j \in \mathbb{N},\\
&(ii) \;\; T_k \to T \text{ in } L^2(\Omega), \quad \nabla T_k \to \nabla T \text{ in } L^2(\Omega),
\end{aligned} \quad (6.11)$$

for $k \to \infty$. Moreover, we get

$$\begin{aligned}
&(i) \;\; u_k \to u \quad a.e. \text{ in } \Omega,\\
&(ii) \;\; e(\nabla T_k^T \nabla u_k) \rightharpoonup e(\nabla T^T \nabla u) \quad \text{weakly in } L^2(\Omega, \mathbb{R}^{2\times 2}_{\text{sym}}),\\
&(iii) \;\; \|\nabla u_k\|_{L^\infty(\Omega)} \le C\varepsilon_k^{-1/8},
\end{aligned} \quad (6.12)$$

for $k \to \infty$ and for the surface energy we obtain

$$\liminf_{k\to\infty} \mathcal{H}^1(J_{y_k}) \ge \frac{1}{2}\sum\nolimits_j P(P_j, \Omega) + \mathcal{H}^1(J_u \setminus \partial P), \quad (6.13)$$

where $\partial P := \bigcup_j \partial^* P_j$.

Here ∂^* denotes the *essential boundary* (see (A.8)). If we drop the condition $\|y\|_\infty \le M$ in the definition of $SBV_M(\Omega)$, then (6.11) only holds for the derivatives of the piecewise rigid motions. In the following we say a triple $(u_k, \mathcal{P}^k, T_k) \in \mathcal{D}$ converges to $(u, \mathcal{P}, T) \in \mathcal{D}_\infty$ and write $(u_k, \mathcal{P}^k, T_k) \to (u, \mathcal{P}, T)$ if (6.10)-(6.13) are satisfied. Of course, the partition of the limiting configuration is not unique as the following simple example shows.

Example 6.2.2. Consider $\Omega = (0,1) \times (0,1)$, $\Omega_1 = (0,1) \times (0, \frac{1}{2})$, $\Omega_2 = (0,1) \times (\frac{1}{2}, 1)$ and

$$y_k = \mathbf{id}\chi_{\Omega_1} + (\mathbf{id} + a\sqrt{\varepsilon_k})\chi_{\Omega_2}$$

for $a \in \mathbb{R}^2$. Then possible alternatives are e.g. (1) $P^1 = \Omega$ with $R_1^1 x + c_1^1 = \mathbf{id}$ or (2) $P_1^2 = \Omega_1$, $P_2^2 = \Omega_2$ with $R_1^2 x + c_1^2 = \mathbf{id}$ and $R_2^2 x + c_2^2 = \mathbf{id} + a\sqrt{\varepsilon_k}$. Letting $u_k^1 = {\varepsilon_k}^{-\frac{1}{2}}(y_k - \mathbf{id})$ and $u_k^2 = {\varepsilon_k}^{-\frac{1}{2}}(y_k - \mathbf{id}\chi_{\Omega_1} - (\mathbf{id} + a\sqrt{\varepsilon_k})\chi_{\Omega_2})$ we obtain in the limit $\varepsilon_k \to 0$ two different configurations:

$$\begin{aligned} u^1 &= 0 \cdot \chi_{\Omega_1} + a\chi_{\Omega_2}, \quad P_1^1 = \Omega, \\ u^2 &= 0, \quad P_1^2 = \Omega_1, P_2^2 = \Omega_2. \end{aligned}$$

Clearly, we can equally well consider an example where we vary the rotations, e.g.

$$y_k(x) = \begin{pmatrix} 1 & 0 \\ 0 & 1 \end{pmatrix} x \ \chi_{\Omega_1}(x) + \begin{pmatrix} \cos a\sqrt{\varepsilon_k} & \sin a\sqrt{\varepsilon_k} \\ -\sin a\sqrt{\varepsilon_k} & \cos a\sqrt{\varepsilon_k} \end{pmatrix} x \ \chi_{\Omega_2}(x)$$

for $a \in \mathbb{R}$.

We now introduce a special subclass of partitions in which uniqueness will be guaranteed. The above example already shows that different partitions are not equivalent in the sense that they may contain a different 'amount of information'. Note that on the various elements of the partition the configuration u is defined separately and the different pieces of the domain are not 'aware of each other'. In particular, the possible discontinuities of u on ∂P do not have any physically reasonable interpretation. On the contrary, in the first example where we did not split up the domain, we gain the jump height as an additional information. The observation that coarser partitions provide more information about the behavior at the jump set motivates the definition of the *coarsest partition.*

Definition 6.2.3. Let $(y_k)_k$ be a given (sub-)sequence as in Theorem 6.2.1.

(i) We say a partition $\mathcal{P}$ of Ω is *admissible* for $(y_k)_k$ and write $\mathcal{P} \in \mathcal{Z}_P((y_k)_k)$ if there are triples $(u_k, \mathcal{P}^k, T_k) \in \mathcal{D}$ for $k \in \mathbb{N}$ as well as u, T such that $(u, \mathcal{P}, T) \in \mathcal{D}_\infty$ and (6.10)-(6.13) hold.

(ii) We say a piecewise rigid motion T is *admissible* for $(y_k)_k$ and $\mathcal{P}$ writing $T \in \mathcal{Z}_T((y_k)_k, \mathcal{P})$ if there are triples $(u_k, \mathcal{P}^k, T_k) \in \mathcal{D}$ for $k \in \mathbb{N}$ as well as u such that $(u, \mathcal{P}, T) \in \mathcal{D}_\infty$ and (6.10)-(6.13) hold.

(iii) We say a configuration u is *admissible* for $(y_k)_k$ and $\mathcal{P}$ and write $u \in \mathcal{Z}_u((y_k)_k, \mathcal{P})$ if there are triples $(u_k, \mathcal{P}^k, T_k) \in \mathcal{D}$ for $k \in \mathbb{N}$ as well as T such that $(u, \mathcal{P}, T) \in \mathcal{D}_\infty$ and (6.10)-(6.13) hold.

(iv) We say a partition $\mathcal{P}$ of Ω is a *coarsest partition* for $(y_k)_k$ if the following holds: The partition is admissible, i.e. $\mathcal{P} \in \mathcal{Z}_P((y_k)_k)$, and for all admissible $u \in \mathcal{Z}_u((y_k)_k, \mathcal{P})$ the corresponding piecewise rigid motions $T_k = \sum_j (R_j^k x + c_j^k)\chi_{P_j^k}$ given by (iii) satisfy

$$\frac{|R_{j_1}^k - R_{j_2}^k| + |c_{j_1}^k - c_{j_2}^k|}{\sqrt{\varepsilon_k}} \to \infty \tag{6.14}$$

for all $j_1, j_2 \in \mathbb{N}$, $j_1 \neq j_2$ and $k \to \infty$.

In Lemma 10.2.2 below we find an equivalent characterization of coarsest partitions being the maximal elements of the partial order on the sets of admissible partitions which is induced by subordination. Loosely speaking, the above definition particularly implies that given a coarsest partition a region of the domain is partitioned into different sets $(P_j)_j$ if and only if the jump height of the approximating sequence u_k tends to infinity on $(\partial^* P_j)_j$.

Recall the definition of the piecewise infinitesimal rigid motions $\mathcal{A}(\mathcal{P})$ in (6.9). We now obtain a unique characterization of the limiting configuration up to piecewise infinitesimal rigid motions.

Theorem 6.2.4. *Let $\varepsilon_k \to 0$ be given. Let $E_{\varepsilon_k}(y_k) \le C$ for a sequence $y_k \in SBV_M(\Omega)$ and let $(y_{k_n})_{n\in\mathbb{N}}$ be a subsequence for which the assertion of Theorem 6.2.1 holds. Then we have the following:*

(i) There is a unique $T \in \mathcal{Z}_T((y_{k_n})_n, \mathcal{P})$ for all $\mathcal{P} \in \mathcal{Z}_P((y_{k_n})_n)$.

(ii) There is a unique coarsest partition $\bar{\mathcal{P}}$ of Ω.

(iii) Given some $u \in \mathcal{Z}_u((y_{k_n})_n, \bar{\mathcal{P}})$ all possible limiting configurations are of the form $u + \nabla T \mathcal{A}(\bar{\mathcal{P}})$, i.e. the limiting configuration is determined uniquely up to piecewise infinitesimal rigid motions.

Remark 6.2.5. Although not stated explicitly, for problems of the form (6.7) we can derive a compactness result similar to Theorem 6.2.1. This allows to solve more general variational problems for fracture mechanics in the realm of linearized elasticity. For technical reasons dealing with energy functionals with the main energy term

$$\int_\Omega |e(\nabla u)(x)|^2\,dx + \mathcal{H}^1(J_u) \tag{6.15}$$

often an a priori L^∞ bound is imposed in the literature (see e.g. [8, 26, 66] or Section 1.6.1) such that compactness results in SBD can be applied. Possible alternatives are to add a term of the form $\int_{J_u} |[u] \odot \nu_u| \, d\mathcal{H}^1$ giving control over the jump height (see e.g. [8]).

Recently, the space of *generalized functions of bounded deformation* was introduced to overcome this difficulty. In this framework it suffices to assume an L^1 bound on the function u similarly as in the compactness results for GSBV, i.e. variational problems for energy functionals of the form (6.15) with an additional term $\|u\|_{L^1(\Omega)}$ are treatable. In fact, in many situations such a lower order term is present, see e.g. [4, 54]. However, there are also applications where the existence of lower order terms can not be expected such as the work at hand which deals with the passage to rescaled configurations. Moreover, in a wide class of problems arising from discrete energies one typically does not have an L^1 bound for the functions as the energies only depend on the relative distance of the material points.

The aforementioned result sheds a new light on this problem. Exploiting condition (6.8)(ii) we may derive a compactness result for energies (6.15) without any extra term by subtracting suitable infinitesimal rigid motions on a partition of the domain.

6.3 Γ-convergence and application to cleavage laws

We now show that the energies E_ε converge to a Griffith functional with linearized elastic energy. Let $Q = D^2W(\mathbf{Id})$ be the Hessian of the stored energy density W at the identity. Define $E : \mathcal{D}_\infty \to [0, \infty)$ by

$$E(u, \mathcal{P}, T) = \int_\Omega \frac{1}{2} Q(e(\nabla T^T \nabla u)) + \frac{1}{2} \sum\nolimits_j P(P_j, \Omega) + \mathcal{H}^1(J_u \setminus \partial P), \qquad (6.16)$$

where as before $\mathcal{P} = (P_j)_j$ and $\partial P = \bigcup_j \partial^* P_j$. The surface energy of the limiting functional has two parts. We call the left part *segmentation energy* and the right part *inner crack energy*. Recall that we say $(u_k, \mathcal{P}^k, T_k) \to (u, \mathcal{P}, T)$ if (6.10)-(6.13) hold.

Theorem 6.3.1. *Let $\varepsilon_k \to 0$. Then E_{ε_k} Γ-converge to E with respect to the convergence given in Theorem 6.2.1, i.e.*

(i) $\Gamma - \liminf$ inequality: For all $(u, \mathcal{P}, T) \in \mathcal{D}_\infty$ and for all sequences $(y_k)_k \subset SBV_M(\Omega)$ and corresponding $(u_k, \mathcal{P}^k, T_k) \in \mathcal{D}$ as given in Theorem 6.2.1 such that $(u_k, \mathcal{P}^k, T_k) \to (u, \mathcal{P}, T)$ we have

$$\liminf_{k \to \infty} E_{\varepsilon_k}(y_k) \geq E(u, \mathcal{P}, T).$$

(ii) *Existence of recovery sequences: For every* $(u, \mathcal{P}, T) \in \mathcal{D}_\infty$ *with* $u \in L^2(\Omega)$ *we find a sequence* $(y_k)_k \subset SBV_M(\Omega)$ *and corresponding* $(u_k, \mathcal{P}^k, T_k) \in \mathcal{D}$ *such that* $(u_k, \mathcal{P}^k, T_k) \to (u, \mathcal{P}, T)$ *and*

$$\lim_{k \to \infty} E_{\varepsilon_k}(y_k) = E(u, \mathcal{P}, T).$$

As a direct consequence we get that the Γ-limit is given by the segmentation energy if we do not pass to rescaled configurations.

Corollary 6.3.2. *Let* $\varepsilon_k \to 0$. *Then* E_{ε_k} *Γ-converge to* E_{seg} *with respect to the* $L^1(\Omega)$*-convergence, where*

$$E_{\text{seg}}(y) = \begin{cases} \frac{1}{2}\sum_j P(P_j, \Omega) & y = T \in \mathcal{R}(\mathcal{P}) \text{ for a Caccioppoli partition } \mathcal{P}, \\ +\infty & \text{else.} \end{cases}$$

Finally, as an application of the above results we return to the investigation of cleavage laws (see Sections 1.3, 1.4, 1.6.2). We consider a special boundary value problem of uniaxial compression/extension. Let $\Omega = (0, l) \times (0, 1)$, $\Omega' = (-\eta, l + \eta) \times (0, 1)$ for $l > 0$, $\eta > 0$ and for $a_\varepsilon \in \mathbb{R}$ define

$$\mathcal{A}(a_\varepsilon) := \{y \in SBV_M(\Omega') : y_1(x) = (1 + a_\varepsilon)x_1 \text{ for } x_1 \leq 0 \text{ or } x_1 \geq l\}.$$

As usual in the theory of SBV functions the boundary values have to be imposed in small neighborhoods of the boundary. In what follows the elastic part of the energy (6.2) still only depends on $y|_\Omega$, whereas the surface energy is given by $\mathcal{H}^1(J_y)$ with $J_y \subset \Omega'$. In particular, jumps on $\{0, l\} \times (0, 1)$ contribute to the energy $E_\varepsilon(y)$ (compare also the discussion before Theorem 1.6.3). The present problem in the framework of continuum fracture mechanics with isotropic surface energies is a slightly simplified model of the version considered in Section 1.6.2.

As a preparation we define $\alpha = \frac{\det Q}{\det \hat{Q}}$, where $\hat{Q}$ arises from Q by deleting the first row and column (see Lemma 2.1.2). Moreover, let $F^a \in \mathbb{R}^{2\times 2}_{\text{sym}}$ be the unique matrix such that $\mathbf{e}_1^T F^a \mathbf{e}_1 = a$ and $Q(F^a) = \inf\{Q(F) : \mathbf{e}_1^T F \mathbf{e}_1 = a\} = \alpha a^2$.

We recall that the proof of the result in Theorem 1.3.1 and the special case in Theorem 1.4.1 fundamentally relied on the application of certain slicing techniques which were not suitable to treat the case of compression. Having general compactness and Γ-convergence results we can now complete the picture found in Section 1.6.2 by extending the results to the case of uniaxial compression.

Theorem 6.3.3. *Suppose* $a_\varepsilon/\sqrt{\varepsilon} \to a \in [-\infty, \infty]$. *The limiting minimal energy is given by*

$$\lim_{\varepsilon \to 0} \inf\{E_\varepsilon(y) : y \in \mathcal{A}(a_\varepsilon)\} = \min\Big\{\frac{1}{2}\alpha l a^2, 1\Big\}. \tag{6.17}$$

Let $a_{\text{crit}} := \sqrt{\frac{2\alpha}{l}}$. *For every sequence* $(y_\varepsilon)_\varepsilon$ *of almost minimizers, up to passing to subsequences, we get* $\varepsilon^{-1/2}(y_\varepsilon(x) - x) \to u(x)$ *for a.e.* $x \in \Omega$, *where*

(i) if $|a| < a_{\text{crit}}$*,* $u(x) = (0, s) + F^a x$ *for* $s \in \mathbb{R}$*,*

(ii) if $|a| > a_{\text{crit}}$*,* $u(x) = \begin{cases} (0, s) & x_1 < p, \\ (la, t) & x_1 > p, \end{cases}$ *for* $s, t \in \mathbb{R}$*,* $p \in (0, l)$*.*

6.4 Overview of the proof

As the proof of the rigidity result is very long and technical, we present here a short overview and highlight the principal strategy. The first idea is to replace the highly nonlinear problem of Theorem 6.1.1 by a linearized version (cf. Theorem 6.1.3) which is easier to treat since (1) the estimate only involves the function itself and not its derivative and (2) the set of infinitesimal rigid motions is a linear space in contrast to $SO(2)$. This estimate will be an essential ingredient to establish the general nonlinear result afterwards.

6.4.1 Korn-Poincaré-type inequality

Essentially, (6.8)(ii) can be seen as a kind of Korn-Poincaré inequality. The classical Korn-Poincaré inequality in BD states that there is a constant C depending only on the domain $\Omega \subset \mathbb{R}^2$ such that

$$\|u - Pu\|_{L^2(\Omega)} \leq C|Eu|(\Omega) \tag{6.18}$$

for all $u \in BD(\Omega, \mathbb{R}^2)$, where P is a linear projection onto the space of infinitesimal rigid motions and $|Eu|$ denotes the total variation of the symmetrized distributional derivative (see Theorem B.4). Assuming that the linearized elastic energy is given by $\|e(\nabla u)\|^2_{L^2(\Omega)} \sim \varepsilon$, the aim is to show that the right hand side in (6.18) is of order ε. Clearly, this cannot be inferred directly by energy bounds since we only have control over the size of the discontinuity set which typically satisfies $\mathcal{H}^1(J_u) \sim 1$. The fundamental idea is to show that the jump height satisfies $[u] \sim \sqrt{\varepsilon}$ in a suitable sense, whence the classical Korn-Poincaré inequality can be applied to obtain the desired result in (6.8)(ii).

However, it turns out that in general there is no hope to control the jump heights in terms of $\sqrt{\varepsilon}$. This can already seen by rather simple configurations where the jump set is given by the boundary of balls (cf. the example in [5] demonstrating that BD does not embed into BV). Counterexamples can even be constructed in a way such that the domain is not disconnected by the jump set into different connected components as we show in Section 7.1. The main strategy of the work at hand is to find bounds on the jump heights after a suitable modification of the jump set and the displacement field whose total energy almost coincides with the original energy. An exact statement of the Korn-Poincaré-type inequality can be found in Theorem 8.0.1 below.

By a density argument we can assume that the jump set is contained in a finite number of rectangle boundaries. (We will call these sets boundary components or cracks in the following.) These boundaries will be altered during an iterative procedure. Clearly, we have to assure that in this process the length of the boundary components does not increase too much. To this end, it is convenient to measure the length of the jump set by a convex combination of the Hausdorff-measure $\mathcal{H}^1$ and the 'diameter' of a crack given by

$$|\Gamma|_\infty = \sqrt{|\pi_1\Gamma|^2 + |\pi_2\Gamma|^2}, \tag{6.19}$$

where Γ denotes the boundary component and π_1, π_2 the orthogonal projections onto the coordinate axes. One of the advantages of $|\cdot|_\infty$ in contrast to $\mathcal{H}^1$ is that due to the strict convexity of $|\cdot|_\infty$ it is often energetically favorable if different cracks are combined to one larger boundary component leading to a simplification of the jump set.

Nevertheless, during the modification process it cannot be avoided that additional cracks are added near original ones. To keep track of this amplification, the boundary components have to be assigned with a weight which indicates if (or: how much) this crack has already been 'used' to introduce another discontinuity set. Now a further difficulty arises from the the fact that during each iteration step of the modification these weights have to be carefully adjusted (see Section 8.2).

The overall aim of the modification is to assure that in a small neighborhood of a boundary component Γ the energy can be controlled. Indeed, it turns out that if the elastic energy exceeds $\varepsilon|\Gamma|_\infty$ or the size of the jump set in a neighborhood is much larger than $|\Gamma|_\infty$, then it is energetically favorable to replace the crack by a larger rectangle and to replace the function u in the interior of the rectangle by an infinitesimal rigid motion (see proof of Theorem 8.4.2a),b)). Moreover, the modification of the jump set occurs not only due to energetic but also due to geometrical reasons. Exploiting the properties of $|\cdot|_\infty$ we can find a finer characterization of the cracks in the neighborhood, e.g. one can show that there are at most two cracks whose size is comparable to $|\Gamma|_\infty$ (see Corollary 8.3.4). Furthermore, we can always find small stripes in the neighborhood which do not intersect the jump set (see Lemma 8.3.5).

Having these properties for the neighborhood of a rectangle Γ and assuming that for all smaller cracks Γ_l we have already established that $[u] \sim \sqrt{|\Gamma_l|_\infty \varepsilon}$, the main technical issue is to derive a trace estimate on the boundary of Γ. Then replacing the function u by an appropriate infinitesimal rigid motion in the interior of the rectangle we will indeed obtain $[u] \sim \sqrt{|\Gamma|_\infty \varepsilon}$ on the boundary of Γ. Consequently, the assertion can be proved using an algorithm which iteratively changes the jump set and determines the trace at boundary components once the required conditions in a neighborhood are fulfilled (see Theorem 8.4.2).

Obviously one expects that the crack opening of small cracks is generically small. In our framework this heuristically follows by a rescaling argument in

(6.18) and the observation that after modification the energy in a neighborhood is bounded by $\sim |\Gamma|_\infty \varepsilon$. However, a rigorous investigation of the trace on the boundary of Γ is very subtle as due to the iterative application of the arguments the involved constants might become arbitrarily large. The proof will be carried out in several steps.

In the first step we assume that in the neighborhood N of Γ only small cracks Γ_l are present. This indeed induces that $|Eu|(N)$ is sufficiently small as on each Γ_l we have already shown $[u] \sim \sqrt{|\Gamma_l|_\infty \varepsilon}$. In general, the idea is to construct thin long paths in N which avoid cracks being to large. We then first measure the distance of the function from an infinitesimal rigid motion only on this path and may apply this result to estimate the distance in the whole set N afterwards (see Section 8.5.2). A major drawback of such a technique is that the constant in (6.18) crucially depends on the domain and explodes for sets getting arbitrarily thin. Consequently, in this context we have to carry out a careful quantitative analysis how the constant in (6.18) depends on the shape of the domain (see Section 7.1).

It turns out, however, that the paths in general cannot be selected in a way such that they only intersect sufficiently small cracks. Nevertheless, it can be shown that boundary components being too large for a direct application of the above ideas occupy only a comparably small region. In this region we then do not use the Korn-Poincaré inequality (Theorem B.4), but circumvent the estimation of the surface energy by a slicing technique. Indeed, by the modification procedure alluded to above we always find small stripes in the neighborhood which do not intersect the jump set at all. The assertion then follows as this exceptional set can then taken arbitrarily small by an iterative application of the slicing method (see Section 8.5.3). We briefly note that such a technique is only employable as we treat a linear problem. This is one of the main reasons why the derivation of Theorem 6.1.3 is easier than the proof of Theorem 6.1.1. (Compare also [25], where the treatment of the linearized version is remarkably easier than the nonlinear problem due to the applicability of a slicing method.)

Finally, one has to face the problem that there are (at most) two other cracks Γ_1, Γ_2 intersecting N being larger than Γ. In particular, (6.18) cannot be directly applied since no estimate of the jump heights at Γ_1 and Γ_2 is available. However, the result can also be established in this case if the elastic and surface energy in the two areas close to Γ, Γ_1 and Γ, Γ_2 is sufficiently small (see Section 8.5.5). In fact, such a smallness assumption can always be inferred by a careful modification of the crack set (see proof of Theorem 8.4.2c) and Section 8.3.2). Finally, we remark that the result crucially depends on the application of a suitable L^2- trace theorem for SBV functions (see Lemma 7.2) which can be established in our framework because of the sufficiently regular jump set. Moreover, it is essential that there are at most two large cracks in a neighborhood. Already with three or four cracks the configurations might be significantly less rigid.

6.4.2 SBD-rigidity

The main estimates in the rigidity result (see (6.6)) do not only provide bounds for the function itself but also for the derivative. The key point is the derivation of an estimate for the symmetric part of the gradient. Using the expansion

$$|e(R^T(\nabla y - \mathbf{Id}))|^2 = \operatorname{dist}^2(\nabla y, SO(2)) + O(|\nabla y - R|^4) \tag{6.20}$$

and recalling that $\|\operatorname{dist}(\nabla y, SO(2))\|^2_{L^2(\Omega)} \sim \varepsilon$ we see that it suffices to establish an estimate of fourth order. Indeed, also in the proof of the geometric rigidity result in nonlinear elasticity (see [49]) one first derives a bound for $\|\nabla y - R\|^4_{L^4(\Omega)}$ to control the symmetric part. The control over the full gradient is then obtained by Korn's inequality.

Clearly, in our framework the result in Theorem B.1 cannot be applied due to the presence of cracks, in particular $\Omega \setminus J_y$ will generically not be a Lipschitz set. Therefore, by a density argument we again first assume that the jump set is contained in a finite number of rectangle boundaries. A careful quantitative analysis shows that the constant in (B.1) depends on the quotient of the diameter of the domain, denoted by k, and the minimal distance of two cracks, denoted by s. In particular, $C = C(k/s) \sim 1$ if $k \sim s$. Provided that $\frac{k}{s}$ is not too large, the principal strategy will be to show that possibly after a modification we get $\|\nabla y - R\|^2_{L^\infty(\Omega)} \le (C(k/s))^{-1}$ which then gives

$$\|e(R^T(\nabla y - \mathbf{Id}))\|^2_{L^2(\Omega)} \le \varepsilon + (C(k/s))^{-1}\|\nabla y - R\|^2_{L^2(\Omega)} \le C\varepsilon \tag{6.21}$$

by (6.20) and Theorem B.1. Of course, in general we cannot suppose that $\frac{k}{s}$ is not large. Moreover, a global rigidity result may fail due to the separation of the domain by the jump set. Consequently, we will apply the presented ideas on a fine partition of the Lipschitz domain Ω consisting of squares with diameter k. This local result will be used to modify the jump set such that the minimal distance of each pair of cracks increases. Then we can repeat the arguments for a larger k. The idea is that after an iterative application of the arguments we obtain an estimate for $k \approx \rho$ which then will provide rigid motions on the connected components of the domain (see (6.5)) with the desired properties.

In Section 9.2 we construct piecewise constant $SO(2)$-valued mappings approximating the deformation gradient. In each square Q of diameter k we may assume that the elastic energy is bounded by $\sim \varepsilon k$ as otherwise it would be energetically favorable to introduce jumps at the boundary of the square and to replace the deformation in the interior by a rigid motion. (The same technique has been used in the proof of the Korn-Poincaé inequality.) Similarly as in [49] we pass to the harmonic part of the deformation (denoted by $\hat{y}$) and obtain by the mean value property

$$\begin{aligned}\|\nabla \hat{y} - R_Q\|^2_{L^\infty(\hat{Q})} &\le Ck^{-2}\|\nabla \hat{y} - R_Q\|^2_{L^2(Q)} \\ &\le C(k/s)k^{-2}\|\operatorname{dist}(\nabla y, SO(2))\|^2_{L^2(Q)} \le C(k/s)k^{-1}\varepsilon\end{aligned} \tag{6.22}$$

for a suitable $R_Q \in SO(2)$, where $\hat{Q} \subset Q$ is a slightly smaller square. Consequently, if we can assure that $\frac{\varepsilon}{k} \le (C(k/s))^{-2}$ we obtain the desired L^∞-bound which allows to derive an estimate of the form (6.21). We note that for this argument we at least have to assume that $k \gg \varepsilon$ which will be denoted as the 'superatomistic regime' (recall the discussion about the signification of ε after (6.2)).

In the subsequent Section 9.2.2 we show that not only the distance of the derivative from a piecewise rigid motion can be controlled but also the distance of the function itself. On the one hand this is essential for (6.6), on the other hand such an estimate is crucial for establishing a modification of the deformation and the jump set. The main idea is to apply the Korn-Poincaré-type inequality on the function $R_Q^T y - \mathbf{id}$. Major difficulties arise from the facts that the rotation R_Q may vary from one square to another and that the inequality derived in Section 8 only provides a local estimate (see formulation in Theorem 8.4.8). Consequently, the arguments have to be repeated for several shifted copies of the fine partition (see Lemma 9.2.3). Moreover, the projections P_Q onto the the space of infinitesimal rigid motions (see (6.18)) have to be combined with the rotations R_Q in a suitable way to obtain appropriate rigid motions, which do not vary too much on adjacent squares (see Lemma 9.2.5).

Having an approximation of the deformation by piecewise rigid motions defined on squares with diameter k, we then are able to modify the function such that the minimal distance $\tilde{s}$ of two cracks of the new configuration satisfies $\tilde{s} \sim k$ (see Lemma 9.3.1). Now we can repeat the above procedure for some larger $\tilde{k}$ such that $\varepsilon/\tilde{k} \le (C(\tilde{k}/\tilde{s}))^{-2}$ is guaranteed and we can repeat the arguments in (6.22).

The strategy is to end up with $k \approx \rho$ after a finite number of iterations. As the number of iteration steps is not bounded but grows logarithmically with $\frac{1}{\varepsilon}$ we have to assure that in each step the surface and the elastic energy do not increase to much. The crucial point is that during the iteration process the coarseness of the partition k grows much faster than the stored elastic energy ε such that the argument in (6.22) may be repeated. The details are given in Theorem 9.4.3. Having an estimate for $k \approx \rho$ it is then not hard to establish the desired result up to a small exceptional set (see Theorem 9.4.2).

Clearly, we cannot assume that initially $s \ge \varepsilon$. In this case the argument in (6.22) can typically not be applied. As a remedy we do not employ the geometric rigidity result directly but first approximate the deformation in each square by an H^1-function, where the distance can be measured by the curl of ∇y. (see Theorem A.1.9 below which was one of the essential ingredients to prove the qualitative result in [25].) We address this problem in Lemma 9.2.2 and subsequently we show that we may modify the configuration such that $\tilde{s} \ge \varepsilon$ (see Theorem 9.4.4).

Finally, by a density argument we can approximate each SBV function by a configuration where the jump set is contained in a finite number of rectangle boundaries (see proof of Theorem 9.4.1). Observe that standard density results as [31] cannot be applied directly in our framework since in general an L^∞ bound for the derivative is not preserved. The problem can be circumvented by using a different approximation introduced in [26] at cost of a non exact approximation of the jump set, which suffices for our purposes.

The rigidity result, which we then have established, only holds up to a small exceptional set as due to the modification of the jump set the deformation might not be defined in the interior of certain rectangles. We emphasize that such an estimate is not enough to obtain good compactness and convergence results, in particular for the convergence of the surface energy further difficulties arise. Therefore, we eventually have to construct a suitable extension to the whole domain. A major challenge is to determine the surface energy correctly, at least for the relaxed functional (6.3). This problem is addressed in Section 9.5.

For small cracks a good extension is already provided by the Korn-Poincaré inequality. Near large cracks we define the extension as a piecewise constant rigid motion such that the jump heights on the new jump sets are sufficiently small (see the proof of Theorem 6.1.1). Consequently, the length of these jumps may possibly be much larger than $\mathcal{H}^1(J_y)$, but due to the small jump height their contribution to (6.3) is considerably small. Whereas the diameter defined in (6.19) was very convenient for the derivation of the linearized estimate, it does not provide the right surface energy for large cracks. Thus, for large boundary components, in particular for the boundary ∂P of the partition $(P_j)_j$, we have to construct an appropriate jump set consisting of Jordan curves which provides the correct crack energy up to a small error (see Lemma 9.5.1).

6.4.3 Compactness and Γ-convergence

The SBD-rigidity estimate turns out to be the fundamental ingredient to derive compactness and Γ-convergence results for functionals of the form (6.2). Theorem 6.2.1 essentially relies on a diagonal sequence argument for $\varepsilon, \rho \to 0$, where ρ denotes the 'error' of the modification obtained by Theorem 6.1.1. The convergence of the partitions and the corresponding rigid motions is based on compactness theorems for Caccioppoli partitions and piecewise constant functions (see Section A.2).

A major challenge for the compactness of the rescaled configurations is the fact that the constant in (6.6) depends on ρ and explodes for $\rho \to 0$. For the symmetric part of the gradient this problem can be bypassed by taking (6.20), (6.6)(iv) and $E_{\varepsilon_k}(y_{\varepsilon_k}) \le C$ into account, which shows that the constant may be chosen independently of ρ. For the function itself, however, the problem is more subtle since a uniform bound cannot be inferred by energies bounds. In particular, generically the limiting configurations are not in L^2, but only finite almost everywhere. The strategy to establish the latter assertion is to show that for fixed ε the functions $(u^\rho_\varepsilon)_\rho$ essentially coincide in a certain sense on the bulk part of the domain (see Lemma 10.1.2) if one chooses the rigid motions in (6.5) in an appropriate way. Afterwards, by a careful analysis we can derive that such a property is preserved in the limit $\varepsilon \to 0$ (see the proof of Theorem 6.2.1).

To establish (6.13) one has to separate the effects arising form the segmentation energy and the inner crack energy. This can be done by employing a structure theorem for Caccioppoli partitions (see Theorem A.2.1). Moreover, the estimate is first carried out in terms of the relaxed functionals (6.3) and drawing ideas from Section 5.1 we conclude that it is also satisfied for E.

In Theorem 6.2.4 the fundamental point is the proof of existence and uniqueness of the coarsest partition. Uniqueness follows from the observation that under the assumption that there are two different coarsest partitions one always can find an even coarser partition. Existence is a more challenging problem. We first give an alternative characterization and identify coarsest partitions as the maximal elements of the partial order on the set of admissible partitions which is induced by subordination. We then show that each chain of the partial order has an upper bound repeating some arguments of the main compactness result. Consequently, the claim is inferred by an application of Zorn's lemma. Finally, the identification of the limiting configurations follows by measuring the difference of admissible triples and taking the linearization formula (6.20) into account.

Eventually, the Γ-convergence result is straightforward since the main estimates, in particular the estimates for the surface energy, were provided in Theorem 6.2.1. To establish Theorem 6.3.3 we follow the proof of Theorem 1.6.5.

Chapter 7

Preliminaries

In this short preparatory chapter we establish a trace theorem and analyze how the constants of the geometric rigidity result and the Korn-Poincaré-inequality depend on the shape of the domain.

7.1 Geometric rigidity and Korn: Dependence on the set shape

In general, the constants of the inequalities stated in Section B depend crucially on the set shape. This will be discussed in detail in this section. As an introductory example we consider the deflection of a thin elastic beam.

Example 7.1.1. Let $U = (0,1)\times(0,\delta)$ and let $y : U \to \mathbb{R}^2$ be given by $y(x_1, x_2) = (x_2+1)(\sin(x_1), \cos(x_1))$. Then

$$\nabla y(x_1, x_2) = \begin{pmatrix} (x_2+1)\cos(x_1) & \sin(x_1) \\ -(x_2+1)\sin(x_1) & \cos(x_1) \end{pmatrix}$$

and therefore $\mathrm{dist}^2(\nabla y, SO(2)) = |\sqrt{\nabla y^T \nabla y} - \mathbf{Id}|^2 = x_2^2$, i.e.

$$\|\,\mathrm{dist}(\nabla y, SO(2))\|^2_{L^2(U)} = \tfrac{1}{3}\delta^3.$$

Let $R_\phi \in SO(2)$, $R_\phi = \begin{pmatrix} \cos\phi & \sin\phi \\ -\sin\phi & \cos\phi \end{pmatrix}$ for $\phi \in [0, 2\pi]$. Then $|\nabla y(x) - R|^2 \geq |\sin(x_1) - \sin\phi|^2 + |\cos(x_1) - \cos\phi|^2$. It is not hard to see that it exists a $C > 0$ such that $\int_0^1 |\nabla y(x) - R|^2\, dx_1 \geq C$ for all $\phi \in [0, 2\pi]$ and $x_2 \in (0, \delta)$. We conclude that

$$\|\nabla y - R\|^2_{L^2(U)} \geq C\delta \geq \frac{C}{\delta^2}\|\,\mathrm{dist}(\nabla y, SO(2))\|^2_{L^2(U)}$$

for all $R \in SO(2)$. A similar argument shows

$$\|y - (R\,\cdot + c)\|^2_{L^2(U)} \geq C\delta \geq \frac{C}{\delta^2}\|\,\mathrm{dist}(\nabla y, SO(2))\|^2_{L^2(U)}$$

for all $R \in SO(2)$ and $c \in \mathbb{R}^2$.

Similar examples can be constructed in the linearized framework for the Korn-Poincaré inequality given in Theorem B.4. As a direct consequence we get that the estimate (6.6) might be wrong without allowing for a small modification of the deformation.

Example 7.1.2. Let $\varepsilon > 0$. Assume without restriction that the set $U = (0,1) \times (0, \varepsilon^{\frac{1}{3}})$ considered above satisfies $\overline{U} \subset \Omega$. Define $y : \Omega \to \mathbb{R}^2$ by $y(x) = \mathbf{id} + \mathbf{e}_2$ for $x \in \Omega \backslash U$ and $y(x) = (x_2+1)(\sin(x_1), \cos(x_1))$ for $x \in U$. Then $y \in SBV^2(\Omega)$ with $J_y = (0,1) \times \{0, \varepsilon^{\frac{1}{3}}\} \cup \{1\} \times (0, \varepsilon^{\frac{1}{3}})$ and $\| \operatorname{dist}(\nabla y, SO(2))\|^2_{L^2(\Omega)} = \frac{\varepsilon}{3}$. However, for all $R \in SO(2)$ and $c \in \mathbb{R}^2$ we have

$$\|\nabla y - R\|^2_{L^2(\Omega)} \geq C\varepsilon^{\frac{1}{3}}, \qquad \|y - (R\, \cdot + c)\|^2_{L^2(\Omega)} \geq C\varepsilon^{\frac{1}{3}}.$$

Although omitted here, a similar estimate can be derived for the symmetric part of the gradient.

For $s > 0$ we partition $\mathbb{R}^2$ up to a set of measure zero into squares $Q^s(p) = p + s(-1,1)^2$ for $p \in I^s := s(1,1) + 2s\mathbb{Z}^2$. Let

$$\mathcal{U}^s := \Big\{ U \subset \mathbb{R}^2 : U = \Big(\bigcup\nolimits_{p \in I} \overline{Q^s(p)} \Big)^\circ : \ \ I \subset I^s \Big\}. \tag{7.1}$$

Here the superscript $\circ$ denotes the interior of a set. In order to quantify how the constants in Theorem B.1 and Theorem B.4 depend on the set shape we will estimate the variation from a square $Q^s(a)$ to a neighboring square $Q^s(b)$, $b = a + 2s\nu$ for $\nu = \pm \mathbf{e}_i$, $i = 1,2$. We first introduce some further notation. For $y \in SBV^2(U)$ with $U \in \mathcal{U}^s$ and for $R \in SO(2)$ we set $u_R = R^T y - \mathbf{id}$, where $\mathbf{id}$ is the identity function. On a cube $Q^s(p) \subset U$ we define for shorthand (we drop the integration variable if no confusion arises)

$$\gamma(p) = \int_{Q^s(p)} \operatorname{dist}^2(\nabla y, SO(2)), \quad \mathcal{E}_R(p) = \int_{Q^s(p)} |\bar{e}_R(\nabla y)| + |D^j y|(Q^s(p) \cap U).$$

Here the symmetric part of the gradient is defined by

$$\bar{e}_R(\nabla y) := e(\nabla u_R) = \frac{R^T \nabla y + (\nabla y)^T R}{2} - \mathbf{Id}, \tag{7.2}$$

where $\mathbf{Id}$ denotes the identity matrix. Moreover, for subsets $V \subset U$, $V \in \mathcal{U}^s$ we write

$$\gamma(V) = \sum\nolimits_{p \in I^s(V)} \gamma(p), \quad \mathcal{E}_R(V) = \int_V |\bar{e}_R(\nabla y)| + |D^j y|(V), \tag{7.3}$$

where $I^s(V) := \{p \in I^s : Q^s(p) \subset V\}$.

We first assume $y \in H^1(U)$ and proceed similarly as in [49]. Applying Theorem B.1 we obtain $R(a), R(b) \in SO(2)$ such that

$$\int_{Q^s(p)} |\nabla y - R(p)|^2 \leq C\gamma(p) \quad \text{for } p = a, b. \tag{7.4}$$

Likewise on the rectangle $Q^s(a,b) := (\overline{Q^s(a)} \cup \overline{Q^s(b)})^\circ$ we obtain $R(a,b) \in SO(2)$ such that

$$\int_{Q^s(a,b)} |\nabla y - R(a,b)|^2 \, dx \leq C \int_{Q^s(a,b)} \operatorname{dist}^2(\nabla y, SO(2)) \leq C(\gamma(a) + \gamma(b)).$$

Combining these estimates we see $|Q^s(p)||R(p) - R(a,b)|^2 \leq C(\gamma(a) + \gamma(b))$ for $p = a, b$ and therefore

$$s^2|R(a) - R(b)|^2 \leq C(\gamma(a) + \gamma(b)). \tag{7.5}$$

More general, we consider a difference quotient with two arbitrary points $a, b \in I^s(U)$. We assume that there is a path $\xi = (\xi_0, \ldots, \xi_m)$ such that

$$\begin{aligned} &\xi_1 = a, \quad \xi_m = b, \\ &\xi_j - \xi_{j-1} = \pm 2s\mathbf{e}_i \text{ for some } i = 1, 2, \quad \forall j = 2, \ldots, m. \end{aligned} \tag{7.6}$$

Then iteratively applying the above estimate (7.5) we obtain

$$s^2|R(a) - R(b)|^2 \, dx \leq Cm \sum\nolimits_{j=1}^m \gamma(\xi_j). \tag{7.7}$$

Now we concern ourselves with the Korn-Poincaré inequality. Assume that $y \in SBV^2(U)$ and recall $u_R = R^T y - \mathbf{id}$ for $R \in SO(2)$. For later purposes, we consider more general rectangles and derive the difference of the deformation on adjacent squares as a special case. Let $b_1, b_2 \in \mathbb{R}^2$, and $B_i = b_i + (-l_i, l_i) \times (-m_i, m_i) \in \mathcal{U}^s$ for $i = 1, 2$, where we assume without restriction that $l_1 \geq m_1 > 0$, $l_2 \geq m_2 > 0$. Suppose that there is a point $b_{12} \in \overline{B_1} \cap \overline{B_2}$.

For given $A_1, A_2, A_{12} \in \mathbb{R}^{2\times 2}_{\text{skew}} = \{G \in \mathbb{R}^{2\times 2} : G^T = -G\}$ and $c_1, c_2, c_{12} \in \mathbb{R}^2$ we set $E_i := \|u_R - (A_i \,\cdot + c_i)\|^2_{L^2(B_i)}$ for $i = 1, 2$ and suppose that

$$\|u_R - (A_{12} \,\cdot + c_{12})\|^2_{L^2(B_1 \cup B_2)} \leq C(E_1 + E_2). \tag{7.8}$$

As above this implies

$$\|(A_i - A_{12}) \cdot + (c_i - c_{12})\|^2_{L^2(B_i)} \leq C(E_1 + E_2) \quad \text{for } i = 1, 2.$$

We let $B_i^- = b_i + (-l_i, 0) \times (-m_i, m_i)$, $B_i^+ = b_i + (0, l_i) \times (-m_i, m_i)$ and for shorthand we write $\hat{A}_i = A_i - A_{12}$, $\hat{c}_i = c_i - c_{12}$. We then derive

$$\begin{aligned} |B_i| \, l_i^2 |\hat{A}_i|^2 &\leq 4\|\hat{A}_i \, l_i \mathbf{e}_1\|^2_{L^2(B_i^-)} = 4\|\hat{A}_i\,(\cdot + l_i\mathbf{e}_1) + \hat{c}_i - \hat{A}_i \,\cdot - \hat{c}_i\|^2_{L^2(B_i^-)} \\ &\leq 8\|\hat{A}_i \,\cdot + \hat{c}_i\|^2_{L^2(B_i^+)} + 8\|\hat{A}_i \,\cdot + \hat{c}_i\|^2_{L^2(B_i^-)} \leq C(E_1 + E_2) \end{aligned} \tag{7.9}$$

for $i = 1, 2$ and therefore

$$|B_1 \cup B_2|(l_1 + l_2)^2 \, |A_1 - A_2|^2 \leq C\kappa(E_1 + E_2), \tag{7.10}$$

where $\kappa = \frac{|B_1 \cup B_2|}{\min_j |B_j|} \left(\frac{l_1 + l_2}{\min_j l_j}\right)^2$. Observe that in the first equality we essentially used the skew symmetry. Since $|y - b_{12}| \leq C(l_1 + l_2)$ for all $y \in B_1 \cup B_2$ we likewise compute

$$\begin{aligned} |B_1 \cup B_2||\hat{A}_i \, b_{12} + \hat{c}_i|^2 &\leq C \frac{|B_1 \cup B_2|}{|B_i|} \|\hat{A}_i \, \cdot + \hat{c}_i\|^2_{L^2(B_i)} + C|B_1 \cup B_2|(l_1 + l_2)^2 |\hat{A}_i|^2 \\ &\leq C\kappa(E_1 + E_2) \end{aligned}$$

for $i = 1, 2$. Employing the triangle inequality we then deduce

$$|B_1 \cup B_2||(A_2 - A_1)\, b_{12} + c_2 - c_1|^2 \leq C\kappa(E_1 + E_2). \tag{7.11}$$

Consider $Z \subset B_1 \cup B_2$, $Z \in \mathcal{U}^s$. Similar arguments yield by (7.10)

$$\begin{aligned} \|(A_2 - A_1) \cdot + c_2 - c_1\|^2_{L^2(Z)} &\leq C\|(A_2 - A_1)b_{12} + c_2 - c_1\|^2_{L^2(Z)} \\ &\quad + C|Z| \max_j l_j^2 |A_1 - A_2|^2 \\ &\leq C \frac{|Z|}{|B_1 \cup B_2|} \kappa(E_1 + E_2) \end{aligned} \tag{7.12}$$

and therefore by the triangle inequality

$$\begin{aligned} \|u_R - (A_1 \; \cdot + c_1)\|^2_{L^2(B_2 \cap Z)} &\leq C\|u_R - (A_2 \; \cdot + c_2)\|^2_{L^2(B_2 \cap Z)} \\ &\quad + C \frac{|Z|}{|B_1 \cup B_2|} \kappa(E_1 + E_2). \end{aligned} \tag{7.13}$$

In particular, employing $Z = B_1 \cup B_2$ and recalling (7.8) we find

$$\|u_R - (A_1 \; \cdot + c_1)\|^2_{L^2(B_1 \cup B_2)} \leq C\kappa(E_1 + E_2). \tag{7.14}$$

Before we treat the case of two adjacent squares we observe that in the above estimates the constants may be refined in the case that $B_1 \subset B_2$ under additional assumptions on the energies. Let $\delta \geq Csl_2^{-1}$. Let $B_2 = (-l_2, l_2) \times (-s, s) \in \mathcal{U}^s$ and $B_1 \subset B_2$, $B_1 \in \mathcal{U}^s$ a general set such that $|B_2 \setminus B_1| \leq \delta |B_2|$. In particular, this implies that the diameter of each connected component of $B_2 \setminus B_1$ is smaller than $C\delta l_2$. Moreover, we assume that for all $Z \subset B_2$, $Z \in \mathcal{U}^s$ one has

$$\|u_R - (A_i \; \cdot + c_i)\|^2_{L^2(B_i \cap Z)} \leq |B_i \cap Z||B_2|^{-1} H_i \tag{7.15}$$

for $H_1, H_2 \geq 0$. Arguing similarly as in (7.10) we find $|A_1 - A_2|^2 \leq C|B_2|^{-1} l_2^{-2}(H_1 + H_2)$. (Observe that the connectedness of B_1 is not necessary. Moreover, the estimate can also be derived if B_2 consists of several connected components.) We write $\tilde{A} = A_1 - A_2$ and $\tilde{c} = c_1 - c_2$ for shorthand. Let $b_0 \in B_1$ and $Q \subset B_1$ be the square containing b_0. Applying a scaled version of Young's inequality and using $s \le C\delta l_2$ we compute

$$\begin{aligned}|Q||\tilde{A}\,b_0 + \tilde{c}|^2 &= \|\tilde{A}\,b_0 + \tilde{c}\|^2_{L^2(Q)} \le (1+\delta)\|\tilde{A}\ \cdot + \tilde{c}\|^2_{L^2(Q)} + (1+\tfrac{1}{\delta})\|\tilde{A}\,(\cdot - b_0)\|^2_{L^2(Q)} \\ &\le (1+\delta)^2\|u_R - (A_1\ \cdot + c_1)\|^2_{L^2(Q)} + \tfrac{C}{\delta}\|u_R - (A_2\ \cdot + c_2)\|^2_{L^2(Q)} \\ &\quad + \tfrac{C}{\delta}|Q|s^2|B_2|^{-1}l_2^{-2}(H_1 + H_2) \\ &\le (1+C\delta)|B_2|^{-1}|Q|H_1 + \tfrac{C}{\delta}|B_2|^{-1}|Q|H_2 + C|B_2|^{-1}|Q|\delta(H_1+H_2) \\ &\le (1+C\delta)|B_2|^{-1}|Q|H_1 + \tfrac{C}{\delta}|B_2|^{-1}|Q|H_2.\end{aligned}$$

Consider some connected $Z \subset B_2 \setminus B_1$, $Z \in \mathcal{U}^s$, and observe that we find some $b_0 \in B_1$ such that $|x - b_0| \le C\delta l_2$. Then repeating the above calculation, again employing Young's inequality, we derive

$$\begin{aligned}\|\tilde{A}\ \cdot + \tilde{c}\|^2_{L^2(Z)} &\le (1+\delta)|Z||\tilde{A}\,b_0 + \tilde{c}|^2 + (1+\tfrac{1}{\delta})|Z|\max_{x\in Z}|x-b_0|^2|\tilde{A}|^2 \\ &\le (1+C\delta)|B_2|^{-1}|Z|H_1 + \tfrac{C}{\delta}|B_2|^{-1}|Z|H_2.\end{aligned}$$

Then Young's inequality yields

$$\begin{aligned}\|u_R - (A_1\ \cdot + c_1)\|^2_{L^2(Z)} &\le (1+\delta)\|\tilde{A}\ \cdot + \tilde{c}\|^2_{L^2(Z)} + (1+\tfrac{1}{\delta})\|u_R - (A_2\ \cdot + c_2)\|^2_{L^2(Z)} \\ &\le (1+C\delta)|B_2|^{-1}|Z|H_1 + \tfrac{C}{\delta}|B_2|^{-1}|Z|H_2. \qquad (7.16)\end{aligned}$$

Finally, it is not hard to see that (7.16) holds for all $Z \subset B_2$, $Z \in \mathcal{U}^s$.

Now assume the special case that $B_1 = Q^s(a)$ and $B_2 = Q^s(b)$, $b = a + 2s\nu$ for $\nu = \pm\mathbf{e}_i$, $i = 1, 2$. Then by Theorem B.4 we obtain $A(a), A(b), A(a,b) \in \mathbb{R}^{2\times 2}_{\text{skew}}$ and $c(a), c(b), c(a,b)$ such that

$$\begin{aligned}&\|u_R - (A(p)\ \cdot + c(p))\|_{L^2(Q^s(p))} \le C\mathcal{E}_R(p) \quad \text{for } p = a, b, \\ &\|u_R - (A(a,b)\ \cdot + c(a,b))\|_{L^2(Q^s(a,b))} \le C\mathcal{E}_R(a,b),\end{aligned} \qquad (7.17)$$

where for shorthand $\mathcal{E}_R(a,b) = \mathcal{E}_R(Q^s(a,b))$. As in this case $\kappa = 8$, (7.10) and (7.12) for $Z = Q^s(a,b)$ yield

$$\begin{aligned}&s^2|A(a) - A(b)| \le C\mathcal{E}_R(a,b), \\ &\|(A(b) - A(a))\ \cdot + c(b) - c(a)\|_{L^2(Q^s(a,b))} \le C\mathcal{E}_R(a,b).\end{aligned}$$

Similarly as in (7.6) we now consider a difference quotient with two arbitrary points $a, b \in I^s(U)$ connected by a path $\xi = (\xi_1, \dots, \xi_m)$. Iterative application of the last estimate yields

$$
\begin{aligned}
&\|(A(b) - A(a)) \cdot + c(b) - c(a))\|_{L^2(Q^s(b))} \\
&\quad \le \sum_{j=2}^{m} \|(A(\xi_j) - A(\xi_{j-1})) \cdot + c(\xi_j) - c(\xi_{j-1}))\|_{L^2(Q^s(b))} \\
&\quad \le \sum_{j=2}^{m} \|(A(\xi_j) - A(\xi_{j-1})) \cdot + c(\xi_j) - c(\xi_{j-1}))\|_{L^2(Q^s(\xi_j))} \\
&\qquad + \sum_{j=2}^{m} 2s|(A(\xi_j) - A(\xi_{j-1}))\,(b - \xi_j)| \\
&\quad \le C\sum_{j=1}^{m} \mathcal{E}_R(\xi_j) + \sum_{j=2}^{m} 2s|A(\xi_j) - A(\xi_{j-1})|m2s \\
&\quad \le Cm\sum_{j=2}^{m} \mathcal{E}_R(\xi_j, \xi_{j-1})
\end{aligned}
\tag{7.18}
$$

and therefore

$$
\begin{aligned}
&\|u_R - (A(a) \cdot + c(a))\|^2_{L^2(Q^s(b))} \le 2\|u_R - (A(b) \cdot + c(b))\|^2_{L^2(Q^s(b))} \\
&\quad + 2\|(A(b) - A(a)) \cdot + c(b) - c(a))\|^2_{L^2(Q^s(b))} \\
&\quad \le Cm^2\Big(\sum_{j=2}^{m} \mathcal{E}_R(\xi_j, \xi_{j-1})\Big)^2 \le Cm^3\sum_{j=2}^{m} \big(\mathcal{E}_R(\xi_j, \xi_{j-1})\big)^2.
\end{aligned}
\tag{7.19}
$$

In the last step we have used Hölder's inequality. We now apply (7.7) and (7.19) to derive a first weak rigidity result and a Korn-Poincaré-type inequality, respectively.

Lemma 7.1.3. *Let $\mu, s > 0$ such that $l := \mu s^{-1} \in \mathbb{N}$. Then there is a constant $C > 0$ independent of μ, s such that for all connected sets $U \in \mathcal{U}^s$, $U \subset (-\mu, \mu)^2$, the following holds:*

(i) For all $y \in H^1(U)$ there is a rotation $R \in SO(2)$ such that

$$\int_U |\nabla y - R|^2 \le C(s^{-2}|U|)^2 \int_U \operatorname{dist}^2(\nabla y, SO(2)) \le Cl^4 \int_U \operatorname{dist}^2(\nabla y, SO(2)).$$

(ii) For all $y \in SBV^2(U)$ and all rotations $R \in SO(2)$ there is an $A \in \mathbb{R}^{2\times 2}_{\mathrm{skew}}$ and $c \in \mathbb{R}^2$ such that

$$\int_U |u_R(x) - (A\,x + c)|^2\,dx \le C(s^{-2}|U|)^3(\mathcal{E}_R(U))^2.$$

(iii) More precisely, for all $V \subset U$, $V \in \mathcal{U}^s$ one has ($⨍_V$ stands for $\frac{1}{|V|}\int$)

$$|U|⨍_V |u_R(x) - (A\,x + c)|^2\,dx \le C(s^{-2}|U|)^3(\mathcal{E}_R(U))^2.$$

Proof. We first show (i). The second inequality is obvious as $|U| \le 4\mu^2$. To see the first inequality we fix $p_0 \in I^s(U)$ and consider an arbitrary $p \in I^s(U)$. As U is connected there is a path $\xi = (\xi_1 = p_0, \ldots, \xi_m = p)$ with $m \le |U|(2s)^{-2}$. We first apply (7.4) on each square and then by (7.7) we obtain

$$\int_{Q^s(p)} |R(p) - R(p_0)|^2 \le C|U|s^{-2} \sum\nolimits_{j=1}^{m} \gamma(\xi_j) \le C|U|s^{-2}\gamma(U).$$

Then setting $R = R(p_0)$ and summing over all $p \in I^s(U)$ we derive

$$\begin{aligned} \int_U |\nabla y - R|^2 &\le C \sum\nolimits_{p \in I^s(U)} \int_{Q^s(p)} \Big(|\nabla y - R(p)|^2 + |R(p) - R(p_0)|^2\Big) \\ &\le C \sum\nolimits_{p \in I^s(U)} (\gamma(p) + |U|s^{-2}\gamma(U)) \le C\#I^s(U)\,|U|s^{-2}\gamma(U) \\ &\le C(|U|s^{-2})^2\gamma(U). \end{aligned}$$

Property (ii) can be proved similarly using (7.19) instead of (7.7). Finally, (iii) is a direct consequence of (ii) since we may replace $\#I^s(U)$ by $\#I^s(V)$ if we only integrate over the set V. □

Remark 7.1.4. (i) The fact that we covered the sets U with squares is not essential. Recalling how we derived (7.5) and (7.10) we could equally well cover U with rectangles $R_i = t_i + (-a_i, a_i) \times (-b_i, b_i)$, where $c_1 a_i \le b_i \le c_2 a_i$ and $c_1 s \le b_i \le c_2 s$ for constants $0 < c_1 < c_2$. The constants in (7.4) and (7.17), respectively, only depend on c_1, c_2 as all the possible shapes are related to $(-s,s)^2$ through bi-Lipschitzian homeomorphisms with Lipschitz constants of both the homeomorphism itself and its inverse bounded (see Section B).

(ii) Let $U = (0,1) \times (0,\delta)$. If we choose $s = \frac{\delta}{2}$, Lemma 7.1.3(i) provides a constant $\sim \delta^{-2}$. Example 7.1.1 shows that this estimate is sharp in the sense that the exponent of δ cannot be improved.

(iii) The argumentation developed in (7.8)-(7.14) remains true if we pass from the linearized to the nonlinear setting, i.e. replacing $\mathbb{R}^{2\times 2}_{\mathrm{skew}}$ by $SO(2)$. This follows from the fact that we essentially use the property that the matrices satisfy $|A\mathbf{e}_1| = |A\mathbf{e}_2|$ (cf. (7.9)). Likewise, we can estimate the difference of rigid motions on chains similarly as in (7.19). This will be used in Section 9.

(iv) Following the above arguments we find that in Lemma 7.1.3(i) one can replace $p = 2$ by any $1 < p < \infty$ replacing l^4 suitably by l^{2p}.

(v) In view of the proof of (i),(ii), in the choice of R and $Ax + c$ we have the freedom to select any of the rotations or infinitesimal rigid motions which are given on each square $Q^s(p) \subset U$ by application of (7.4) or (7.17), respectively.

7.2 A trace theorem in SBV^2

By the trace theorem for BD functions (Theorem B.5) one can control the L^1-norm of the function on the boundary. In our framework we may establish a trace theorem in L^2 for SBV^2 functions if the jump set is sufficiently regular: Let $Q_\mu = (-\mu,\mu)^2$ and recall the definition of $SBV^2(Q_\mu)$ in Section A.1. We suppose that some $y \in SBV^2(Q_\mu)$ or the corresponding $u_R = R^T y - \mathbf{id}$, $R \in SO(2)$, respectively, satisfies $J_{u_R} = \bigcup_j \Gamma_j \cap Q_\mu$, where $\Gamma_i = \partial R_i$ for rectangles $R_i = (a_1^i, a_2^i) \times (b_1^i, b_2^i) \subset \mathbb{R}^2$ (note that for the application we have in mind we do not require that the rectangles are subsets of Q_μ.). Clearly, as $u_R \in H^1(Q_\mu \setminus J_{u_R})$ the trace is well defined in L^2. More precisely, we have the following statement.

Lemma 7.2.1. *Let $R \in SO(2)$ and $\mu > 0$. There is a constant $C > 0$ such that for all $u_R \in SBV^2(Q_\mu)$ with $J_{u_R} = \bigcup_{j=1}^n \Gamma_j \cap Q_\mu$, where $\Gamma_j = \partial R_j$, one has*

$$\begin{aligned}\int_{\partial Q_\mu} |u_R|^2 \, d\mathcal{H}^1 \le C\mu \|e(\nabla u_R)\|^2_{L^2(Q_\mu)} + \tfrac{C}{\mu}\|u_R\|^2_{L^2(Q_\mu)} \\ + C\sum\nolimits_{j=1}^n \mathcal{H}^1(\Gamma_j) \sum\nolimits_{j=1}^n \Big((\mathcal{H}^1(\Gamma_j))^{-1} \int_{\Gamma_j \cap Q_\mu} |[u_R]|^2 \, d\mathcal{H}^1 \Big).\end{aligned} \tag{7.20}$$

Proof. Let $Q_\mu = (-\mu,\mu)^2$ and $u_R \in SBV^2(Q_\mu)$ with $J_{u_R} = \bigcup_{j=1}^n \Gamma_j \cap Q_\mu$. In what follows we drop the subscripts μ and R for notational convenience. First by approximation of Sobolev functions on Lipschitz sets (see, e.g., [40, Section 4.2]) we may assume that $u_R|_{R_j}$ is smooth for $j = 0, \ldots, n$, where $R_0 = Q \setminus \bigcup_{j=1}^n R_j$. We only consider the part $\partial' Q = (-\mu,\mu) \times \{\mu\}$ of the boundary. Let $\pi_x = \{x\} \times \mathbb{R}$ and compute for the second component u_2 by a slicing argument in $\mathbf{e}_2$-direction:

$$\begin{aligned}\int_{-\mu}^{\mu}\int_{-\mu}^{\mu} |u_2(x,\mu) - u_2(x,y)|^2 \, dx\, dy &= \int_{-\mu}^{\mu}\int_{-\mu}^{\mu} \Big| \int_y^{\mu} D_2 u_2(x,t)\, dt \Big|^2 dx\, dy \\ &\le C \int_{-\mu}^{\mu}\int_{-\mu}^{\mu} \Big(\mu \int_{-\mu}^{\mu} |\partial_2 u_2(x,t)|^2 \, dt + \Big(\sum_{z \in J_u \cap \pi_x} |[u](z)| \Big)^2 \Big) dx\, dy \\ &\le C\mu^2 \|e(\nabla u)\|^2_{L^2(Q)} + C\mu \int_{-\mu}^{\mu} \Big(\sum_{z \in J_u \cap \pi_x} |[u](z)| \Big)^2 dx.\end{aligned}$$

In the second step we have used Hölder's inequality. We now estimate the term on the right side. As Γ_j is a rectangle, except for two x-values there are exactly two points $t_j^1, t_j^2 \in \mathbb{R}$ such that $\Gamma_j \cap \pi_x = \{(x,t_j^1),(x,t_j^2)\}$ if $\Gamma_j \cap \pi_x \neq \emptyset$. We write $|\Gamma_j|_{\mathcal{H}} = \mathcal{H}^1(\Gamma_j)$, $|S|_{\mathcal{H}} = \sum_j \mathcal{H}^1(\Gamma_j)$ for shorthand. Letting $z_j^{k,x} = (x, t_j^k) \in \mathbb{R}^2$ and setting $|[u](z_j^{k,x})| = 0$ if $z_j^{k,x} \notin Q \cap \Gamma_j$, we then obtain by the discrete version of Jensen's inequality

$$\int_{-\mu}^{\mu}\Big(\sum_{z\in J_u\cap\pi_x}|[u](z)|\Big)^2\,dx = 4\int_{-\mu}^{\mu}\Big(\sum\nolimits_j\sum\nolimits_{k=1,2}\frac{|\Gamma_j|_{\mathcal{H}}}{2|S|_{\mathcal{H}}}\,|[u](z_j^{k,x})|\frac{|S|_{\mathcal{H}}}{|\Gamma_j|_{\mathcal{H}}}\Big)^2\,dx$$
$$\leq 4\int_{-\mu}^{\mu}\sum\nolimits_j\sum\nolimits_{k=1,2}\frac{|\Gamma_j|_{\mathcal{H}}}{2|S|_{\mathcal{H}}}\Big(|[u](z_j^{k,x})|\frac{|S|_{\mathcal{H}}}{|\Gamma_j|_{\mathcal{H}}}\Big)^2\,dx$$
$$\leq 2|S|_{\mathcal{H}}\sum\nolimits_j\Big(|\Gamma_j|_{\mathcal{H}}^{-1}\int_{\Gamma_j\cap Q}|[u]|^2\,d\mathcal{H}^1\Big).$$

Consequently, letting E be the right hand side of (7.20) we derive

$$\int_{\partial' Q}|u_2|^2\,d\mathcal{H}^1 \leq \frac{C}{\mu}\Big(\int_{-\mu}^{\mu}\int_{-\mu}^{\mu}|u_2(x,\mu)-u_2(x,y)|^2\,dx\,dy + \|u\|^2_{L^2(Q)}\Big) \leq CE.$$

The same argument with slicing in the directions $\xi_1 = \frac{1}{\sqrt{2}}(1,-1)$ and $\xi_1 = \frac{1}{\sqrt{2}}(-1,-1)$ yields

$$\int_{\partial_1' Q}|u\cdot\xi_1|^2\,d\mathcal{H}^1 \leq CE, \qquad \int_{\partial_2' Q}|u\cdot\xi_2|^2\,d\mathcal{H}^1 \leq CE,$$

where $\partial_1' Q = (-\mu,0)\times\{\mu\}$ and $\partial_2' Q = (0,\mu)\times\{\mu\}$. The claim now follows by combination of the previous estimates. □

Chapter 8

A Korn-Poincaré-type inequality

This section is devoted to the derivation of a Korn-Poincaré-type inequality which roughly speaking emerges from the inequality in Theorem B.4 by replacing $|E^j u_R|(\Omega)$ by $\sqrt{\varepsilon}\mathcal{H}^1(J_{u_R})$, where $\varepsilon \sim \int_\Omega |e(\nabla u_R)|^2$. The main strategy will be to show that the jump height on J_{u_R} is at most of order $O(\sqrt{\varepsilon})$ from which the claim will follow by application of Theorem B.4. The examples in Section 7.1, however, show that this might not be possible in general. Therefore, in Sections 8.2, 8.3 we first introduce a suitable modification scheme to alter the jump set. Section 8.5 then contains the main technical estimates for the analysis of the jump height. In Section 8.4 we combine the previously established results and present an algorithm which iteratively modifies the jump set such that the estimates on the jump height may be applied.

We first formulate the main result of this chapter. For $\mu > 0$ let $Q_\mu = (-\mu, \mu)^2$ and by $\mathrm{diam}(R)$ denote the diameter of a rectangle $R \subset Q_\mu$.

Theorem 8.0.1. *Let $\varepsilon > 0$ and $h_* > 0$ sufficiently small. Then there is a constant $C = C(h_*)$ and a universal constant $\bar{c} > 0$ such that for all $u \in SBD^2(Q_\mu, \mathbb{R}^2) \cap L^2(Q_\mu, \mathbb{R}^2)$ the following holds: We get pairwise disjoint, paraxial rectangles $R_1, \ldots, R_n$ with*

$$\sum\nolimits_{j=1}^{n} \mathrm{diam}(R_j) \le (1 + \bar{c}h_*)\big(\mathcal{H}^1(J_u) + \varepsilon^{-1}\|e(\nabla u)\|^2_{L^2(Q_\mu)}\big)$$

such that for $E := \bigcup_{j=1}^n R_j$ and the square $\tilde{Q} = (-\tilde{\mu}, \tilde{\mu})^2$ with $\tilde{\mu} = \max\{\mu - 2\sum_j \mathrm{diam}(R_j), 0\}$ we have $|E| \le \bar{c}(\sum_j \mathrm{diam}(R_j))^2$ and

$$\|u(x) - (A\,x + c)\|^2_{L^2(\tilde{Q}\setminus E)} \le C\mu^2 \|e(\nabla u)\|^2_{L^2(Q_\mu)} + C\mu^2 \varepsilon \mathcal{H}^1(J_u)$$

for some $A \in \mathbb{R}^{2\times 2}_{\mathrm{skew}}$ and $c \in \mathbb{R}^2$.

This Korn-Poincaré-type estimate is in the spirit of the Poincaré inequality in SBV due to De Giorgi, Carriero, Leaci (see [39]) with the difference that we do not truncate the function (which is forbidden in the SBD framework), but provide an exceptional set $E = \bigcup_j R_j$, where the estimate does not hold. This set is associated to the parts of Q_μ being detached from the bulk part of Q_μ by J_u. In contrast to the recently established estimate in [24], Theorem 8.0.1 provides an exceptional set with a rather simple geometry. Most notably we have control over $\mathcal{H}^1(\partial E)$ which allows to apply compactness results for GSBD functions (see Theorem A.1.3). Moreover, for $h_* \ll 1$ and $\varepsilon \gg \|e(\nabla u)\|^2_{L^2(Q_\mu)}(\mathcal{H}^1(J_u))^{-1}$ we obtain a fine estimate on the sum of the diameter of the rectangles which will be essential in the energy estimates in Chapter 9.

8.1 Preparations

Let $Q_\mu = (-\mu,\mu)^2$ and recall the definition of $\mathcal{U}^s$ in (7.1). We will concern ourselves with subsets $V \subset Q_\mu$ of the form

$$\mathcal{V}^s := \{V \subset Q_\mu : V = Q_\mu \setminus \bigcup_{i=1}^m X_i, \ \ X_i \in \mathcal{U}^s, \ X_i \text{ pairwise disjoint}\} \quad (8.1)$$

for $s > 0$. Note that each set in $V \in \mathcal{V}^s$ coincides with a set $U \in \mathcal{U}^s$ up to subtracting a set of zero Lebesgue measure, i.e. $U \subset V$, $\mathcal{L}^2(V \setminus U) = 0$. The essential difference of V and the corresponding U concerns the connected components of the complements $Q_\mu \setminus V$ and $Q_\mu \setminus U$. Observe that one may have $Q_\mu \setminus \bigcup_{i=1}^m X_i = Q_\mu \setminus \bigcup_{i=1}^{\hat{m}} \hat{X}_i$ with $(X_1,\dots,X_m) \neq (\hat{X}_1,\dots,\hat{X}_{\hat{m}})$, e.g. by combination of different sets (see Figure 8.1). In such a case we will regard $V_1 = Q_\mu \setminus \bigcup_{i=1}^m X_i$ and $V_2 = Q_\mu \setminus \bigcup_{i=1}^{\hat{m}} \hat{X}_i$ as different elements of $\mathcal{V}^s$. For the whole chapter we will tacitly assume that all considered sets are elements of $\mathcal{V}^s$ for some small, fixed $s > 0$.

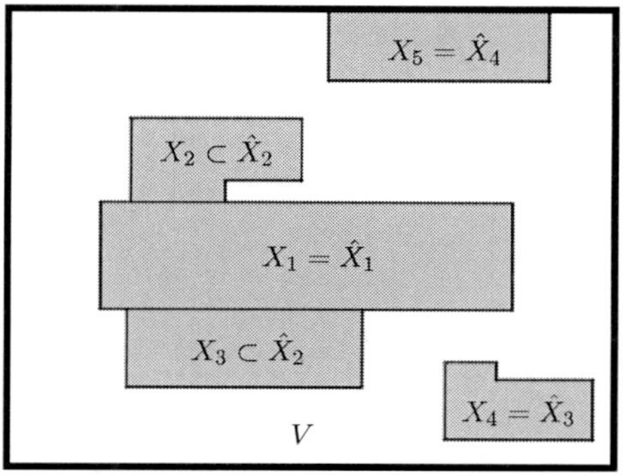

Figure 8.1: The square Q_μ with a subset V. The set V has two representations $V_1 = Q_\mu \setminus \bigcup_{i=1}^5 X_i$ and $V_2 = Q_\mu \setminus \bigcup_{i=1}^4 \hat{X}_i$, where $\hat{X}_2 = X_2 \cup X_3$, which are regarded as different elements of $\mathcal{V}^s$. The corresponding set $U \in \mathcal{U}^s$ arises from V by subtracting the black boundary lines $\bigcup_{i=1}^5 \partial X_i$.

Let $W \in \mathcal{V}^s$ and arrange the components $X_1, \dots, X_m$ of the complement such that $\partial X_i \subset Q_\mu$ for $1 \le i \le n$ and $\partial X_i \cap \partial Q_\mu \neq \emptyset$ otherwise. Define $\Gamma_i(W) = \partial X_i$ for $i = 1, \dots, n$. In the following we will often refer to these sets as *boundary components*. Note that $\bigcup_{i=1}^n \Gamma_i(W)$ might not cover $\partial W \cap Q_\mu$ completely if $n < m$. We frequently drop the subscript and write $\Gamma(W)$ or just Γ if no confusion arises. In addition to the Hausdorff-measure $|\Gamma|_{\mathcal{H}} = \mathcal{H}^1(\Gamma)$ (we will use both notations) we define the 'diameter' of a boundary component by $|\Gamma|_\infty := \sqrt{|\pi_1\Gamma|^2 + |\pi_2\Gamma|^2}$, where π_1, π_2 denote the orthogonal projections onto the coordinate axes.

Note that by definition of $\mathcal{V}^s$ (in contrast to the definition of $\mathcal{U}^s$) two components in $(\Gamma_i)_i$ might not be disjoint. Therefore, we choose an (arbitrary) order $(\Gamma_i)_{i=1}^n = (\Gamma_i(W))_{i=1}^n$ of the boundary components of W, introduce

$$\Theta_i = \Theta_i(W) = \Gamma_i \setminus \bigcup\nolimits_{j<i} \Gamma_j \tag{8.2}$$

for $i = 1, \dots, n$ and observe that the boundary components $(\Theta_i)_i$ are pairwise disjoint. With a slight abuse of notation we define

$$|\Theta_i|_\infty = |\Gamma_i|_\infty.$$

Again we will often drop the subscript if we consider a fixed boundary component. We now introduce a convex combination of $|\cdot|_\infty$ and $|\cdot|_{\mathcal{H}}$. For an $h_* > 0$ to be specified below we set

$$\Theta|_* = h_*|\Theta|_{\mathcal{H}} + (1 - h_*)|\Theta|_\infty. \tag{8.3}$$

For sets $W \in \mathcal{V}^s$ we then define

$$\|W\|_Z = \sum\nolimits_{j=1}^n |\Theta_j(W)|_Z$$

for $Z = \mathcal{H}, \infty, *$. Note that $\|W\|_\infty, \|W\|_{\mathcal{H}}$ and thus also $\|W\|_*$ are independent of the specific order which we have chosen in (8.2). Indeed, for $\|W\|_\infty$ this is clear as $|\Theta_i|_\infty = |\Gamma_i|_\infty$, for $\|W\|_{\mathcal{H}}$ it follows from the fact that $\|W\|_{\mathcal{H}} = \mathcal{H}^1(\bigcup_{i=1}^n \Gamma_i)$.

Before we introduce the modification procedure we collect some elementary properties of $|\cdot|_*$.

Lemma 8.1.1. *Let $W \subset Q_\mu$. Let $\Gamma = \Gamma(W)$ be a boundary component with $\Gamma = \partial X$ and let $\Theta \subset \Gamma$ be the corresponding set defined in* (8.2). *Moreover, let $V \in \mathcal{U}^s$ be a rectangle with $\overline{V} \cap \overline{X} \neq \emptyset$. Suppose that h_* is sufficiently small. Then*

(i) $|\Gamma|_* \ge |\partial R(\Gamma)|_*$ *if Γ is connected, where $R(\Gamma)$ denotes the smallest (closed) rectangle such that $\Gamma \subset R(\Gamma)$,*

(ii) $|\Theta|_* = |\Gamma|_* \Leftrightarrow |\Theta|_{\mathcal{H}} = |\Gamma|_{\mathcal{H}}$,

(iii) $|\partial(X \setminus \overline{V})|_\infty \le |\Theta|_\infty$ *and* $|\Theta \setminus \overline{V}|_{\mathcal{H}} \le |\Theta|_{\mathcal{H}}$,

(iv) $|\partial(V \cup X)|_* \leq |\partial V|_* + |\Gamma|_*$,

(v) $|\partial(V \cup X)|_* \geq |\partial R(V \cup X)|_*$ *if* $\overline{V} \cup \overline{X}$ *is connected, where* $R(V \cup X)$ *denotes the smallest rectangle such that* $V \cup X \subset R(V \cup X)$.

Now assume that $\Gamma = \partial R$ *for a rectangle* $R \in \mathcal{U}^s$. *Then*

(vi) $\frac{1}{\sqrt{2}}|\Gamma|_{\mathcal{H}} \leq 2|\Gamma|_\infty \leq |\Gamma|_{\mathcal{H}}$,

(vii) $|\partial(V \cup R)|_* \leq |\partial V|_* + \frac{1}{2}|\Gamma|_*$ *provided that* $\Gamma \setminus \overline{V}$ *is not connected and* $|\Gamma|_\infty \leq c|\partial V|_\infty$ *for a constant* $c > 0$ *sufficiently small.*

Proof. If Γ is connected, we obtain $|\Gamma|_{\mathcal{H}} \geq |\partial R(\Gamma)|_{\mathcal{H}}$ and $|\Gamma|_\infty = |\partial R(\Gamma)|_\infty$. This yields (i) and likewise we obtain (v). Assertions (ii)-(iv) follow directly from the definition of $|\cdot|_*$, where in (ii) we particularly use $|\Theta|_\infty = |\Gamma|_\infty$. Claim (vi) is elementary. To see (vii) we assume without restriction $V = (-a, a) \times (0, b)$ and $\pi_1\Gamma = (-d, d)$ with $d > a$ as well as $\pi_2\Gamma \subset (0, b)$. An elementary calculation yields $|\partial(V \cup R)|_\infty = \sqrt{(2d)^2 + b^2} \leq b + \frac{(2d)^2}{2b} \leq b + \frac{2d}{4} \leq |\partial V|_\infty + \frac{1}{4}|\Gamma|_\infty$. Here we used that $4d \leq b$ for c small enough. As $|\partial(V \cup R)|_{\mathcal{H}} \leq |\partial V|_{\mathcal{H}} + |\Gamma|_{\mathcal{H}}$ the claim now follows from (8.3) and (vi) if we choose h_* small enough. □

The properties stated here will be exploited frequently and we will not always refer to this lemma.

One method of the modification procedure below will be the 'combination' of different boundary components by adding additional sets to the original boundary (see case c) in the proof of Theorem 8.4.2). To keep track of the components we already 'used' to modify the boundary, it is convenient to introduce a weight $\omega_{\min} \leq \omega(\Gamma_j) \leq 1$ for all $\Gamma_j = \Gamma_j(W)$ with $\frac{1}{2} \leq \omega_{\min} < 1$ to be specified below. We define $|\Theta_j|_{Z,\omega} = \omega(\Gamma_j)|\Theta_j|_Z$ and likewise a weighted version of $\|\cdot\|_Z$ by setting

$$\|W\|_{Z,\omega} := \sum\nolimits_j \omega(\Gamma_j)|\Theta_j|_Z \tag{8.4}$$

for $Z = \mathcal{H}, \infty, *$. For $Z = *$ we write for shorthand $|\cdot|_\omega = |\cdot|_{*,\omega}$ and $\|\cdot\|_\omega = \|\cdot\|_{*,\omega}$. We briefly note that in contrast to $\|\cdot\|_*$, the value of (8.4) depends on the order given in (8.2) and therefore we will always consider a specific order of the boundary components in the following.

8.2 Modification of sets

For $\lambda \geq 0$ and fixed small $v > 0$ let $\mathcal{W}^s_\lambda \subset \mathcal{V}^s$ be the subset consisting of the sets $W \in \mathcal{V}^s$ with a corresponding weight ω and an ordering of the boundary components $(\Gamma_i)_{i=1}^n$ such that the following properties are satisfied:

$$\begin{array}{lll} (i) & \Theta_i \subset \partial R_i,\ \Gamma_i \subset \overline{R_i} \ \text{ for a rectangle } R_i & \forall\, \Gamma_i:\ \omega(\Gamma_i) < 1, \\ (ii) & |\partial R_i|_* \le \omega_{\min}^{-1}\omega(\Gamma_i)|\Theta_i|_* & \forall\, \Gamma_i:\ \omega(\Gamma_i) < 1, \\ (iii) & R_i \setminus X_j \text{ is connected for all } j = 1,\dots,n & \forall\, \Gamma_i:\ \omega(\Gamma_i) < 1, \\ (iv) & \omega(\Gamma_i) = 1 & \forall\, \Gamma_i:\ |\Gamma_i|_\infty \ge 19\upsilon\lambda, \\ (v) & \Gamma_i = \Theta_i = \partial R_i \text{ for a rectangle } R_i & \forall\, \Gamma_i:\ \omega(\Gamma_i) = 1. \end{array} \tag{8.5}$$

Observe that (iv),(v) imply that boundary components larger than $19\upsilon\lambda$ are always rectangular and pairwise disjoint. In particular, $\mathcal{W}_0^s$ consists of the sets where all boundary components are rectangular. By an elementary argumentation taking (8.3), (8.5)(i),(ii) into account and recalling $\omega_{\min} \ge \frac{1}{2}$, $h_* \ll 1$, we observe

$$|\Gamma_i|_\infty \le |\partial R_i|_\infty \le C|\Gamma_i|_\infty \quad \forall\, \Gamma_i:\ \omega(\Gamma_i) < 1, \tag{8.6}$$

i.e. the diameter of Γ_i and the corresponding rectangle R_i are comparable.

Consider a set $W = Q_\mu \setminus \bigcup_{i=1}^m X_i \in \mathcal{W}_\lambda^s$, $\lambda \ge 0$, and a rectangle $V \in \mathcal{U}^s$ with $|\partial V|_\infty \ge \lambda$ and $\overline{V} \subset Q_\mu$. We define the modification

$$\tilde{W} = Q_\mu \setminus \bigcup\nolimits_{i=0}^m \tilde{X}_i, \tag{8.7}$$

where $\tilde{X}_i = X_i \setminus \overline{V}$ for $i = 1,\dots,m$ and $\tilde{X}_0 = V$. We observe that $\tilde{W} = (W \setminus V) \cup \partial V$ (as a subset of $\mathbb{R}^2$). Therefore, for shorthand we will write $\tilde{W} = (W \setminus V) \cup \partial V$ to indicate the element of $\mathcal{V}^s$ which is given by (8.7).

We have the following boundary components of $\tilde{W}$: First let $\Gamma_0(\tilde{W}) = \partial V$ (it is convenient to start with index 0) and for $j \ge 1$ we have by construction $\Gamma_j(\tilde{W}) = \partial(X_j \setminus \overline{V})$. Observe that some boundary components may be empty and therefore reordering the indices we let $(\Gamma_j(\tilde{W}))_{j=1}^{\tilde{n}}$ for $\tilde{n} \le n$ be the nonempty boundary components. Clearly, for each $\Gamma_j(\tilde{W})$, $j \ge 1$, there is exactly one corresponding $\partial X_{i_j} = \Gamma_{i_j}(W)$ such that $\Gamma_j(\tilde{W}) = \partial(X_{i_j} \setminus \overline{V})$. (This mapping is injective.) We order the components of $\tilde{W}$ such that $1 \le j_1 < j_2$ if and only if $i_{j_1} < i_{j_2}$, i.e. we preserve the ordering of W.

We now define the corresponding subsets as in (8.2) and obtain $\Theta_0(\tilde{W}) = \partial V$ as well as $\Theta_j(\tilde{W}) = \Theta_{i_j}(W) \setminus \overline{V}$ for $j \ge 1$. Moreover, we choose the same corresponding rectangles as given for W by (8.5)(i), i.e. for $\Gamma_j(\tilde{W})$ with $\omega(\Gamma_j(\tilde{W})) < 1$ we define $R_j(\tilde{W}) = R_{i_j}(W)$.

From now on for notational convenience we may assume that $i_j = j$ for all $j \ge 1$. We obtain the following 'new' weights: Set $\omega(\Gamma_0(\tilde{W})) = 1$ and for $j \ge 1$

$$\omega(\Gamma_j(\tilde{W})) = \begin{cases} 1 & \text{if } \omega(\Gamma_j(W)) = 1, \\ \min\left\{\frac{|\Theta_j(W)|_*}{|\Theta_j(\tilde{W})|_*}\omega(\Gamma_j(W)), 1\right\} & \text{else.} \end{cases} \tag{8.8}$$

We note that

$$\omega(\Gamma_j(\tilde{W})) \geq \omega(\Gamma_j(W)) \quad \text{and} \quad \omega(\Gamma_j(\tilde{W}))|\Theta_j(\tilde{W})|_* \leq \omega(\Gamma_j(W))|\Theta_j(W)|_* \quad (8.9)$$

for all $j \geq 1$. To see this, it suffices to show $|\Theta_j(\tilde{W})|_* \leq |\Theta_j(W)|_*$. This follows from Lemma 8.1.1(iii) and the observation that by construction (recall in particular (8.2)) we have $\Gamma_j(\tilde{W}) = \partial(X_j \setminus \overline{V})$ and $\Theta_j(\tilde{W}) = \Theta_j(W) \setminus \overline{V}$.

Note that $\tilde{W}$ might not be an element of $\mathcal{W}^s_\lambda$. We now show, however, that $\tilde{W}$ can be modified to a set in $\mathcal{W}^s_\lambda$.

Lemma 8.2.1. *Let $\lambda \geq 0$ and $W \in \mathcal{W}^s_\lambda$. Let $\tilde{W} = (W \setminus V) \cup \partial V$ for a rectangle $V \in \mathcal{U}^s$ with $|\partial V|_\infty \geq \lambda$ and $\overline{V} \subset Q_\mu$. Then there is another rectangle $V' \in \mathcal{U}^s$ with $\overline{V} \subset \overline{V'} \subset Q_\mu$ such that $U := (W \setminus V') \cup \partial V' \in \mathcal{W}^s_\lambda$ and*

$$\|U\|_\omega \leq \|\tilde{W}\|_\omega, \quad (8.10)$$

where for both sets $\tilde{W}$, U we adjusted the weights as in (8.8).

Proof. Without restriction we can assume $\overline{V} \cap W \neq \emptyset$ as otherwise there is nothing to show. We first see that $\tilde{W}$ clearly satisfies (8.5)(i),(iv). (Recall that in (i) we take the same rectangles as for the boundary components of W.) To see (8.5)(ii) it suffices to note that for a given $\Theta_j(\tilde{W})$ with $\omega(\Gamma_j(\tilde{W})) < 1$, (8.8) implies $\omega(\Gamma_j(\tilde{W}))|\Theta_j(\tilde{W})|_* = \omega(\Gamma_j(W))|\Theta_j(W)|_*$. Possibly (8.5)(iii) or (8.5)(v) are violated, i.e. there are $\Gamma_{j_i}(\tilde{W})$, $i = 1, \ldots, k$, with $\omega(\Gamma_{j_i}(\tilde{W})) = 1$ such that $\Gamma_{j_i}(\tilde{W})$ is not rectangular or $\Gamma_{j_i}(\tilde{W}) \neq \Theta_{j_i}(\tilde{W})$ or there are sets $\Theta_{j_i}(\tilde{W})$, $i = k+1, \ldots, l$, such that for the corresponding rectangles R_{i_j} given by (8.5)(i) one has that $R_{j_i} \setminus X$ is disconnected for a suitable component X. Note that $\partial V \cap \Gamma_{j_i}(\tilde{W}) \neq \emptyset$ for $i = 1, \ldots, l$ as $W \in \mathcal{W}^s_\lambda$. So it remains to modify $\tilde{W}$ iteratively to obtain a set satisfying (8.5)(iii) and (8.5)(v).

Set $W_0 = \tilde{W}$ and $V_0 = V$. Assume $W_i = (W \setminus V_i) \cup \partial V_i \subset \tilde{W}$ has been constructed, where $V_i \in \mathcal{U}^s$ is a rectangle with $V \subset V_i$. Moreover, suppose that (8.10) holds replacing U by W_i and that W_i satisfies (8.5)(i),(ii),(iv) and

$$\Gamma(W_i) \cap \partial V_i \neq \emptyset \quad \text{for all } \Gamma(W_i) \in \mathcal{F}_i \quad (8.11)$$

for the boundary component $\partial V_i = \Gamma_0(W_i)$ with $\omega(\Gamma_0(W_i)) = 1$. Here $\mathcal{F}_i = \mathcal{F}^1_i \cup \mathcal{F}^2_i$, where $\mathcal{F}^1_i$ denotes the set of the not rectangular boundary components $\Gamma(W_i)$ with $\omega(\Gamma(W_i)) = 1$ and $\mathcal{F}^2_i$ denotes the set of boundary components for which the corresponding rectangle is disconnected. Observe that $|\partial V_i|_\infty \geq \lambda$ as $|\partial V|_\infty \geq \lambda$. If now $W_i \in \mathcal{W}^s_\lambda$ (i.e $\mathcal{F}^1_i = \mathcal{F}^2_i = \emptyset$), we stop and set $U = W_i$. Otherwise, we choose $\hat{\Gamma} \in \mathcal{F}_i$. If $\hat{\Gamma} \in \mathcal{F}^1_i$ we let $V_{i+1} \in \mathcal{U}^s$ be the smallest (closed) rectangle containing V_i and $\hat{\Gamma}$. By Lemma 8.1.1(v) we get $|\partial V_{i+1}|_* \leq |\partial(V_i \cup \hat{X})|_*$, where $\hat{X}$ is the component of $Q_\mu \setminus W_i$ corresponding to $\hat{\Gamma}$. Now by Lemma 8.1.1(iv) and (8.5)(v) we obtain

$$|\partial V_{i+1}|_* \leq |\partial(V_i \cup \hat{X})|_* \leq |\partial V_i|_* + |\hat{\Gamma}|_* = |\partial V_i|_\omega + |\hat{\Theta}|_\omega$$

for $\hat{\Theta} \subset \hat{\Gamma}$ as given by (8.2). If $\hat{\Gamma} \in \mathcal{F}_i^2 \setminus \mathcal{F}_i^1$ we let $V_{i+1} \in \mathcal{U}^s$ be the smallest rectangle containing V_i and $\partial\hat{R}$, where $\hat{R}$ is the corresponding rectangle given by (8.5)(i). By (8.5)(ii) and $\omega_{\min} \geq \frac{1}{2}$ we derive $|\partial\hat{R}|_* \leq 2|\hat{\Theta}|_\omega$. Moreover, the fact that (8.5)(iii) holds for W, is violated for W_i and $W_i = (W \setminus V_i) \cup \partial V_i$ implies $\hat{R} \setminus V_i$ is disconnected. As $|\hat{\Gamma}|_\infty \leq 19v|\partial V_i|_\infty$ by (8.5)(iv) and thus $|\partial\hat{R}|_\infty \leq Cv|\partial V_i|_\infty$ by (8.6), Lemma 8.1.1(v),(vii) then yields for v sufficiently small

$$|\partial V_{i+1}|_* \leq |\partial(V_i \cup \hat{R})|_* \leq |\partial V_i|_* + \tfrac{1}{2}|\partial\hat{R}|_* \leq |\partial V_i|_\omega + |\hat{\Theta}|_\omega.$$

Let $W_{i+1} = (W_i \setminus V_{i+1}) \cup \partial V_{i+1}$ (recall (8.7)) and adjust the weights of the boundary components of W_{i+1} as in (8.8). Recall that for all $\Gamma_j(W_{i+1})$ with $\Gamma_j(W_{i+1}) \neq \partial V_{i+1}$ we find a (unique) corresponding $\Gamma_j(W_i)$ with $\Gamma_j(W_i) \neq \partial V_i, \hat{\Gamma}$. By (8.9) we then derive

$$\begin{aligned}\|W_{i+1}\|_\omega &= |\partial V_{i+1}|_* + \sum\nolimits_{\Gamma_j(W_{i+1}) \neq \partial V_{i+1}} \omega(\Gamma_j(W_{i+1}))|\Theta_j(W_{i+1})|_* \\ &\leq |\partial V_i|_\omega + |\hat{\Theta}|_\omega + \sum\nolimits_{\Gamma_j(W_i) \neq \partial V_i, \hat{\Gamma}} \omega(\Gamma_j(W_i))|\Theta_j(W_i)|_* = \|W_i\|_\omega.\end{aligned} \tag{8.12}$$

Consequently, (8.10) still holds and arguing as before we see that W_{i+1} satisfies (8.5)(i),(ii),(iv). Moreover, (8.5) (iii),(v) can only be violated if (8.11) holds with $\mathcal{F}_{i+1} \neq \emptyset$. We now continue with iteration step $i+1$ and observe that after a finite number of steps i^* we find a rectangle $V_{i^*} \supset V$ and a set $W_{i^*} = (W \setminus V_{i^*}) \cup \partial V_{i^*} \in \mathcal{W}_\lambda^s$ as in each step the number of boundary components decreases. Define $V' = V_{i^*}$ and $U = (W \setminus V') \cup \partial V'$.

Note that U and W_{i^*} coincide as sets in $\mathbb{R}^2$, but the weights have been obtained in a different way. Therefore, to see (8.10) it remains to show $\omega(\Gamma_j(W_{i^*})) = \omega(\Gamma_j(U))$ for all boundary components Γ_j. For $\Gamma_j = \partial V'$ it suffices to recall that $\omega(\partial V') = \omega(\partial V_i) = 1$ for all $1 \leq i \leq i^*$. If $\Gamma_j \cap \partial V' = \emptyset$ it follows from the fact that Γ_j has not been changed during the modification procedure. Otherwise, as $W_{i^*} \in \mathcal{W}_\lambda^s$ and thus boundary components of W_{i^*} with weight 1 are pairwise disjoint (see (8.5)(v)), we know that $\omega(\Gamma_j(W_{i^*})) < 1$. Let $\Theta_j \subset \Gamma_j$ as given in (8.2) and let $\Theta_j(\tilde{W})$ be the component corresponding to Θ_j. Then by iterative application of (8.9) we get $\omega(\Gamma_j(\tilde{W})) \leq \omega(\Gamma_j(W_{i^*})) < 1$ and thus using iteratively (8.8) we find

$$\omega(\Gamma_j(W_{i^*}))|\Theta_j|_* = \omega(\Gamma_j(W_{i^*-1}))|\Theta_j(W_{i^*-1})|_* = \ldots = \omega(\Gamma_j(\tilde{W}))|\Theta_j(\tilde{W})|_*.$$

Consequently, again employing (8.8) we derive $\omega(\Gamma_j(W_{i^*})) = \frac{|\Theta_j(\tilde{W})|_*}{|\Theta_j|_*}\omega(\Gamma_j(\tilde{W})) = \omega(\Gamma_j(U))$, as desired. □

As a direct consequence of the above result, we get that sets in $\mathcal{V}^s$ can be modified such that the boundary components have rectangular form.

Corollary 8.2.2. *Let $W \in \mathcal{V}^s$ with connected boundary components. Then there is a subset $U \subset W$ such that $|W \setminus U| \leq c\|U\|_\infty^2$ for some $c > 0$ and all boundary components of U are rectangular and pairwise disjoint. Moreover, we have*

$$\|U\|_* \leq \|W\|_*.$$

In particular, if we introduce a weight ω corresponding to U by $\omega(\Gamma_j(U)) = 1$ for all j and define an (arbitrary) ordering of the boundary components we obtain $U \in \mathcal{W}_0^s$.

Proof. We follow the lines of the previous proof. Set $W_0 = W$ and assume $W_i \subset W$ has been constructed with $\|W_i\|_* \leq \|W\|_*$. If $W_i \in \mathcal{W}_0^s$ we stop, otherwise we find a component $\Gamma = \Gamma(W_i)$ which is not rectangular. Let $W_{i+1} = (W_i \setminus R(\Gamma)) \cup \partial R(\Gamma)$, where $R(\Gamma)$ is the smallest closed rectangle which contains Γ and all components Γ_j with $\Gamma_j \cap \Gamma \neq \emptyset$. Using Lemma 8.1.1(i) we clearly have $|\partial R(\Gamma)|_* \leq |\Gamma|_*$. As in the previous proof, in particular by (8.12), we then get $\|W_{i+1}\|_* \leq \|W_i\|_* \leq \|W\|_*$. We now continue with iteration step $i+1$ and note that we find the desired set U after a finite number of iterations.

Let $(\Gamma_i)_i$ be the boundary components of U with corresponding sets $(X_i)_i$. It is elementary to see that $W \setminus U \subset \bigcup_i X_i$ and thus by the isoperimetric inequality we conclude $|W \setminus U| \leq \sum_i |X_i| \leq C \sum_i |\Gamma_i|^2 \leq C\|U\|_\infty^2$. □

8.3 Neighborhoods of boundary components

Consider $W \in \mathcal{W}_\lambda^s$, $\lambda \geq 0$. In this section we concern ourselves with neighborhoods of a boundary component $\Gamma = \Gamma(W)$ with $\omega(\Gamma) = 1$ and $|\Gamma|_\infty \geq \lambda$. This implies that Γ has rectangular shape by (8.5)(v). We begin with a rectangular neighborhood and show that essentially the neighborhood can contain at most two other 'large' boundary components. Afterwards we will introduce a dodecagonal neighborhood. The main condition which will allow us to investigate properties of the neighborhoods will be the following minimality condition for $\|\cdot\|_\omega$: We require

$$\|\tilde{W}\|_\omega \geq \|W\|_\omega \quad \text{for all rectangles } V \in \mathcal{U}^s \text{ with } \Gamma \subset \overline{V} \subset Q_\mu, \tag{8.13}$$

where $\tilde{W} = (W \setminus V) \cup \partial V$ (recall (8.7)) and the weights are adjusted as in (8.8). In Section 8.4 we will see that (8.13) is one of the necessary conditions such that a trace estimate on Γ can be established (see Theorem 8.4.1). On the contrary, if (8.13) is violated for some $\tilde{W}$, we will show that it is convenient to replace W by $\tilde{W}$ (see case a) in the proof of Theorem 8.4.2).

Without restriction let $\Gamma = \partial X$ with $X = (-l_1, l_1) \times (-l_2, l_2)$ for $0 < l_2 \leq l_1$ and $l_1, l_2 \in s\mathbb{N}$.

8.3.1 Rectangular neighborhood

This section is devoted to the definition and properties of rectangular neighborhoods of Γ. As the technical proofs in this part are in principle not relevant to understand the proof of the main result in Section 8.4, they may be omitted on first reading. The essential points in this section are the definition of the neighborhood $N^t(\Gamma)$ (cf. Figure 8.2), the choice of the size of the neighborhoods (see (8.14) and (8.26)) and the properties that the length of ∂W in $N^t(\Gamma)$ can be controlled (see Lemma 8.3.1) as well as that there are at most two other 'large' boundary components (see Corollary 8.3.4 and Figure 8.3). Moreover, Lemma 8.3.5 shows that up to two small exceptional sets one can find a covering of the neighborhood (see Figure 8.4) such that on each element the projection $\|\cdot\|_\pi$ (see (8.18)) can be controlled which will be essential for a slicing argument in the proof of Theorem 8.4.1.

For $t \in s\mathbb{N}$ with $t \ll l_1$ we set

$$N^t(\Gamma) := (-t - l_1, l_1 + t) \times (-t - l_2, l_2 + t) \setminus \overline{X},$$
$$N^t_{j,\pm}(\Gamma) := N^t(\Gamma) \cap \{\pm x_j \geq l_j\} \quad \text{for } j = 1, 2.$$

(in the following we will use $\pm$ for shorthand if something holds for sets with index $+$ and $-$.) We drop Γ in the brackets if no confusion arises.

We cover $N^t_{2,\pm}$ up to a set of measure 0 with disjoint translates of a 'quasi square' $(0, \tilde{t}) \times (0, t)$, $\frac{\tilde{t}}{t} \approx 1$. If $l_2 \geq \frac{t}{2}$ we cover $N^t_{1,\pm} \setminus (N^t_{2,-} \cup N^t_{2,+})$ with translates of the rectangle $(0, t) \times (0, a)$ with $\frac{1}{2}t \leq a \leq t$. By $E^t_{\pm,\pm}$ we denote the four squares in the corners whose boundaries contain the points $(\pm l_1, \pm l_2)$, respectively. For $l_2 < \frac{t}{2}$ we cover each $N^t_{1,\pm}$ by itself, i.e. by a translate of the rectangle $(0, t) \times (0, 2t + 2l_2)$. For convenience we will often refer to these sets as 'squares' in the following. We number the squares by $Q^t_0, Q^t_1, \ldots, Q^t_n = Q^t_0$ such that $\overline{Q^t_j} \cap \overline{Q^t_{j+1}} \neq \emptyset$ for $j = 0, \ldots, n-1$ and let $J^t = \{Q^t_1, \ldots, Q^t_n\}$.

For shorthand we define $\bar{\tau} = v|\Gamma|_\infty$ for $0 < v \ll 1$ and we will assume that (possibly by passing to a smaller s)

$$\bar{\tau} = v|\Gamma|_\infty \in s\mathbb{N} \quad \text{and} \quad \bar{\tau} \gg s. \tag{8.14}$$

This assures that all the neighborhoods we consider below can be chosen as elements of $\mathcal{U}^s$. Let $(\Gamma_j)_j = (\Gamma_j(W))_j$ be the boundary components of W and $(\Theta_j)_j$ the corresponding subsets defined by (8.2). Let $(R_j)_j$ be the associated rectangles as given in (8.5)(i) and (8.5)(v), respectively. We will always add a subscript to avoid a mix up with Γ.

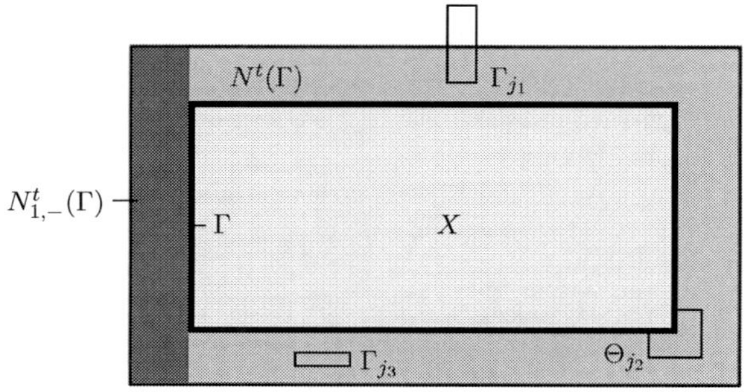

Figure 8.2: Neigborhood $N^t(\Gamma)$ with other small boundary components. The part $N^t_{1,-}(\Gamma)$ is colored in dark grey.

Lemma 8.3.1. *Let $\lambda \geq 0$ and $c > 0$. Let $W \in \mathcal{W}^s_\lambda$ and let Γ be a boundary component with $\omega(\Gamma) = 1$ and $|\Gamma|_\infty \geq \lambda$. Assume that* (8.13) *holds. Then there is a constant $C = C(c)$ such that*

$$\begin{aligned} &(i) \quad |\partial W \cap N^t|_{\mathcal{H}} \leq C\tfrac{t}{h_*} &&\text{for all} \quad t \geq c\bar{\tau}, \\ &(ii) \quad |\Gamma_j \cap N^t|_{\mathcal{H}} \leq C\tfrac{t}{h_*} &&\text{for all} \quad t \in s\mathbb{N} \quad \text{and all } \Gamma_j \text{ with } \omega(\Gamma_j) = 1. \end{aligned}$$

Proof. (i) Let $V = (-l_1-\hat{t}, l_1+\hat{t}) \times (-l_2-\hat{t}, l_2+\hat{t}) \in \mathcal{U}^s$, where $\hat{t} = 2\max\{t, 19C\bar{\tau}\}$ with the constant C from (8.6). Define $\tilde{W} = (W \setminus V) \cup \partial V$ and adjust the weights as in (8.8). It is not hard to see that $|\partial V|_* \leq |\Gamma|_* + 8\hat{t}$.

Let $\mathcal{F}$ be the set of boundary components having nonempty intersection with N^t and let $\mathcal{G} \subset \mathcal{F}$ be the subset satisfying $\omega(\Gamma_j(W)) = 1$ for $\Gamma_j(W) \in \mathcal{G}$. By (8.5)(iv) and (8.6) we find $\Theta_j(W) \subset \partial R_j \subset V$ for $\Gamma_j(W) \in \mathcal{F} \setminus \mathcal{G}$. Recall that due to the choice of $\hat{t}$ for all $\Gamma_j(\tilde{W}) \in \mathcal{F}$ with $\Gamma_j(\tilde{W}) \neq \partial V$ we find a (unique) corresponding $\Gamma_j(W) \in \mathcal{G}$ with $\Gamma_j(W) \neq \partial V, \Gamma$. (Without restriction we take the same index.) For $\Gamma_j(W) \in \mathcal{G}$ it is elementary to see that $|\Theta_j(\tilde{W})|_* \leq |\Gamma_j(W)|_* - h_*|\Gamma_j(W) \cap V|_{\mathcal{H}}$. Consequently, using $\omega(\partial V) = \omega(\Gamma) = 1$ we derive

$$\begin{aligned} \|\tilde{W}\|_\omega &= |\partial V|_* + \sum\nolimits_{\Gamma_j(\tilde{W}) \neq \partial V} \omega(\Gamma_j(\tilde{W}))|\Theta_j(\tilde{W})|_* \\ &\leq |\Gamma|_\omega + 8\hat{t} + \sum\nolimits_{\Gamma_j(W) \in \mathcal{G}} (|\Gamma_j(W)|_* - h_*|\Gamma_j(W) \cap V|_{\mathcal{H}}) \\ &\quad + \sum\nolimits_{\Gamma_j(W) \notin \mathcal{F}} \omega(\Gamma_j(W))|\Theta_j(W)|_* \\ &\leq \|W\|_\omega + 8\hat{t} - \sum_{\Gamma_j(W) \in \mathcal{G}} h_*|\Gamma_j(W) \cap V|_{\mathcal{H}} - \sum_{\Gamma_j(W) \in \mathcal{F} \setminus \mathcal{G}} \omega(\Gamma_j(W))|\Theta_j(W)|_* \\ &\leq \|W\|_\omega + 8\hat{t} - \omega_{\min} h_* |\partial W \cap N^t|_{\mathcal{H}}. \end{aligned}$$

For the components not being in $\mathcal{F}$ we proceeded as in (8.12). Since $\|W\|_\omega \le \|\tilde{W}\|_\omega$ by condition (8.13) and $\omega_{\min} \ge \frac{1}{2}$, we find $|\partial W \cap N^t|_{\mathcal{H}} \le C\frac{\hat{t}}{h_*} \le C\frac{t}{h_*}$, where in the last step we used $t \ge c\bar{\tau}$.

(ii) We argue as in (i) with the difference that we set $\hat{t} = t$ and $\mathcal{F} = \mathcal{G} = \{\Gamma_j\}$. Then repeating the above calculation we obtain

$$\|\tilde{W}\|_\omega \le \|W\|_\omega + 8\hat{t} - h_*|\Gamma_j \cap N^t|_{\mathcal{H}},$$

where for all other components we proceeded as in (8.12). We conclude by employing $\|W\|_\omega \le \|\tilde{W}\|_\omega$. □

We now analyze the components intersecting N^t more precisely. In particular, we will show that at most two large boundary components lie in the neighborhood of Γ (see Corollary 8.3.4). The properties can be established by exploiting elementary geometric arguments and essential ideas of the procedure are exemplarily illustrated in Figure 8.3.

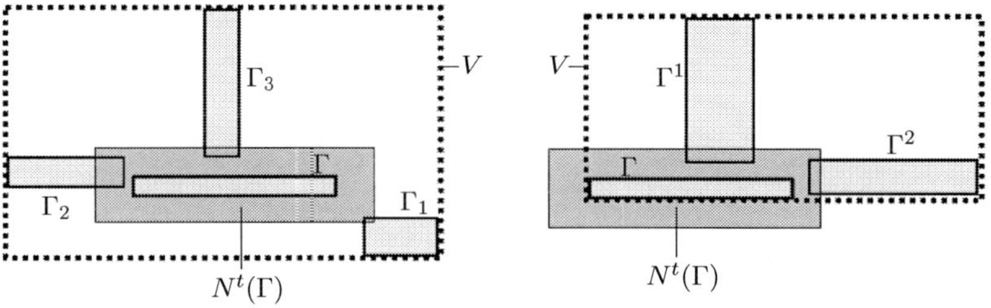

Figure 8.3: The left picture shows three boundary components $\Gamma_1, \Gamma_2, \Gamma_3$ intersecting $N^t(\Gamma)$. Below we argue that such a configuration violates (8.13) for $\tilde{W} = (W \setminus V) \cup \partial V$, where V is dotted rectangle. Indeed, one might have $|\partial V|_{\mathcal{H}} > |\Gamma|_{\mathcal{H}} + \sum_{k=1,2,3} |\Gamma_k|_{\mathcal{H}}$, but we can show that one always has $|\partial V|_\infty < |\Gamma|_\infty + \sum_{k=1,2,3} |\Gamma_k|_\infty$, whereby we obtain $|\partial V|_* < |\Gamma|_* + \sum_{k=1,2,3} |\Gamma_k|_*$ for h_* sufficiently small. Likewise, we can control the position of the at most two large components Γ^1, Γ^2 in $N^t(\Gamma)$: A configuration depicted on the right, where Γ^1, Γ^2 do not intersect opposite parts of $N^t(\Gamma)$, violates (8.13) for the dotted rectangle V.

We first introduce a coarser covering of N^t: Let $c\bar{\tau} \le t \le C\bar{\tau}$. Let $\mathcal{Y}^t$ be the union of connected sets Y having the form $Y = \left(\bigcup_{i=j}^k \overline{Q_i^t}\right)^\circ$ for $Q_i^t \in J^t$. Cover each set $N_{2,\pm}^t$ with seven sets $Y_{2,\pm}^j$ such that

$$|Y_{2,\pm}^j| \ge \bar{C}t|\Gamma|_\infty, \qquad \tfrac{1}{8}\bar{C}t|\Gamma|_\infty \le |Y_{2,\pm}^j \cap Y_{2,\pm}^{j+1}| \le \tfrac{1}{4}\bar{C}t|\Gamma|_\infty \tag{8.15}$$

for a constant $\bar{C} > 0$. If $l_2 \ge \frac{l_1}{2}$ we proceed likewise for $N_{1,\pm}^t$ passing possibly to a smaller constant $\bar{C}$. If $l_2 < \frac{l_1}{2}$ we cover $N_{1,\pm}^t$ by itself. Denote the covering by $\mathcal{C}^t = \mathcal{C}^t(\Gamma) = \{Y_1^t, \dots Y_m^t\}$ and order the sets in a way that $Y_i^t \cap Y_{i+1}^t \neq \emptyset$ for all $i = 1, \dots, m$, where by convention $Y_i^t = Y_{i \,\mathrm{mod} m}^t$. In particular, (8.15) implies $Y_{i \,\mathrm{mod} m}^t \cap Y_{j \,\mathrm{mod} m}^t = \emptyset$ for $|i - j| \geq 2$.

This construction implies that for υ sufficiently small

$$R_j \cap Y_i^t \neq \emptyset \quad \Rightarrow \quad R_j \cap Y_{i+l}^t = \emptyset \quad \text{for } |l| \geq 3 \tag{8.16}$$

for all R_j and $i = 1, \dots, n$. To see this, we first observe that

$$|\partial R_j \cap N^t|_{\mathcal{H}} \leq C t h_*^{-1}. \tag{8.17}$$

Indeed, if $|\Gamma_j|_\infty < 19\bar{\tau}$, we obtain $|\partial R_j|_\infty \leq C\bar{\tau}$ by (8.6) and thus $|\partial R_j|_{\mathcal{H}} \leq 2\sqrt{2}|R_j|_\infty \leq Ct$. Otherwise, recalling $|\Gamma|_\infty \geq \lambda$, by (8.5)(iv),(v) we have $\partial R_j = \Gamma_j$ and thus employing Lemma 8.3.1(ii) we get $|\partial R_j \cap N^t|_{\mathcal{H}} \leq C t h_*^{-1}$.

If now $\mathrm{dist}(Y_i^t, Y_{i+l}^t) \geq \bar{C}|\Gamma|_\infty$ for some $|l| \geq 3$, (8.16) follows as $|\partial R_j \cap N^t|_{\mathcal{H}} \leq C t h_*^{-1} \ll \bar{C}|\Gamma|_\infty$ for υ small enough (depending on h_*).

On the other hand, suppose $\mathrm{dist}(Y_i^t, Y_{i+l}^t) \leq \bar{C}|\Gamma|_\infty$. This is only possible in the case $l_2 \leq \frac{l_1}{2}$ if (up to interchanging $+$ and $-$) $Y_i^t \subset N_{2,+}^t \setminus (N_{1,-}^t \cup N_{1,+}^t)$, $Y_{i+l}^t \subset N_{2,-}^t$ and $\mathrm{dist}(Y_i^t, N_{1,\pm}^t) \geq c|\Gamma|_\infty$ or $\mathrm{dist}(Y_{i+l}^t, N_{1,\pm}^t) \geq c|\Gamma|_\infty$. Now assume that (8.16) was wrong. Then by (8.15) and $|\partial R_j \cap N^t|_{\mathcal{H}} \leq C t h_*^{-1} \ll \bar{C}|\Gamma|_\infty$ this would imply $R_j \cap N_{2,\pm}^t \neq \emptyset$ and $R_j \cap (N_{1,-}^t \cup N_{1,+}^t) = \emptyset$. But then we would get that $R_j \setminus X$ is not connected which contradicts (8.5)(iii).

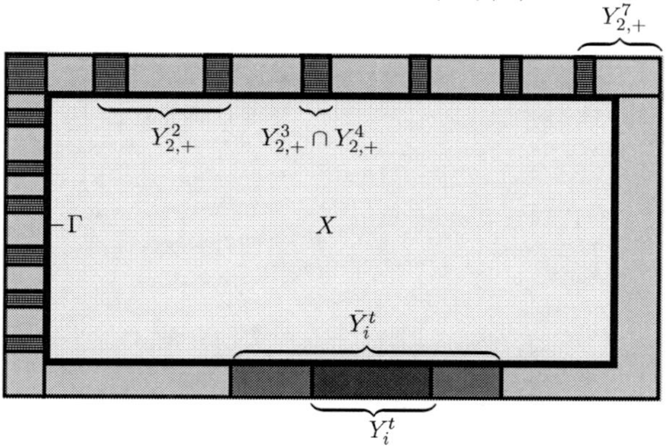

Figure 8.4: On the upper left side of the neigborhood $N^t(\Gamma)$ one can see elements of the partition $\mathcal{C}^t(\Gamma)$ (which are not necessarily of the same size). The sets where two elements overlap are striped. In the lower part an element Y_i^t and the corresponding enlarged set $\bar{Y}_i^t$ are highlighted.

For $Y \subset N^t$ we set $\mathcal{R}(Y) = \{R_j : R_j \cap Y \neq \emptyset\}$ and define

$$|\partial R_j|_\pi = \min\{|\partial R_j|_\infty,\ t - \max_{i=1,2} \mathrm{dist}(\pi_i R_j, \pi_i \Gamma)\} \tag{8.18}$$

for all $R_j \in \mathcal{R}(N^t)$. It is obvious that $|\partial R_j|_\pi \leq |\partial R_j|_\infty$. For a set $Y \subset N^t$ we then define $\|Y\|_\pi = \sum_{R_j \in \mathcal{R}(Y)} |\partial R_j|_\pi$. The projection $\|\cdot\|_\pi$ is one essential object we will need to apply a slicing argument in the investigation of the jump heights in Section 8.5.

Remark 8.3.2. We have already introduced the (small) parameters $h_*, 1-\omega_{\min}, \upsilon$. In the following sections we will additionally consider q, r. The subsequent lemmas will hold if we choose the involved parameters sufficiently small. To avoid confusion about the relation of the different parameters, we state at this point that the parameters can be chosen in the order $h_*, q, 1-\omega_{\min}, r, \upsilon$. In what follows, we will not always repeat the relation of the parameters for convenience.

We now show that we can control $\|\cdot\|_\pi$ in a suitable way. For that purpose, for a set $Y_i^t \in \mathcal{C}^t$ we define

$$\bar{Y}_i^t = \overline{\bigcup_{|l|\leq 1} Y_{i+l}^t}.$$

Lemma 8.3.3. *Let $\lambda \geq 0$, $W \in \mathcal{W}_\lambda^s$. Let Γ be a boundary component with $\omega(\Gamma) = 1$ and $|\Gamma|_\infty \geq \lambda$. Assume that* (8.13) *holds. Let $c\bar{\tau} \leq t \leq C\bar{\tau}$. If we choose h_*, υ and $1-\omega_{\min}$ small enough, there are two sets $Y^1, Y^2 \in \mathcal{C}^t$ such that $\|Y^t\|_\pi \leq \frac{19}{20}t$ for all $Y^t \in \mathcal{C}^t$ with $Y^t \cap (\bar{Y}^1 \cup \bar{Y}^2) = \emptyset$.*
Additionally, if $\|Y^1\|_\pi, \|Y^2\|_\pi \geq \frac{19}{20}t$, then $\bar{Y}^1 \cup \bar{Y}^2$ intersects both $N_{1,+}^t$ and $N_{1,-}^t$ or both $N_{2,+}^t$ and $N_{2,-}^t$. If $l_2 \leq \frac{l_1}{2}$, then $\bar{Y}^1 \cup \bar{Y}^2$ intersects $N_{1,+}^t$ and $N_{1,-}^t$.

We briefly remark that by similar arguments the additional statement can also be proved without the extra assumption $\|Y^1\|_\pi, \|Y^2\|_\pi \geq \frac{19}{20}t$. We omit the proof of this fact here as we will not need it in the following.
Proof. For convenience we drop the superscript t in the following proof. We proceed in two steps:

In a) we first show that it is not possible that there are three sets $Y^1, Y^2, Y^3 \in \mathcal{C}$ such that $\bar{Y}^k \cap Y^l = \emptyset$ if $k \neq l$ and $\|Y^k\|_\pi > \frac{19}{20}t$ for $k, l = 1, 2, 3$. Provided that a) is proven we can then select the two desired sets Y^1, Y^2 as follows:

(1) If $\|Y\|_\pi \leq \frac{19}{20}t$ for all $Y \in \mathcal{C}$, we can choose arbitrary sets Y^1, Y^2 satisfying the additional condition. Otherwise, we can assume that there is some Y^* with $\|Y^*\|_\pi > \frac{19}{20}t$.

(2) If $\|Y\|_\pi \leq \frac{19}{20}t$ for all $Y \in \mathcal{C}^t$ with $Y \cap \bar{Y}^* = \emptyset$, we set $Y^1 = Y^*$ and choose Y^2 arbitrarily such that the additional condition holds.

(3) Otherwise, we set $Y^1 = Y^*$ and choose Y^2 with $\|Y^2\|_\pi > \frac{19}{20}t$ and $Y^2 \cap \bar{Y}^* = \emptyset$. Now a) indeed shows that $\|Y\|_\pi \leq \frac{19}{20}t$ for all $Y \in \mathcal{C}^t$ with $Y \cap (\bar{Y}^1 \cup \bar{Y}^2) = \emptyset$.

In step b) we concern ourselves with the additional assertions on the position of $\bar{Y}^1 \cup \bar{Y}^2$ in case (3).

a) Suppose that there are three sets $Y^1, Y^2, Y^3 \in \mathcal{C}$ such that $\bar{Y}^k \cap Y^l = \emptyset$ if $k \neq l$ and $\|Y^k\|_\pi > \frac{19}{20}t$ for $k, l = 1, 2, 3$. First note that the assumption implies that if e.g. $Y^1 = Y_i$, then $Y^2, Y^3 \notin \{Y_{i-2}, \dots, Y_{i+2}\}$. Let V be the smallest rectangle containing Γ and the sets $\mathcal{R} := \bigcup_{k=1}^3 \mathcal{R}(Y^k)$. Define $\tilde{W} = (W \setminus V) \cup \partial V$ (recall (8.7)). Similarly as in (8.12) we intend to estimate $\|\tilde{W}\|_\omega$. To this end, we have to control the difference of $|\partial R_j|_*$ and $|\Theta_j|_\omega$ for $R_j \in \mathcal{R}$. By (8.5)(ii),(iv),(v) and (8.6) we have

$$|\Theta_j|_\omega = \omega(\Gamma_j)|\Theta_j|_* \geq \begin{cases} \omega_{\min}|\partial R_j|_* & |\partial R_j|_\infty \leq 19Cv\lambda, \\ |\partial R_j|_* & \text{else}, \end{cases}$$

with the constant C from (8.6). For notational convenience we define $\|W\|_{\omega,\mathcal{R}} = \|W\|_\omega + \sum_{R_j \in \mathcal{R}}(|\partial R_j|_* - |\Theta_j|_\omega) = \sum_{R_j \notin \mathcal{R}} |\Theta_j|_\omega + \sum_{R_j \in \mathcal{R}} |\partial R_j|_*$. We get

$$\begin{aligned} \|W\|_{\omega,\mathcal{R}} &\leq \|W\|_\omega + \sum\nolimits_{R_j \in \mathcal{R},\, |\partial R_j|_\infty < 19Cv\lambda} (\omega_{\min}^{-1} - 1)|\Theta_j|_\omega \\ &\leq \|W\|_\omega + (\omega_{\min}^{-1} - 1)\mathcal{H}^1(N^{t+C\bar{\tau}} \cap \partial W) \\ &\leq \|W\|_\omega + C\tfrac{t}{h_*}(\omega_{\min}^{-1} - 1). \end{aligned} \tag{8.19}$$

In the second step we used $|\cdot|_* \leq |\cdot|_\mathcal{H}$ and in the last step we applied Lemma 8.3.1(i). We will show below that

$$|\partial V|_\infty \leq |\Gamma|_\infty + \sum\nolimits_{R_j \in \mathcal{R}} |\partial R_j|_\infty - \tfrac{1}{50}t. \tag{8.20}$$

Moreover, it is not hard to see that $|\partial V|_\mathcal{H} \leq |\Gamma|_\mathcal{H} + \sum_{R_j \in \mathcal{R}} |\partial R_j|_\mathcal{H} + 8t$. Then recalling $\|W\|_{\omega,\mathcal{R}} = \sum_{R_j \notin \mathcal{R}} |\Theta_j|_\omega + \sum_{R_j \in \mathcal{R}} |\partial R_j|_*$ and arguing as in (8.12) we get $\|\tilde{W}\|_\omega - \|W\|_{\omega,\mathcal{R}} \leq -(1 - h_*)\frac{t}{50} + h_* 8t$. Consequently, for h_* small enough we get $\|\tilde{W}\|_\omega - \|W\|_{\omega,\mathcal{R}} < -\frac{t}{100}$ and thus by (8.19) we derive $\|\tilde{W}\|_\omega - \|W\|_\omega < 0$ for $1 - \omega_{\min}$ sufficiently small (with respect to h_*). This gives a contradiction to (8.13) and concludes the proof of a).

We now proceed to show (8.20). Assume $V = (-a_{1,-} - l_1, l_1 + a_{1,+}) \times (-a_{2,-} - l_2, l_2 + a_{2,+})$ and select (not necessarily pairwise different) $R_{k,\pm} \in \mathcal{R}$ such that $\pm(l_k + a_{k,\pm}) \in \pi_k \partial R_{k,\pm}$ for $k = 1, 2$. (If $a_{k,\pm} = 0$ then $R_{k,\pm} = \emptyset$.) We find by (8.18)

$$|\pi_k R_{k,\pm}| \geq a_{k,\pm} - \operatorname{dist}(\pi_k R_{k,\pm}, \pi_k \Gamma) \geq a_{k,\pm} - t + |\partial R_{k,\pm}|_\pi. \tag{8.21}$$

We suppose for the moment that $R_{k,+} \neq R_{k,-}$ for $k = 1, 2$. (In particular, this implies that three rectangles never coincide.) At the end of the proof we will briefly indicate how the following arguments can be adapted to the general case. We first assume that two rectangles coincide, e.g. $R = R_{1,-} = R_{2,-}$. By (8.21) and an elementary computation we obtain

$$\sqrt{a_{1,-}^2 + a_{2,-}^2} \leq |\partial R|_\infty + \sqrt{2}(t - |\partial R|_\pi) \leq |\partial R|_\infty + \sqrt{2}t - |\partial R|_\pi.$$

Otherwise, if e.g. $R_{1,-} \neq R_{2,-}$, again applying (8.21) we get

$$\begin{aligned}\sqrt{a_{1,-}^2 + a_{2,-}^2} &\leq \sqrt{(t + (|\partial R_{1,-}|_\infty - |\partial R_{1,-}|_\pi))^2 + (t + (|\partial R_{2,-}|_\infty - |\partial R_{2,-}|_\pi))^2} \\ &\leq \sqrt{2}t + |\partial R_{1,-}|_\infty - |\partial R_{1,-}|_\pi + |\partial R_{2,-}|_\infty - |\partial R_{2,-}|_\pi.\end{aligned}$$

Consequently, we obtain

$$F \leq 2\sqrt{2}t + \sum\nolimits_{k,\pm} (|\partial R_{k,\pm}|_\infty - |\partial R_{k,\pm}|_\pi), \tag{8.22}$$

where each rectangle is only counted once in the sum and

$$F = \sqrt{a_{1,-}^2 + a_{2,-}^2} + \sqrt{a_{1,+}^2 + a_{2,+}^2} \quad \text{or} \quad F = \sqrt{a_{1,-}^2 + a_{2,+}^2} + \sqrt{a_{1,+}^2 + a_{2,-}^2}.$$

Moreover, note that by assumption and (8.16) each boundary component R_j intersects at most one of the three sets Y^k. Therefore, as $|\cdot|_\infty \geq |\cdot|_\pi$ we obtain

$$\begin{aligned}|\partial V|_\infty &\leq |\Gamma|_\infty + F \leq |\Gamma|_\infty + 2\sqrt{2}t + \sum\nolimits_{R_j \in \mathcal{R}} (|\partial R_j|_\infty - |\partial R_j|_\pi) \\ &= |\Gamma|_\infty + \sum\nolimits_{R_j \in \mathcal{R}} |\partial R_j|_\infty + 2\sqrt{2}t - \sum\nolimits_{k=1}^{3} \|Y^k\|_\pi.\end{aligned} \tag{8.23}$$

As $\sum_k \|Y^k\|_\pi \geq 3 \cdot \frac{19}{20}t$ this gives (8.20).

b) Suppose that $Y^1, Y^2 \in \mathcal{C}$ with $\|Y^1\|_\pi, \|Y^2\|_\pi \geq \frac{19}{20}t$ have been chosen according to case (3) above. We show that $\bar{Y}^1 \cup \bar{Y}^2$ intersect both $N_{1,+}$ and $N_{1,-}$ or both $N_{2,+}$ and $N_{2,-}$ ($N_{1,\pm}$ if $l_2 \leq \frac{l_1}{2}$). Let V be the smallest rectangle containing Γ and the sets $\mathcal{R} := \bigcup_{k=1}^{2} \mathcal{R}(Y^k)$. Set $\tilde{W} = (W \setminus V) \cup \partial V$. As before we define for convenience $\|W\|_{\omega,\mathcal{R}} = \|W\|_\omega + \sum_{R_j \in \mathcal{R}} (|\partial R_j|_* - |\Theta_j|_\omega)$ and note that (8.19) holds. Observe that by assumption and (8.16) each boundary component R_j intersects at most one of the two sets Y^k.

(i) First we assume $\bar{Y}_1 \cup \bar{Y}_2$ intersects at most two adjacent parts of the neighborhood, e.g. $(\bar{Y}_1 \cup \bar{Y}_2) \cap (N_{1,+}^t \cup N_{2,+}^t) = \emptyset$. This implies $V = (-a_1 - l_1, l_1) \times (-a_2 - l_2, l_2)$. Selecting (not necessarily different) R_1, R_2 such that $-l_k - a_k \in \pi_k \partial R_k$ for $k = 1, 2$ and proceeding as in (8.22) we obtain

$$\sqrt{a_1^2 + a_2^2} \leq \sqrt{2}t + \sum\nolimits_{k=1,2} (|R_k|_\infty - |R_k|_\pi)$$

and therefore

$$|\partial V|_\infty \leq |\Gamma|_\infty + \sum\nolimits_{R_j \in \mathcal{R}} |\partial R_j|_\infty + \sqrt{2}t - \sum\nolimits_{k=1,2} \|Y^k\|_\pi. \tag{8.24}$$

Moreover, we have $|\partial V|_\mathcal{H} \leq |\Gamma|_\mathcal{H} + 4t + \sum_{R_j \in \mathcal{R}} |\partial R_j|_\mathcal{H}$ which together with $\sum_k \|Y^k\|_\pi \geq 2 \cdot \frac{19}{20}t$ implies $\|\tilde{W}\|_\omega - \|W\|_{\omega,\mathcal{R}} \leq -\frac{t}{100}$ for h_* small enough. Recalling (8.19) we again obtain a contradiction to (8.13) for $1 - \omega_{\min}$ sufficiently small.

(ii) We finally show the additional statement that $\bar{Y}_1 \cup \bar{Y}_2$ intersects $N^t_{1,\pm}$ in the case $l_2 \leq \frac{l_1}{2}$. Assume without restriction that $(\bar{Y}_1 \cup \bar{Y}_2) \cap N^t_{1,+}(\Gamma) = \emptyset$. Then we have $V = (-a_1 - l_1, l_1) \times (-a_- - l_2, l_2 + a_+)$ and select rectangles R_1, R_-, R_+ as before. (For the moment we assume $R_{k,+} \neq R_{k,-}$ for $k = 1, 2$.) Similarly as in a), we find

$$\sqrt{a_1^2 + a_-^2} \leq \sqrt{2}t + \sum\nolimits_{k=1,-} (|\partial R_k|_\infty - |\partial R_k|_\pi), \quad a_+ \leq t + |\partial R_+|_\infty - |\partial R_+|_\pi.$$

Then

$$|\partial V|_\infty = \max\nolimits_{c\in[0,1]} \left(\sqrt{1-c^2}|\pi_1\Gamma| + c|\pi_2\Gamma| + \sqrt{1-c^2}a_1 + c(a_+ + a_-)\right).$$

We define $f(c) = \frac{2}{\sqrt{5}}(\sqrt{1-c^2} + \frac{1}{2}c)$ for $c \geq \frac{1}{\sqrt{5}}$ and $f(c) = 1$ else. As $l_2 \leq \frac{l_1}{2}$ an elementary argument yields $\sqrt{1-c^2}|\pi_1\Gamma| + c|\pi_2\Gamma| \leq f(c)|\Gamma|_\infty$. Using $|\Gamma|_\infty \geq (Cv)^{-1}t$ we then obtain

$$\begin{aligned} |\partial V|_\infty &\leq \max\nolimits_{c\in[0,1]} \left(f(c)|\Gamma|_\infty + \sqrt{2}t + ct + \sum\nolimits_{k=1,\pm} (|\partial R_k|_\infty - |\partial R_k|_\pi)\right) \\ &\leq |\Gamma|_\infty + t\max\nolimits_{c\in[0,1]} r(c) + \sum\nolimits_{R_j\in\mathcal{R}} |\partial R_j|_\infty - \sum\nolimits_{k=1,2} \|Y^k\|_\pi, \end{aligned} \tag{8.25}$$

where $r(c) = (f(c)-1)(Cv)^{-1} + \sqrt{2} + c$. A computation yields $r(c) \leq \frac{19}{10} - \frac{1}{100}$ for $c \leq \sqrt{\frac{9}{40}}$ as $f \leq 1$. Otherwise we have $\max_{[\sqrt{\frac{9}{40}},1]}(f(c)-1) < 0$ and thus for v sufficiently small we also obtain $r(c) \leq \frac{19}{10} - \frac{1}{100}$ for $c \in [\sqrt{\frac{9}{40}}, 1]$. Moreover, we have $|\partial V|_{\mathcal{H}} \leq |\Gamma|_{\mathcal{H}} + 6t + \sum_{R_j\in\mathcal{R}} |\partial R_j|_{\mathcal{H}}$ which together with $\sum_k \|Y^k\|_\pi \geq 2 \cdot \frac{19}{20}t$ implies $\|\tilde{W}\|_\omega - \|W\|_\omega < 0$ for $h_*, 1 - \omega_{\min}$ small enough. This again yields the desired contradiction.

To finish the proof we briefly indicate how to proceed if e.g. $R_{2,-} = R_{2,+}$. This may happen in the cases a) and b)ii) above if $l_2 \ll l_1$. In this case we reduce the problem to the above treated situation by applying a translation argument: We replace R_j by $R_j' := R_j - a_{2,+}\mathbf{e}_2$ for all $R_j \in \mathcal{R}$ as well as V by $V' := V - a_{2,+}\mathbf{e}_2$. Then we may set $R_{2,+} = \emptyset$ and can repeat the arguments above to derive (8.23) and (8.25), respectively, for V' and $R'_{k,\pm}$. But then (8.23) and (8.25) also hold for the original sets V and $R_j \in \mathcal{R}$ as $|V'|_* = |V|$ and $|R'_{k,\pm}|_* = |R_{k,\pm}|_*$. Consequently, we may then proceed as before and can employ (8.19) to derive a contradiction to (8.13). □

As a corollary we obtain that at most two large boundary components lie in the neighborhood of Γ.

Corollary 8.3.4. *Let $\lambda \geq 0$, $W \in \mathcal{W}^s_\lambda$. Let Γ be a boundary component with $\omega(\Gamma) = 1$ and $|\Gamma|_\infty \geq \lambda$. Assume that (8.13) holds. Let $\bar{\tau} \leq \bar{t} \leq C\bar{\tau}$. Then for h_*, v and $1 - \omega_{\min}$ small enough there are at most two boundary components Γ_1*

and Γ_2 with $|\Gamma_i|_\infty \geq 19\bar{t}$ having nonempty intersection with $N^{\bar{t}}$.
If Γ_1, Γ_2 exist, $\Gamma_1 \cup \Gamma_2$ intersects both $N^{\bar{t}}_{1,+}$ and $N^{\bar{t}}_{1,-}$ or both $N^{\bar{t}}_{2,+}$ and $N^{\bar{t}}_{2,-}$. Additionally, if $l_2 \leq \frac{l_1}{2}$ then $\Gamma_1 \cup \Gamma_2$ intersects both $N^{\bar{t}}_{1,+}$, $N^{\bar{t}}_{1,-}$ and $|\pi_1\Gamma_k| \geq \frac{1}{2}|\pi_2\Gamma_k|$ for $k = 1, 2$.

We remark that the additional statement $|\pi_1\Gamma_k| \geq \frac{1}{2}|\pi_2\Gamma_k|$ also holds if only one Γ_k exists.
Proof. Let $\bar{\tau} \leq \bar{t} \leq C\bar{\tau}$ be given and assume that there are three components Γ_k, $k = 1, 2, 3$, intersecting $N^{\bar{t}}$ with $|\Gamma_k|_\infty \geq 19\bar{t}$. By (8.5)(iv),(v), (8.14) and $\bar{t} \geq \bar{\tau}$ we see that Γ_k are rectangular with $\omega(\Gamma_k) = 1$ for $k = 1, 2, 3$. Set $t = 20\bar{t}$ and recalling (8.18) we observe that $|\Gamma_k|_\pi \geq 19\bar{t} = \frac{19}{20}t$. We now may follow the lines of the proof of Lemma 8.3.3 with the essential difference that we replace the set of rectangles $\mathcal{R} = \bigcup_{k=1}^3 \mathcal{R}(Y^k)$ (see beginning of step a)) by $\mathcal{R} = \{\Gamma_1\} \cup \{\Gamma_2\} \cup \{\Gamma_3\}$ and in (8.23) we replace $\sum_{k=1}^3 \|Y^k\|_\pi$ by $\sum_{k=1}^3 |\Gamma_k|_\pi$. Noting that $\sum_{k=1}^3 |\Gamma_k|_\pi \geq 3 \cdot \frac{19}{20}t$ we again obtain a contradiction to (8.13) and thus there are at most two large components Γ_k, $k = 1, 2$, in $N^{\bar{\tau}}$. Likewise, we can proceed to determine the possible position of the two sets.

It remains to show that $|\pi_1\Gamma_k| < \frac{1}{2}|\pi_2\Gamma_k|$ leads to a contradiction if $l_2 \leq \frac{l_1}{2}$. Let V be the smallest rectangle containing Γ, Γ_k and derive similarly as in (8.25)

$$\begin{aligned}|\partial V|_\infty &\leq \max\nolimits_{c\in[0,1]} \left(\sqrt{1-c^2}|\pi_1\Gamma| + c|\pi_2\Gamma| + \sqrt{1-c^2}(\bar{\tau} + |\pi_1\Gamma_k|) + c(\bar{\tau} + |\pi_2\Gamma_k|)\right)\\ &\leq \max\nolimits_{c\in[0,1]} \left(f(c)|\Gamma|_\infty + \sqrt{2}\bar{\tau} + \sqrt{1-c^2}|\pi_1\Gamma_k| + c|\pi_2\Gamma_k|)\right),\end{aligned}$$

where we used $\sqrt{1-c^2}|\pi_1\Gamma| + c|\pi_2\Gamma| \leq f(c)|\Gamma|_\infty$ due to the fact that $|\pi_2\Gamma| \leq \frac{1}{2}|\pi_2\Gamma|$. Likewise, we use the assumption $|\pi_1\Gamma_k| < \frac{1}{2}|\pi_2\Gamma_k|$ to find $\sqrt{1-c^2}|\pi_1\Gamma_k| + c|\pi_2\Gamma_k| \leq f(\sqrt{1-c^2})|\Gamma_k|_\infty$ and thus obtain

$$|\partial V|_\infty \leq |\Gamma|_\infty + |\Gamma_k|_\infty + \sqrt{2}\bar{\tau} + \max_{c\in[0,1]} \left((f(c)-1)\upsilon^{-1}\bar{\tau} + (f(\sqrt{1-c^2})-1)19\bar{t}\right),$$

where we used $|\Gamma|_\infty \geq \upsilon^{-1}\bar{\tau}$ and $|\Gamma|_\infty \geq 19\bar{t}$. Again separating the cases $c \geq \sqrt{\frac{9}{40}}$, where $\max_{[\sqrt{\frac{9}{40}},1]}(f(c)-1) < 0$, and $c \leq \sqrt{\frac{9}{40}}$, where $(f(\sqrt{1-c^2})-1)19\bar{\tau} \leq -3\bar{t}$, we obtain for υ small enough $|\partial V|_\infty \leq |\Gamma|_\infty + |\Gamma_k|_\infty - \bar{\tau}$. As $|\partial V|_\mathcal{H} \leq |\Gamma|_\mathcal{H} + 4\bar{t} + |\Gamma_k|_\mathcal{H}$ we derive $\|\tilde{W}\|_\omega - \|W\|_\omega < 0$ for h_* small enough, where $\tilde{W} = (W \setminus V) \cup \partial V$. This gives a contradiction to (8.13) and finishes the proof. $\square$

We now use Corollary 8.3.4 for $\bar{t} = \bar{\tau}$ to find (at most) two Γ_i, $i = 1, 2$, with $|\Gamma_1|_\infty, |\Gamma_2|_\infty \geq 19\bar{\tau}$ intersecting $N^{\bar{\tau}}$. We can choose

$$\tfrac{1}{800}\bar{\tau} \leq \tau \leq \tfrac{1}{2}\bar{\tau} \tag{8.26}$$

in such a way that the neighborhood $N^\tau = N^\tau(\Gamma)$ satisfies $\Gamma_i \cap N^{\tau/20} \neq \emptyset$ or $\Gamma_i \cap N^\tau = \emptyset$ for $i = 1, 2$: If $\Gamma_1, \Gamma_2 \cap N^{\bar{\tau}/800} \neq \emptyset$ choose $\tau = \frac{\bar{\tau}}{2}$, if $\Gamma_1, \Gamma_2 \cap N^{\bar{\tau}/800} = \emptyset$ choose $\tau = \frac{\bar{\tau}}{800}$, otherwise choose either $\tau = \frac{\bar{\tau}}{2}$ or $\tau = \frac{1}{40}\bar{\tau}$. For shorthand we set

$N = N^\tau(\Gamma)$, $N_{j,\pm} = N^\tau(\Gamma)_{j,\pm}$, $J = J^\tau = \{Q_0, \dots, Q_n\}$, $\mathcal{Y} = \mathcal{Y}^\tau$ and $\mathcal{C} = \mathcal{C}^\tau = \{Y_1, \dots, Y_m\}$ (recall the constructions before (8.14) and Lemma 8.3.3).

We are now in a position to formulate the main lemma of this section.

Lemma 8.3.5. *Let $\lambda \geq 0$, $W \in \mathcal{W}^s_\lambda$. Let Γ be a boundary component with $\omega(\Gamma) = 1$ and $|\Gamma|_\infty \geq \lambda$. Assume that (8.13) holds. Choosing h_*, v and $1 - \omega_{\min}$ small enough we obtain sets $K_1, K_2 \in \mathcal{Y}$ with $|K_j| \leq C\frac{\tau^2}{h_*}$, $j = 1, 2$, and $\operatorname{dist}(K_1, K_2) \geq c|\Gamma|_\infty$ for some $c > 0$ small enough such that*

(i) The covering $\{\hat{Y}_1, \dots, \hat{Y}_k\}$ of $N \setminus (K_1 \cup K_2)$ consisting of the connected components of $\{Y \setminus (K_1 \cup K_2) : Y \in \mathcal{C}\}$, satisfies $\|\hat{Y}_i\|_\pi \leq \frac{19}{20}\tau$ for all $i = 1, \dots, k$.

(ii) $\Gamma_i \cap N \subset K_1 \cup K_2$ for all components Γ_i with $|\Gamma_i|_\infty \geq 19\bar{\tau}$.

Proof. By Lemma 8.3.3 we obtain that there are two sets $Y^1, Y^2 \in \mathcal{C}$ with $\bar{Y}^1 \cap \bar{Y}^2 = \emptyset$ such that $\|Y\|_\pi \leq \frac{19}{20}\tau$ for $Y \in \mathcal{C}$ with $Y \cap (\bar{Y}^1 \cup \bar{Y}^2) = \emptyset$. We only construct the set K_1. Choose $Y_i = Y^1$ and set $S_l = Y_{i+l} \in \mathcal{C}$ for $|l| \leq 3$. In particular, we have $S_l \cap S_0 \neq \emptyset$ for $l = -1, 1$ and $\|S_l\|_\pi \leq \frac{19}{20}\tau$ for $l = -3, 3$. Set $S = \bigcup_{l=-2}^2 S_l$.

Arguing as in (8.24) or (8.25) for $t = \tau$, respectively, depending on whether S is contained in at most two adjacent parts of the neighborhood or S intersects three parts of the neighborhood (possible for $l_2 \leq \frac{l_1}{2}$), we derive $|\partial V|_\infty \leq |\Gamma|_\infty + \sum_{R_j \in \mathcal{R}(S)} |\partial R_j|_\infty + (\frac{19}{10} - \frac{1}{100})\tau - \|S\|_\pi$, where V is the smallest rectangle containing Γ and $\mathcal{R}(S)$. (Note that in the above calculation we possibly have to repeat the translation argument indicated at the end of the proof of Lemma 8.3.5.) Thus, arguing as in the proof of Lemma 8.3.3, in particular taking (8.19) and condition (8.13) into account, we find

$$0 \leq \|\tilde{W}\|_\omega - \|W\|_\omega \leq (1 - h_*)\tfrac{19}{10}\tau - (1 - h_*)\|S\|_\pi \tag{8.27}$$

for h_*, $1 - \omega_{\min}$ small enough, where $\tilde{W} = (W \setminus V) \cup \partial V$. We now construct the set K_1 and the corresponding (at most) two connected components T_1, T_2 of $S \setminus K_1$ by distinction of the two following cases:

a) If there is some R_j with $|\partial R_j|_\pi \geq \frac{19}{20}\tau$ we choose $K_1 \in \mathcal{Y}$ as the smallest set such that $R_j \cap N \subset K_1$. Then the (at most) two connected components T_1, T_2 of $S \setminus K_1$ satisfy $\|T_i\|_\pi \leq \frac{19}{20}\tau$ by (8.27). Using (8.17) we derive that $|K_1| \leq C\frac{\tau^2}{h_*}$, as desired.

b) Otherwise, we choose K_1 as follows. Assume $S = (\bigcup_{i=1}^{n'} \overline{Q_i})^\circ$ for $Q_i \in J$ and let $k \in \{0, \dots, n'\}$ be the index (if existent) such that $\|(\bigcup_{i=1}^{k} \overline{Q_i})^\circ\|_\pi \leq \frac{19}{20}\tau$ and $\|(\bigcup_{i=1}^{k+1} \overline{Q_i})^\circ\|_\pi > \frac{19}{20}\tau$. Now define $T_1 = (\bigcup_{i=1}^{k} \overline{Q_i})^\circ$ and choose $K_1 = (\bigcup_{i=k+1}^{l} \overline{Q_i})^\circ$ for l large enough such that $|K_1| \geq \bar{c}\frac{\tau^2}{h_*}$. Finally, let $T_2 = S \setminus (\overline{T_1 \cup K_1})$ and observe that for $\bar{c}$ large enough also $\|T_2\|_\pi \leq \frac{19}{20}\tau$ by (8.27) and (8.17) since each rectangle can intersect at most one of the sets T_1, T_2.

Let S_l^1, S_l^2 be the connected components of $S_l \setminus K_1$ for $l = -2, -1, 0, 1, 2$. Both cases a),b) above imply $\|S_l^i \setminus K_1\|_\pi \le \max_{k=1,2} \|T_k\|_\pi \le \frac{19}{20}\tau$ for $l = -2, \dots, 2$, $i = 1, 2$, which gives assertion (i). Assertion (ii) follows from the construction of the set K_1 and definition (8.26). Indeed, if $\Gamma_i \cap N \neq \emptyset$, then $\Gamma_i \cap N^{\tau/20}$ and thus recalling (8.18) we find $|\Gamma_i|_\pi \ge \frac{19}{20}\tau$ and then $\Gamma_i \cap N \subset \bar{Y}^1 \cup \bar{Y}^2$. Finally, $\mathrm{dist}(K_1, K_2) \ge c|\Gamma|_\infty$ follows directly from the fact that in the case $\|Y^1\|_\pi, \|Y^2\|_\pi \ge \frac{19}{20}\tau$ the set $\bar{Y}^1 \cup \bar{Y}^2$ intersects both $N_{1,+}$ and $N_{1,-}$ or both $N_{2,+}$ and $N_{2,-}$ ($N_{1,\pm}$ if $l_2 \le \frac{l_1}{2}$). □

8.3.2 Dodecagonal neighborhood

We now introduce neighborhoods of Γ which in general have dodecagonal shape and differ from $N^t(\Gamma)$ near the corners of Γ. These neighborhoods will be essential in the modification algorithm below (see Section 8.4.2) as we have to treat the modification near the corners of a boundary component with special care. For $t > 0$ we define

$$\hat{M}^t(\Gamma) = \bigcup_{i=1,2} \{x \in N^t(\Gamma) : \ |x_i + l_i| \ge q h_*^{-1} t, |x_i - l_i| \ge q h_*^{-1} t\} \qquad (8.28)$$

for $q \gg 1$ to be specified below. Moreover, for $\tilde{l} = l_1 + \min\{t, q^{-1} h_* l_2\}$ let

$$M^t(\Gamma) := \mathrm{co}(\hat{M}^t(\Gamma) \cup \Gamma \cup (\tilde{l}, 0) \cup (-\tilde{l}, 0)) \cap N^t(\Gamma), \qquad (8.29)$$

where $\mathrm{co}(\cdot)$ denotes the convex hull of a set. Observe that $M^t(\Gamma) \supset \hat{M}^t(\Gamma)$ and that $M^t(\Gamma)$, $\hat{M}^t(\Gamma)$ differ by some triangles. Moreover, the shape of $M^t(\Gamma)$ is dodecagonal for $l_2 > q h_*^{-1} t$ and decagonal otherwise, cf. Figure 8.7. For shorthand we write $M = M^\tau(\Gamma)$ and $\hat{M} = \hat{M}^\tau(\Gamma)$ for a choice of τ satisfying (8.26). For later reference we also define

$$M_k^t(\Gamma) = M^t(\Gamma) \cap (N_{k,+}^t(\Gamma) \cup N_{k,-}^t(\Gamma)) \quad \text{for } k = 1, 2. \qquad (8.30)$$

Recall the definition in (8.14). Let $K_1, K_2 \in \mathcal{Y}$ be the sets constructed in Lemma 8.3.5. Let $\Gamma_m = \Gamma_m(W)$ be another boundary component satisfying $\Gamma_m \cap K \neq \emptyset$ for some $K \in \{K_1, K_2\}$ and $|\Gamma_m|_\infty \ge \frac{q^2 \bar{\tau}}{h_*}$ with q given in (8.28). For q large enough we have $|\Gamma_m|_\infty \ge 19\bar{\tau}$ and thus $\omega(\Gamma_m) = 1$ by (8.5)(iv). Moreover, (8.26) implies that Γ_m is one of the (at most) two rectangular boundary components given by Corollary 8.3.4. By the choice in (8.26), K is constructed in case a) of the proof of Lemma 8.3.5 and therefore it is not hard to see that K is contained in one of the sets $N_{j,\pm}$, $j = 1, 2$. Let $X_m \in \mathcal{U}^s$ be the corresponding component of $Q_\mu \setminus W$. We now treat two different cases depending on whether K is near a corner of Γ or not:

(I) Assume $K \cap \hat{M} \neq \emptyset$. As K is contained in one of the sets $N_{j,\pm}$, $j = 1, 2$ we assume e.g. $K \subset N_{1,-}$. As $|K| \le C\frac{\tau^2}{h_*}$ by Lemma 8.3.5, we find $|\pi_2 \Gamma_m| \le C\frac{\tau}{h_*}$ and thus $|\pi_1\Gamma_m| \gg |\pi_2\Gamma_m|$. Consequently, for q sufficiently large we have $|\pi_1\Gamma_m| \gg \bar{\tau}$ which implies

$$\Gamma_m \cap \{-l_1 - 21\bar{\tau}\} \times \mathbb{R} \neq \emptyset. \tag{8.31}$$

Let $Q_1, Q_2 \in J$ be the neighboring squares of K, i.e. $Q_i \cap K = \emptyset$ and $\partial K \cap \partial Q_i \neq \emptyset$ for $i = 1, 2$. Let $\Psi = (\overline{Q_1 \cup K \cup Q_2} \setminus X_m)^\circ$ and observe that $\Psi \subset N_{1,-}$ as $K \cap \hat{M} \neq \emptyset$. By (8.31) the set $\Psi = \Psi_1 \cup \Psi_2 \cup \Psi_3$ decomposes into three rectangles, where (up to translation and sets of measure zero) $\Psi_1 = (0, \tau) \times (0, \tau + a_1)$, $\Psi_2 = (0, \psi) \times (0, \hat{\psi})$ and $\Psi_3 = (0, \tau) \times (0, \tau + a_3)$ for $-\frac{1}{2}\tau \leq a_1, a_3 \leq \tau$. (Recall the construction of K in the proof of Lemma 8.3.5 a).) Furthermore, let

$$\Phi = \{x \in Q_\mu : \mathrm{dist}(x, \Psi) \leq 20\bar{\tau}\}.$$

Before we go on with case (II) we state two observations. We say that two sets are C-Lipschitz equivalent if they are related through a bi-Lipschitzian homeomorphism with Lipschitz constants of both the homeomorphism itself and its inverse bounded by C.

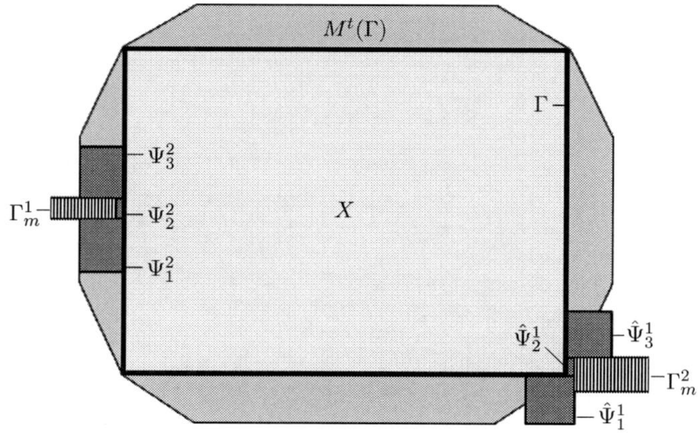

Figure 8.5: Neigborhood $M^t(\Gamma)$ with two other boundary components Γ_m^1, Γ_m^2 (the interiors X_m^1, X_m^2 are striped) and corresponding neighborhoods $\hat{\Psi}^1$ and Ψ^2.

Lemma 8.3.6. *Let Γ, Γ_m with $\omega(\Gamma_m) = \omega(\Gamma) = 1$ and $|\Gamma|_\infty \geq \lambda$, $|\Gamma_m|_\infty \geq q^2 \frac{\bar{\tau}}{h_*}$ be given. In the situation of (I) the following holds:*

(i) Let $V \in \mathcal{U}^s$ be the smallest rectangle containing X and X_m. Then $\Phi \subset V$.

(ii) $\hat{\psi} \leq C\frac{\psi}{h_*}$. In particular, there is a suitable set $\Psi_2 \subset \Psi_2^* \subset \Psi$ such that each set Ψ_1, Ψ_2^*, Ψ_3 is $C(h_*)$-Lipschitz equivalent to a square.

Proof. (i) As $K \cap \hat{M} \neq \emptyset$ we get that $\Phi \subset N^* := N_{1,-}^{21\bar{\tau}} \setminus (N_{2,+}^{21\bar{\tau}} \cup N_{2,-}^{21\bar{\tau}})$ if we again choose q large enough. By (8.31) we have $\partial N^* \cap \Gamma_m \neq \emptyset$. Therefore, the smallest rectangle V containing Γ and Γ_m satisfies $N^* \subset \overline{V}$ which gives the assertion.

(ii) By Lemma 8.3.1(ii) we obtain $\hat{\psi} \leq |\Gamma_m \cap N^{2\psi}|_{\mathcal{H}} \leq C\frac{\psi}{h_*}$. If also $\hat{\psi} \geq \frac{h_*}{C}\psi$ we set $\Psi_2^* = \Psi_2$, otherwise we choose some $\Psi_2^* \supset \Psi_2$ with $|\pi_2 \Psi_2^*| = \psi$. □

(II) Assume now $K \cap \hat{M} = \emptyset$. (i) We first treat the case $l_2 \gg \frac{\tau}{h_*}$ and similarly as in (I) suppose without restriction that $K \subset N_{1,-}$. Again let Q_1, Q_2 be the neighboring squares of K and set $\hat{\Psi} = (\overline{Q_1 \cup K \cup Q_2 \setminus X_m})^\circ$. If $Q_j \subset N_{1,-}$ for $j = 1, 2$ the set $\hat{\Psi}$ decomposes as before in (I).

Otherwise, we may assume that e.g. $Q_1 \subset N_{2,-} \setminus N_{1,-}$. Observe that then $K_1, Q_2 \subset N_{1,-}$ as $|K_1| \leq C\frac{\tau^2}{h_*}$ and $l_2 \gg \frac{\tau}{h_*}$. As indicated in Figure 8.5, the set $\hat{\Psi}$ contains three rectangles $\hat{\Psi}_1, \hat{\Psi}_2, \hat{\Psi}_3$, where (up to translation and sets of measure zero) $\hat{\Psi}_1 = (0, \tau + \psi) \times (0, \tau)$, $\hat{\Psi}_2 = (0, \psi) \times (0, \hat{\psi})$ and $\hat{\Psi}_3 = (0, \tau) \times (0, \tau + a_3)$ for $0 \leq a_3 \leq \tau$. Note that $\hat{\psi} = 0$ is possible and that an argumentation as in Lemma 8.3.6 yields $\hat{\psi} \leq C\frac{\psi}{h_*}$. Now let

$$\Psi_j = \hat{\Psi}_j \setminus (M^{21\bar{\tau}}(\Gamma) \cup M^{21\bar{\tau}_m}(\Gamma_m)),\ j = 1, 2, 3, \qquad \Psi = \left(\bigcup\nolimits_{j=1}^{3} \overline{\Psi_j}\right)^\circ,$$

where $\bar{\tau}_m = v|\Gamma_m|_\infty$. Furthermore, let $\Phi = \{x \in Q_\mu : \mathrm{dist}(x, \Psi) \leq 20\bar{\tau}\}$.

(ii) We finally treat the case that l_2 is small with respect to l_1 (i.e. $l_2 \leq C\frac{\tau}{h_*}$) which particularly implies that $M^{21\bar{\tau}}(\Gamma)$ is decagonal. Suppose without restriction that $K \subset N_{1,-}$. If $K \cap N_{2,+} = \emptyset$ or $K \cap N_{2,-} = \emptyset$ we may proceed as before in (II)(i). Otherwise, the set $\hat{\Psi} \supset \hat{\Psi}_1 \cup \hat{\Psi}_2 \cup \hat{\Psi}_3$ contains three rectangles, where (up to translation and sets of measure zero) $\hat{\Psi}_1 = (0, \tau + \psi) \times (0, \tau)$, $\hat{\Psi}_2 = (0, \psi) \times (0, 2l_2)$ and $\hat{\Psi}_3 = (0, \tau + \psi) \times (0, \tau)$ (cf. Figure 8.6). The same argumentation as in Lemma 8.3.6(ii) yields $2l_2 \leq C\frac{\psi}{h_*}$. We let $\Psi_j = \hat{\Psi}_j \setminus M^{21\bar{\tau}}(\Gamma)$ for $j = 1, 2, 3$. Observe that in contrast to case (II)(i) we only subtract the set $M^{21\bar{\tau}}(\Gamma)$. We now have the following properties.

Lemma 8.3.7. *Let Γ, Γ_m with $\omega(\Gamma_m) = \omega(\Gamma) = 1$ and $|\Gamma|_\infty \geq \lambda$, $|\Gamma_m|_\infty \geq q^2\frac{\bar{\tau}}{h_*}$ be given. In the situation of (II) the following holds:*

(i) Let $V \in \mathcal{U}^s$ be the smallest rectangle containing X and X_m. Then we have $\Phi \cap \{x : x_1 \geq -l_1 - \psi\} \cap M^{21\bar{\tau}_m}(\Gamma_m) \subset V$.

(ii) In the cases (II)(i),(ii) we have $\hat{\psi} \leq C\frac{\psi}{h_*}$ and $2l_2 \leq C\frac{\psi}{h_*}$, respectively. Moreover, there is a suitable set $\Psi_2 \subset \Psi_2^* \subset \Psi$ such that each set Ψ_1, Ψ_2^*, Ψ_3 is $C(h_*)$-Lipschitz equivalent to a square.

Proof. (i) It suffices to note that $\{x : x_2 \geq -l_1 - \psi\} \cap M^{21\bar{\tau}_m}(\Gamma_m) \subset [-l_1 - \psi, \infty) \times \pi_2\Gamma_m$ and $\pi_1\Phi \subset (-\infty, l_1]$ (cf. Figure 8.6).

(ii) The bounds on $\hat{\psi}$ and l_2 were already discussed above. As in the proof of Lemma 8.3.6(ii) we can choose $\hat{\Psi}_2^* \supset \hat{\Psi}_2$ such that $\hat{\Psi}_2^*$ is $C(h_*)$-Lipschitz equivalent to a square. Let $\Psi_2^* = \hat{\Psi}_2^* \setminus (M^{21\bar{\tau}}(\Gamma) \cup M^{21\bar{\tau}_m}(\Gamma_m))$ or $\Psi_2^* = \hat{\Psi}_2^* \setminus M^{21\bar{\tau}}(\Gamma)$, respectively, depending on the cases (II)(i) and (II)(ii). For q sufficiently large in (8.28) it is elementary to see that Ψ_1, Ψ_2^*, Ψ_3 are $C(h_*)$-Lipschitz equivalent to a square. □

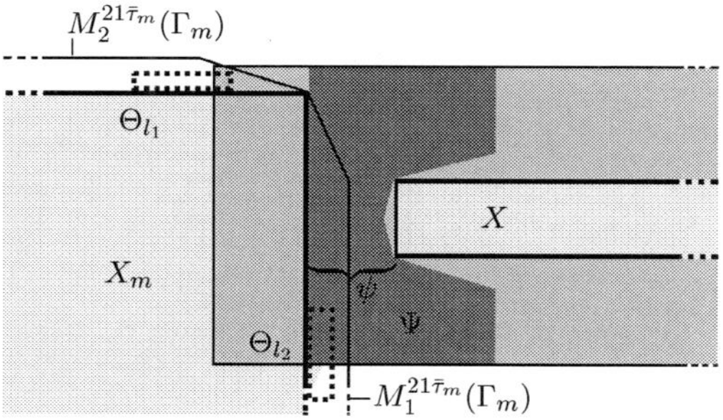

Figure 8.6: Sketch of Ψ (grey) in the case (II)(ii), where only parts of the boundary components Γ,Γ_m are depicted. In particular $M^{21\bar{\tau}_m}(\Gamma_m) \cap \Psi \neq \emptyset$ and $M^{21\bar{\tau}}(\Gamma) \cap \Psi = \emptyset$. Also note that $M^{21\bar{\tau}_m}(\Gamma_m)$ is dodecagonal, whereas $M^{21\bar{\tau}}(\Gamma)$ is decagonal. Moreover, for later reference (see proof of Lemma 8.4.5) we have also drawn two boundary components $\Theta_{l_1}, \Theta_{l_2} \subset M^{21\bar{\tau}_m}(\Gamma_m)$ in dashed lines.

8.4 Proof of the Korn-Poincaré-inequality

This section is devoted to the main proof of Theorem 8.0.1. We concern ourselves with functions $y \in H^1(W)$ on $W \in \mathcal{V}^s$, $W \subset Q_\mu$ (recall (7.1), (8.1)). In the following we will again omit to write $\mathcal{V}^s$. Let $R \in SO(2)$ and define $u = u_R = R^T y - \mathbf{id}$. For shorthand we set $\alpha(U) = \|\bar{e}_R(\nabla y)\|^2_{L^2(U)} = \|e(\nabla u)\|^2_{L^2(U)}$ for $U \subset W$ (cf. (7.2)).

As a further preparation, we define $H(W) \supset W \in \mathcal{V}^s$ as the 'variant of W without holes'. Arrange the components $X_1, \ldots, X_m$ such that $\partial X_i \subset Q_\mu$ for $1 \leq i \leq n$ and $\partial X_i \cap \partial Q_\mu \neq \emptyset$ otherwise. We set

$$H(W) = W \cup \bigcup\nolimits_{j=1}^n X_j. \tag{8.32}$$

The main idea will be to analyze the trace of u at the boundary components. Therefore, we will have to change the set W iteratively. We first introduce further conditions for the neighborhood of a boundary component which allows us to apply a trace estimate. Then we present the main modification algorithm. Afterwards, the proof of Theorem 8.0.1 will be straightforward by employing Theorem B.4.

8.4.1 Conditions for boundary components and trace estimate

Recall definition (8.5) and assume that in an iteration step $W_i \in \mathcal{W}^s_\lambda$ for $\lambda \geq 0$ with the corresponding weight ω and a specific ordering of the boundary components $(\Gamma(W_i)_j)_{j=1}^n$ is given. Consider $\Gamma = \Gamma(W_i)$ with $|\Gamma|_\infty \geq \lambda$ and recall that $\Gamma = \Theta$ is rectangular by (8.5)(v). Let $\hat{N} = N^{2\hat{\tau}}(\Gamma)$, where

$$\hat{\tau} = q^2 \bar{\tau} h_*^{-1} = q^2 \upsilon h_*^{-1} |\Gamma|_\infty \ll |\Gamma|_\infty \tag{8.33}$$

with q from (8.28) and $\bar{\tau}$ as defined in (8.14). Recall that $\hat{\tau}$ is the least length of boundary components considered in Section 8.3.2. The latter inequality holds if we choose υ sufficiently small with respect to q. For $\varepsilon > 0$ and for $D = D(h_*)$ sufficiently large we require

$$\alpha(\hat{N} \cap W_i) + \varepsilon |\partial W_i \cap \hat{N}|_{\mathcal{H}} \leq D \varepsilon \hat{\tau}. \tag{8.34}$$

Moreover, let Ψ^j and ψ^j, $j = 1, 2$, be defined as in Section 8.3.2 (I),(II) corresponding to the sets K_j, $j = 1, 2$, provided by Lemma 8.3.5. We introduce the condition

$$\alpha(\Psi^j \cap W_i) + \varepsilon |\partial W_i \cap \Psi^j|_{\mathcal{H}} \leq D(1 - \omega_{\min})^{-1} \varepsilon \psi^j \tag{8.35}$$

for $j = 1, 2$, where $D = D(h_*)$ as in (8.34). For $\eta \geq 0$ we let $\mathcal{T}_\eta(W_i)$ be the set of $\Gamma_l(W_i)$ satisfying $|\Gamma_l(W_i)|_\infty \leq \eta$ and

$$N^{2\hat{\tau}_l}(\partial R_l) \subset H(W_i),$$

where $\hat{\tau}_l = q^2 \upsilon |\Gamma_l|_\infty h_*^{-1}$ (cf. (8.33)) and R_l is the corresponding rectangle given in (8.5)(i) or (8.5)(v), respectively. Moreover, recalling again the definition of $\Theta_l(W_i) \subset \Gamma_l(W_i) = \partial X_l$ in (8.2) we define

$$S_\lambda(W_i) = \bigcup\nolimits_{\Gamma_l(W_i) \in \mathcal{S}_\lambda(W_i)} \Theta_l(W_i), \tag{8.36}$$

where $\mathcal{S}_\lambda(W_i) = \{\Gamma_l : |\Gamma_l|_\infty > \lambda\} \cup \{\Gamma_l : \omega(\Gamma_l) = 1,\ N^{2\hat{\tau}_l}(\partial R_l) \not\subset H(W_i)\}$. (Note that by (8.5)(iv) all components of $\mathcal{S}_\lambda(W_i)$ have weight 1.)

We assume that for all $\Gamma_l(W_i) \in \mathcal{T}_{\hat{\tau}}(W_i)$ there are $A_l \in \mathbb{R}^{2\times 2}_{\text{skew}}$, $c_l \in \mathbb{R}^2$ such that for the extension $\bar{u} \in SBV(W_{\hat{\tau}})$, $W_{\hat{\tau}} := W_i \cup \bigcup_{\Gamma_l(W_i) \in \mathcal{T}_{\hat{\tau}}(W_i)} X_l$, defined by

$$\bar{u}(x) = \begin{cases} A_l\, x + c_l & x \in X_l \ \text{ for } \ \Gamma_l(W_i) \in \mathcal{T}_{\hat{\tau}}(W_i) \\ u(x) & \text{else,} \end{cases} \tag{8.37}$$

we have the trace estimate

$$\begin{aligned} \int_{\Theta_l(W_i)} |\bar{u}(x) - (A_l\, x + c_l)|^2 \, d\mathcal{H}^1(x) &= \int_{\Theta_l(W_i)} |[\bar{u}](x)|^2 \, d\mathcal{H}^1(x) \\ &\leq C_* \frac{\varepsilon^4}{\upsilon} |\Theta_l(W_i)|_*^2 \end{aligned} \tag{8.38}$$

for some $C_* = C_*(h_*) > 0$ sufficiently large. (The left hand side has to be understood as the trace of $\bar{u}|_{W_{\hat{\tau}} \setminus X_l}$ on $\Theta_l(W_i)$.)

We now state that under suitable conditions also $\Gamma = \Gamma(W_i)$ with $|\Gamma|_\infty \geq \lambda$ satisfies an estimate similar to (8.38).

Theorem 8.4.1. *Let $\upsilon, h_*, \varepsilon, \omega_{\min} > 0$ and $\lambda > 0$. Then there is a constant $\hat{C} = \hat{C}(h_*) > 0$ such that for υ sufficiently small (depending on h_* and $\omega_{\min}$) the following holds: For all $W_i \in \mathcal{W}^s_\lambda$, for all $u \in H^1(W_i)$ and boundary components $\Gamma = \Gamma(W_i)$ with $|\Gamma|_\infty \geq \lambda$ such that (8.13), (8.34), (8.35), $N^{2\hat{\tau}}(\Gamma) \subset H(W_i)$ hold and (8.38) is satisfied for $\mathcal{T}_{\hat{\tau}}(W_i)$ one has (in the sense of traces)*

$$\int_\Gamma |\bar{u}(x) - (A\,x - c)|^2 \, d\mathcal{H}^1(x) \leq \Big(\hat{C} + \frac{C_*}{2}\Big)\frac{\varepsilon}{\upsilon^4}|\Gamma|_*^2 \tag{8.39}$$

for suitable $A \in \mathbb{R}^{2\times 2}_{\text{skew}}$, $c \in \mathbb{R}^2$.

As the proof of this assertion is very technical and involves several steps we postpone it to Section 8.5.

8.4.2 Modification algorithm

We now show that we may modify the set W iteratively such that successively we find a component Γ which satisfies the conditions (8.13), (8.34), (8.35) and (8.38) such that Theorem 8.4.1 can be applied. Define $\mathcal{W}^s = \bigcup_{\lambda \geq 0} \mathcal{W}^s_\lambda$.

Theorem 8.4.2. *Let $\varepsilon > 0$ and $h_* \geq \sigma > 0$ sufficiently small. Let $C_1 = C_1(\sigma, h_*) \geq 1$ large, $0 < C_2 = C_2(\sigma, h_*) < 1$ small enough and let $c > 0$ be a universal constant. For all $W \in \mathcal{V}^s$ with connected boundary components and $u \in H^1(W)$ there is a set $U \in \mathcal{W}^{C_2 s}$ with $|U \setminus W| = 0$ and an extension $\bar{u}$ defined by*

$$\bar{u}(x) = \begin{cases} A_l\,x + c_l & x \in X_l \quad \textit{for all } \Gamma_l(U) \textit{ with } N^{2\hat{\tau}_l}(\partial R_l) \subset H(U), \\ u(x) & \textit{else,} \end{cases} \tag{8.40}$$

such that for all $\Gamma_l(U)$ with $N^{2\hat{\tau}_l}(\partial R_l) \subset H(U)$

$$\int_{\Theta_l(U)} |[\bar{u}](x)|^2 \, d\mathcal{H}^1(x) \leq C_1 \varepsilon |\Theta_l(U)|_*^2. \tag{8.41}$$

Moreover, one has $|W \setminus U| \leq c\|U\|_\infty^2$ and

$$\varepsilon\|U\|_* + \alpha(U) \leq (1+\sigma)(\varepsilon\|W\|_* + \alpha(W)). \tag{8.42}$$

Remark 8.4.3. In the proofs of Theorem 8.4.2 and Theorem 8.4.1 we will see that the constants $C_i = C_i(\sigma, h_*)$ have polynomial growth in σ: We find $z \in \mathbb{N}$ large enough such that $C_1(\sigma, h_*) \leq C(h_*)\sigma^{-z}$ and $C_2(\sigma, h_*) \geq C(h_*)\sigma^z$.

The proof relies on an iterative modification procedure. First choose

$$C_2 = Cv \tag{8.43}$$

for C small enough and consider $W \in \mathcal{V}^s$ as an element of $\mathcal{V}^{C_2 s}$ such that (8.14) is satisfied for all boundary components $\Gamma_l(W)$. From now on we will always tacitly assume that all involved sets lie in $\mathcal{V}^{C_2 s}$ and write $\mathcal{W}_\lambda$ instead of $\mathcal{W}_\lambda^{C_2 s}$. In the proof below we will show that C_2 is in fact a constant only depending on h_* and σ.

We set $W_0 = \hat{W}$, where $\hat{W}$ is the modification constructed in Corollary 8.2.2. Choosing an ordering of the boundary components and setting $\omega(\Gamma_j(W_0)) = 1$ for all j we obtain $W_0 \in \mathcal{W}_0$. Moreover, we let $\lambda_0 = 0$, $B_0^0 = \emptyset$. Assume that $\lambda_0 \leq \ldots \leq \lambda_i$ and that $W_i \subset \ldots \subset W_0$, $W_j \in \mathcal{W}_{\lambda_j}$, are given (the inclusion holds up to sets of negligible $\mathcal{L}^2$-measure) as well as $\{B_k^j : k = 0, \ldots, j\}$ for $j = 0, \ldots, i$. In each iteration step the sets B_k^j, $k = 0, \ldots, j$, will describe the set where we already 'used' the 'energy lying in the set' to modify W.

Suppose that in an iteration step i the following conditions are satisfied:

$$\varepsilon \|W_i\|_\omega + \alpha(W_i) \leq \varepsilon \|W\|_* + \alpha(W) + h_*(1 - \omega_{\min}) \sum\nolimits_{j=0}^{i} \alpha(B_j^i) \tag{8.44}$$

as well as

$$\begin{aligned} (i) \quad & \text{Each } x \in Q_\mu \text{ lies in at most two different } B_{j_1}^i,\ B_{j_2}^i \text{ and} \\ & \text{each } x \in W_i \text{ lies in at most one } B_j^i,\ j, j_1, j_2 \in \{0, \ldots, i\}, \\ (ii) \quad & \text{Either } \Theta_l(W_i) \subset B_j^i \text{ for some } 0 \leq j \leq i \text{ or} \\ & \Gamma_l(W_i) \in \mathcal{G}_i := \Big\{\Gamma_l : \Gamma_l \cap \bigcup\nolimits_{j=0}^{i} B_j^i = \emptyset,\ \omega(\Gamma_l) = 1\Big\} \ \text{ for all } \Gamma_l(W_i), \\ (iii) \quad & \text{Each } B_j^i \text{ with } B_j^i \cap W_i \neq \emptyset, \ \text{ satisfies } B_j^i \cap W_i \subset M_k^{\eta_l^i}(\Gamma_l(W_i)), \\ & \qquad \text{for some } \Gamma_l(W_i) \in \mathcal{G}_i \text{ and } k \in \{1, 2\},\ j = 0, \ldots, i. \end{aligned} \tag{8.45}$$

Here $\eta_l^i := 21v \min\{|\Gamma_l(W_i)|_\infty, \lambda_i\} = \min\{21\bar{\tau}_l, 21v\lambda_i\}$ and the neighborhood M_k was defined in (8.30). Moreover, recalling (8.36) we suppose

$$\begin{aligned} \alpha(N^{\hat{\tau}_l}(\Gamma_l(W_i)) \cap W_i) + \varepsilon |N^{\hat{\tau}_l}(\Gamma_l(W_i)) \cap (\partial W_i \setminus S_{\lambda_i}(W_i))|_{\mathcal{H}} \leq D\varepsilon \hat{\tau}_l \\ \text{for all } \Gamma_l(W_i) \in \mathcal{G}_i \cap \mathcal{T}_{\lambda_i}(W_i) \end{aligned} \tag{8.46}$$

where D is defined as in (8.34). Furthermore, recalling (8.37) we assume that there is an extension $\bar{u}_i \in SBV^2(W_{\lambda_i})$ such that all boundary components $\Gamma_l(W_i) \in \mathcal{T}_{\lambda_i}(W_i)$ satisfy

$$\int_{\Theta_l(W_i)} |\bar{u}_i(x) - (A_l\, x - c_l)|^2 \, d\mathcal{H}^1 \leq \hat{C} \sum_{n=0}^{i} \Big(\frac{2}{3}\Big)^n \frac{\omega(\Gamma_l(W_i))^2}{\hat{\omega}_i(\Gamma_l(W_i))^2} \frac{\varepsilon}{v^4} |\Theta_l(W_i)|_*^2 \tag{8.47}$$

for $A_l \in \mathbb{R}^{2\times 2}_{\rm skew}$, $c_l \in \mathbb{R}^2$, where $\hat{\omega}_i(\Gamma_l(W_i)) := 1 - \frac{1-\omega_{\min}}{2}\#\{j = 0,\ldots,i : \Theta_l(W_i) \subset B^i_j\}$ and $\hat{C}$ is the constant from (8.39). In particular, this implies that (8.38) is satisfied if we replace C_* by $\hat{C}\omega_{\min}^{-2}\sum_{n=0}^{i}(2/3)^n$ as $\hat{\omega}_i(\Gamma_l(W_i)) \ge \omega_{\min}$ (see (8.45)(i), (ii)). Finally, we assume

$$\begin{aligned} &(i)\ \ \omega(\Gamma_l(W_i)) \ge \hat{\omega}_i(\Gamma_l(W_i)), && \forall\, \Gamma_l, \\ &(ii)\ \ |\partial R_l(W_i)|_* \le \omega(\Gamma_l(W_i))(\hat{\omega}_i(\Gamma_l(W_i)))^{-1}|\Theta_l(W_i)|_*, && \forall\, \Gamma_l :\ \omega(\Gamma_l) < 1. \end{aligned} \tag{8.48}$$

The second condition is a refinement of (8.5)(ii).

Recall that $\partial W_i \cap Q_\mu \subset W_i$ by definition (see (8.1)). This particularly implies that each $\Theta_l(W_i)$ is contained in at most one set B^i_j (see (8.45)(i),(ii)). Before we give the proof of Theorem 8.4.2 we first observe that the above stated properties are preserved under modification.

Lemma 8.4.4. *Let $\varepsilon > 0$ and $\lambda \ge 0$. Let $W_i \in \mathcal{W}_\lambda$, $\bar{u}_i$ and $\{B^i_j : j = 0,\ldots,i\}$ be given such that (8.44)-(8.48) hold for W_i, $\bar{u}_i$ and λ (replace λ_i by λ). For a rectangle $\overline{V} \subset Q_\mu$ with $|\partial V|_\infty > \lambda$, let $\tilde{W}_i = (W_i \setminus V) \cup \partial V$ and assume that (recall (8.7), (8.8))*

$$\varepsilon\|\tilde{W}_i\|_\omega + \alpha(\tilde{W}_i) \le \varepsilon\|W_i\|_\omega + \alpha(W_i).$$

Let $W_{i+1} \in \mathcal{W}_\lambda$ be the set given by Lemma 8.2.1 and define $B^{i+1}_j = B^i_j \setminus S_\lambda(W_{i+1})$ for $j = 0,\ldots,i$ (recall (8.36)) and $B^{i+1}_{i+1} = \emptyset$ as well as $\bar{u}_{i+1} = \bar{u}_i$. Then (8.44)-(8.48) hold for W_{i+1}, $\bar{u}_{i+1}$ and λ (replace λ_{i+1} by λ).

Proof. Let $\tilde{W} = (W_i \setminus V) \cup \partial V$ for some rectangle V with $|V|_\infty > \lambda$ and choose $V' \supset V$ such that $W_{i+1} := (W_i \setminus V') \cup \partial V' \in \mathcal{W}_\lambda$ as in Lemma 8.2.1. Then by assumption and (8.10) we have $\varepsilon\|W_{i+1}\|_\omega + \alpha(W_{i+1}) \le \varepsilon\|W_i\|_\omega + \alpha(W_i)$, where we adjust the weights as described in (8.8). As $|B^i_j \setminus B^{i+1}_j| = 0$ for all $j = 0,\ldots,i$, (8.44) is trivially satisfied.

Clearly, (8.45)(i) still holds as $B^{i+1}_j \subset B^i_j$ for all $j = 0,\ldots,i$. Moreover, $\Gamma_l(W_{i+1}) \cap \bigcup_{j=0}^{i+1} B^{i+1}_j = \emptyset$ for all $\Gamma_l(W_{i+1}) \in \mathcal{S}_\lambda(W_{i+1})$ by definition. Consequently, $\mathcal{S}_\lambda(W_{i+1}) \subset \mathcal{G}_{i+1}$ and to confirm (8.45)(ii) it suffices to consider the components not lying in $\mathcal{S}_\lambda(W_{i+1})$. Let $\Gamma_l(W_{i+1}) \notin \mathcal{S}_\lambda(W_{i+1})$. Using that $\partial V' \in \mathcal{S}_\lambda(W_{i+1})$ and arguing similarly as in Section 8.2 (see remark after (8.7)) we find a (unique) corresponding $\Gamma_l(W_i)$ (for notational convenience we use the same index) such that $\Theta_l(W_{i+1}) = \Theta_l(W_i) \setminus \overline{V'}$. If $\Gamma_l(W_i) \in \mathcal{G}_i$ we immediately get $\Gamma_l(W_{i+1}) \in \mathcal{G}_{i+1}$ by (8.9) and the fact that the sets $(B^i_j)^i_{j=1}$ do not become larger. Consequently, we can assume that $\Theta_l(W_i) \subset B^i_j$ for some j and it now remains to show $\Theta_l(W_{i+1}) \subset B^{i+1}_j$. To see this, it suffices to observe $\Theta_l(W_{i+1}) = \Theta_l(W_i) \setminus \overline{V'} \subset \Theta_l(W_i)$ and $\Theta_l(W_{i+1}) \cap S_\lambda(W_{i+1}) = \emptyset$, where the latter holds as the sets $(\Theta_j(W_{i+1}))_j$ are pairwise disjoint (see (8.2)).

Due to the modification procedure (see the construction of V' in the proof of Lemma 8.2.1), for all $\Gamma_l(W_i) \in \mathcal{G}_i$ we find a $\Gamma_j(W_{i+1}) \in \mathcal{G}_{i+1}$ such that $\Gamma_l(W_i) \subset$

$\overline{X_j(W_{i+1})}$, where $\partial X_j(W_{i+1}) = \Gamma_j(W_{i+1})$). In fact, one can choose either $\Gamma_l(W_i)$ itself or $\partial V'$. (Note that both are elements of $\mathcal{G}_{i+1}$.) Therefore, $M_k^{\eta_l^i}(\Gamma_l(W_i)) \subset M_k^{\eta_j^{i+1}}(\Gamma_j(W_{i+1}))$ for $k = 1, 2$ and thus condition (8.45)(iii) holds.

To confirm (8.48)(i) we first observe that $\hat{\omega}_{i+1}(\partial V') = \omega(\partial V') = 1$. Moreover, we see that $\hat{\omega}_{i+1}(\Gamma_l(W_{i+1})) = \hat{\omega}_i(\Gamma_l(W_i))$ for all $\Gamma_l(W_{i+1}) \neq \partial V'$, where $\Gamma_l(W_i)$ is the unique corresponding component. In fact, $\hat{\omega}_{i+1}(\Gamma_l(W_{i+1})) \geq \hat{\omega}_i(\Gamma_l(W_i))$ follows immediately from the definition of $(B_j^{i+1})_{j=1}^{i+1}$ and the reverse inequality follows from the observation that $\Theta_l(W_i) \subset B_j^i$ implies $\Theta_l(W_{i+1}) \subset B_j^{i+1}$ (see above). Now this together with the fact that $\omega(\Gamma_l(W_{i+1})) \geq \omega(\Gamma_l(W_i))$ (see (8.9)) yields (8.48)(i).

We now show that (8.47) remains true. Similarly as before we find for all $\Gamma_l(W_{i+1}) \in \mathcal{T}_\lambda(W_{i+1})$ a (unique) corresponding $\Gamma_l(W_i)$. If $\Gamma_l(W_{i+1}) \cap \partial V' = \emptyset$, then $\Theta_l(W_{i+1}) = \Theta_l(W_i)$ and there is nothing to show since $\Gamma_l(W_i) \in \mathcal{T}_\lambda(W_i)$ due to the fact that $|\Gamma_l(W_i)|_\infty = |\Gamma_l(W_{i+1})|_\infty \leq \lambda$ and $\omega(\Gamma_l(W_i))$, $\hat{\omega}_i(\Gamma_l(W_i))$ remain unchanged. Otherwise, $\omega(\Gamma_l(W_{i+1})) < 1$ by (8.5)(v) and thus

$$\omega(\Gamma_l(W_{i+1}))|\Theta_l(W_{i+1}))|_* = \omega(\Gamma_l(W_i))|\Theta_l(W_i)|_* \tag{8.49}$$

by (8.8), which together with the fact that $\bar{u}_{i+1} = \bar{u}_i$ and $\hat{\omega}_{i+1}(\Gamma_l(W_{i+1})) = \hat{\omega}_i(\Gamma_l(W_i))$ implies (8.47). To see that $|\Gamma_l(W_i)|_\infty \leq \lambda$ also holds in this case (and thus $\Gamma_l(W_i) \in \mathcal{T}_\lambda(W_i)$) we note that $|\Gamma_l(W_{i+1})|_\infty \leq 19\upsilon\lambda$ by (8.5)(iv) and therefore (8.49) together with (8.5)(i) and (8.6) implies $|\Gamma_l(W_i)|_\infty \leq \lambda$ for υ small enough.

Observe that by the same argument as in (8.49) property (8.48)(ii) is satisfied. (Recall that in the modification procedure we never change the rectangles ∂R_l.)

Finally, (8.46) holds. Indeed, for a given $\Gamma_l(W_{i+1}) \in \mathcal{T}_\lambda(W_{i+1}) \cap \mathcal{G}_{i+1}$ we deduce $\Gamma_l(W_{i+1}) \cap \partial V' = \emptyset$ by (8.5)(v) and thus $\Gamma_l(W_{i+1}) = \Gamma_l(W_i)$, where $\Gamma_l(W_i)$ is the corresponding component of W_i. The assertion now follows from the i-th iteration step of (8.46). In fact, for the left part it suffices to recall $W_{i+1} \subset W_i$. For the right part we note $S_\lambda(W_{i+1}) = \partial V' \cup (S_\lambda(W_i) \cap \partial W_{i+1}) \supset S_\lambda(W_i) \cap \partial W_{i+1}$ (again recall that we did not change the rectangles ∂R_l) and $\partial W_{i+1} \setminus \partial W_i \subset \partial V' \subset S_\lambda(W_{i+1})$ which then yields $\partial W_{i+1} \setminus S_\lambda(W_{i+1}) \subset \partial W_i \setminus S_\lambda(W_i)$ by an elementary computation. □

We are now in a position to prove Theorem 8.4.2.

Proof of Theorem 8.4.2. Using Corollary 8.2.2 we first see that (8.44)-(8.48) hold for $W_0 = \hat{W}$ and $\lambda_0 = 0$, $B_0^0 = \emptyset$. Assume that $W_i \in \mathcal{W}_{\lambda_i}$, λ_i, $\{B_j^i : j = 0, \dots, i\}$ and $\bar{u}_i$ have already been constructed and that (8.44)-(8.48) hold.

If now all $\Gamma_l(W_i)$ with $N^{2\hat{\tau}_l}(\partial R_l(W_i)) \subset H(W_i)$ satisfy $|\Gamma_l(W_i)|_\infty \leq \lambda_i$ we stop and set $U = W_i$. We observe that in this case (8.41) holds for $C_1 = \hat{C}\sum_{n=0}^\infty (2/3)^n \omega_{\min}^{-2} \upsilon^{-4}$ by (8.47). Otherwise, there is some smallest $\Gamma = \Gamma(W_i)$ with respect to $|\cdot|_\infty$ satisfying $|\Gamma|_\infty > \lambda_i$ and $N^{2\hat{\tau}}(\Gamma) \subset H(W_i)$. To simplify the exposition, we will suppose that the choice of Γ is unique. At the end of the proof we briefly indicate the necessary changes if there are several components of the same size.

Choose $\omega_{\min} \geq \sqrt{\frac{3}{4}}$. We observe that $\mathcal{T}_{\hat{\tau}} \subset \mathcal{T}_{\lambda_i}$ for $\hat{\tau}$ as defined in (8.33). Indeed, for $\hat{\tau} \leq \lambda_i$ it is obvious and for $\hat{\tau} > \lambda_i$ it follows from the choice of Γ with respect to $|\cdot|_\infty$. Thus, by (8.47) we get that (8.38) is satisfied replacing C_* by $\frac{4}{3}\hat{C}\sum_{n=0}^{i}(2/3)^n$. If Γ additionally fulfills (8.13), (8.34) and (8.35), we may apply Theorem 8.4.1. Therefore, recalling $\omega(\Gamma) = 1$ we get that for suitable $A \in \mathbb{R}^{2\times 2}_{\text{skew}}$, $c \in \mathbb{R}^2$

$$\begin{aligned}\int_\Gamma |\bar{u}_i(x) - (Ax+c)|^2\,dx &\leq \Big(\hat{C} + \frac{1}{2}\cdot\frac{4}{3}\hat{C}\sum\nolimits_{n=0}^{i}\Big(\frac{2}{3}\Big)^n\Big)\frac{\varepsilon}{\upsilon^4}|\Gamma|_*^2 \\ &\leq \hat{C}\sum\nolimits_{n=0}^{i+1}\Big(\frac{2}{3}\Big)^n\frac{\omega(\Gamma)^2}{\hat{\omega}(\Gamma)^2}\frac{\varepsilon}{\upsilon^4}|\Gamma|_*^2.\end{aligned}$$

Thus, (8.47) holds, as desired. We define $\bar{u}_{i+1}(x) = Ax + c$ for $x \in X$ and $\bar{u}_{i+1} = \bar{u}_i$ else, where $\partial X = \Gamma$. Moreover, we set $W_{i+1} = W_i$, $\lambda_{i+1} = |\Gamma|_\infty$, $B_j^{i+1} = B_j^i$ for $j = 0, \dots, i$ and $B_{i+1}^{i+1} = \emptyset$. Clearly, (8.44)-(8.48) still hold due to choice of Γ with respect to $|\cdot|_\infty$. In particular, for (8.47) we note that $\mathcal{T}_{\lambda_{i+1}}(W_{i+1}) = \mathcal{T}_{\lambda_i}(W_i)\cup\{\Gamma\}$. Likewise, (8.46) is fulfilled by (8.34) and the fact that $S_{\lambda_{i+1}}(W_{i+1}) = S_{\lambda_i}(W_i)$ (recall (8.36)). Moreover, (8.45)(iii) still holds as $\lambda_{i+1} \geq \lambda_i$. As also (8.5) is satisfied for λ_{i+1}, we get $W_{i+1} \in \mathcal{W}_{\lambda_{i+1}}$. We continue with the next iteration step.

Otherwise (a) (8.13), (b) (8.34) or (c) (8.35) is violated.

In case (a) we find some $V \supsetneq \Gamma$ such that setting $\tilde{W}_i = (W_i \setminus V) \cup \partial V$, we get $\|\tilde{W}_i\|_\omega \leq \|W_i\|_\omega$ and therefore

$$\varepsilon\|\tilde{W}_i\|_\omega + \alpha(\tilde{W}_i) \leq \varepsilon\|W_i\|_\omega + \alpha(W_i).$$

Here we adjusted the weights as in (8.8). Let $\lambda_{i+1} = |\Gamma|_\infty$. It is not hard to see that $W_i \in \mathcal{W}_{\lambda_{i+1}}$ satisfies (8.44)-(8.48) also for λ_{i+1}. In fact, (8.47) follows from the choice of Γ with respect to $|\cdot|_\infty$ and the fact that $|\partial V|_\infty > \lambda_{i+1}$. For the other properties we may argue as before. Now Lemma 8.4.4 yields a set $W_{i+1} \in \mathcal{W}_{\lambda_{i+1}}$ with $W_{i+1} \subset \tilde{W}_i$ as well as $(B_j^{i+1})_{j=1}^{i+1}$ and $\bar{u}_{i+1} = \bar{u}_i$ such that (8.44)-(8.48) hold for W_{i+1}, $\bar{u}_{i+1}$ and λ_{i+1}. We now continue with the next iteration step.

In case b) set $\tilde{W}_i = (W_i \setminus V) \cup \partial V$, where V is the smallest rectangle containing $N^{4\hat{\tau}}(\Gamma)$. Observe that $|\partial V|_* \leq |\Gamma|_* + C\hat{\tau}$. Choosing $D = D(h_*) \geq \frac{2C}{h_*} \geq \frac{C}{h_*\omega_{\min}}$ and arguing as in the proof of Lemma 8.3.1 we derive

$$\begin{aligned}\varepsilon\|\tilde{W}_i\|_\omega + \alpha(\tilde{W}_i) &\leq \varepsilon\|W_i\|_\omega + \alpha(W_i) + C\varepsilon\hat{\tau} - \alpha(V \cap W_i) \\ &\quad - \varepsilon h_*\omega_{\min}|\partial W_i \cap \hat{N}|_{\mathcal{H}} \leq \varepsilon\|W_i\|_\omega + \alpha(W_i).\end{aligned}$$

As usual we adjusted the weights as in (8.8). We now may proceed as in case (a) and then continue with the next iteration step.

Finally, consider case (c). Let $\Psi_i = \Psi^j$ and $\psi_i = \psi^j$, where Ψ^j is a set such that (8.35) is violated. As derived above in Section 8.3.2, we find a boundary component $\Gamma_m = \Gamma_m(W_i)$ with $|\Gamma_m|_\infty \geq \hat{\tau}$, $\omega(\Gamma_m) = 1$. Moreover, there is a rectangle $T \subset Q_\mu$ with $|\partial T|_{\mathcal{H}} \leq 4\psi_i$ and $\overline{T} \cap \Gamma \neq \emptyset$, $\overline{T} \cap \Gamma_m \neq \emptyset$ (cf. Figure 8.8).

Let $\mathcal{A}_i \subset (\Gamma_l(W_i))_l \setminus \{\Gamma, \Gamma_m\}$ be the boundary components with $\Theta_l(W_i) \cap \Psi_i \neq \emptyset$ or, if $\Gamma_l(W_i) = \Theta_l(W_i) \in \mathcal{G}_i$, with $M^{\eta_l^i}(\Gamma_l(W_i)) \cap \Psi_i \neq \emptyset$. We now define an additional set B^i_{i+1}, where we will 'use the energy' to modify W_i. Let

$$B^i_{i+1} = \Big((\Psi_i \cap W_i) \cup \bigcup\nolimits_{\Gamma_l(W_i)\in\mathcal{A}_i} \Theta_l(W_i)\Big) \setminus \bigcup\nolimits_{B^i_j \in \mathcal{B}_i} B^i_j,$$

where $\mathcal{B}_i := \{B^i_j : B^i_j \cap W_i \subset M^{\eta_l^i}(\Gamma_l(W_i)) \text{ for some } \Gamma_l(W_i) \in \mathcal{A}_i \cap \mathcal{G}_i\}$. In the definition of B^i_{i+1} it is essential to subtract the set on the right hand side such that we will be able to assure (8.45)(i).

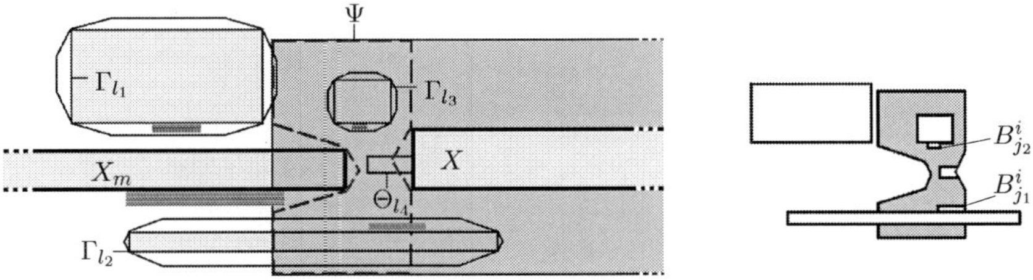

Figure 8.7: On the left side Ψ (the set surrounded by the dashed grey line) and parts of Γ, Γ_m are sketched. Observe that $M^{21\bar{\tau}_m}(\Gamma_m) \cap \Psi = M^{21\bar{\tau}}(\Gamma) \cap \Psi = \emptyset$. Moreover, the picture includes several boundary components with corresponding dodecagonal or decagonal neighborhoods as well as four striped sets $B^i_{j_1}, \ldots, B^i_{j_4}$. On the right hand side the resulting B^i_{i+1} is drawn, where ∂B^i_{i+1} is black and the interior $(B^i_{i+1})^\circ$ is grey. Observe that in general only parts of ∂B^i_{i+1} are contained in B^i_{i+1}.

Note that by (8.45) we have for all $\Gamma_l(W_i) \in \mathcal{A}_i$ either $\Theta_l(W_i) \subset B^i_{i+1}$ or $\Theta_l(W_i) \cap B^i_{i+1} = \emptyset$ depending on whether $\Theta_l(W_i) \cap \bigcup_{B^i_j \in \mathcal{B}_i} B^i_j = \emptyset$ or $\Theta_l(W_i) \subset B^i_j \in \mathcal{B}_i$. Moreover, the components $\Gamma_l(W_i) \notin \mathcal{A}_i$ clearly satisfy $\Theta_l(W_i) \cap B^i_{i+1} = \emptyset$. Denote by $\tilde{\mathcal{A}}_i \subset \mathcal{A}_i$ the boundary components completely contained in B^i_{i+1} and observe that $\mathcal{G}_i \cap \mathcal{A}_i \subset \tilde{\mathcal{A}}_i$.

By (8.52) below we obtain $|\Gamma_l(W_i)|_\infty < 19\bar{\tau}$ for all $\Gamma_l(W_i) \in \mathcal{A}_i$ which by (8.33) for v sufficiently small implies $N^{2\hat{\tau}_l}(\partial R_l) \subset N^{2\hat{\tau}}(\Gamma) \subset H(W_i)$. Moreover, as $|\Gamma_l(W_i)|_\infty < |\Gamma|_\infty$, by the choice of Γ with respect to $|\cdot|_\infty$ we obtain $|\Gamma_l(W_i)|_\infty \leq \lambda_i$ for all $\Gamma_l(W_i) \in \mathcal{A}_i$ and thus $\mathcal{A}_i \subset \mathcal{T}_{\lambda_i}(W_i)$. As $\mathcal{T}_{\lambda_i}(W_i) \cap \mathcal{S}_{\lambda_i}(W_i) = \emptyset$, this also yields $\Psi_i \cap S_{\lambda_i}(W_i) = \emptyset$. This together with (8.46) shows that $\gamma(M^{21\bar{\tau}_l}(\Gamma_l(W_i))) \leq$

$D\varepsilon\hat{\tau}_l$ for all $\Gamma_l(W_i) \in \mathcal{A}_i \cap \mathcal{G}_i$, where

$$\gamma(A) := \alpha(A \cap W_i \cap \Psi_i) + \varepsilon|A \cap (\partial W_i \cap \Psi_i)|_{\mathcal{H}}$$

for $A \subset \mathbb{R}^2$. Observe that $\eta_l^i = 21\bar{\tau}_l$ as $|\Gamma_l(W_i)|_\infty \le \lambda_i$ for all $\Gamma_l(W_i) \in \mathcal{A}_i$. Using the definition of $\mathcal{B}_i$, (8.33), (8.46) and recalling $D = D(h_*)$ we find for v small enough (with respect to h_*)

$$\begin{aligned}\sum\nolimits_{B_j^i \in \mathcal{B}_i} \gamma(B_j^i) &\le \sum\nolimits_{\Gamma_l(W_i) \in \mathcal{A}_i \cap \mathcal{G}_i} \gamma(M^{21\bar{\tau}_l}(\Gamma_l(W_i))) \\ &\le \sum\nolimits_{\Gamma_l(W_i) \in \tilde{\mathcal{A}}_i} \varepsilon|\Gamma_l(W_i)|_\infty \le \varepsilon|B_{i+1}^i \cap \partial W_i|_{\mathcal{H}}.\end{aligned}$$

In the first step we used that $B_{j_1}^i \cap B_{j_2}^i \cap W_i = \emptyset$ for $j_1 \neq j_2$ by (8.45)(i) and $\partial W_i \cap Q_\mu \subset W_i$. The last step follows from the definition of $\tilde{\mathcal{A}}_i$. Recall that The fact that (8.35) is violated and the definition of B_{i+1}^i then imply

$$\begin{aligned}D(1-\omega_{\min})^{-1}\varepsilon\psi_i &< \alpha(\Psi_i \cap W_i) + \varepsilon|\partial W_i \cap \Psi_i|_{\mathcal{H}} \\ &\le \alpha(\Psi_i \cap W_i) + \varepsilon|\partial W_i \cap \Psi_i|_{\mathcal{H}} - \sum\nolimits_{B_j^i \in \mathcal{B}_i} \gamma(B_j^i) \\ &\quad + \varepsilon|\partial W_i \cap B_{i+1}^i|_{\mathcal{H}} \\ &\le \alpha(B_{i+1}^i) + 2\varepsilon|\partial W_i \cap B_{i+1}^i|_{\mathcal{H}}.\end{aligned} \tag{8.50}$$

We adjust the weights for components in $\tilde{\mathcal{A}}_i$: Let $W_i^* = W_i$ and $\omega(\Gamma_l(W_i^*)) = \omega(\Gamma_l(W_i)) - \frac{1-\omega_{\min}}{2}$ for $\Gamma_l(W_i^*) = \Gamma_l(W_i) \in \tilde{\mathcal{A}}_i$ and $\omega(\Gamma_l(W_i^*)) = \omega(\Gamma_l(W_i))$ otherwise. (The set as a subset of $\mathbb{R}^2$ is left unchanged, we have only changed the weights of the boundary components.) This implies

$$\|W_i^*\|_\omega \le \|W_i\|_\omega - \tfrac{1}{2}h_*(1-\omega_{\min})|\partial W_i \cap B_{i+1}^i|_{\mathcal{H}}. \tag{8.51}$$

We briefly note that (8.47), (8.48) are still satisfied for W_i^* if we replace $\hat{\omega}_i$ by $\hat{\omega}_{i+1}^*$, where $\hat{\omega}_{i+1}^*(\Gamma_l(W_i^*)) := 1 - \frac{1-\omega_{\min}}{2}\#\{j = 0, \dots, i+1 : \Theta_l(W_i) \subset B_j^i\}$. Indeed, as $\omega_i(\Gamma_l(W_i)) \ge \hat{\omega}_i(\Gamma_l(W_i))$ by (8.48) we find

$$\frac{\omega_i(\Gamma_l(W_i^*))}{\hat{\omega}_{i+1}^*(\Gamma_l(W_i^*))} = \frac{\omega_i(\Gamma_l(W_i)) - (1-\omega_{\min})/2}{\hat{\omega}_i(\Gamma_l(W_i)) - (1-\omega_{\min})/2} \ge \frac{\omega_i(\Gamma_l(W_i))}{\hat{\omega}_i(\Gamma_l(W_i))} \ge 1$$

for $\Gamma_l(W_i) \in \tilde{\mathcal{A}}_i$. (Observe that $\hat{\omega}_{i+1}^*$ may slightly differ from the desired $\hat{\omega}_{i+1}$ as given in (8.47). Below we will see, however, that the properties are still satisfied for $\hat{\omega}_{i+1}$.)

We set $\tilde{W}_i = (W_i^* \setminus V) \cup \partial V$, where $\overline{V}$ is the smallest rectangle containing Γ, Γ_m and T. As usual we define $\omega(\partial V) = 1$ and adjust the other weights as in (8.8). We then derive by (8.50) and (8.51)

$$\begin{aligned}\|\tilde{W}_i\|_\omega &\le \|W_i^*\|_\omega + |\partial T|_{\mathcal{H}} \le \|W_i\|_\omega - \tfrac{1}{2}h_*(1-\omega_{\min})|\partial W_i \cap B_{i+1}^i|_{\mathcal{H}} + |\partial T|_{\mathcal{H}} \\ &\le \|W_i\|_\omega + h_*(1-\omega_{\min})\tfrac{1}{4\varepsilon}\alpha(B_{i+1}^i) - \tfrac{1}{4}h_* D\psi_i + |\partial T|_{\mathcal{H}},\end{aligned}$$

where for the other boundary components not involved we proceeded as in (8.12). Recall $|\partial T|_{\mathcal{H}} \le 4\psi_i$. Now choosing $D \ge \frac{16}{h_*}$ we conclude

$$\varepsilon\|\tilde{W}_i\|_\omega + \alpha(\tilde{W}_i) \le \varepsilon\|W_i\|_\omega + \alpha(W_i) + h_*(1-\omega_{\min})\alpha(B^i_{i+1}).$$

Define $\lambda_{i+1} = |\Gamma|_\infty$ and $\bar{u}_{i+1} = \bar{u}_i$. Observe that $W_i^* \in \mathcal{W}_{\lambda_i}$. In fact, (8.5)(iv) follows from the definition of the weights and (8.52) below. Moreover, (8.5)(ii) is a consequence of (8.48)(ii). As before, following the proof of Lemma 8.4.4, we find a set $W_{i+1} = (W_i^* \setminus V') \cup \partial V' \in \mathcal{W}_{\lambda_{i+1}}$ for a rectangle $V' \supset V$ with $W_{i+1} \subset \tilde{W}_i$ and $B^{i+1}_j = B^i_j \setminus S_{\lambda_{i+1}}(W_{i+1})$ for $j = 0,\dots,i+1$ such that (8.44), (8.46) hold and (8.47), (8.48) are satisfied for $\hat{\omega}_{i+1}$. Observe that (8.45) does not follow from Lemma 8.4.4 as $B^{i+1}_{i+1} \neq \emptyset$. We postpone the proof of (8.45) to Lemma 8.4.5 below. We now continue with the next iteration step.

In each iteration step either the number of components satisfying (8.47) increases or the volume of W_i decreases by at least $(2C_2 s)^2$. Consequently, after a finite number of steps, denoted by i^*, we find a set $U = W_{i^*} \in \mathcal{W}^s_{\lambda_U}$, $\lambda_U \ge 0$, satisfying (8.47) for all boundary components $\Gamma_l(U)$ with $N^{2\hat{\tau}}(\partial R_l(U)) \subset H(U)$. Let $\bar{u} = \bar{u}_{i^*}$. Then (8.41) holds for all such boundary components as $\sum_n (2/3)^n < \infty$, $\hat{\omega}(\Gamma_l(U)) \ge \omega_{\min}$ for all $\Gamma_l(U)$ by (8.45)(i) and since v can be chosen in dependence of h_*. Similarly, by (8.45)(i) we find $\sum_{j=0}^{i^*} \alpha(B^{i^*}_j) \le 2\alpha(W)$ and by (8.48) we get $\omega(\Gamma_l(U)) \ge \omega_{\min}$ for all $\Gamma_l(U)$. Setting $\sigma = 2(1-\omega_{\min})$, by (8.44) and $h_* \le 1$ we conclude

$$\varepsilon\|U\|_* + \alpha(U) \le (1 - \tfrac{1}{2}\sigma)^{-1}\varepsilon\|W\|_* + (1+\sigma)\alpha(W) \le (1+\sigma)(\varepsilon\|W\|_* + \alpha(W)).$$

As v is chosen in dependence of h_* and σ, the constant C_2 in (8.43) depends only on h_* and σ. Finally, the property $|W \setminus U| \le c\|U\|^2_\infty$ relies on the isoperimetric inequality and can be derived as in Corollary 8.2.2.

It remains to indicate the necessary changes if in some iteration step i the choice of Γ is not unique. If there are several components $\Gamma_1,\dots,\Gamma_m$ with $\lambda_{i+1} := |\Gamma_j|_\infty > \lambda_i$ for $j = 1,\dots,m$ we choose an order such that $\Gamma_1,\dots,\Gamma_{m'}$, $m' \le m$, are the components satisfying (8.13), (8.34), (8.35). We now apply Theorem 8.4.1 successively on each Γ_j, $j = 1,\dots,m'$, and replace $\mathcal{T}_{\lambda_{i+1}}(W_{i+1})$ in (8.46), (8.47) by $\mathcal{T}^j_{\lambda_{i+1}}(W_{i+1}) := \mathcal{T}_{\lambda_i}(W_i) \cup \bigcup_{k=1}^j \{\Gamma_k\}$. For each Γ_j, $j = m'+1,\dots,m$, we proceed as in one of the cases a) - c) and let $\mathcal{T}^j_{\lambda_{i+1}}(W_{i+1}) := \mathcal{T}_{\lambda_i}(W_i) \cup \bigcup_{k=1}^{m'} \{\Gamma_k\}$ in (8.46), (8.47). □

It remains to show (8.45) in case c).

Lemma 8.4.5. *If in the i-th iteration step of the above modification procedure case c) is applied, then* (8.45) *holds for W_{i+1}.*

Proof. We first show that

$$|\Gamma_l(W_i)|_\infty < 19\bar{\tau} \quad \text{and} \quad \operatorname{dist}(\Gamma_l(W_i), \Psi_i) \le \bar{\tau} \quad \text{for all } \Gamma_l(W_i) \in \mathcal{A}_i. \qquad (8.52)$$

For sets $\Gamma_l = \Gamma_l(W_i)$ intersecting $\Psi_i \subset N^{\bar{\tau}}(\Gamma)$ this is clear by construction of Ψ_i and Corollary 8.3.4. (We can assume that property (8.13) holds and Corollary 8.3.4 is applicable as otherwise we would have applied case (a).) Now assume $\Gamma_l \cap N^{\bar{\tau}}(\Gamma) = \emptyset$ but $M^{\eta_l^i}(\Gamma_l) \cap \Psi_i \neq \emptyset$ for $\Gamma_l \in \mathcal{G}_i$, which implies $\mathrm{dist}(\Gamma_l, \Psi_i) < \eta_i^l \leq 21 \upsilon \lambda_i \leq 21\bar{\tau}$. In particular, this yields $\Gamma_l \cap N^{22\bar{\tau}}(\Gamma) \neq \emptyset$. Recall that $\Gamma_m \cap \overline{\Psi}_1 \neq \emptyset$ and $|\Gamma_m|_\infty \geq \hat{\tau} \geq 19 \cdot 22\bar{\tau}$ for q large enough. Therefore, applying Corollary 8.3.4 for $\bar{t} = 22\bar{\tau}$ we derive that that $|\Gamma_l|_\infty \leq 19 \cdot 22\bar{\tau}$. Repeating the above arguments we obtain $\mathrm{dist}(\Gamma_l, \Psi_i) \leq 21\eta_i^l = 21\bar{\tau}_l \leq \upsilon \cdot 21 \cdot 19 \cdot 22\bar{\tau} < \frac{\bar{\tau}}{2}$ for υ small enough due to the choice of Γ with respect to $|\cdot|_\infty$. This gives the second part of (8.52). Moreover, we have $\Gamma_l \cap N^{\bar{\tau}}(\Gamma) \neq \emptyset$ as $M^{21\bar{\tau}_l}(\Gamma_l) \cap N^{\tau}(\Gamma) \neq \emptyset$ and $\tau \leq \frac{\bar{\tau}}{2}$ by (8.26). This, however, gives a contradiction to the assumption and thus $\Gamma_l \cap N^{\bar{\tau}}(\Gamma) \neq \emptyset$. Then the first part of (8.52) follows again from Corollary 8.3.4.

We now show that (8.45) holds for W_{i+1}. Note that by Lemma 8.4.4 we have $W_{i+1} = (W_i^* \setminus V') \cup \partial V'$ for a rectangle V' which contains Γ, Γ_m and T. Moreover, recall that $W_i = W_i^*$ only differ by the definition of the weights.

First of all, to see (8.45)(ii) it suffices to show that either $\Theta_l(W_i^*) \subset B_j^i$ for some $0 \leq j \leq i+1$ or $\Theta_l(W_i^*) \cap \bigcup_{j=0}^{i+1} B_j^i = \emptyset$ and $\omega(\Gamma_l(W_i^*)) = 1$. In fact, we can then follow the argumentation in the proof of Lemma 8.4.4 to obtain the desired property also for the sets $B_j^{i+1} = B_j^i \setminus S_{\lambda_{i+1}}(W_{i+1})$, $j = 0, \ldots, i+1$.

Recall that $\Theta_l(W_i^*) \subset B_{i+1}^i$ or $\Theta_l(W_i^*) \cap B_{i+1}^i = \emptyset$ for all $(\Gamma_l(W_i^*))_l$. Thus, if $\Theta_l(W_i^*) \not\subset B_j^i$ for some $0 \leq j \leq i+1$ we find $\Theta_l(W_i^*) \cap \bigcup_{j=0}^{i+1} B_j^i = \emptyset$ by (8.45)(ii) for iteration step i. This particularly implies $\Gamma_l(W_i^*) \notin \tilde{\mathcal{A}}_i$ as $\Theta_l(W_i^*) \cap B_{i+1}^i = \emptyset$. Again by (8.45)(ii) and the construction of the weights in (8.51) we then get $\omega(\Gamma_l(W_i^*)) = 1$, as desired.

We concern ourselves with (8.45)(i). First, the assertion is clear for $x \in W_{i+1} \setminus W_i$ as $W_{i+1} \setminus W_i \subset \partial V'$ and $\partial V' \in S_{\lambda_{i+1}}(W_{i+1})$. For $x \notin W_{i+1} \setminus W_i$ it is enough to show the property for $(B_j^i)_{j=1}^{i+1}$ since $B_j^{i+1} \subset B_j^i$ for $j = 0, \ldots, i+1$. As $\bigcup_{l=1}^n \Theta_l(W_i) \subset W_i$ and thus $B_{i+1}^i \subset W_i$, it is elementary to see that it suffices to confirm $B_{i+1}^i \cap \bigcup_{j=0}^i B_j^i \subset W_i \setminus W_{i+1}$. Recall that $\Gamma, \Gamma_m \notin \mathcal{A}_i$. By the definition of B_{i+1}^i we have

$$B_{i+1}^i \cap \bigcup\nolimits_{j=0}^i B_j^i \subset B_{i+1}^i \cap \big(f(M^{21\bar{\tau}}(\Gamma)) \cup f(M^{21\bar{\tau}_m}(\Gamma_m))\big),$$

where $f(A) = A$ if $A \cap \Psi_i \neq \emptyset$ and $f(A) = \emptyset$ else for $A \subset \mathbb{R}^2$. (The possible different cases can be seen in Figure 8.5, 8.6, 8.7.) To see this, let $\mathcal{A}^* \subset \{\Gamma, \Gamma_m\}$ such that the boundary component is contained in $\mathcal{A}^*$ if the corresponding neighborhood intersects Ψ_i. Observe that if $B_j^i \cap B_{i+1}^i \neq \emptyset$, then by (8.45)(iii) (for W_i) we get $B_j^i \cap B_{i+1}^i \subset B_j^i \cap W_i \subset M^{\eta_l^i}(\Gamma_l)$ for some $\Gamma_l \in \mathcal{G}_i \setminus \tilde{\mathcal{A}}_i = \mathcal{G}_i \setminus \mathcal{A}_i$. (The last equality follows from $\mathcal{G}_i \cap \mathcal{A}_i \subset \tilde{\mathcal{A}}_i$.) On the other hand, by (8.45)(ii),(iii) we derive that each $\Gamma_l(W_i) \in \mathcal{A}_i$ with $\Theta_l(W_i) \subset B_j^i$ satisfies $\Gamma_l(W_i) \notin \mathcal{G}_i$, $\Theta_l(W_i) \cap \Psi_i \neq \emptyset$ and thus $\Theta_l(W_i) \cap M^{\eta_l^i}(\Gamma_l) = \emptyset$ for all $\Gamma_l \in \mathcal{G}_i \setminus (\mathcal{A}_i \cup \mathcal{A}^*)$. Likewise, we get

$\Psi_i \cap M^{\eta_l^i}(\Gamma_l) = \emptyset$ for all $\Gamma_l \in \mathcal{G}_i \setminus (\mathcal{A}_i \cup \mathcal{A}^*)$ and thus $(B_j^i \cap B_{i+1}^i) \cap M^{\eta_l^i}(\Gamma_l) = \emptyset$ for all $\Gamma_l \in \mathcal{G}_i \setminus (\mathcal{A}_i \cup \mathcal{A}^*)$. This implies $B_j^i \cap B_{i+1}^i \subset M^{\eta_l^i}(\Gamma_l)$ for some $\Gamma_l \in \mathcal{A}^*$.

Setting $\Phi := \{x \in Q_\mu : \mathrm{dist}(x, \Psi_i) \le 20\bar{\tau}\}$ and recalling (8.52) we then find

$$B_{i+1}^i \cap \bigcup_{j=0}^{i} B_j^i \subset \Phi \cap \big(f(M^{21\bar{\tau}}(\Gamma)) \cup f(M^{21\bar{\tau}_m}(\Gamma_m))\big).$$

We now differ the cases (I) and (II) as considered in Lemma 8.3.6, 8.3.7. In case (I) we get $\Phi \cap W_{i+1} = \emptyset$ as by Lemma 8.3.6(i) the rectangle V' satisfies $\Phi \subset V'$. In (II)(i) the assertion follows as $M^{21\bar{\tau}}(\Gamma), M^{21\bar{\tau}_m}(\Gamma_m) \cap \Psi_i = \emptyset$. Finally, in (II)(ii) it suffices to derive

$$B_{i+1}^i \cap \bigcup_{j=0}^{i} B_j^i \subset \Phi \cap \{x : x_1 \ge -l_1 - \psi\} \cap M^{21\bar{\tau}_m}(\Gamma_m), \qquad (8.53)$$

where without restriction we treat the case $\Gamma_m \cap N(\Gamma) \subset N_{1,-}(\Gamma)$. Then Lemma 8.3.7(i) gives $\Phi \cap \{x : x_1 \ge -l_1 - \psi\} \cap M^{21\bar{\tau}_m}(\Gamma_m) \subset V'$ which finishes the proof of (8.45)(i). To see (8.53), first note that $f(M^{21\bar{\tau}}(\Gamma)) = \emptyset$ and $\Psi_i \subset \{x : x_1 \ge -l_1 - \psi\}$. Consequently, recalling (8.45)(ii),(iii) if the assertion was wrong, there would be some $\Gamma_l(W_i) \in \mathcal{A}_i \setminus \mathcal{G}_i$ which satisfies $\Theta_l(W_i) \subset M^{21\bar{\tau}_m}(\Gamma_m)$ and $\Theta_l(W_i) \cap \{x : x_1 < -l_1 - \psi\} \neq \emptyset$. Again by (8.45)(iii) we then get $\Theta_l(W_i) \subset M_2^{21\bar{\tau}_m}(\Gamma_m)$ (see Figure 8.6) and therefore $\Theta_l(W_i) \cap \Psi_i = \emptyset$. This implies $\Gamma_l(W_i) \notin \mathcal{A}_i$ and yields a contradiction.

Finally we show (8.45)(iii). It suffices to consider B_{i+1}^{i+1} as for the other sets the property follows from Lemma 8.4.4. Without restriction we set $V' = (-v_1, v_1) \times (-v_2, v_2)$. We first observe that $\Phi \setminus \overline{V'} \subset N^{21\bar{\lambda}}(\partial V')$, where $\bar{\lambda} = v\lambda_{i+1}$. This is a consequence of the definition of Φ and the fact that $\bar{\lambda} = \bar{\tau}$. We may assume that $|\pi_1 \Gamma_m| \ge \frac{1}{2}|\pi_2 \Gamma_m|$. In fact, if $l_2 \le \frac{l_1}{2}$ this follows from Corollary 8.3.4, if $l_2 \ge \frac{l_1}{2}$ then l_1, l_2 are comparable and the assumption holds possibly after a rotation of the components by $\frac{\pi}{2}$. Recalling $|\Gamma_m|_\infty \ge \hat{\tau}$ we thus obtain $|\pi_1 \Gamma_m| \ge \frac{1}{\sqrt{5}}\hat{\tau} = \frac{q^2 \bar{\tau}}{\sqrt{5} h_*}$. Recall that $|\pi_1 \Gamma_m \cap \pi_1 \Gamma| \le C\frac{\bar{\tau}}{h_*}$ by Lemma 8.3.1(ii).

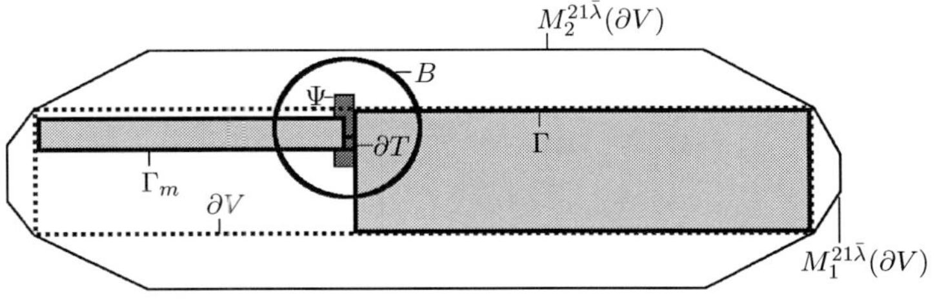

Figure 8.8: Sketch of the components Γ, Γ_m, ∂T and in dashed lines the corresponding rectangle V (which in this example coincides with V'). The ball B is chosen large enough such that $\Phi \subset B$. (The proportions were adapted for illustration purposes.)

We now find for all $x \in \Phi \setminus V'$ with $x_1, x_2 \geq 0$

$$\frac{v_1 - x_1}{x_2 - v_2} \geq \frac{\min\{|\pi_1\Gamma|, |\pi_1\Gamma_m|\} - C\bar{\tau}h_*^{-1}}{21\bar{\tau}} \geq \frac{q^2\bar{\tau}(\sqrt{5}h_*)^{-1} - C\bar{\tau}h_*^{-1}}{21\bar{\tau}} \geq qh_*^{-1}$$

for q sufficiently large and may proceed likewise for $\pm x_1, \pm x_2 \geq 0$. Thus, upon recalling (8.28) and (8.29), we obtain $B^i_{i+1} \cap W_{i+1} \subset \Phi \setminus \overline{V'} \subset M_2^{21\bar{\lambda}}(\partial V')$. As $v|\partial V'|_\infty \geq \bar{\lambda}$, $\omega(\partial V') = 1$ and $\partial V' \cap \bigcup_{j=0}^{i+1} B_j^{i+1} = 0$ (since $\partial V' \subset S_{\lambda_{i+1}}(W_{i+1})$) we finally obtain (8.45)(iii). □

Remark 8.4.6. (i) During the modification process in Theorem 8.4.2 the components $X_{n+1}(W), \dots, X_m(W)$ at the boundary of Q_μ might be changed and the corresponding components of U are given by $X_j(U) = X_j(W) \setminus \overline{H(U)}$ for $j = n+1, \dots, m$. In particular, we observe $|\partial X_j(U)|_* \leq |\partial X_j(W)|_*$ arguing as in Lemma 8.1.1.

(ii) In general, the components of the set U might not be connected as they can be separated by other components during the modification process. However, by application of Corollary 8.2.2 we obtain a set $U' \subset U$ with $\|U'\|_* \leq \|U\|_*$ and $|U \setminus U'| \leq C\|U'\|^2_\infty \leq C\mu\|U'\|_\infty$ such that all components of U' are pairwise disjoint and rectangular and thus particularly connected. Moreover, recalling the modification process (cf. Section 8.2) we find that for each $\Gamma(U)$ the corresponding rectangle $R(U)$ given by (8.5) is contained in a component of U'.

8.4.3 Proof of the main theorem

We now are in a position to prove our Korn-Poincaré-type inequality. For later purpose in Section 9 we split the proof into three steps and begin with a corollary of Theorem 8.4.2. In what follows, we will frequently employ (8.41) and in doing so we apply the inequalities

$$|\Theta|_{\mathcal{H}} \leq C|\Theta|_* \leq C|\Gamma|_\infty \leq C|\Gamma|_{\mathcal{H}}, \qquad |\Theta|_* \leq |\partial R|_* \leq |\partial R|_{\mathcal{H}} \tag{8.54}$$

for a boundary component $\Theta \subset \Gamma$ and the corresponding rectangle R given by (8.5). The properties follow from (8.5)(i), (8.6) and Lemma 8.1.1(vi)). Moreover, we observe that for $W \in \mathcal{V}^s$ and a subset $A \subset Q_\mu$ one has $\sum_{\Gamma_l(W)} |\Gamma_l(W) \cap A|_{\mathcal{H}} \leq 2|\partial W \cap A|$. Recall the definition of $\mathcal{W}^{C_2 s}$ in Theorem 8.4.2 as well as (7.3) and (A.1).

Corollary 8.4.7. *Let $\varepsilon, \mu, h_* > 0$. Let $U \subset Q_\mu = (-\mu, \mu)^2$, $U \in \mathcal{W}^{C_2 s}$ and $u_R \in H^1(U)$. Assume there is a square $\tilde{Q} = (-\tilde{\mu}, \tilde{\mu})^2 \subset Q_\mu$ such that (8.41) is satisfied for all components $\Theta_l(U)$ having nonempty intersection with $\tilde{Q}$, where $\bar{u}_R$ is the extension of u_R defined in (8.37). Then there is a universal constant C such that*

$$|E\bar{u}_R|(\tilde{Q})^2 \le (\mathcal{E}_R(\tilde{Q}))^2 \le C\tilde{\mu}^2 \|e(\nabla u_R)\|^2_{L^2(U\cap\tilde{Q})} + CC_1\mu\varepsilon|\partial U \cap \tilde{Q}|_{\mathcal{H}}|\partial U \cap Q_\mu|_{\mathcal{H}},$$

where C_1 is the constant in Theorem 8.4.2.

Proof. Recall $\alpha(V) = \|e(\nabla \bar{u}_R)\|^2_{L^2(V)}$ for $V \subset U$. Note that by Hölder's inequality we have

$$|E\bar{u}_R|(V)^2 \le (\mathcal{E}_R(V))^2 \le C|V|\alpha(V) + C\Big(\int_{\overline{V}\cap J_{\bar{u}_R}} |[u_R]|\, d\mathcal{H}^1\Big)^2. \qquad (8.55)$$

for $V \subset U$. Moreover, observe that $J_{\bar{u}_R} \cap \tilde{Q} = \bigcup_l \Theta_l(U) \cap \tilde{Q}$. We now derive by (8.41) and (8.54)

$$\begin{aligned}
(\mathcal{E}_R(\tilde{Q}))^2 &\le C\tilde{\mu}^2\alpha(\tilde{Q}) + C\Big(\sum\nolimits_{\Theta_l(U)\cap\tilde{Q}\neq\emptyset} |\Theta_l(U)\cap\tilde{Q}|^{\frac{1}{2}}_{\mathcal{H}} \|[\bar{u}_R]\|_{L^2(\Theta_l(U))}\Big)^2 \\
&\le C\tilde{\mu}^2\alpha(\tilde{Q}) + CC_1\varepsilon|\partial U \cap \tilde{Q}|_{\mathcal{H}} \sum\nolimits_l |\Gamma_l(U)|^2_\infty \\
&\le C\tilde{\mu}^2\alpha(\tilde{Q}\cap U) + CC_1\mu\varepsilon|\partial U \cap \tilde{Q}|_{\mathcal{H}}|\partial U \cap Q|_{\mathcal{H}}.
\end{aligned}$$

In the second and third step we employed Hölder's inequality. In the last step we used $\alpha(\tilde{Q}\setminus U)$ by (8.40) as well as $|\Gamma_l(U)|_\infty \le 2\mu$ and $\sum_l |\Gamma_l(U)|_\infty \le C|\partial U \cap Q|_{\mathcal{H}}$. □

We now formulate and prove the main theorem of this chapter first in terms of sets $W \in \mathcal{V}^s$. The assertion follows combining Theorem 8.4.2, Corollary 8.4.7 and Theorem B.4.

Theorem 8.4.8. *Let $\varepsilon, \mu > 0$ and $h_* > 0$ sufficiently small. There is a constant $C = C(h_*)$ and a universal constant $\bar{c} > 0$ such that for all sets $W \subset Q_\mu = (-\mu,\mu)^2$, $W \in \mathcal{V}^s$ with connected boundary components, and all $y \in H^1(W)$, $R \in SO(2)$, the following holds: There is a set $U \in \mathcal{W}^{C_2 s}$ with $|U \setminus W| = 0$, $|W \setminus U| \le \bar{c}|\partial U \cap Q_\mu|^2_\infty$ and*

$$\varepsilon\|U\|_* + \|e(\nabla u_R)\|^2_{L^2(U)} \le (1+h_*)\big(\varepsilon\|W\|_* + \|e(\nabla u_R)\|^2_{L^2(W)}\big)$$

such that for the square $\tilde{Q} = (-\tilde{\mu},\tilde{\mu})^2$ with $\tilde{\mu} = \max\{\mu - 2|\partial U \cap Q_\mu|_{\mathcal{H}}, 0\}$ we have

$$\|u_R(x) - (A\,x + c)\|^2_{L^2(U\cap\tilde{Q})} \le C\mu^2\|e(\nabla u_R)\|^2_{L^2(U\cap\tilde{Q})} + C\mu\varepsilon|\partial U \cap Q|^2_{\mathcal{H}}$$

for some $A \in \mathbb{R}^{2\times 2}_{\rm skew}$ and $c \in \mathbb{R}^2$.

Proof. Choose $\sigma = \sigma(h_*) \le h_*$ and apply Theorem 8.4.2 to get a set $U \in \mathcal{W}^{C_2 s}$ with $|U \setminus W| = 0$ satisfying (8.42). We can assume that $\tilde{\mu} = \mu - 2|\partial U \cap Q_\mu|_{\mathcal{H}} > 0$ as otherwise there is nothing to show. By definition of $\tilde{Q}$ and (8.33) it is not hard to see that every boundary component $\Gamma_l(U)$ with $\Theta_l(U) \cap \tilde{Q} \neq \emptyset$ fulfills

$N^{2\hat{n}}(\partial R_l(U)) \subset H(U)$ and therefore (8.41) holds. The claim now follows from Theorem B.4 and Corollary 8.4.7. □

We can now finally give the proof of Theorem 8.0.1.

Proof of Theorem 8.0.1. Let $u \in SBD^2(Q_\mu, \mathbb{R}^2) \cap L^2(Q_\mu, \mathbb{R}^2)$ be given. Following the arguments in the proof of Theorem 9.4.1 below (see (9.110)), in particular covering J_u with rectangles, we find a set $W \in \mathcal{V}^s$ with connected boundary components such that $\|W\|_* \le (1+\bar{c}h_*)\mathcal{H}^1(J_u)$ and a modification $\tilde{u} \in SBD^2(Q_\mu, \mathbb{R}^2)$ with $\|\tilde{u} - u\|^2_{L^2(Q_\mu)} + \|e(\nabla\tilde{u}) - e(\nabla u)\|^2_{L^2(Q_\mu)} \le \delta$ for $\delta > 0$ arbitrarily small such that $\tilde{u} \in H^1(W)$.

We now apply Theorem 8.4.8 on $\tilde{u}$ and W for $R = \mathbf{Id}$ and $y = \mathbf{id} + \tilde{u}$. Up to a modification by applying Corollary 8.2.2 we can assume that the components of U are pairwise disjoint rectangles $R_1, \ldots, R_n$ with $\sum_j |R_j|_* \le \|U\|_*$ which yields $\sum_j |R_j|_\infty \le (1 + \bar{c}h_*)\|U\|_* \le (1 + \bar{c}h_*)\big(\mathcal{H}^1(J_u) + \varepsilon^{-1}\|e(\nabla\tilde{u})\|^2_{L^2(W)}\big)$. Finally, as we may assume $\|U\|_* \le C\mu$ (otherwise $\tilde{Q} = \emptyset$) we conclude

$$\begin{aligned}
\|u(x) - (A\,x + c)\|^2_{L^2(U\cap\tilde{Q})} &\le C\delta + \|\tilde{u}(x) - (A\,x + c)\|^2_{L^2(U\cap\tilde{Q})} \\
&\le C\delta + C\mu^2\|e(\nabla\tilde{u})\|^2_{L^2(U\cap\tilde{Q})} + C\mu\varepsilon\|U\|^2_* \\
&\le C(1+\mu^2)\delta + C\mu^2\|e(\nabla u)\|^2_{L^2(Q_\mu)} + C\mu^2\varepsilon\mathcal{H}^1(J_u).
\end{aligned}$$

As δ was arbitrary, we obtain the desired estimate. □

8.5 Trace estimates for boundary components

This section is entirely devoted to the proof of Theorem 8.4.1. We start with some preliminary estimates including an approximation of u_R by a piecewise infinitesimal rigid motion. Here we also discuss the passage from an estimate in the neighborhood to a trace estimate. Afterwards the proof is performed in several steps. We will first assume that in a neighborhood of Γ only small boundary components are present (Step 1). Then we suppose that we have a bound on the projection $\|\cdot\|_\pi$ (recall definition (8.18)) which will allow us to apply a slicing method in the regions of the domain where too large boundary components exist (Step 2). In this context, we have to be particularly careful at the corners of Γ (Step 3). Finally, we present the general proof taking into account the possible existence of sets Ψ^1, Ψ^2 discussed in Section 8.3.2 (Step 4). At this point, the trace theorem we derived in Section 7.2 will play an essential role.

8.5.1 Preliminary estimates

Assume $h_*, q, \omega_{\min} > 0$ have been chosen in the previous section (in this order, see Remark 8.3.2). The parameter $\upsilon > 0$ considered before is not assumed to be already chosen, but will be specified below. Moreover, let $r > 0$ such that $r(1-\omega_{\min})^3 \geq v$. This implies $v = v(h_*, q, \omega_{\min}, r)$. Moreover, we will show $r = r(h_*, q)$ and recalling that $\sigma = 2(1-\omega_{\min})$ (see proof of Theorem 8.4.2) as well as using $q = q(h_*)$ we will find $v \sim C(h_*)\sigma^3$ (cf. Remark 8.4.3).

Let $\varepsilon > 0$, $\lambda \geq 0$ and let $y \in H^1(W)$ for $W = W_i \in \mathcal{W}^s_\lambda$. (We drop the subscript i in the following.) Again we drop the subscript $R \in SO(2)$ and write $u = R^T y - \mathbf{id}$ instead of u_R. Recall $\alpha(U) = \|e(\nabla u)\|^2_{L^2(U)}$ for $U \subset W$. Let $\Gamma = \Gamma(W)$ with $|\Gamma|_\infty \geq \lambda$ and the corresponding neighborhoods $N(\Gamma) = N^\tau(\Gamma)$ and $\hat{N}(\Gamma) = N^{2\hat{\tau}}(\Gamma)$ be given. In addition, we define $\tilde{N}(\Gamma) = N(\Gamma) \setminus (X^1 \cup X^2)$, where $\partial X^1 = \Gamma^1$, $\partial X^2 = \Gamma^2$ are the boundary components satisfying $|\Gamma^i|_\infty \geq \hat{\tau} = q^2 h_*^{-1} v |\Gamma|_\infty$ and $\Gamma_i \cap N(\Gamma) \neq \emptyset$, see Corollary 8.3.4 (note that $X_1, X_2 = \emptyset$ is possible). As before, for shorthand we will write N, $\hat{N}$ and $\tilde{N}$ if no confusion arises. By Remark 7.1.4(i) it is not restrictive to assume that $J = J(\Gamma)$ as defined before equation (8.14) consists of (almost) squares. Suppose that (8.13), (8.34), (8.35) and (8.38) for $\mathcal{T}_{\hat{\tau}}(W)$ hold. Assume that $N^{2\hat{\tau}} \subset H(W)$.

Note that the inclusion $W \cap N \subset \tilde{N}$ may be strict due to boundary components Γ_l with $|\Gamma_l|_\infty < \hat{\tau}$ having nonempty intersection with N. Observe that by (8.5)(iv),(v) and (8.6) for q large we have $|\partial R_l|_\infty < \hat{\tau}$ for these Γ_l, where R_l is the corresponding rectangle given by (8.5)(i),(v). Then for v sufficiently small we get

$$N^{2\hat{\tau}_l}(\partial R_l) \subset N^{2\hat{\tau}}(\Gamma) \subset H(W) \tag{8.56}$$

and thus $\Gamma_l \in \mathcal{T}_{\hat{\tau}}(W)$. Consequently, we can extend u as an SBV function from $N \cap W$ to $\tilde{N}$ as defined in (8.37). For convenience we denote this extension still by u. Clearly, one has $\alpha(\tilde{N}) = \alpha(W \cap N)$.

We now begin with some preliminary estimates. Recall definition (7.3) as well as (8.55) and write $\mathcal{E}$ instead of $\mathcal{E}_R$. First assume $\tilde{N} = N$. We apply Theorem B.4 on each $Q \in J$ recalling that the constant is invariant under rescaling of the domain: This yields functions $\bar{A} : N \to \mathbb{R}^{2\times 2}_{\mathrm{skew}}$, $\bar{c} : N \to \mathbb{R}^2$ being constant on each $Q \in J$ such that by (8.34) and (8.38) we obtain

$$\begin{aligned}
\int_N |u(x) - (\bar{A}\,x + \bar{c})|^2\,dx &\leq C \sum\nolimits_{Q\in J} (\mathcal{E}(Q))^2 \leq C(\mathcal{E}(N))^2 \\
&\leq Cv|\Gamma|^2_\infty \alpha(N) + C\Big(\sum\nolimits_l |\Theta_l \cap N|^{1/2}_{\mathcal{H}} \|[u]\|_{L^2(\Theta_l)}\Big)^2 \\
&\leq Cv^2|\Gamma|^3_\infty \varepsilon + CC_* v^{-4}\varepsilon |\partial W \cap N|_{\mathcal{H}} \Big(\sum\nolimits_l |\Theta_l|_*\Big)^2 \\
&\leq Cv^2|\Gamma|^3_\infty \varepsilon + CC_* v^{-4}\varepsilon |\partial W \cap \hat{N}|^3_{\mathcal{H}} \leq C(1+C_*)v^{-1}|\Gamma|^3_\infty \varepsilon
\end{aligned} \tag{8.57}$$

for some $C = C(h_*, q)$. In the third step we employed Hölder's inequality and $|N| \leq Cv|\Gamma|^2_\infty$. In the penultimate step we used $\sum_l |\Theta_l|_* \leq C\sum_l |\Gamma_l|_{\mathcal{H}} \leq C|\partial W \cap \hat{N}|_{\mathcal{H}}$ by (8.54) and (8.56). The constants used in this section may as usual vary from line to line but are always independent of the parameters r, $\omega_{\min}$ and v.

In the general case, recall the definition of Ψ^1 and Ψ^2 in Section 8.3.2. By Lemma 8.3.6(ii) and Lemma 8.3.7(ii) it is not restrictive to assume that Ψ^i_1, $\Psi^{i,*}_2$, Ψ^i_3 are squares for $i = 1,2$. Similarly as in the previous estimate we obtain functions $\bar{A} : N \to \mathbb{R}^{2\times 2}_{\rm skew}$, $\bar{c} : N \to \mathbb{R}^2$ being constant on each $Q \in J$ with $Q \cap (\Psi^1 \cup \Psi^2) = \emptyset$ and Ψ^i_j, $i = 1,2$, $j = 1,2,3$, such that

$$\begin{aligned} &(i) \int_{\tilde{N}\setminus(\Psi^1\cup\Psi^2)} |u(x) - (\bar{A}\,x + \bar{c})|^2\,dx \le C(1+C_*)v^{-1}|\Gamma|^3_\infty\varepsilon, \\ &(ii) \int_{\Psi^i_j} |u(x) - (\bar{A}\,x + \bar{c})|^2\,dx \le C(1+C_*)v^{-3}|\Gamma|^2_\infty\psi^i\varepsilon, \end{aligned} \tag{8.58}$$

To see (ii), we apply Theorem B.4 on the sets Ψ^i_1, $\Psi^{i,*}_2$, Ψ^i_3 and follow the lines of the previous estimate to obtain that the left hand side is bounded by $Cv|\Gamma|^2_\infty\alpha(\Psi^i_j) + CC_*v^{-4}\varepsilon|\partial W \cap \Psi^i_j|_{\mathcal{H}}|\partial W \cap \hat{N}|^2_{\mathcal{H}}$. We then use $\alpha(\Psi^i_j) \le D\varepsilon(1-\omega_{\min})^{-1}\psi^i \le D\varepsilon v^{-1}\psi^i$ and $|\partial W \cap \Psi^i|_{\mathcal{H}} \le Cv^{-1}\psi^i$ by (8.34), (8.35).

The goal will be to replace the functions $\bar{A}$, $\bar{c}$ in (8.58) by constants $A \in \mathbb{R}^{2\times 2}_{\rm skew}$ and $c \in \mathbb{R}^2$ such that

$$\begin{aligned} &(i) \int_{\tilde{N}\setminus(\Psi^1_2\cup\Psi^2_2)} |u(x) - (A\,x + c)|^2\,dx \le C(1+rC_*)v^{-3}|\Gamma|^3_\infty\varepsilon, \\ &(ii) \int_{\Psi^i_2} |u(x) - (A\,x + c)|^2\,dx \le C(1+rC_*)v^{-4}|\Gamma|^2_\infty\psi^i\varepsilon, \end{aligned} \tag{8.59}$$

for $i = 1,2$ and for $r(1-\omega_{\min})^3 \ge v$. Then the trace theorem applied on each square (if J or Ψ^j, $j = 1,2$, consist also of rectangles, they can be covered by possibly overlapping squares) implies the assertion:

To satisfy the assumptions of Lemma 7.2.1, the jump set has to be the union of rectangle boundaries. Therefore, we extend $J_u \cap \tilde{N}$ to $\tilde{J}_u = \bigcup_l \partial R_l \cap \tilde{N}$ by $[u](x) = 0$ for $x \in \tilde{J}_u \setminus J_u$, where R_l are the corresponding rectangles given in (8.5)(i). We observe that by (8.5)(ii) and (8.54) we get

$$\sum\nolimits_l |\partial R_l|_{\mathcal{H}} \le C\sum\nolimits_l |\Theta_l|_*, \quad \sum\nolimits_l |\partial R_l|^{-1}_{\mathcal{H}}|\Theta_l|^2_* \le C\sum\nolimits_l |\Theta_l|_*. \tag{8.60}$$

Note that by every boundary component Γ_l is contained in at most $C(h_*)$ different squares (see Lemma 8.3.1). Then by Lemma 7.2.1, either for $\mu \sim v|\Gamma|_\infty$ or $\mu \sim \psi^i$, (8.59) and (8.38) we obtain for v small enough

$$\begin{aligned} &\int_\Gamma |u(x) - (A\,x + c)|^2\,d\mathcal{H}^1(x) \\ &\le Cv|\Gamma|_\infty\alpha(\tilde{N}) + CC_*\sum\nolimits_l |\partial R_l|_{\mathcal{H}}\sum\nolimits_l |\partial R_l|^{-1}_{\mathcal{H}}\varepsilon v^{-4}|\Theta_l|^2_* \\ &\quad + C(v|\Gamma|_\infty)^{-1}\|u - (A\,\cdot + c)\|^2_{L^2(\tilde{N}\setminus(\Psi^1_2\cup\Psi^2_2))} + \sum\nolimits_{i=1}^2 C(\psi^i)^{-1}\|u - (A\,\cdot + c)\|^2_{L^2(\Psi^i_2)} \\ &\le C(1+rC_*)v^{-2}|\Gamma|^2_\infty\varepsilon + C(1+rC_*)v^{-4}|\Gamma|^2_\infty\varepsilon \le C(1+rC_*)\varepsilon v^{-4}|\Gamma|^2_*, \end{aligned}$$

where for the first two terms we proceeded similarly as in (8.57), also taking (8.60) into account. Finally, choosing $\hat{C} = C = C(h_*, q)$ and $r = r(h_*, q)$ small enough (i.e. also v small enough) such that $r\hat{C} \leq \frac{1}{2}$ we get (8.39), as desired. Consequently, it suffices to establish (8.59).

8.5.2 Step 1: Small boundary components

We first treat the case that only small components Γ_l lie in N. For $1 > r \geq v > 0$ define $T = \lfloor \log_r(v) \rfloor$ and let

$$\mathcal{S}_t(Q) = \{\Gamma_l : \Gamma_l \cap Q \neq \emptyset,\ v^4 r^{-2t}|\Gamma|_\infty < |\Gamma_l|_\infty \leq v^4 r^{-2t-2}|\Gamma|_\infty\}$$

for all $t \in \mathbb{N}$ and $\mathcal{S}_0(Q) = \{\Gamma_l : \Gamma_l \cap Q \neq \emptyset, |\Gamma_l|_\infty \leq v^4 r^{-2}|\Gamma|_\infty\}$.

Lemma 8.5.1. *Theorem 8.4.1 holds under the additional assumption that there is some $\frac{T}{4} + 2 \leq t \leq \frac{T}{2} - 1$ such that $\bigcup_{s>t} \mathcal{S}_s(Q) = \emptyset$ for all $Q \in J$ and $\sum_Q \#\mathcal{S}_t(Q) \leq v^{-3} r^{2t+3}$.*

Proof. We first observe that the assumption implies $\tilde{N} = N$. Let $\frac{T}{4}+2 \leq t \leq \frac{T}{2}-1$ with the above properties be given and write $\hat{v} = v^2 r^{-\frac{2t+1}{2}}$ for shorthand. For later we note that

$$v^{\frac{7}{4}} r^{-\frac{3}{2}} \leq \hat{v} \leq v^{\frac{3}{2}} \sqrt{r}. \tag{8.61}$$

We cover N with squares $\hat{Q}(\xi) = \hat{Q}^{\hat{v}|\Gamma|_\infty}(\xi)$ of length $2\hat{v}|\Gamma|_\infty$ and midpoint ξ. (If the sets in $J = J(\Gamma)$ constructed in Section 8.3.1 are not perfect squares, the sets $\hat{Q}(\xi)$ shall be chosen appropriately. The difference in the possible shapes, however, does not affect the following estimates by Remark 7.1.4(i).) We will now consider a rectangular path, i.e. a path $\xi = (\xi_0, \dots \xi_n = \xi_0)$ of square midpoints intersecting all $Q \in J$ such that there are indices i_1, i_2, i_3 with $\xi_j - \xi_{j-1} = \pm 2\hat{v}|\Gamma|_\infty \mathbf{e}_1$ for all $0 \leq j \leq i_1$, $i_2 \leq j \leq i_3$ and $\xi_j - \xi_{j-1} = \pm 2\hat{v}|\Gamma|_\infty \mathbf{e}_2$ else. Observe that the number of squares in a path satisfies $n \leq C\hat{v}^{-1}$ and that we can find $\sim v\hat{v}^{-1}$ disjoint rectangular paths in N. Consequently, by assumption and (8.34) we can find at least one rectangular path $P := \overline{\bigcup_j \hat{Q}(\xi_j)}$ such that

$$\alpha(P) \leq C\hat{v}\varepsilon|\Gamma|_\infty, \quad \sum_{\Gamma_l \in \hat{\mathcal{S}}(P)} |\Gamma_l|_\infty \leq C\hat{v}|\Gamma|_\infty, \quad \#\hat{\mathcal{S}}^t(P) \leq C\tfrac{\hat{v}}{v} v^{-3} r^{2t+3} \tag{8.62}$$

for some sufficiently large constant $C = C(h_*, q)$, where $\hat{\mathcal{S}}(P) = \{\Gamma_l : \Gamma_l \cap P \neq \emptyset\}$ and $\hat{\mathcal{S}}^t(P) = \hat{\mathcal{S}}(P) \cap \bigcup_Q \mathcal{S}_t(Q)$. Here we used that each $\Gamma_l \in \bigcup_{Q \in J} \bigcup_{s \leq t} \mathcal{S}_s(Q)$ intersects at most four adjacent squares $\hat{Q}$ because $|\partial R_l|_\infty \leq C\hat{v}^2 r^{-1}|\Gamma|_\infty \ll \hat{v}|\Gamma|_\infty$ by (8.6), (8.61) and $\Gamma_l \subset \overline{R_l}$ (see (8.5)(i)). Observe that the above path can be chosen in the way that also $|P \cap Q| \geq C\frac{\hat{v}}{v}|Q|$ for $Q = E_{\pm,\pm}$, where $E_{\pm,\pm}$

denote the squares in the corners of N, i.e. $(\pm l_1, \pm l_2) \cap \overline{E_{\pm,\pm}} \neq \emptyset$. This implies $|P \cap Q| \geq C\frac{\hat{v}}{v}|Q|$ for all $Q \in J$. It is convenient to write the above estimate in the form

$$v^4 r^{-2t-1\pm 1} = \hat{v}^2 r^{\pm 1}, \quad \#\hat{\mathcal{S}}^t(P) v^4 r^{-2t-2} \leq C\hat{v}r. \tag{8.63}$$

We now apply Lemma 7.1.3(ii) with $s = \hat{v}|\Gamma|_\infty$ and $|V| = |P| \sim \hat{v}|\Gamma|_\infty^2$. Recall that we get $\|[u]\|_{L^1(\Theta_l)} \leq \sqrt{|\Theta_l|_{\mathcal{H}}}\|[u]\|_{L^2(\Theta_l)} \leq C\sqrt{C_*\varepsilon v^{-4}}|\Theta_l|_*^{3/2}$ by (8.38), (8.54) and Hölder's inequality. Arguing similarly as in (8.57) we find $A \in \mathbb{R}^{2\times 2}_{\rm skew}$ and $c \in \mathbb{R}^2$ such that by (8.56), (8.62), (8.63) and (8.54)

$$\begin{aligned}
\int_P |u(x) - (A\,x + c)|^2\,dx &\leq C\hat{v}^{-3}(\mathcal{E}(P))^2 \\
&\leq C\hat{v}^{-3}\hat{v}|\Gamma|_\infty^2\alpha(P) + CC_*\hat{v}^{-3}\frac{\varepsilon}{v^4}\Big(\sum\nolimits_{\Gamma_l\in\hat{\mathcal{S}}(P)}(|\Theta_l|_*)^{3/2}\Big)^2 \\
&\leq C\hat{v}^{-1}|\Gamma|_\infty^3\varepsilon + CC_*\hat{v}^{-1}r^{-1}|\Gamma|_\infty\frac{\varepsilon}{v^4}\Big(\sum\nolimits_{\Gamma_l\in\hat{\mathcal{S}}^t(P)}|\Gamma_l|_\infty\Big)^2 \\
&\quad + CC_*\hat{v}^{-1}r|\Gamma|_\infty\frac{\varepsilon}{v^4}\Big(\sum\nolimits_{\Gamma_l\in\hat{\mathcal{S}}(P)\setminus\hat{\mathcal{S}}^t(P)}|\Gamma_l|_\infty\Big)^2 \\
&\leq C\hat{v}^{-1}|\Gamma|_\infty^3\varepsilon + CC_*\hat{v}r\frac{\varepsilon}{v^4}|\Gamma|_\infty^3 + CC_*\hat{v}r\frac{\varepsilon}{v^4}|\Gamma|_\infty^3.
\end{aligned} \tag{8.64}$$

Observing that $\hat{v}^{-1} \leq \hat{v}v^{-4}$ by definition of $\hat{v}$, we derive

$$\int_P |u(x) - (A\,x + c)|^2\,dx \leq C(1 + C_*r)\,\hat{v}\frac{\varepsilon}{v^4}|\Gamma|_\infty^3 =: F. \tag{8.65}$$

We now pass from an estimate on P to an estimate on N. For later purpose in Section 8.5.3, we consider general subsets $V \subset N$ consisting of squares in J. Then repeating (8.64) we obtain by Lemma 7.1.3(iii) $\|u(x) - (A\,x + c)\|^2_{L^2(P\cap V)} \leq |P \cap V||P|^{-1}CF$.

Since $|P \cap Q| \geq C\frac{\hat{v}}{v}|Q|$ for all $Q \in J$ we find $\frac{|V|}{|N|} \geq C\frac{|V\cap P|}{|P|} \geq Cv$. Therefore, by (8.57) (recall that $\tilde{N} = N$) and $v^2 \leq \hat{v}r$ (see (8.61)) we also have

$$\int_V |u(x) - (\bar{A}\,x + \bar{c})|^2\,dx \leq |V \cap P||P|^{-1}CF \leq C|V||N|^{-1}F.$$

We apply (7.14) on each $Q \subset V$ with $B_1 = P \cap Q$, $B_2 = Q$ noting that $\bar{A}, \bar{c}$ are constant on each square. (Although $P\cap Q$ is not a rectangle if $Q = E_{\pm,\pm}$, we can still argue as in (7.14) since $P\cap Q$ consists of two rectangles.) As $|P\cap Q| \geq C\frac{\hat{v}}{v}|Q|$ for all $Q \in J$ we obtain

$$\|u(x) - (A\,x + c)\|^2_{L^2(Q)} \leq C\frac{v}{\hat{v}}\Big(\int_Q |u(x) - (\bar{A}\,x + \bar{c})|^2\,dx + \int_{P\cap Q}|u(x) - (A\,x + c)|^2\,dx\Big)$$

and thus summing over all $Q \subset V$ we derive

$$|N| \fint_V |u(x) - (A\,x + c)|^2\,dx \leq C\hat{v}^{-1}vF = C(1 + C_*r)\,\frac{\varepsilon}{v^3}|\Gamma|_\infty^3. \tag{8.66}$$

Consequently, setting $V = N$, (8.59)(i) is established, as desired. □

8.5.3 Step 2: Subset with small projection of components

The next step will be the case that $\|\cdot\|_\pi$ is not too large. For that purpose, recall (8.18) and the definition of $\mathcal{Y}$ (see before (8.15)). Consider some $U \in \mathcal{Y}$ with $|U| \geq Cv|\Gamma|_\infty^2$. Moreover, by $\mathcal{Y}'$ we denote the set of subsets of N consisting of squares in J. (In contrast to $\mathcal{Y}$ the connectedness of the sets is not required.) In this section we show that for all $Z \subset U$, $Z \in \mathcal{Y}'$, one has

$$|U| \fint_Z |u(x) - (A_U\, x + c_U)|^2\, dx \leq C(1 + rC_*)v^{-3}|\Gamma|_\infty^3 \varepsilon \qquad (8.67)$$

for $A_U \in \mathbb{R}^{2\times 2}_{\text{skew}}$, $c_U \in \mathbb{R}^2$. Recall that $E_{\pm,\pm}$ denote the squares at the corners of Γ (see construction before (8.14)).

Lemma 8.5.2. *Let $r \geq v > 0$. Let $U \in \mathcal{Y}$ with $|U| \geq Cv|\Gamma|_\infty^2$ and $U \cap E_{\pm,\pm} = \emptyset$ be given and assume that $\|U\|_\pi \leq \frac{19}{20}\tau$. Then there is a subset $U' \subset U$, $U' \in \mathcal{Y}'$, with $|U \setminus U'| \leq Cr|U|$ such that* (8.67) *holds for all $Z \subset U'$, $Z \in \mathcal{Y}'$.*

Proof. Let $U \in \mathcal{Y}$ be given with $\|U\|_\pi \leq \frac{19}{20}\tau$ and assume without restriction $U \subset N_{2,+} \setminus (N_{1,-} \cup N_{1,+})$. By the choice of τ in (8.26) we obtain that all Γ_l having nonempty intersection with U satisfy $|\Gamma_l|_\infty < 19\bar{\tau}$. In particular, this implies $U \cap \tilde{N} = U$. Let $(\partial R_l)_l$ be the rectangles corresponding to $(\Gamma_l)_l$ as given by (8.5)(i),(v). We first prove that there is a $\frac{T}{4} + 2 \leq t \leq \frac{T}{2} - 1$ such that $\sum_{Q \subset U} \#\mathcal{S}_t(Q) \leq v^{-3}r^{2t+3}$ as in the assumption of Lemma 8.5.1. If the claim were false, we would have (assume without restriction that $T \in 4\mathbb{N}$)

$$\begin{aligned} |\partial W \cap N^{(1+19C)\bar{\tau}}|_{\mathcal{H}} &\geq C \sum\nolimits_{t=\frac{T}{4}+2}^{\frac{T}{2}-1} \sum\nolimits_{Q\subset U} \#\mathcal{S}_t(Q) v^4 r^{-2t} |\Gamma|_\infty \geq CTvr^3|\Gamma|_\infty \\ &\geq C\log_r(v) r^3 v |\Gamma|_\infty \gg v|\Gamma|_\infty \end{aligned} \qquad (8.68)$$

for v small enough (with respect to $r = r(h_*, q)$) giving a contradiction to (8.34). In the first step we used that $|\partial R_l|_\infty \leq 19C\bar{\tau}$ by (8.6) which implies $\Gamma_l \subset N^{(1+19C)\bar{\tau}}(\Gamma)$ and assures that Γ_l intersects only a uniformly bounded number of different squares $Q \subset U$ (independently of r, v). As before, we define $\hat{v} = v^2 r^{-\frac{2t+1}{2}}$ for shorthand.

As in the previous proof we will select a path in U with certain properties. Recalling (8.18) it is not hard to see that $|\pi_2(R_l \cap U)| \leq |\partial R_l|_\pi$. As by assumption $\|U\|_\pi \leq \frac{19}{20}\tau$, we find a set $S \subset (l_2, l_2 + \tau)$ being the union of intervals $2k's + (-s, s), k' \in \mathbb{Z}$, with $|S| \geq \frac{\tau}{20}$ such that the stripe $\hat{U} = U \cap (\mathbb{R} \times S)$ satisfies $\partial W \cap \hat{U} = \emptyset$. We cover U by k horizontal paths $\mathcal{P} = (P_i)_i$, $i = 1, \ldots, k$ consisting of $\hat{Q}(\xi) = \hat{Q}^{\hat{v}|\Gamma|_\infty}(\xi)$, i.e. $k = \lceil (2\hat{v}|\Gamma|_\infty)^{-1}\tau \rceil$ as $|\pi_2 U| = \tau$. We can find a subset $\hat{\mathcal{P}}_1 \subset \mathcal{P}$ with $\#\hat{\mathcal{P}}_1 \geq c_1 k$ for c_1 small enough such that

$$\Gamma_l \cap P_i = \emptyset \quad \text{for all } |\Gamma_l|_\infty \geq \bar{C}^2 \hat{v}|\Gamma|_\infty \text{ and } P_i \in \hat{\mathcal{P}}_1 \qquad (8.69)$$

and $|S \cap \pi_2 \bigcup_{P_i \in \hat{\mathcal{P}}_1} P_i| \geq \frac{\tau}{21}$, if $\bar{C}$ is chosen sufficiently large. Indeed, for $\{\Gamma_l : |\pi_2 \Gamma_l| \geq \bar{C}\hat{v}|\Gamma|_\infty\}$ this follows by an elementary argument. On the other hand, by (8.6) we see that each component in $\mathcal{G} := \{\Gamma_l : |\Gamma_l|_\infty \geq \bar{C}^2\hat{v}|\Gamma|_\infty, |\pi_2\Gamma_l| \leq \bar{C}\hat{v}|\Gamma|_\infty\}$ intersects at most $\sim \frac{\bar{C}\hat{v}|\Gamma|_\infty}{\hat{v}|\Gamma|_\infty} = \bar{C}$ different $P_i \in \mathcal{P}$ and thus using (8.34) $\mathcal{G}$ intersects at most $\bar{C}\frac{C\tau}{\bar{C}^2\hat{v}|\Gamma|_\infty} \sim \frac{C\tau}{\hat{v}|\Gamma|_\infty \bar{C}} \ll \frac{\tau}{|\Gamma|_\infty \hat{v}}$ different $P_i \in \mathcal{P}$.

Moreover, it is not hard to see that there is a subset $\hat{\mathcal{P}}_2 \subset \hat{\mathcal{P}}_1$ with $\#\hat{\mathcal{P}}_2 \geq c_2 k$ for c_2 sufficiently small such that

$$|\pi_2(P_i \cap \hat{U})| \geq \tfrac{1}{22} 2\hat{v}|\Gamma|_\infty \quad \text{for all } \; P_i \in \hat{\mathcal{P}}_2. \tag{8.70}$$

Recall that we have already found a $\frac{T}{4}+2 \leq t \leq \frac{T}{2}-1$ such that $\sum_{Q \subset U} \#\mathcal{S}_t(Q) \leq v^{-3}r^{2t+3}$. Using (8.69) we can now choose a path $P = \overline{\bigcup_j \hat{Q}(\xi_j)} \in \hat{\mathcal{P}}_2$ such that (8.62) is satisfied possibly passing to a larger constant $C > 0$ depending on $\bar{C}$. (The essential difference to the argument developed in (8.62) is the fact that every boundary component may intersect not only four squares but a number depending on $\bar{C}$.) Observe that $n \sim \hat{v}^{-1}$, where n denotes the number of squares in the path P. Recall $\hat{\mathcal{S}}(P) = \{\Gamma_l : \Gamma_l \cap P \neq \emptyset\}$ and let

$$\hat{\mathcal{S}}^t_>(P) = \{\Gamma_l : \Gamma_l \cap P \neq \emptyset, \Gamma_l \in \bigcup_{Q \in J} \bigcup_{s>t} \mathcal{S}_s(Q)\}. \tag{8.71}$$

Moreover, define $\mathcal{K} = \{\hat{Q} = \hat{Q}(\xi_j) : \hat{Q} \cap \Gamma_l = \emptyset \text{ for all } \Gamma_l \in \hat{\mathcal{S}}^t_>(P)\}$. By (8.62) it is elementary to see that $\#\hat{\mathcal{S}}^t_>(P) \leq C\hat{v}|\Gamma|_\infty (v^4 r^{-2t-2}|\Gamma|_\infty)^{-1} = C\hat{v}^{-1}r$. Consequently, as by (8.69) every $\Gamma_l \in \hat{\mathcal{S}}^t_>(P)$ intersects only a uniformly bounded number of adjacent sets, we find

$$n - \#\mathcal{K} \leq C\#\hat{\mathcal{S}}^t_>(P) \leq C\hat{v}^{-1}r. \tag{8.72}$$

Consider two squares $\hat{Q}(a), \hat{Q}(b) \in \mathcal{K}$ and the path $(\xi_0 = a, \xi_1, \ldots, \xi_m = b)$. Define $D = \overline{\bigcup_{j=0}^m \hat{Q}(\xi_j)}$. Without restriction we assume $\hat{Q}_0 := \hat{Q}(a) = \mu(-1,1)^2$ and $\hat{Q}_m = \hat{Q}(b) = \mu((2m,0)+(-1,1)^2)$, where for shorthand we write $\mu = \hat{v}|\Gamma|_\infty$. We will now derive an estimate of the form (7.18). First of all, Theorem B.4 (see also (7.17)), Theorem B.5 and a rescaling argument show

$$\|u - (A^i \cdot + c^i)\|_{L^1(\partial \hat{Q}_i)} \leq C\mathcal{E}(\hat{Q}_i) \tag{8.73}$$

for $A^i \in \mathbb{R}^{2\times 2}_{\text{skew}}$, $c^i \in \mathbb{R}^2$, $i = 0, m$, and a constant independent of μ. For shorthand let $\mathcal{E} = \mathcal{E}(\hat{Q}_0) + \mathcal{E}(\hat{Q}_m)$ and define $\hat{\alpha}(D) = \int_D |e(\nabla u)|$. We claim that

$$\begin{aligned} &\mu^2|a^0 - a^m| + \mu|c^0_1 - c^m_1| \leq C\mathcal{E} + C\hat{\alpha}(D), \\ &\mu|c^0_2 - c^m_2| \leq Cm\mathcal{E} + Cm\hat{\alpha}(D), \end{aligned} \tag{8.74}$$

where c^i_j denotes the j-th component of c^i, $i = 0, m$, and a^0, a^m are defined such that $A^i = \begin{pmatrix} 0 & a^i \\ -a^i & 0 \end{pmatrix}$. By (8.70) we find two (measurable) sets $B_1, B_2 \subset \mu(-1,1)$

such that $|B_j| \geq \frac{\mu}{44}$, $\operatorname{dist}(B_1, B_2) \geq \frac{\mu}{22}$ and $E := \mu(-1, 2m+1) \times B_1 \cup B_2 \subset W$. We apply a slicing argument in the first coordinate direction and obtain

$$\begin{aligned}\int_{B_1 \cup B_2} &|u_1(\mu(2m-1), y) - u_1(\mu, y)|\, dy \\ &\leq \int_{B_1 \cup B_2} \Big| \int_{\mu}^{\mu(2m-1)} \partial_1 u_1(t, y)\, dt \Big|\, dy \leq C\hat{\alpha}(E).\end{aligned} \tag{8.75}$$

This together with (8.73) and the triangle inequality yields

$$\|(a^0 - a^m)\, \cdot + (c_1^0 - c_1^m)\|_{L^1(B_1 \cup B_2)} \leq C\mathcal{E} + C\hat{\alpha}(E).$$

Choose $f : B_1 \to \mathbb{R}$ such that $\mathbf{id} + f : B_1 \to B_2$ is piecewise constant and bijective. Thanks to $|B_1| \geq \frac{\mu}{44}$ and $f(y) \geq \frac{\mu}{22}$ for $y \in B_1$ we derive

$$\begin{aligned}\mu^2 |a^0 - a^m| \leq C\|(a^0 - a^m)\, f(\cdot)\|_{L^1(B_1)} &\leq C\|(a^0 - a^m)\, \cdot + (c_1^0 - c_1^m)\|_{L^1(B_1)} \\ &+ C\|(a^0 - a^m)\,(\cdot + f(\cdot)) + (c_1^0 - c_1^m)\|_{L^1(B_1)} \leq C\mathcal{E} + C\hat{\alpha}(E)\end{aligned}$$

and likewise $\mu|c_1^0 - c_1^m| \leq C\mathcal{E} + C\hat{\alpha}(E)$. This gives the first bound in (8.74) since $E \subset D$. Analogously, we slice in $\zeta = \mu(2m-2, c)$ direction for $0 < c < 1$. By (8.62) we find $|\pi_2(\partial W \cap P_i)| \leq C\mu$. Consequently, choosing c small enough and recalling (8.70), we find a set $B_3 \subset \mu(-1, 1-c)$ with $|B_3| \geq \frac{1}{24} 2\hat{v}|\Gamma|_\infty = \frac{\mu}{12}$ such that $\{\mu\} \times B_3 + [0,1]\zeta \subset W$. Letting $\bar{\zeta} = \frac{\zeta}{|\zeta|}$ we get

$$\int_{B_3} |u(\mu, y) \cdot \bar{\zeta} - u((\mu, y) + \zeta) \cdot \bar{\zeta}|\, dy \leq C\hat{\alpha}(E)$$

similarly to (8.75) and thus, using (8.73) and the fact that $A^m \bar{\zeta} \cdot \bar{\zeta} = 0$, we derive

$$\int_{B_3} |(A^0 - A^m)\,(\mu, y)^T \cdot \bar{\zeta} + (c^0 - c^m) \cdot \bar{\zeta}|\, dy \leq C\mathcal{E} + C\hat{\alpha}(E).$$

This together with first part of (8.74) then leads to

$$\mu|c_2^0 - c_2^m + (a^m - a^0)\mu| \frac{c}{|\zeta|} \leq C\hat{\alpha}(E) + C\mathcal{E}$$

and implies the second part of (8.74) as $E \subset D$. Summarizing, (8.74) yields

$$\|c^0 - c^m + (A^0 - A^m)\, \cdot\|_{L^2(\hat{Q}(a))} \leq Cm(\mathcal{E}(\hat{Q}_0) + \mathcal{E}(\hat{Q}_m)) + Cm\hat{\alpha}(D), \tag{8.76}$$

which is an estimate of the form (7.18) with the difference that $\mathcal{E}_R$ is replaced by the elastic part of the energy $\hat{\alpha}$ in squares not contained in $\mathcal{K}$. We briefly note that in (8.76) we can replace $\hat{Q}(a)$ by $\hat{Q}(b)$ due to (8.74).

Define $\tilde{P} = \overline{\bigcup_{\hat{Q} \in \mathcal{K}} \hat{Q}}$. Recall that the essential point for the derivation of Lemma 7.1.3 was an estimate of the form (7.18), (7.19). Consequently, arguing similarly as in Lemma 7.1.3(ii) for $s = \hat{v}|\Gamma|_\infty$ and $|V| = |\tilde{P}| \sim \hat{v}|\Gamma|^2_\infty$ and derive

$$\|u - (A\ \cdot + c)\|^2_{L^2(\tilde{P})} \leq C\hat{v}^{-3}((\mathcal{E}(\tilde{P}))^2 + (\hat{\alpha}(P))^2) \tag{8.77}$$

for suitable $A \in \mathbb{R}^{2\times 2}_{\text{skew}}$ and $c \in \mathbb{R}^2$. Recall the definition of $\mathcal{K}$ (cf. (8.71)) and note that $\Gamma_l \cap \tilde{P} = \emptyset$ for all $\Gamma_l \in \hat{\mathcal{S}}^t_>$. Proceeding as in (8.64) and (8.65), in particular using (8.62) and (8.63), it is not hard to see that

$$\|u - (A\ \cdot + c)\|^2_{L^2(\tilde{P})} \leq C(1 + C_* r)\, \hat{v} \frac{\varepsilon}{v^4} |\Gamma|^3_\infty. \tag{8.78}$$

Note that the difference to the estimate in the proof of Lemma 8.5.1 is that due to the above slicing argument it suffices to consider the elastic part of the energy in the connected components of $P \setminus \tilde{P}$. Now let $J' \subset J$ be the set of squares such that $|Q \cap \tilde{P}| \geq Cv\hat{v}|\Gamma|^2_\infty \geq C\frac{\hat{v}}{v}|Q|$ for all $Q \in J'$. Setting $U' = \left(\bigcup_{Q \in J'} \overline{Q}\right)^\circ \in \mathcal{Y}'$ it is not hard to see that $|U \setminus U'| \leq Cr|U|$ for r small enough as $|P \setminus \tilde{P}| \leq Cr|P|$ by (8.72). Let $Z \subset U'$, $Z \in \mathcal{Y}'$. As before in Lemma 8.5.1, applying Lemma 7.1.3(iii) instead of Lemma 7.1.3(ii), (8.78) yields

$$\|u - (A\ \cdot + c)\|^2_{L^2(\tilde{P} \cap Z)} \leq C|\tilde{P} \cap Z||\tilde{P}|^{-1}(1 + C_* r)\, \hat{v} \frac{\varepsilon}{v^4} |\Gamma|^3_\infty.$$

Then applying (7.14), (8.57) and arguing as in (8.66) we derive

$$|U'| \fint_Z |u(x) - (A\,x + c)|^2\, dx \leq C(1 + C_* r) \frac{\varepsilon}{v^3} |\Gamma|^3_\infty.$$

As $|U \setminus U'| \leq Cr|U|$, this gives (8.67), as desired. □

The next step will be to replace U' by U in Lemma 8.5.2. To this end, we will apply the above arguments iteratively.

Lemma 8.5.3. *Let $r \geq v > 0$. Let $U \in \mathcal{Y}$ with $|U| \geq Cv|\Gamma|^2_\infty$ and $U \cap E_{\pm,\pm} = \emptyset$ be given and assume that $\|U\|_\pi \leq \frac{19}{20}\tau$. Then (8.67) holds.*

Proof. Define $U_1 = U'$ and $J_1 = J'$ as given in Lemma 8.5.2. Assume that $U_i = \left(\bigcup_{Q \in J_i} \overline{Q}\right)^\circ$, $J_i \subset J$, with $U_1 \subset \ldots \subset U_i$ is given such that for $\bar{C} > 0$ sufficiently large

$$|U \setminus U_i| \leq \bar{C} r^i |U| \tag{8.79}$$

and for all $Z \subset U_i$, $Z \in \mathcal{Y}'$, one has

$$\int_Z |u - (A\,x + c)|^2\, dx \leq |Z||U|^{-1} C \prod_{j=0}^{i-1} \left(1 + \bar{C} r^{\frac{j}{8}}\right) G, \tag{8.80}$$

where $A \in \mathbb{R}^{2\times 2}_{\text{skew}}$, $c \in \mathbb{R}^2$ as given by Lemma 8.5.2 and $G := (1 + C_* r)\frac{\varepsilon}{v^3}|\Gamma|^3_\infty$.

Observe that (8.79), (8.80) hold for $i = 1$ by Lemma 8.5.2. We now pass from i to $i+1$ and suppose $i \le T + 2$. First, it is not restrictive to assume that $|U \setminus U_i| \ge \bar{C} r^{i+1}|U|$ for $\bar{C} > 0$ as above since otherwise we may set $U_{i+1} = U_i$. We cover $U \setminus U_i$ with pairwise disjoint, connected sets $N_i^1, \dots, N_i^m \in \mathcal{Y}$, such that

$$\tfrac{1}{2} r^{\frac{i}{8}} |N_i^k| \le |N_i^k \setminus U_i| \le 2 r^{\frac{i}{8}} |N_i^k| \tag{8.81}$$

for all $k = 1, \dots, m$. This can be done in the following way: Let $V_0 = U = (\bigcup_{j=1}^n \overline{Q_j})^\circ$. First, to construct $\tilde{N}_i^1$ let l_1, l_2 be the smallest and largest index, respectively, such that $Q_l \subset U \setminus U_i$ and choose $l = l_1$ if $l_1 < n - l_2$ and $l = l_2$ otherwise. Then add neighbors $Q_{l-1}, Q_{l+1} \subset U$, $Q_{l-2}, Q_{l+2} \subset U$, $\dots$ until $|\tilde{N}_i^1 \setminus U_i| \le 2 r^{\frac{i}{8}} |\tilde{N}_i^1|$ holds. (I.e. the right inequality in (8.81) is satisfied.) This is possible due to the fact that $|V_0 \setminus U_i| \le \bar{C} r^i |V_0| \le \frac{1}{2} r^{\frac{i}{8}} |V_0|$ by (8.79) for r sufficiently small. Then note that also $r^{\frac{i}{8}} |\tilde{N}_i^1| \le |\tilde{N}_i^1 \setminus U_i|$ holds, in particular the left inequality in (8.81) is fulfilled. We now define V_1 as the connected component of $V_0 \setminus \tilde{N}_i^1$ which is not completely contained in U_i. (If both are contained in U_i we have finished.) We repeat the procedure on sets V_j to define $\tilde{N}_i^j$, $1 \le j \le k$, satisfying (8.81), where k is the smallest index such that $|V_k \setminus U_i| > \frac{1}{2} r^{\frac{i}{8}} |V_k|$. We now define $N_i^j = \tilde{N}_i^j$ for $j < k$ and $N_i^k := \tilde{N}_i^k \cup V_k$.

It remains to show that also N_i^k satisfies (8.81). Recall $|V_{k-1} \setminus U_i| \le \frac{1}{2} r^{\frac{i}{8}} |V_{k-1}|$. As due to the choice of l and the fact that $r^{\frac{i}{8}} |\tilde{N}_i^{k-1}| \le |\tilde{N}_i^{k-1} \setminus U_i|$ we have $|V_k| \ge \frac{1}{2}|V_{k-1}| - |\tilde{N}_i^{k-1}|$ and $|V_k \setminus U_i| = |V_{k-1} \setminus U_i| - |\tilde{N}_i^{k-1} \setminus U_i| \le \frac{1}{2} r^{\frac{i}{8}} |V_{k-1}| - r^{\frac{i}{8}} |\tilde{N}_i^{k-1}|$, we find $|V_k \setminus U_i| \le r^{\frac{i}{8}} |V_k|$. This together with (8.81) for $\tilde{N}_i^k$ implies the desired property for N_i^k.

Let $N_i = \bigcup_{k=1}^m N_i^k$. Similarly as in (8.68) we find some $\frac{T}{4} + 2 \le t \le \frac{T}{2} - 1$ such that for $t_i = t + \frac{9}{8} \cdot \frac{i}{2}$ we have $\sum_{Q \subset N_i} \#\mathcal{S}_{t_i}(Q) \le v^{-3} r^{2t_i + 3}$. Again set $\hat{v} = v^2 r^{-\frac{2t+1}{2}}$. Arguing as in (8.70) we can find a horizontal path P_i consisting of $\hat{Q}(\xi_j) = \hat{Q}^{\hat{v}|\Gamma|_\infty}(\xi_j)$, $j = 1, \dots, n_i$, and lying in N_i such that (8.62), (8.69) and (8.70) are satisfied replacing t by t_i. By (8.79) and (8.81) we obtain

$$\bar{C} r^{i+1} \hat{v}^{-1} \le n_i \le C \bar{C} r^{i - \frac{i}{8}} \hat{v}^{-1}. \tag{8.82}$$

Clearly, in general the path P_i is not connected. Define $\hat{\mathcal{S}}^{t_i}_>(P_i)$ and $\mathcal{K}_i$ similarly as in (8.71). By (8.62) it is elementary to see that

$$\#\hat{\mathcal{S}}^{t_i}_>(P_i) \le C \hat{v} |\Gamma|_\infty (v^4 r^{-2t_i - 2} |\Gamma|_\infty)^{-1} \le C \hat{v}^{-1} r^{\frac{9}{8} i} r = C \hat{v}^{-1} r^{\frac{9}{8} i + 1}.$$

Therefore, letting $\tilde{P}_i = \overline{\bigcup_{\hat{Q} \in \mathcal{K}_i} \hat{Q}} \subset P_i$ we find by (8.69) (cf. (8.72))

$$n_i - \#\mathcal{K}_i \le C \hat{v}^{-1} r^{\frac{9}{8} i + 1} \quad \text{and} \quad |P_i| - |\tilde{P}_i| \le C r^{\frac{9}{8} i + 1} \hat{v} |\Gamma|^2_\infty. \tag{8.83}$$

We now repeat the slicing arguments above on each N_i^k and obtain expressions similar to (8.76). As before, applying Lemma 7.1.3(ii) we get (cf. (8.77))

$$\|u-(A^k\,\cdot+c^k)\|^2_{L^2(\tilde{P}_i\cap N_i^k)}\le C(n^k)^3(\mathcal{E}(\tilde{P}_i\cap N_i^k)+\hat{\alpha}(P_i\cap N_i^k))^2$$

for suitable $A^k\in\mathbb{R}^{2\times 2}_{\text{skew}}$ and $c^k\in\mathbb{R}^2$, where as before $\hat{\alpha}(D)=\int_D|e(\nabla u)|$ for $D\subset N$. Here n^k denotes the number of squares forming the path $P_i\cap N_i^k$, particularly $n_i=\sum_{k=1}^m n^k$. We observe that by (8.62) the estimate in (8.63) can now be replaced by

$$v^4r^{-2t_i-1\pm1}=\hat{v}^2r^{-\frac{9}{8}i\pm1},\qquad \#\hat{\mathcal{S}}^{t_i}(P)v^4r^{-2t_i-2}\le C\hat{v}r.$$

Consequently, recalling $n_i\le C\bar{C}r^{\frac{7}{8}i}\hat{v}^{-1}$ by (8.82), $t_i=t+\frac{9}{8}\cdot\frac{i}{2}$ and following the arguments in (8.64), (8.77) and (8.78) we obtain

$$\begin{aligned}&\sum\nolimits_k(n^k)^{-1}\|u-(A^k\,\cdot+c^k)\|^2_{L^2(N_i^k\cap\tilde{P}_i)}\\&\quad\le C\sum\nolimits_k(n^k)^2(\mathcal{E}(N_i^k\cap\tilde{P}_i)+\hat{\alpha}(N_i^k\cap P_i))^2\le Cn_i^2(\mathcal{E}(\tilde{P}_i)+\hat{\alpha}(P_i))^2\\&\quad\le C\hat{v}r^{\frac{7}{4}i}\hat{v}^{-3}(\mathcal{E}(\tilde{P}_i)+\hat{\alpha}(P_i))^2\le C\hat{v}r^{\frac{7}{4}i}r^{-\frac{9}{8}i}F\le C\hat{v}r^{\frac{i}{2}}F,\end{aligned}$$

where F was defined in (8.65). Observe that in the calculation the additional $r^{-\frac{9}{8}i}$ in front of F occurs as in (8.63) $v^4r^{-2t-1\pm1}$ was replaced by $v^4r^{-2t-1\pm1}r^{-\frac{9}{8}i}$. Moreover, the above estimate can be repeated applying Lemma 7.1.3(iii) instead of Lemma 7.1.3(ii): For $Z\subset N_i$, $Z\in\mathcal{Y}'$, we obtain

$$\sum\nolimits_k(n^k)^{-1}|N_i^k\cap\tilde{P}_i|\fint_{N_i^k\cap\tilde{P}_i\cap Z}|u(x)-(A^k\,x+c^k)|^2\,dx\le C\hat{v}r^{\frac{i}{2}}F.$$

Define $J_i^k\subset J$ such that $|Q\cap(\tilde{P}_i\cap N_i^k)|\ge Cv\hat{v}|\Gamma|^2_\infty\ge C\frac{\hat{v}}{v}|Q|$ for $Q\in J_i^k$ and set $\hat{N}_i^k=\bigcup_{Q\in J_i^k}Q$. Assume $\hat{N}_i^k\cap Z\neq\emptyset$ which implies $|\hat{N}_i^k\cap Z|\ge v^2|\Gamma|^2_\infty$. Observe $\hat{v}\ge v^{\frac{7}{4}}r^{-\frac{3}{2}}\ge v^2r^{-\frac{i}{4}-1}$ by (8.61) and the fact that $i\le T+2$. As $|N_i^k|(n^k)^{-1}\le C\hat{v}v|\Gamma|^2_\infty$, we find by (8.57)

$$\begin{aligned}&\sum\nolimits_k(n^k)^{-1}|\hat{N}_i^k|\fint_{\hat{N}_i^k\cap Z}|u(x)-(\bar{A}\,x+\bar{c})|^2\,dx\\&\quad\le C\hat{v}v^{-1}\int_U|u(x)-(\bar{A}\,x+\bar{c})|^2\,dx\le C\hat{v}v^{-1}\,v^3(\hat{v}r)^{-1}F\le C\hat{v}r^{\frac{i}{4}}F.\end{aligned}$$

Again arguing as in (8.66), in particular applying (7.14), we derive

$$\sum\nolimits_k(n^k)^{-1}|\hat{N}_i^k|\fint_{Z\cap\hat{N}_i^k}|u(x)-(A^k\,x+c^k)|^2\,dx\le C\hat{v}r^{\frac{i}{4}}\,\hat{v}^{-1}vF=Cr^{\frac{i}{4}}vF.\quad(8.84)$$

We set $U_i^k=\hat{N}_i^k$ if $|N_i^k\setminus\hat{N}_i^k|\le r^{\frac{i}{8}}|N_i^k|$ and $U_i^k=\emptyset$ else for all $k=1,\dots,m$. We now estimate the difference between A,c given in (8.80) and A^k,c^k for $k=$

$1, \dots, m$. Consider U_i^k such that $U_i^k = \hat{N}_i^k$. Then $|U_i^k| \ge (1 - r^{\frac{i}{8}})|N_i^k|$ and thus by (8.81) we have

$$|U_i^k \cap U_i| \ge |N_i^k \cap U_i| - |N_i^k \setminus \hat{N}_i^k| \ge (1 - Cr^{\frac{i}{8}})|N_i^k| \ge (1 - Cr^{\frac{i}{8}})|U_i^k|$$

for r sufficiently small and some $C > 0$. Consequently, we are in the position to apply (7.16) for $B_2 = U_i^k$, $B_1 = U_i^k \cap U_i$ and $s = \frac{T}{2}$, $\delta = Cr^{\frac{i}{8}}$, where we observe $\delta \ge Cs|\pi_1(U_i^k)|^{-1}$ by (8.81). (Recall the remark in Section 7.1 that B_2 does not have to be connected.) Set $\bar{C}_i = C\prod_{j=0}^{i-1}\left(1 + \bar{C}r^{\frac{j}{8}}\right)$ (cf. (8.80)). Using (8.80) and (8.84), in particular recalling that the sets $(\hat{N}_i^k)_k$ are pairwise disjoint, we find for $Z \subset U$, $Z \in \mathcal{Y}'$

$$\begin{aligned}
\|u - (A\ \cdot + c)\|^2_{L^2(U_i^k \cap U_i \cap Z)} &\le |U_i^k \cap U_i \cap Z||U_i^k|^{-1}H_1^k, \\
\|u - (A^k\ \cdot + c^k)\|^2_{L^2(U_i^k \cap Z)} &\le |U_i^k \cap Z||U_i^k|^{-1}H_2^k,
\end{aligned}$$

where $H_1^k = |U_i^k||U|^{-1}\bar{C}_i G$ and $H_2^k = Cn^k r^{\frac{i}{4}} vF$. Therefore, (7.16) yields

$$\begin{aligned}
\|u - (A\ \cdot + c)\|^2_{L^2(U_i^k \cap Z)} \le\ & |U_i^k \cap Z||U_i^k|^{-1}(1 + Cr^{\frac{i}{8}})|U_i^k||U|^{-1}\bar{C}_i G \\
& + |U_i^k \cap Z||U_i^k|^{-1}Cr^{-\frac{i}{8}}n^k r^{\frac{i}{4}} vF.
\end{aligned} \tag{8.85}$$

For shorthand we write $U^* = \left(\bigcup_{k=1}^m \overline{U_i^k}\right)^\circ$ and define $U_{i+1} = (\overline{U_i} \cup \overline{U^*})^\circ$. We recall $N_i = \bigcup_{k=1}^m N_i^k$ as constructed in (8.81). We claim

$$|N_i \setminus U^*| \le Cr^{i+1}|U| \tag{8.86}$$

and postpone the proof of this assertion to the end of the proof. Then (8.82) for $\bar{C}$ sufficiently large implies $|U^*| \ge |N_i| - Cr^{i+1}|U| \ge cn_i\hat{v}|U|$. As for $U_i^k \neq \emptyset$ we have $|N_i^k| \le (1 - r^{\frac{i}{8}})^{-1}|U_i^k|$, it is not hard to see that $\frac{|U^*|}{|U_i^k|} \le C\frac{n_i}{n^k}$ and thus $n^k|U_i^k|^{-1} \le C|U^*|^{-1}n_i \le C|U|^{-1}\hat{v}^{-1}$. Let $V \subset U_{i+1}$, $V \in \mathcal{Y}'$. Now by (8.85), the fact that the sets U_i^k are pairwise disjoint and $F \le C\hat{v}v^{-1}G$ we derive

$$\begin{aligned}
\sum\nolimits_k \|u - (A\ \cdot + c)\|^2_{L^2(U_i^k \cap V)} &\le |U^* \cap V||U|^{-1}\big(\bar{C}_i(1 + Cr^{\frac{i}{8}})G + Cr^{\frac{i}{8}}G\big) \\
&\le |U^* \cap V||U|^{-1}\bar{C}_{i+1}G,
\end{aligned}$$

The last estimate follows for $\bar{C}$ sufficiently large. By (8.80) we now conclude for $V \subset U_{i+1}$

$$\begin{aligned}
\|u - (A\ \cdot + c)\|^2_{L^2(V)} &= \|u - (A\ \cdot + c)\|^2_{L^2(V \setminus U^*)} + \sum\nolimits_k \|u - (A\ \cdot + c)\|^2_{L^2(U_i^k \cap V)} \\
&\le |V \setminus U^*||U|^{-1}\bar{C}_i G + |U^* \cap V||U|^{-1}\bar{C}_{i+1}G \\
&\le |V||U|^{-1}\bar{C}_{i+1}G.
\end{aligned}$$

This yields (8.80). To see (8.79) for $i + 1$, we apply (8.86) to obtain $|U \setminus U_{i+1}| \le |(U \setminus U_i) \setminus N_i| + |N_i \setminus U^*| \le 0 + \bar{C}r^{i+1}|U|$. Here we used that $U \setminus U_i \subset N_i$.

Finally, we choose $i_* \le T+2$ large enough such that $|U \setminus U_{i_*}| \le \bar{C} r v |U| \ll (v|\Gamma|_\infty)^2$ for r sufficiently small which implies $U_{i_*} = U$. Consequently, thanks to (8.80), (8.67) holds.

It remains to show (8.86). First, by (8.83) and the construction of $\hat{N}_i^k$ we have $|\bigcup_k N_i^k \setminus \bigcup_k \hat{N}_i^k| = \sum_k |N_i^k \setminus \hat{N}_i^k| \le C r^{\frac{9}{8}i+1} v |\Gamma|_\infty^2 \le C r^{\frac{9}{8}i+1} |U|$. Therefore, it suffices to prove

$$\sum\nolimits_k |\hat{N}_i^k \setminus U_i^k| \le r^{-\frac{i}{8}} \sum\nolimits_k |N_i^k \setminus \hat{N}_i^k| \tag{8.87}$$

as then we conclude $|N_i \setminus U^*| \le \sum_k |N_i^k \setminus \hat{N}_i^k| + \sum_k |\hat{N}_i^k \setminus U_i^k| \le C r^{i+1} |U|$.

To see (8.87) we observe that if $\hat{N}_i^k \neq U^k$, then $|\hat{N}_i^k| \le |N_i^k| < r^{-\frac{i}{8}} |N_i^k \setminus \hat{N}_i^k|$. Consequently, we calculate $\sum_k |\hat{N}_i^k \setminus U_i^k| = \sum_{k: U_i^k = \emptyset} |\hat{N}_i^k| \le r^{-\frac{i}{8}} \sum_{k: U_i^k = \emptyset} |N_i^k \setminus \hat{N}_i^k| \le r^{-\frac{i}{8}} \sum_k |N_i^k \setminus \hat{N}_i^k|$, as desired. □

Remark 8.5.4. We briefly note that the previous proof shows that the assertion of Lemma 8.5.3 holds for $U \in \mathcal{Y}$ with $U \cap E_{\pm,\pm} = \emptyset$ of arbitrary size. In fact, we can choose $0 \le i_0 \le T+2$ such that $C r^{i_0+1} v |\Gamma|_\infty^2 < |U| \le C r^{i_0} v |\Gamma|_\infty^2$ and begin the induction in (8.79), (8.80) not for $i = 0$, but for $i = i_0$. For the first step $i = i_0$ we do not apply Lemma 8.5.2, but follow the lines of the proof of Lemma 8.5.3 for one single set $N_{i_0}^1 = U$.

We now drop the assumption that $U \in \mathcal{Y}$ does not intersect a corner of Γ.

Corollary 8.5.5. *Let $r \ge v > 0$. Let $U \in \mathcal{Y}$ be given and assume that $\|U\|_\pi \le \frac{19}{20}\tau$. Then* (8.67) *holds.*

Proof. Assume without restriction $E_{+,+} \subset U$ and define $U' = U \setminus E_{+,+}$. Using Lemma 8.5.3 and Remark 8.5.4 we find

$$|U'| \fint_Z |u(x) - (A\,x + c)|^2 \, dx \le CG$$

for $Z \subset U'$, $Z \in \mathcal{Y}'$, where $G := (1 + C_* r)\frac{\varepsilon}{v^3}|\Gamma|_\infty^3$. Let $Q \in J$, $Q \subset U'$ such that $\partial Q \cap E_{+,+} \neq \emptyset$. Setting $Z = Q$ in the above inequality and arguing as in (8.57) we find $\hat{A} \in \mathbb{R}^{2\times 2}_{\rm skew}$, $\hat{c} \in \mathbb{R}^2$ such that

$$\int_Q |u(x) - (A\,x + c)|^2 \, dx \le CvG, \quad \int_{Q \cup E_{+,+}} |u(x) - (\hat{A}\,x + \hat{c})|^2 \, dx \le C v^2 r^{-1} G.$$

Applying (7.14) on $B_1 = Q$ and $B_2 = Q \cup E_{+,+}$ we find $\|u - (A\,\cdot + c)\|^2_{L^2(Q \cup E_{+,+})} \le CvG$ as $v \le r$. Now it is not hard to see that (8.67) is satisfied. □

8.5.4 Step 3: Neighborhood with small projection of components

Recall the covering $\mathcal{C}$ of the neighborhood N introduced in (8.15). We now treat the case that $\|U\|_\pi$ is small for all $U \in \mathcal{C}$. It is essential that adjacent elements of the covering overlap sufficiently. Therefore, we introduce another covering $\hat{\mathcal{C}} \subset \mathcal{C}$ as follows. First assume $l_2 \geq \frac{l_1}{2}$. If some $U \in \mathcal{C}$ intersects only one of the four sets $N_{j,\pm}$, $j = 1,2$, we let $U \in \hat{\mathcal{C}}$. Then eight sets $U^1_{\pm,\pm}, U^2_{\pm,\pm}$ remain where $U^i_{\pm,\pm} \cap E_{\pm,\pm} \neq \emptyset$ and $U^i_{\pm,\pm} \subset N_{i,-} \cup N_{i,+}$ for $i = 1,2$. As before $E_{\pm,\pm}$ denote the sets at the corners of Γ. Add the four sets $U^1_{\pm,\pm} \cup U^2_{\pm,\pm}$ to $\hat{\mathcal{C}}$. If $l_2 < \frac{l_1}{2}$ we proceed likewise with the only difference that instead of $U^1_{\pm,\pm} \cup U^2_{\pm,\pm}$ we add the two sets $U^2_{k,+} \cup U^1_{k,+} \cup U^1_{k,-} \cup U^2_{k,-}$, $k = +,-$, to $\hat{\mathcal{C}}$. (Note that by definition of $\mathcal{C}$ we have $U^1_{\pm,+} = U^1_{\pm,-} = N_{1,\pm}$ in this case.)

Lemma 8.5.6. *Theorem 8.4.1 holds under the additional assumption that $\|U\|_\pi \leq \frac{19}{20}\tau$ for all $U \in \mathcal{C}$.*

Proof. It suffices to show that for all $U \in \hat{\mathcal{C}}$ there are $A_U \in \mathbb{R}^{2\times 2}_{\mathrm{skew}}$, $c_U \in \mathbb{R}^2$ such that

$$|U| \fint_Z |u(x) - (A_U\, x + c_U)|^2 \, dx \leq C(1 + rC_*)v^{-3}|\Gamma|^3_\infty \varepsilon \tag{8.88}$$

holds for all $Z \subset U$, $Z \in \mathcal{Y}'$. Indeed, the desired result then follows from the construction of the covering $\hat{\mathcal{C}}$ and the arguments developed in Section 7.1: Write $\hat{\mathcal{C}} = \{U_1, \ldots, U_n\}$ with $U_{i-1} \cap U_i \neq \emptyset$ for all $i = 1, \ldots, n$, where $U_0 = U_n$. Now let

$$\mathcal{D} = \{U_i \setminus \overline{U_{i-1} \cup U_{i+1}} : i = 0, \ldots, n-1\} \cup \{U_i \cap U_{i+1} : i = 0, \ldots, n-1\},$$

where $U_{-1} = U_{n-1}$. Note that the elements in $\mathcal{D}$ are pairwise disjoint. We write $\mathcal{D} = \{V_1, \ldots, V_m\}$ such that $\partial V_{i-1} \cap \partial V_i \neq \emptyset$ for $i = 1, \ldots, m$, where $V_0 = V_m$. By (8.15) and the definition of the 'combined sets' in $\hat{\mathcal{C}}$, we find $|V_i| \sim v|\Gamma|^2_\infty$. Clearly, (8.88) also holds for all $V_i \in \mathcal{D}$ for corresponding infinitesimal rigid motions as each set is contained in an element of $\hat{\mathcal{C}}$. We can now estimate the difference of the infinitesimal rigid motions of $B_1 = V_{i-1}$ and $B_2 = V_i$, $i = 1, \ldots, m$, proceeding as in (7.10), (7.12) and (7.14). Here it is essential to observe that assumption (7.8) is satisfied as $B_1 \cup B_2 \subset U$ for some $U \in \hat{\mathcal{C}}$ and so (8.88) may be applied. We now obtain (8.59) following the argument in (7.18), (7.19) replacing the squares $(Q(\xi_j))_j$ by the elements of $\mathcal{D}$ and noting that $\#\mathcal{D}$ is uniformly bounded independently of v.

More general, taking (8.88) and (7.13) into account, we have even shown that

$$|N| \fint_V |u(x) - (A\, x + c)|^2 \, dx \leq C(1 + rC_*)v^{-3}|\Gamma|^3_\infty \varepsilon \tag{8.89}$$

for all $V \subset N$, $V \in \mathcal{Y}'$.

It remains to establish (8.88) for $U \in \hat{\mathcal{C}}$. By assumption and Lemma 8.5.3 the assertion is clear if $U \cap E_{\pm,\pm} = \emptyset$ as then particularly $U \in \mathcal{C}$. Therefore, we first let $l_2 \geq \frac{l_1}{2}$ and assume that e.g. $U \cap E_{+,+} \neq \emptyset$. The necessary changes for the case $l_2 \leq \frac{l_1}{2}$ are indicated at the end of the proof.

As in (8.68) we find $\frac{T}{4} + 2 \leq t \leq \frac{T}{2} - 1$ such that $\sum_{Q \subset U} \#\mathcal{S}_t(Q) \leq v^{-3} r^{2t+3}$. Again let $\hat{v} = v^2 r^{-\frac{2t+1}{2}}$. As before, the main strategy will be to construct a suitable path in U. Let $(\Gamma_l)_l$ be the boundary components such that the corresponding rectangles $(\partial R_l)_l$ given by (8.5)(i) and (8.5)(v), respectively, satisfy $\partial R_l \cap U \neq \emptyset$ and $|\partial R_l|_\pi \neq |\partial R_l|_\infty$. Let $V_l \subset \overline{N}$ be the smallest rectangle containing $R_l \cap N$ and $(l_1 + \tau, l_2 + \tau)$. We partition $(V_l)_l$ into $\mathcal{V}_1$ and $\mathcal{V}_2$ depending on whether $|\pi_1 V_l| \leq |\pi_2 V_l|$ or $|\pi_1 V_l| > |\pi_2 V_l|$. Recalling (8.18) it is not hard to see that $|\pi_j V_l| = |R_l|_\pi$ for $V_l \in \mathcal{V}_j$ for $j = 1, 2$. Let $a_j = \inf\{s \in \mathbb{R} : s \in \pi_j V_l \text{ for a } V_l \in \mathcal{V}_j\}$ and define the stripes

$$A_1 = (-\infty, a_1) \times (-\infty, a_2) \cap N_{1,+} \cap U, \quad A_2 = (-\infty, a_1) \times (-\infty, a_2) \cap N_{2,+} \cap U.$$

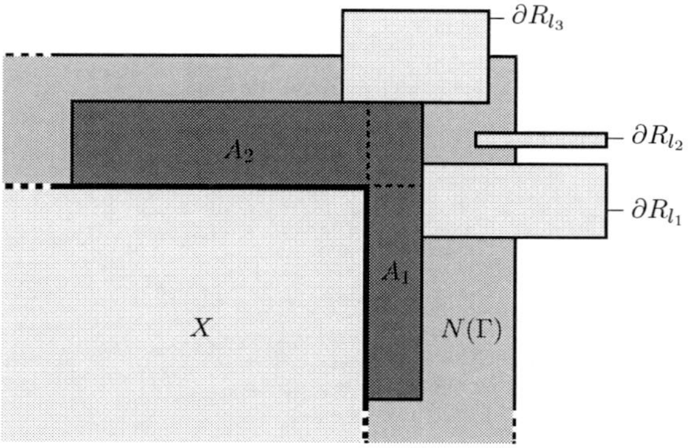

Figure 8.9: Sketch of a part of $N(\Gamma)$ containing $U^1_{+,+} \cup U^2_{+,+}$. The sets A_1, A_2 are highlighted in grey.

As by assumption $\|U\|_\pi \leq \frac{19}{20}\tau$ for all $U \in \mathcal{C}$, we find sets $S_j \subset (l_j, a_j)$ with $|S_j| \geq \frac{\tau}{20}$ such that the stripes $\hat{A}_1 = A_1 \cap (S_1 \times \mathbb{R}) \in \mathcal{U}^s$ and $\hat{A}_2 = A_2 \cap (\mathbb{R} \times S_2) \in \mathcal{U}^s$ satisfy $\partial W \cap \hat{A}_j = \emptyset$ for $j = 1, 2$. Moreover, observe that $|a_j - l_j| \geq \frac{\tau}{20}$ for $j = 1, 2$. We cover A_1 by vertical paths $\mathcal{P}_1 = (P^1_i)_i$, $i = 1, \dots, k_1$, and A_2 by horizontal paths $\mathcal{P}_2 = (P^2_i)_i$, $i = 1, \dots, k_2$, consisting of squares $\hat{Q}^{\hat{v}|\Gamma|_\infty}(\xi) = \hat{Q}(\xi)$, i.e. $k_j = \lceil (2\hat{v}|\Gamma|_\infty)^{-1}(a_j - l_j) \rceil$. As in (8.70) it is not hard to see that there are subsets $\hat{\mathcal{P}}_j \subset \mathcal{P}_j$ with $\#\hat{\mathcal{P}}_j \geq ck_j \geq cv\hat{v}^{-1}$ for $c > 0$ sufficiently small such that (8.69) and (8.70) hold for all $P^j_i \in \hat{\mathcal{P}}_j$, $j = 1, 2$. We can now choose

$P^j \in \hat{\mathcal{P}}_j$, $j = 1, 2$, such that (8.62) is satisfied possibly passing to a larger constant. Moreover, this can be done in a way that $Q^* := P^1 \cap P^2$ satisfies

$$\begin{aligned} &\sum\nolimits_{\Gamma_k \cap Q^* \neq \emptyset} |\Gamma_k|_\infty \le \tilde{C} v^{-1} \hat{v}^2 |\Gamma|_\infty, \\ &Q^* \cap \Gamma_k = \emptyset \quad \text{for all } \Gamma_k : |\Gamma_k| \ge \tilde{C} \hat{v}^2 v^{-1} |\Gamma|_\infty, \end{aligned} \tag{8.90}$$

for $\tilde{C} > 0$ sufficiently large. To see the latter, note that we have $\sim \tau^2(\hat{v}|\Gamma|_\infty)^{-2} = v^2 \hat{v}^{-2}$ possibilities to combine paths in $\hat{\mathcal{P}}_1$, $\hat{\mathcal{P}}_2$ such that (8.62) hold. Moreover, we also have $\sim \tau^2(\hat{v}|\Gamma|_\infty)^{-2} = v^2\hat{v}^{-2}$ possibilities to combine paths in $\hat{\mathcal{P}}_1$, $\hat{\mathcal{P}}_2$ such that Q^* additionally has empty intersection with all Γ_k satisfying $|\Gamma_k|_\infty \ge \tilde{C}\hat{v}^2 v^{-1}|\Gamma|_\infty$. This follows from (8.69) and the fact that by (8.34) we derive

$$\#\{\hat{Q} : \exists \Gamma_k : \tilde{C}\hat{v}^2 v^{-1}|\Gamma|_\infty \le |\Gamma_k|_\infty \le C\hat{v}|\Gamma|_\infty, \Gamma_k \cap \hat{Q} \neq \emptyset\} \le C\tilde{C}^{-1} v^2 \hat{v}^{-2}.$$

Since all other components Γ_k intersect at most four adjacent squares $\hat{Q}$, using again (8.34) we can select Q^* such that also $\sum_{\Gamma_k \cap Q^* \neq \emptyset} |\Gamma_k|_\infty \le Cv|\Gamma|_\infty v^{-2}\hat{v}^2$ holds.

Let $P = \hat{P}^1 \cup Q^* \cup \hat{P}^2$, where $\hat{P}^j$, $j = 1, 2$, is the connected component of $P^j \setminus Q^*$ not completely contained in $E_{+,+}$. Denote the midpoints of the squares in P by $(\xi_1, \dots, \xi_n)$. Recall the definition of $\hat{\mathcal{S}}^t_>$ in (8.71) and let

$$\mathcal{K} = \{\hat{Q} = \hat{Q}(\xi_j) : \hat{Q} \cap \Gamma_l = \emptyset \quad \text{for all} \quad \Gamma_l \in \hat{\mathcal{S}}^t_>(P)\} \cup \{Q^*\}.$$

Consider two sets $\hat{Q}(a), \hat{Q}(b) \in \mathcal{K}$ and the path $(\xi_0 = a, \dots, \xi_m = b)$. We can repeat the slicing method of the previous proofs and end up with an estimate of the form (cf. (8.76))

$$\|c_a - c_b + (A_a - A_b) \cdot \|_{L^2(\hat{Q}(a))} \le Cm\Big(\mathcal{E}(\hat{Q}(a)) + \mathcal{E}(\hat{Q}(b)) + \mathcal{E}(\hat{Q}^*) + \hat{\alpha}(D)\Big)$$

for suitable $A_a, A_b \in \mathbb{R}^{2\times 2}_{\text{skew}}$, $c_a, c_b \in \mathbb{R}^2$, where $D = \bigcup_{j=0}^m \hat{Q}(\xi_j)$. In fact, if $\hat{Q}(a), \hat{Q}(b) \subset \hat{P}^j$ for some $j = 1, 2$, this follows immediately. Otherwise, we apply the arguments leading to (8.74) on each pair $\hat{Q}(a), Q^*$ and $Q^*, \hat{Q}(b)$ and employ the triangle inequality.

Defining $\tilde{P} = \overline{\bigcup_{\hat{Q}\in\mathcal{K}} \hat{Q}}$ and arguing as in (8.77), we then obtain

$$\|u-(A\cdot+c)\|^2_{L^2(\tilde{P})} \le C\hat{v}^{-3}((\mathcal{E}(\tilde{P}))^2+(\alpha(P))^2) \le C(1+C_*r)\frac{\varepsilon\hat{v}}{v^4}|\Gamma|^3_\infty + C\frac{1}{\hat{v}^3}(\mathcal{E}(\hat{Q}^*))^2$$

for some $A \in \mathbb{R}^{2\times 2}_{\text{skew}}$ and $c \in \mathbb{R}^2$. In the last step we proceeded as in (8.78) (see also (8.64)), observing that the paths $\tilde{P}$ defined here and in the proof of Lemma 8.5.2 differ essentially by the square Q^*. By (8.61) we get $\hat{v}^3 v^{-3} \le \hat{v}r$ and using

(8.90) as well as (8.38) we derive (cf. (8.64))

$$
\begin{aligned}
\hat{\upsilon}^{-3}(\mathcal{E}(\hat{Q}^*))^2 &\le C\hat{\upsilon}^{-1}|\Gamma|_\infty^3\varepsilon + C\hat{\upsilon}^{-3}C_*\frac{\varepsilon}{\upsilon^4}\Big(\sum\nolimits_{\Gamma_l\cap Q^*\neq\emptyset}(|\Theta_l|_*)^{3/2}\Big)^2\\
&\le C\hat{\upsilon}^{-1}|\Gamma|_\infty^3\varepsilon + C|\Gamma|_\infty\hat{\upsilon}^{-1}\upsilon^{-1}C_*\frac{\varepsilon}{\upsilon^4}\Big(\sum\nolimits_{\Gamma_l\cap Q^*\neq\emptyset}|\Gamma_l|_\infty\Big)^2 \qquad (8.91)\\
&\le C(1+C_*r)\,\hat{\upsilon}\frac{\varepsilon}{\upsilon^4}|\Gamma|_\infty^3.
\end{aligned}
$$

Consequently, we have re-derived (8.78). Proceeding as in Lemma 8.5.2 we obtain a set $U' \subset U$ with $|U\setminus U'| \le Cr|U|$ such that

$$|U| \fint_Z |u(x) - (A\,x + c)|^2\,dx \le C(1+C_*r)\frac{\varepsilon}{\upsilon^3}|\Gamma|_\infty^3$$

for $Z \subset U'$, $Z \in \mathcal{Y}'$. On the other hand, by Corollary 8.5.5 we find $A_j \in \mathbb{R}^{2\times 2}_{\mathrm{skew}}$, $c_j \in \mathbb{R}^2$ for $j = 1,2$ such that

$$|U \cap N_{j,+}| \fint_{Z\cap N_{j,+}} |u(x) - (A_j\,x + c_j)|^2\,dx \le C(1+C_*r)\frac{\varepsilon}{\upsilon^3}|\Gamma|_\infty^3 \qquad (8.92)$$

for all $Z \subset U$, $Z \in \mathcal{Y}'$. Now (8.89) follows by applying (7.13) on $B_1 = U' \cap N_{j,+}$ and $B_2 = U \cap N_{j,+}$.

Finally, the essentially difference in the treatment of the case $l_2 \le \frac{l_1}{2}$ is that in the construction of the path P one has to choose two sets Q_1^*,Q_2^* where the path changes its direction. Following the above arguments it is not hard to see that these sets can be selected so that the required conditions are satisfied. Note that in the derivation of (8.92) we then exploit that Corollary 8.5.5 also holds for sets which are much smaller than $\upsilon|\Gamma|_\infty^2$. □

8.5.5 Step 4: General case

We are eventually in a position to give the proof of Theorem 8.4.1. We briefly remark that the following proof crucially depends on the trace theorem established in Lemma 7.2 and the fact that there are at most two large cracks in the neighborhood of Γ.

Proof of Theorem 8.4.1. In the general situation we possibly have $N \neq \tilde{N} = N \setminus (X_1 \cup X_2)$. Let $\hat{\mathcal{C}}$ be the covering considered in Lemma 8.5.6. Let K_1, K_2 with $\mathrm{dist}(K_1, K_2) \ge c|\Gamma|_\infty$ be the sets given by Lemma 8.3.5 and let $\tilde{\mathcal{C}}$ be the covering of $N \setminus (K_1 \cup K_2)$ consisting of the connected components of the sets $U \setminus (K_1 \cup K_2)$, $U \in \hat{\mathcal{C}}$. To simplify the exposition we prefer to present first a special case where K_1, K_2 have the form $K_- := K_1 = (-\tau - l_1, -l_1) \times (-\tau, \tau)$ and $K_+ := K_2 = (l_1, l_1 + \tau) \times (-\tau, \tau)$. Moreover, we suppose that the sets $\Psi^\pm$ associated to boundary components larger than $\hat{\tau}$ – if they exist at all – have the form $\Psi^\pm = \Psi_1^\pm \cup \Psi_2^\pm \cup \Psi_3^\pm$, where $\Psi_1^\pm = (\pm l_1, \pm(l_1 + \tau)) \times (\psi^\pm, 2\tau)$,

$\Psi_2^\pm = (\pm l_1, \pm(l_1 + \psi^\pm)) \times (-\psi^\pm, \psi^\pm)$ and $\Psi_3^\pm = (\pm l_1, \pm(l_1 + \tau)) \times (-2\tau, -\psi^\pm)$. Here $\psi^\pm$ denote the corresponding values to $\Psi^\pm$ (see Section 8.3.2).

If Ψ^- or Ψ^+ do not exist, we set $\Psi_2^- = K_1$, $\Psi_2^+ = K_2$, respectively, and let $\Psi_j^\pm$, $j = 1, 3$, be the adjacent squares. In addition, we then define $\psi^\pm = \tau$. We will treat both cases simultaneously in the following.

This special case already covers the fundamental ideas of the proof as the arguments essentially rely on the property that $\mathrm{dist}(K_1, K_2) \geq c|\Gamma|_\infty$ and the fact that the shapes of all sets are comparable (through homeomorphisms with constants depending on h_*) to squares. We will indicate the necessary adaptions for the general case at the end of the proof.

Let $N'_\pm = N \cap \{\pm x_2 \geq 0\} \setminus (K_1 \cup K_2)$. By Lemma 8.3.5 the assumptions of Lemma 8.5.6 are satisfied on each set N'_+ and N'_-. Consequently, there are $A_\pm \in \mathbb{R}^{2\times 2}_{\mathrm{skew}}$ and $c_\pm \in \mathbb{R}^2$ such that for all $V_\pm \subset N'_\pm$, $V_\pm \in \mathcal{Y}$, one has

$$|N'_\pm| \fint_{V_\pm} |u(x) - (A_\pm\, x - c_\pm)|^2\, dx \leq C(1 + C_* r)\frac{\varepsilon}{\upsilon^3}|\Gamma|^3_\infty =: G \tag{8.93}$$

by (8.89). We let $\Xi_0^+ = (l_1, l_1 + \psi^+) \times \{0\}$, $\Xi_0^- = (-l_1 - \psi^-, -l_1) \times \{0\}$ and without restriction (possibly after a small translation in $\mathbf{e}_2$-direction) we can assume $\mathcal{H}^1(\Xi_0^\pm \cap \partial W) = 0$. The goal is to show

$$\int_{\Xi_0^\pm} |(A_+ - A_-)\, x + (c_+ - c_-)|^2\, d\mathcal{H}^1(x) \leq C\frac{\psi^\pm}{\upsilon|\Gamma|^2_\infty}G. \tag{8.94}$$

We prove this only for Ξ_0^-. As a preparation let $\tilde{\Psi}_1^- = (-\frac{\tau}{2} - l_1, -l_1) \times (\psi^-, \frac{3}{2}\tau)$, $\tilde{\Psi}_3^- = (-\frac{\tau}{2} - l_1, -l_1) \times (-\frac{3}{2}\tau, -\psi^-)$ and $\tilde{\Psi}^- = \big(\overline{\tilde{\Psi}_1^-} \cup \overline{\Psi_2^-} \cup \overline{\tilde{\Psi}_3^-}\big)^\circ$. We observe

$$\sum\nolimits_{\Gamma_l \cap \tilde{\Psi}^- \neq \emptyset} |\Gamma_l|_{\mathcal{H}} \leq C(1 - \omega_{\min})^{-1}\psi^- \tag{8.95}$$

for $C = C(h_*, q)$. If $\psi^- \geq c(1 - \omega_{\min})\tau$ this follows from (8.34) and the fact that $\Gamma_l \subset \hat{N} = N^{2\hat{\tau}}(\Gamma)$ for all Γ_l with $\Gamma_l \cap \tilde{\Psi}^- \neq \emptyset$ (recall $|\Gamma_l|_\infty \leq \hat{\tau}$ by the construction in Section 8.3.2). If $\psi^- \leq c(1 - \omega_{\min})\tau$, by (8.35) we obtain $|\Psi^- \cap \partial W|_{\mathcal{H}} \leq D(1 - \omega_{\min})^{-1}\psi^- \ll \tau$ taking $c > 0$ sufficiently small. Thus, we can assume that $\Gamma_l \subset \Psi^-$ if $\Gamma_l \cap \tilde{\Psi}^- \neq \emptyset$. This implies $\sum_{\Gamma_l \cap \tilde{\Psi}_i^- \neq \emptyset} |\Gamma_l|_{\mathcal{H}} \leq C|\Psi^- \cap \partial W|_{\mathcal{H}}$ and gives the assertion.

Recall that $\upsilon \leq r(1 - \omega_{\min})^3$ (see beginning of Section 8.5.1). Applying Theorem B.4 we obtain by (8.35), (8.38), (8.95) and the fact that $\psi^- \leq \upsilon|\Gamma|_\infty$ (cf. also (8.57) for a similar estimate)

$$\begin{aligned}
&\int_{\tilde{\Psi}_i^-} |u(x) - (A_i\, x + c_i)|^2\, dx \\
&\quad \leq C|\tilde{\Psi}_i^-|\alpha(\tilde{\Psi}_i^-) + CC_*\varepsilon\upsilon^{-4}|\partial W \cap \tilde{\Psi}_i^-|\big(\sum\nolimits_{\Gamma_l \cap \tilde{\Psi}_i^- \neq \emptyset} |\Gamma_l|_\infty\big)^2 \\
&\quad \leq C\upsilon^2|\Gamma|^2_\infty(1 - \omega_{\min})^{-1}\varepsilon\psi^- + CC_*(1 - \omega_{\min})^{-3}\varepsilon\upsilon^{-2}|\Gamma|^2_\infty\psi^- \\
&\quad \leq C(1 + C_* r)\upsilon^{-3}|\Gamma|^2_\infty\psi^-\varepsilon
\end{aligned} \tag{8.96}$$

for $A_i \in \mathbb{R}^{2\times 2}_{\rm skew}$ and $c_i \in \mathbb{R}^2$, $i = 1, 3$. Likewise using particularly (8.95) and $|\Psi_2^-| \le C(\psi^-)^2$ we get

$$\int_{\Psi_2^-} |u(x) - (A_2\, x + c_2)|^2 \, dx \le C(1 + C_* r) v^{-4} |\Gamma|_\infty (\psi^-)^2 \varepsilon \tag{8.97}$$

for $A_2 \in \mathbb{R}^{2\times 2}_{\rm skew}$ and $c_2 \in \mathbb{R}^2$. By (8.93) for $V_\pm = \Psi_1^- \setminus K_1, \Psi_3^- \setminus K_1$ we see that

$$\|u - (A_+\, x + c_+)\|^2_{L^2(\Psi_1^- \setminus K_1)} + \|u - (A_-\, x + c_-)\|^2_{L^2(\Psi_3^- \setminus K_1)} \le CvG.$$

Applying (7.10) and (7.11) on $B_1 = \Psi_i^- \setminus K_1$, $B_2 = \tilde{\Psi}_i^-$ $i = 1, 3$, we then derive by (8.96) employing $\psi^- \le v|\Gamma|_\infty$

$$\begin{aligned} \tau^2 |A_+ - A_1|^2 + |c_+ - c_1 + (A_+ - A_1)\, b_+|^2 \le C(v|\Gamma|^2_\infty)^{-1} G, \\ \tau^2 |A_- - A_3|^2 + |c_- - c_3 + (A_- - A_3)\, b_-|^2 \le C(v|\Gamma|^2_\infty)^{-1} G, \end{aligned} \tag{8.98}$$

where $b_- = (-l_1, -\tau)^T$ and $b_+ = (-l_1, \tau)^T$. Furthermore, Lemma 7.2.1, (8.35), (8.38), (8.60), (8.95) and (8.96) yield

$$\begin{aligned} &\int_{\partial \tilde{\Psi}_i^-} |u(x) - (A_i\, x + c_i)|^2 \, d\mathcal{H}^1(x) \\ &\quad \le C(v|\Gamma|_\infty)^{-1} \|u - (A_i\, \cdot + c_i)\|^2_{L^2(\tilde{\Psi}_i^-)} + Cv|\Gamma|_\infty \alpha(\tilde{\Psi}_i^-) \\ &\qquad + CC_* \varepsilon v^{-4} \sum\nolimits_{\Gamma_l \cap \tilde{\Psi}_i^- \neq \emptyset} |\Theta_l|_* \sum\nolimits_{\Gamma_l \cap \tilde{\Psi}_i^- \neq \emptyset} |\Theta_l|_* \\ &\quad \le C(1 + rC_*) v^{-4} |\Gamma|_\infty \psi^- \varepsilon \le C\psi^- (v|\Gamma|^2_\infty)^{-1} G \end{aligned} \tag{8.99}$$

for $i = 1, 3$, where we tacitly assumed that all boundary components are rectangular (cf. discussion after (8.59)). In the penultimate step we again used $\psi^- \le v|\Gamma|_\infty$ and $v \le r(1 - \omega_{\min})^3$. Likewise, we get

$$\int_{\partial \Psi_2^- \cup \Xi_0^-} |u(x) - (A_2\, x + c_2)|^2 \, d\mathcal{H}^1(x) \le C\psi^- (v|\Gamma|^2_\infty)^{-1} G, \tag{8.100}$$

where we replaced $(v|\Gamma|_\infty)^{-1}$ by $(\psi^-)^{-1}$ and used (8.97) instead of (8.96). Observe that (8.100) is well defined in the sense of traces since $\mathcal{H}^1(\Xi_0^- \cap \partial W) = 0$. Define $\Xi_\pm^- = (-\frac{\psi^-}{2} - l_1, -l_1) \times \{\pm \psi^-\}$ and note that $\Xi_+^- \subset \partial \Psi_2^- \cap \partial \tilde{\Psi}_1^-$ and $\Xi_-^- \subset \partial \Psi_2^- \cap \partial \tilde{\Psi}_3^-$. Again up to a small translation in $\mathbf{e}_2$-direction we may suppose $\mathcal{H}^1(\Xi_\pm^- \cap \partial W) = 0$. Combining the estimates (8.98), (8.99) for $i = 1, 3$ we obtain

$$\int_{\Xi_+^-} |u(x) - (A_+\, x + c_+)|^2 \, d\mathcal{H}^1 + \int_{\Xi_-^-} |u(x) - (A_-\, x + c_-)|^2 \, d\mathcal{H}^1 \le C\psi^- (v|\Gamma|^2_\infty)^{-1} G.$$

Using once more the techniques provided in Section 7.1 we may estimate the difference of $A_\pm$, A_2 and $c_\pm$, c_2 on the boundaries $\Xi_\pm^-$ (replace the sets B_1, B_2 in

(7.10), (7.11) by the surfaces $\Xi_\pm^-$) and obtain by (8.100) an expression similar to (8.98):

$$(\psi^-)^2|A_\pm - A_2|^2 + |c_\pm - c_2 + (A_\pm - A_2)\, b_2|^2 \le C(v|\Gamma|_\infty^2)^{-1}G, \tag{8.101}$$

where $b_2 = (-l_1, 0)^T$. Together with (8.100) this leads to

$$\int_{\Xi_0^-} |u(x) - (A_\pm\, x + c_\pm)|^2 \, d\mathcal{H}^1(x) \le C\psi^-(v|\Gamma|_\infty^2)^{-1}G$$

and then by the triangle inequality we derive

$$\int_{\Xi_0^-} |(A_+ - A_-)\, x + (c_+ - c_-)|^2 \, d\mathcal{H}^1(x) \le C\psi^-(v|\Gamma|_\infty^2)^{-1}G.$$

This gives the desired estimate (8.94). From (8.94) applied on both sets, Ξ_0^- and Ξ_0^+, we deduce

$$-C(l_1 v)^2|(A_+ - A_-)\,\mathbf{e}_1|^2 + |-(A_+ - A_-)\, l_1\mathbf{e}_1 + (c_+ - c_-)|^2 \le C(v|\Gamma|_\infty^2)^{-1}G$$

and

$$-C(l_1 v)^2|(A_+ - A_-)\,\mathbf{e}_1|^2 + |(A_+ - A_-)\, l_1\mathbf{e}_1 + (c_+ - c_-)|^2 \le C(v|\Gamma|_\infty^2)^{-1}G.$$

Combining these two estimates we find for v sufficiently small $l_1^2|A_+ - A_-|^2 = 2l_1^2|(A_+ - A_-)\mathbf{e}_1|^2 \le C(v|\Gamma|_\infty^2)^{-1}G$ and then also $|c_+ - c_-|^2 \le C(v|\Gamma|_\infty^2)^{-1}G$. (This is the step where we fundamentally use $\operatorname{dist}(K_1, K_2) \ge c|\Gamma|_\infty$.) We choose $A = A_-$ and $c = c_-$. Recalling the definition of G and $|N| \le v|\Gamma|_\infty^2$ we obtain by (8.93) for $V_\pm = N'_\pm$

$$\int_{N'_+ \cup N'_-} |u(x) - (A\, x - c)|^2 \, dx \le C(1 + C_* r)\frac{\varepsilon}{v^3}|\Gamma|_\infty^3,$$

which together with the estimates (8.96) and (8.98) gives (8.59)(i). Finally, (8.101) yields $(\psi^-)^2|A - A_2|^2 \le C(v|\Gamma|_\infty^2)^{-1}G$ and $|c - c_2 + (A - A_2)\,(-l_1, 0)^T|^2 \le C(v|\Gamma|_\infty^2)^{-1}G$. Then by (8.97) and the fact that $|\Psi_2^-| \le C(\psi^-)^2 \le C\psi^- v|\Gamma|_\infty$ we conclude

$$\int_{\Psi_2^-} |u(x) - (A\, x + c)|^2 \, dx \le C(1 + rC_*)v^{-3}|\Gamma|_\infty^2 \psi^- \varepsilon$$

giving (8.59)(ii). The estimate for Ψ_2^+ follows analogously.

It remains to briefly indicate the necessary adaptions for the general case. The main differences are (i) the shape of the sets $\Psi_i^\pm, i = 1, 2, 3$ and (ii) the position of the sets K_1, K_2. For (i) we observe that $\Psi_i^\pm, i = 1, 3$, are $C(h_*)$-Lipschitz equivalent to a square by Lemma 8.3.6(ii) and Lemma 8.3.7(ii) whereby (8.96) can still be derived (cf. Remark 7.1.4(i)). (Note that the sets are even related by affine mappings.) Likewise, an estimate of the form (8.97) can be derived for sets $(\Psi_2^{\pm})^* \supset \Psi_2^{\pm}$ which have been constructed in Section 8.3.2. Moreover, although not stated explicitly in Section 7.2, the trace estimate used in (8.99), (8.100) can also be applied for sets being an affine transformation of a square. The rest of the arguments concerning the difference of infinitesimal rigid motions (see (8.98), (8.101)) remains unchanged. For (ii) we observe that in the derivation of $|A_+ - A_-|^2 \le Cl_1^{-2}(v|\Gamma|^2_\infty)^{-1}G$ we fundamentally used that $\mathrm{dist}(K_1, K_2) \sim l_1$, but the exact position of the sets K_1, K_2 was not essential. $\square$

We briefly explain Remark 8.4.3. At the beginning of Section 8.5.1 we have already observed that $v \sim C(h_*)\sigma^3$. Now the property for C_2 follows immediately (see (8.43)). For C_1 we use (8.47) and the fact that $\hat{C} = \hat{C}(h_*)$ (see end of Section 8.5.1).

We close this section with an estimate for the skew symmetric matrices involved in the above results.

Lemma 8.5.7. *Let be given the situation of Theorem 8.4.2 for a function* $u = u_{\bar{R}} = \bar{R}^T y - \mathbf{id}$, *where* $y \in H^1(W)$ *and* $\bar{R} \in SO(2)$. *Let* $V \subset Q_\mu$ *be a rectangle and let* $\mathcal{F}(V)$ *be the boundary components* $(\Gamma_l)_l = (\Gamma_l(U))_l$ *satisfying* $N^{\hat{\tau}_l}(\partial R_l) \subset V$ *and* (8.41). *Then there is a* $C_3 = C_3(\sigma, h_*)$ *such that*

$$\sum\nolimits_{\Gamma_l \in \mathcal{F}(V)} |X_l|^2_\infty |A_l|^p \le C_3\big(\|\nabla y - \bar{R}\|^p_{L^p(V\cap W)} + (\varepsilon s^{-1})^{\frac{p}{2}-1}\varepsilon|\partial U \cap V|_{\mathcal{H}}\big)$$

for $p = 2, 4$, *where* $X_l \subset Q_\mu$, $A_l \in \mathbb{R}^{2\times 2}_{\mathrm{skew}}$ *is given in* (8.40).

Remark 8.5.8. Similarly as in Remark 8.4.3 we note that the constant $C_3 = C_3(\sigma, h_*)$ has polynomial growth in σ, i.e. $C_3(\sigma, h_*) \le C(h_*)\sigma^{-z}$ for some $z \in \mathbb{N}$.

Proof. Let $p = 2, 4$. Consider a component $\Gamma = \Gamma_l(U)$ with corresponding rectangle R and X with $\partial X = \Gamma$. It suffices to show $|R|^2_\infty|A|^p \le C_3\big(\|\nabla y - \bar{R}\|^p_{L^p(\tilde{N})} + (\varepsilon s^{-1})^{\frac{p}{2}-1}\varepsilon|\Gamma|_{\mathcal{H}}\big)$ for this component, where $\tilde{N} = N^{\hat{\tau}}(\partial R) \setminus \bigcup_{\Gamma_l \in \mathcal{I}(\Gamma)} N^{\hat{\tau}_l}(\partial R_l)$ and $\mathcal{I}(\Gamma) = \{\Gamma_l : |\Gamma_l|_\infty \le |\Gamma|_\infty\}$. Then the assertion follows by summation over all components and the fact that $|X|_\infty \le |R|_\infty$.

As Γ satisfies (8.41), we observe that we applied Theorem 8.4.1 on ∂R in some iteration step, in particular (8.59) is satisfied. Choose $U \in \mathcal{C}$ with $U \subset N^{2,+}$ as considered in Lemma 8.5.3. By assumption we find a set $S \subset (l_2, l_2 + \tau)$ with $|S| \ge \frac{1}{2}\frac{\tau}{20}$ such that for $T = (\mathbb{R} \times S) \cap U$ we have $T \cap \bigcup_{\Gamma_l \in \mathcal{I}(\Gamma)} N^{\hat{\tau}_l}(\partial R_l) = \emptyset$ by (8.33). It is not restrictive to assume that S is connected as otherwise we follow the subsequent arguments for every connected component of S. Recall $|\Gamma|_\infty \le |\partial R|_\infty \le C|\Gamma|_\infty$ by (8.6). The Poincaré inequality and a rescaling argument imply

$$\int_T |u(x) - \hat{c}|^2\,dx \le C|T|^{1-\frac{2}{p}}|\Gamma|^2_\infty\|\nabla y - \bar{R}\|^2_{L^p(T)}$$

for a constant $\hat{c} \in \mathbb{R}^2$ and $p = 2, 4$. This together with (8.59) yields

$$\int_T |A\,x + c - \hat{c}|^2\,dx \le C(v|\Gamma|_\infty^2)^{1-\frac{2}{p}}|\Gamma|_\infty^2\|\nabla y - \bar{R}\|^2_{L^p(T)} + C\hat{C}\sum_{n=0}^\infty \Big(\frac{2}{3}\Big)^n v^{-3}|\Gamma|^3_\infty\varepsilon.$$

For the constant in the latter part see below (8.47). Arguing as in (7.10) we find

$$|T||\Gamma|^2_\infty|A|^2 \le C\int_T |A\,x + c - \hat{c}|^2\,dx.$$

Thus, by $|T| \ge Cv|\Gamma|^2_\infty$ and an elementary calculation we derive

$$|R|^2_\infty|A|^4 \le C|\Gamma|^2_\infty|A|^4 \le Cv^{-1}\|\nabla y - \bar{R}\|^4_{L^4(T)} + Cv^{-8}\varepsilon^2.$$

As $|\partial R|_* \ge s$, we obtain $|\Gamma|_{\mathcal{H}} \ge C|\partial R|_* \ge Cs$ by (8.5)(ii). Choose C_3 large enough and recall $T \subset N^{\bar{\tau}}(\Gamma) \setminus \bigcup_{\Gamma_l \in \mathcal{I}(\Gamma)} N^{\bar{\tau}_l}(\partial R_l) \subset V$ as well as the fact that $v \ge C(h_*)\sigma^3$. This yields

$$|R|^2_\infty|A|^4 \le C|\Gamma|^2_\infty|A|^4 \le C_3\|\nabla y - \bar{R}\|^4_{L^4(\tilde{N})} + C_3\varepsilon s^{-1}\varepsilon|\Gamma|_{\mathcal{H}}.$$

giving the claim for $p = 4$. Likewise, for $p = 2$ we deduce

$$\begin{aligned}|R|^2_\infty|A|^2 \le C|\Gamma|^2_\infty|A|^2 &\le Cv^{-1}\|\nabla y - \bar{R}\|^2_{L^2(T)} + Cv^{-4}\varepsilon|\Gamma|_\infty \\ &\le C_3\|\nabla y - \bar{R}\|^2_{L^2(\tilde{N})} + C_3\varepsilon|\Gamma|_{\mathcal{H}}.\end{aligned}$$

□

Chapter 9

Quantitative SBD-rigidity

This section is devoted to the proof of Theorem 6.1.1. We will first establish a local rigidity result measuring the distance of the deformation from a piecewise rigid motion in the H^1-norm. At this point the Korn-Poincaré-type inequality established in the last chapter is essential. Afterwards, we show that in regions where only small boundary components are present such a local estimate can be used to replace the original deformation by an H^1-function. Observing that in the modification process the least length of the boundary components increases, we can then apply this estimate iteratively to obtain the rigidity result up to a small set. Finally, by constructing a suitable extension we find that the result holds on the whole domain.

9.1 Preparations

Before we start with a local rigidity estimate, we recall some definitions and introduce further notions. Given a Lipschitz domain $\Omega \subset \mathbb{R}^2$ choose μ_0 so large that $\overline{\Omega} \subset Q_{\mu_0} = (-\mu_0, \mu_0)^2$. Recall the point set $I^s = s(1,1) + 2s\mathbb{Z}^2$, $s > 0$, introduced in Section 7.1 and the definitions of $\mathcal{U}^s, \mathcal{V}^s$ in (7.1), (8.1) with respect to the square Q_{μ_0}. We define additional partitions. Set $z_1 = (0,0)$, $z_2 = (1,0)$, $z_3 = (0,1)$, $z_4 = (1,1)$ and let $I^s_i = sz_i + 2s\mathbb{Z}^2$ as well as $Q^s_i(p) = p + s(-1,1)^2$ for $p \in I^s_i$, $i = 1, \dots, 4$. Moreover, for $U \subset \Omega$ let

$$I^s_i(U) = \{p \in I^s_i : Q^s_i(p) \subset U\}$$

for $i = 1, \dots, 4$. For shorthand we also write $I^s = I^s_4$ and $Q^s = Q^s_4$.

We let Ω^k be the largest set in $\mathcal{V}^{\bar{c}k}$ satisfying $\Omega^k \subset \{x \in \Omega : \mathrm{dist}(x, \partial\Omega) \geq \bar{c}k\}$ for $k \geq 0$ for some $\bar{c} \geq \sqrt{2}$ large enough. For sets $W \subset \Omega^k$, $W \in \mathcal{V}^s$, we assume that one component in definition (8.1) is given by $X = Q_{\mu_0} \setminus \Omega^k$. In particular, all other components $X_1, \dots, X_n$ satisfy $\partial X_i \subset Q_{\mu_0}$ as $\overline{\Omega} \subset Q_{\mu_0}$. We choose an (arbitrary) order of $(\Gamma_j)_{j=1,\dots,n}$ and similarly to (8.2) we define

$$\Theta_i = \Theta_i(W) = \Gamma_i \setminus \bigcup_{j<i} \Gamma_j \tag{9.1}$$

for $i = 1, \dots, n$. Recall the definition of $\|\cdot\|_X$, $X = *, \infty, \mathcal{H}$, in (8.3). As in Section 8.1 we set $\|W\|_* = \sum_{j=1}^n |\Theta_j(W)|_*$. Moreover, let $\mathcal{V}^s_{\mathrm{con}} \subset \mathcal{V}^s$ be the subset consisting of the sets where all $\overline{X_1}, \dots, \overline{X_n}$ are connected. In this chapter we will again modify sets frequently. This will be done in the way described in (8.7).

In the last chapter a general strategy was the combination of different cracks to larger ones in order to simplify the jump set. In this context we have seen that $|\cdot|_\infty$ is a good measure for the diameter of boundary components. However, it turns out that for large surfaces of general shape this object may not be adequate. Therefore, in what follows we have to assure that boundary components do not become too large. For $0 < s \le \lambda \le k$ we introduce

$$\mathcal{V}^s_{(\lambda,k)} := \{W \in \mathcal{V}^s_{\mathrm{con}} : 2\lambda \le \max\{|\pi_1\Gamma_j(W)|, |\pi_2\Gamma_j(W)|\} \le 2k \ \text{ for all } \Gamma_j(W)\}.$$

By definition we have $\max\{|\pi_1\Gamma_j(W)|, |\pi_2\Gamma_j(W)|\} \ge 2s$ for all $\Gamma_j(W)$ and therefore we write for shorthand $\mathcal{V}^s_k = \mathcal{V}^s_{(s,k)}$.

Although we have to avoid that boundary components become to large, it is essential to combine small components. To this end, it is convenient to alter configurations on sets of negligible measure.

Lemma 9.1.1. *Let $t \ge 2k$, $t' > 0$ and $W \in \mathcal{V}^s_t$.*

(i) Then there is a set $\tilde{W} \in \mathcal{V}^s_t$ with $\tilde{W} \subset W$, $|W \setminus \tilde{W}| = 0$ and $\|\tilde{W}\|_* \le \|W\|_*$ such that

$$\Gamma_{j_1}(\tilde{W}) \cap \Gamma_{j_2}(\tilde{W}) = \emptyset \qquad \text{if } |\Gamma_{j_i}(\tilde{W})|_\infty \le k \ \text{ for } i = 1, 2. \tag{9.2}$$

(ii) Then there is a set $U \in \mathcal{V}^s_{t+k}$ with $U \subset W$, $|W \setminus U| = 0$ and $\|U\|_* \le \|W\|_*$ such that

$$\Gamma(U) \cap \Gamma_j(U) = \emptyset \qquad \text{for all } \Gamma_j(U) \neq \Gamma(U) \tag{9.3}$$

for all $\Gamma(U)$ with $|\Gamma(U)|_\infty \le k$.

Proof. (i) The strategy is to combine iteratively different boundary components. Clearly, if $|\Gamma_{j_i}(W)|_\infty \le k$ for $i = 1, 2$ with $\Gamma_{j_1}(W) \cap \Gamma_{j_2}(W) \neq \emptyset$ we may replace W by $W' = W \setminus (\overline{X_{j_1} \cup X_{j_2}})^\circ$ and note that $W' \in \mathcal{V}^s_t$ as well as $|W \setminus W'| = 0$ and $\|W'\|_* \le \|W\|_*$ similarly as in Lemma 8.1.1. (Recall that $\partial X_{j_i} = \Gamma_{j_1}(W)$ for $i = 1, 2$.) We proceed in this way until we obtain a set $\tilde{W} \in \mathcal{V}^s_t$ with $|W \setminus \tilde{W}| = 0$ and $\|\tilde{W}\|_* \le \|W\|_*$ such that (9.2) holds.

(ii) We apply (i) and then proceed to combine two components $\Gamma_{j_1}(\tilde{W}), \Gamma_{j_2}(\tilde{W})$ if $\Gamma_{j_1}(\tilde{W}) \cap \Gamma_{j_2}(\tilde{W}) \neq \emptyset$ and $\min\{|\Gamma_{j_1}(\tilde{W})|_\infty, |\Gamma_{j_2}(\tilde{W})|_\infty\} \le k$. Arguing as before we end up with a set U satisfying $|W \backslash U| = 0$, $\|U\|_* \le \|W\|_*$ and (9.3). It remains to show that $U \in \mathcal{V}^s_{t+k}$. Consider some $\Gamma(U) = \partial X$ with $|\Gamma(U)|_\infty > k$ and observe that there are $\Gamma(\tilde{W}) = \partial X'$ with $|\Gamma(\tilde{W})|_\infty > k$ and $\Gamma_{j_i}(\tilde{W}) = \partial X_{j_i}$, $i = 1, \dots, m$, with $|\Gamma_{j_i}(\tilde{W})|_\infty \le k$, $\Gamma_{j_{i_1}}(\tilde{W}) \cap \Gamma_{j_{i_2}}(\tilde{W}) = \emptyset$ for $i_1 \neq i_2$ and $\Gamma_{j_i}(\tilde{W}) \cap \Gamma(\tilde{W}) \neq \emptyset$ such that $\overline{X} = \overline{X' \cup \bigcup_{i=1}^m X_{j_i}}$. But this implies $|\pi_i \Gamma(U)| \le 2k + |\pi_i \Gamma(\tilde{W})| \le 2k + 2t$ for $i = 1, 2$, as desired. □

In what follows we often modify sets by subtracting rectangular neighborhoods of boundary components. In this context it is particularly important that the components remain connected and do not become too large. By $\triangle$ we denote the symmetric difference of two sets.

Lemma 9.1.2. *Let $k, t, t' > 0$ with $t, t' \le Ck$ and $\nu \ge 0$. Let $V \subset \Omega^k$ with $V \in \mathcal{V}^s_{\mathrm{con}}$.*

(i) Assume that for each component $X_j = X_j(V)$, $j = 1, \dots, n$, there is a rectangle $Z_j \in \mathcal{U}^s$ with $X_j \subset Z_j$, $|\pi_i \partial Z_j| \le |\pi_i \partial X_j| + \nu |\partial X_j|_\infty$ for $i = 1, 2$ and $\max_{i=1,2} |\pi_i \partial Z_j| \le 2t'$ for all $j = 1, \dots, n$. Moreover, assume that $Z_{j_1} \setminus Z_{j_2}$ or $Z_{j_2} \setminus Z_{j_1}$ is connected for all $1 \le j_1 < j_2 \le n$. Then there is a set $U \in \mathcal{V}^s_{t'}$, $U \subset \Omega^k$, with $\bigcup_{j=1}^n \overline{X_j(U)} = \bigcup_{j=1}^n \overline{Z_j} \cap \Omega^k$ and $\|U\|_* \le (1 + c\nu)\|V\|_*$ for a universal constant $c > 0$.

(ii) In addition let $V' \in \mathcal{V}^s_t$ be given and define $\hat{W} = V' \setminus \bigcup_{j=1}^n Z_j$. Then there is a set $W \in \mathcal{V}^{s/2}_{t+2t'}$ with $|W \setminus \hat{W}| = 0$, $|\hat{W} \setminus W| \le ct\|V'\|_*$ and $\|W\|_* \le (1 + c\nu)\|V\|_* + \|V'\|_*$.

Proof. (i) Let $V \subset \Omega^k$ with components $(X_j)_{j=1}^n$ and rectangles $(Z_j)_{j=1}^n$ be given. It suffices to show the following: There are connected, pairwise disjoint $(X'_j)_{j=1}^n$ with $X'_j \subset Z_j$, $\bigcup_{j=1}^n \overline{X'_j} = \bigcup_{j=1}^n \overline{Z_j}$ and

$$|\bigcup\nolimits_{j=1}^n \partial X'_j|_{\mathcal{H}} \le \sum\nolimits_{j=1}^n |\Theta_j(V)|_{\mathcal{H}} + c\nu \sum\nolimits_{j=1}^n |\Gamma_j|_{\mathcal{H}}. \tag{9.4}$$

Moreover, we have $X'_j = R_j \backslash \overline{(A^1_j \cup A^2_j)}$. Here $R_j \in \mathcal{U}^s$ is a rectangle and $A^i_j \in \mathcal{U}^s$, $i = 1, 2$, are (if nonempty) unions of rectangles whose closure intersect the corner $c^i_j \in \partial R_j$, where c^1_j, c^2_j are adjacent corners of R_j.

Then the claim of the lemma follows for $U = \Omega^k \setminus \bigcup_{j=1}^n X'_j$. Indeed, to see $\|U\|_* \le (1 + c\nu)\|V\|_*$ we first observe $\sum_j |\partial X'_j|_\infty \le \sum_j |\partial Z_j|_\infty \le (1 + c\nu) \sum_j |\partial X_j|_\infty$. Moreover, by (9.4) we get

$$\|U\|_{\mathcal{H}} \le |\bigcup\nolimits_{j=1}^n \partial X'_j|_{\mathcal{H}} \le (1 + c\nu)\|V\|_{\mathcal{H}} = (1 + c\nu)|\bigcup\nolimits_{j=1}^n \partial X_j|_{\mathcal{H}}. \tag{9.5}$$

In the first inequality we also used $|\partial X'_j|_\infty \le |\partial Z_j|_\infty \le Ck$ and $\Omega^k \in \mathcal{V}^{\bar{c}k}$ for $\bar{c} \gg 1$. (Arguments of this form will be used frequently in the following and from now on we will omit the details.) Finally, we conclude $U \in \mathcal{V}^s_{t'}$ as $\max_{i=1,2} |\pi_i \partial Z_j| \le 2t'$ for $j = 1, \dots, n$.

We prove the above assertion by induction. Clearly, the claim holds for $n = 1$ for $X'_1 = Z_1$. Now assume the assertion holds for sets with at most $n-1$ components and consider $V \subset \Omega^k$ with components $(X_j)_{j=1}^n$ and corresponding $(Z_j)_{j=1}^n$. Without restriction we assume that $\max_{x \in \overline{Z_n}} x_2 = \max_{x \in \bigcup_{j=1}^n \overline{Z_j}} x_2$. By hypothesis we obtain pairwise disjoint, connected sets X''_j, $j = 1, \ldots, n-1$, fulfilling the above properties, in particular $\bigcup_{j=1}^{n-1} \overline{X''_j} = \bigcup_{j=1}^{n-1} \overline{Z_j}$.

Given $Z_n = (z_1^1, z_1^2) \times (z_2^1, z_2^2)$ we set $\tilde{Z}_n = (z_1^1, z_1^2) \times (z_2^1, z_2^2]$. For $j = 1, \ldots, n-1$ let $Z'_{j,i} \in \mathcal{U}^s$ be the largest rectangle in Z_n satisfying $Z_j \cap Z_n \subset Z'_{j,i} \subset \bigcup_{j=1}^{n-1} \overline{Z_j}$ with $z_1^i \in \overline{Z'_{j,i}}$ for $i = 1, 2$. If $Z'_{j,i} \neq \emptyset$ for some i, we let $Z'_j = Z'_{j,i}$, otherwise we set $Z'_j = Z_j \cap Z_n$. (Note that $Z'_{j,1} = Z'_{j,2}$ if $Z'_{j,1}, Z'_{j,2} \neq \emptyset$.)

Let $J_0 \subset \{1, \ldots, n-1\}$ such that $Z_j \cap Z_n = \emptyset$ for $j \in J_0$. Let $J_1 \subset \{1, \ldots, n-1\} \setminus J_0$ such that $(\overline{Z'_j} \setminus Z_n) \cap \{z_1^1, z_1^2\} = \emptyset$ for $j \in J_1$ and $J_2 \subset \{1, \ldots, n-1\} \setminus J_0$ such that $\tilde{Z}_n \setminus Z'_j$ is a rectangle for $j \in J_2$. (Observe that $J_1 \cap J_2 = \emptyset$.) Let $J_3 = \{1, \ldots, n-1\} \setminus (J_0 \cup J_1 \cup J_2)$. Define $X'_n = Z_n \setminus \bigcup_{j \in J_2 \cup J_3} \overline{Z'_j}$. Moreover, we let $X'_j = X''_j$ for $j \in J_0 \cup J_2 \cup J_3$ and $X'_j = X''_j \setminus \overline{X'_n}$ for $j \in J_1$. Clearly, by construction the sets are pairwise disjoint and fulfill $\bigcup_{j=1}^n \overline{X'_j} = \bigcup_{j=1}^n \overline{Z_j}$.

Moreover, we observe that the sets are connected and have the special shape given above. In fact, for $j \in J_0 \cup J_2 \cup J_3$ this is clear. For X'_n we first note that $J_3 = J_3^1 \dot\cup J_3^2$, where $\overline{Z'_j}$ intersects the lower right and the lower left corner of Z_n for $j \in J_3^1$ and $j \in J_3^2$, respectively. (It cannot happen that $\overline{Z'_j}$ intersects only the other corners due to the choice of Z_n.) We observe $X'_n = R_n \setminus \overline{(A_n^1 \cup A_n^2)}$ is connected, where $R_n = Z_n \setminus \bigcup_{j \in J_2} \overline{Z'_j}$ and $A_n^i = \bigcup_{j \in J_3^i} Z'_j$ for $i = 1, 2$.

Finally, to see the properties for $j \in J_1$ we first observe that $S_j := Z_j \setminus \overline{X'_n}$ is a rectangle. In fact, otherwise due to the special shape of X'_n it is elementary to see that $(\overline{Z'_j} \setminus Z_n) \cap \{z_1^1, z_1^2\} \neq \emptyset$ and thus $j \notin J_1$. We get $X'_j = X''_j \cap S_j = (R_j \cap S_j) \setminus \overline{(A_j^1 \cup A_j^2)}$ is connected and $X'_j = \hat{R}_j \setminus \overline{(\hat{A}_j^1 \cup \hat{A}_j^2)}$, where $\hat{R}_j = S_j$ and $\hat{A}_j^i = A_j^i \cap S_j$ for $i = 1, 2$.

It remains to confirm (9.4). We first observe that

$$\sum\nolimits_{j=1}^n |\Theta_j(V)|_{\mathcal{H}} = \tfrac{1}{2} \sum\nolimits_{j=1}^n |\Gamma_j|_{\mathcal{H}} + \tfrac{1}{2} \big|\partial\big(\bigcup\nolimits_{j=1}^n \overline{X_j}\big)\big|_{\mathcal{H}}. \tag{9.6}$$

(Recall that different boundary components may have nonempty intersection.) Similarly, for the components $(X'_j)_j$ we find

$$\big|\bigcup\nolimits_{j=1}^n \partial X'_j\big|_{\mathcal{H}} = \tfrac{1}{2} \sum\nolimits_{j=1}^n |\partial X'_j|_{\mathcal{H}} + \tfrac{1}{2} \big|\partial\big(\bigcup\nolimits_{j=1}^n \overline{X'_j}\big)\big|_{\mathcal{H}}.$$

We now treat the two terms on the right separately. By $T_j \in \mathcal{U}^s$ we denote the smallest rectangle containing X_j and observe that $|\partial T_j|_\infty = |\Gamma_j|_\infty$, $|\partial T_j|_{\mathcal{H}} \le |\Gamma_j|_{\mathcal{H}}$. Recall $|\partial Z_j|_{\mathcal{H}} \le |\partial T_j|_{\mathcal{H}} + c\nu |\partial T_j|_\infty \le (1 + c\nu)|\Gamma_j|_{\mathcal{H}}$ for $j = 1, \ldots, n$. Due to the special shape of the components X'_j we find $|\partial X'_j|_{\mathcal{H}} \le |\partial Z_j|_{\mathcal{H}}$ and thus

$$\sum\nolimits_{j=1}^n |\partial X'_j|_{\mathcal{H}} \le (1 + c\nu) \sum\nolimits_{j=1}^n |\Gamma_j|_{\mathcal{H}}. \tag{9.7}$$

Moreover, it is elementary to see that we can find a connected set $\tilde{X}_j \supset X_j$ such that $\tilde{\Gamma}_j := \partial \tilde{X}_j$ satisfies $|\tilde{\Gamma}_j|_{\mathcal{H}} \le (1+c\nu)|\Gamma_j|_{\mathcal{H}}$ and $Z_j \in \mathcal{U}^s$ is the smallest rectangle containing $\tilde{X}_j$. By a projection argument it is then not hard to see that

$$\begin{aligned} |\partial(\bigcup\nolimits_{j=1}^n \overline{X_j'})|_{\mathcal{H}} = |\partial(\bigcup\nolimits_{j=1}^n \overline{Z_j})|_{\mathcal{H}} &\le |\partial(\bigcup\nolimits_{j=1}^n \overline{\tilde{X}_j})|_{\mathcal{H}} \\ &\le |\partial(\bigcup\nolimits_{j=1}^n \overline{X_j})|_{\mathcal{H}} + c\nu \sum |\Gamma_j|_{\mathcal{H}}. \end{aligned}$$

Consequently, we derive by (9.6) and (9.7)

$$\begin{aligned} |\bigcup\nolimits_{j=1}^n \partial X_j'|_{\mathcal{H}} &\le \tfrac{1}{2}\sum\nolimits_{j=1}^n |\Gamma_j|_{\mathcal{H}} + \tfrac{1}{2}|\partial(\bigcup\nolimits_{j=1}^n \overline{X_j})|_{\mathcal{H}} + c\nu \sum\nolimits_{j=1}^n |\Gamma_j|_{\mathcal{H}} \\ &= \sum\nolimits_{j=1}^n |\Theta_j(V)|_{\mathcal{H}} + c\nu \sum\nolimits_{j=1}^n |\Gamma_j|_{\mathcal{H}}, \end{aligned}$$

as desired.

(ii) Let $(Y_j)_{j=1}^{n'}$ be the components of V' and let $T_j \in \mathcal{U}^s$ be the smallest rectangle containing Y_j. It is elementary to see that $T_{j_1} \setminus T_{j_2}$ is connected for $1 \le j_1, j_2 \le n'$. Thus, by (i) we obtain pairwise disjoint, connected sets $(Y_j')_j$ with $\bigcup_j \overline{Y_j'} = \bigcup_j \overline{T_j}$ and define $V'' = \Omega^k \setminus \bigcup_{j=1}^{n'} Y_j'$. By (i) for $\nu = 0$ we then also obtain $\|V''\|_* \le \|V'\|_*$. Moreover, the isoperimetric inequality yields $|V' \setminus V''| \le ct\|V'\|_*$ since $|\partial T_j|_\infty \le 2\sqrt{2}t$ for all $j = 1, \ldots, n'$.

Let $(X_j')_{j=1}^n$ and $U \in \mathcal{V}^s_{t'}$ as given in (i). We define $W' = (U \setminus \bigcup_{j=1}^{n'} Y_j') \cup \bigcup_{j=1}^{n'} \partial Y_j'$. Clearly, we have $|W' \setminus \hat{W}| = 0$, $|\hat{W} \setminus W'| \le ct\|V'\|_*$ and $\|W'\|_* \le (1+c\nu)\|V\|_* + \|V'\|_*$ arguing similarly as in Lemma 8.1.1. Observe that possibly $W' \notin \mathcal{V}^s_{\mathrm{con}}$ as the components $(X_j')_{j=1}^n$ of U may have become disconnected. Thus, we now construct a set $W \in \mathcal{V}^{s/2}_{\mathrm{con}}$ with $|W' \triangle W| = 0$.

By $R_j \in \mathcal{U}^s$ we denote the smallest rectangle such that $X_j' \subset R_j$ for $j = 1, \ldots, n$ and observe $\bigcup_j \overline{R_j} = \bigcup_j \overline{X_j'}$. To simplify the exposition we assume that each of the components $(X_j')_j$ has become disconnected as otherwise we do not have to alter the boundary component in the modification procedure described below. Moreover, we can suppose that for each pair Y_{j_1}', X_{j_2}', $1 \le j_1 \le n'$, $1 \le j_2 \le n$, with $R_{j_2} \setminus Y_{j_1}'$ is not disconnected we have $X_{j_2}' \setminus Y_{j_1}'$ is not disconnected. In fact, otherwise we can pass to some $Y_{j_1}^* \subset Y_{j_1}'$ with $|\partial Y_{j_1}^*|_* \le |\partial Y_{j_1}'|_*$ such that $X_{j_2}' \setminus Y_{j_1}^*$ is not disconnected and $\bigcup_j \overline{Y_j'} \cup \bigcup_j \overline{X_j'} = \bigcup_j \overline{Y_j^*} \cup \bigcup_j \overline{X_j'}$.

We now proceed by induction. Let $W_0 = V''$ and $T_j^0 = Y_j'$ for $j = 1, \ldots, n'$. Assume there are pairwise disjoint, connected sets $T_j^{l-1} \in \mathcal{U}^{\frac{s}{2}}$, $j = 1, \ldots, n'$ such that

$$(i)\ \bigcup\nolimits_{j=1}^{n'} \overline{T_j^{l-1}} = \bigcup\nolimits_{j=1}^{n'} \overline{Y_j'} \cup \bigcup\nolimits_{j=1}^{l-1} \overline{X_j'}, \quad (ii)\ T_{j_1}^{l-1} \cap \overline{X_{j_2}'} = T_{j_1}^0 \cap \overline{X_{j_2}'} \qquad (9.8)$$

for all $1 \le j_1 \le n'$, $l \le j_2 \le n$. Moreover, assume that the set $W_{l-1} := \Omega^k \setminus \bigcup_j T_j^{l-1}$

satisfies $\|W_{l-1}\|_\infty \le \sum_j |\partial T_j^0|_\infty + \sum_{i=1}^{l-1} |\partial X_i'|_\infty$ and

$$\|W_{l-1}\|_{\mathcal{H}} \le |\bigcup_j \partial T_j^0|_{\mathcal{H}} + |\bigcup_{i=1}^{l-1} \partial X_i' \setminus \bigcup_j T_j^0|_{\mathcal{H}} + \frac{1}{2}\sum_{i=1}^{l-1} |\partial X_i' \cap \bigcup_j T_j^0|_{\mathcal{H}}. \quad (9.9)$$

We now construct W_l. Let $J^l \subset \{1,\ldots,n'\}$ such that $T_j^{l-1} \cap X_l' \neq \emptyset$ with $J^l = J_1^l \dot\cup J_2^l$, where $j \in J_2^l$ if and only if $R_l \setminus T_j^{l-1}$ is disconnected.

If $j \in J_1^l$, we define $T_j^l = T_j^{l-1} \setminus \hat{X}_l'$, where $\hat{X}_l' \in \mathcal{U}^{\frac{s}{2}}$ is the largest set with $\overline{\hat{X}_l'} \subset X_l'$. It is not hard to see that $|\partial T_j^l|_\infty \le |\partial T_j^{l-1}|_\infty$ for all $j \in J_1^l$ and $|\partial T_j^l|_{\mathcal{H}} \le |\partial T_j^{l-1} \setminus X_l'|_{\mathcal{H}} + \frac{1}{2}|\partial T_j^{l-1} \cap X_l'|_{\mathcal{H}} + \frac{1}{2}|\partial X_l' \cap T_j^{l-1}|_{\mathcal{H}}$. As each $x \in \mathbb{R}^2$ is contained in at most two different ∂T_j^{l-1}, we find $\sum_{j\in J_1^l} \frac{1}{2}|\partial T_j^{l-1} \cap X_l'|_{\mathcal{H}} \le |\bigcup_{j \in J_1^l} \partial T_j^{l-1} \cap X_l'|_{\mathcal{H}}$. Therefore, taking the union over all components we derive

$$|\bigcup_{j\in J_1^l} \partial T_j^l \cup \bigcup_{j \notin J_1^l} \partial T_j^{l-1}|_{\mathcal{H}} \le |\bigcup_j \partial T_j^{l-1}|_{\mathcal{H}} + \tfrac{1}{2}|\partial X_l' \cap \bigcup_{j\in J_1^l} T_j^0|_{\mathcal{H}}. \quad (9.10)$$

Here we used (9.8)(ii) and the fact that the sets $(T_j^{l-1})_j$ are pairwise disjoint. Observe that the above construction together with (9.8)(ii) and the special shape of T_j^0 (see proof of (i)) implies that the sets T_j^l, $j \in J_1^l$, are connected. Moreover, (9.8)(ii) holds for $j_1 \in J_1^l$.

We define $\tilde{X}_l' = X_l' \setminus \bigcup_{j \in J_1^l} \overline{T_j^l} \in \mathcal{U}^{\frac{s}{2}}$. Due to the fact that $\hat{X}_l' \neq \emptyset$ we observe that the number of connected components of the sets $X_l' \setminus \bigcup_{j\in J_2^l} T_j^{l-1}$ and $\tilde{X}_l' \setminus \bigcup_{j\in J_2^l} T_j^{l-1}$ coincide. Therefore, letting $A_1,\ldots,A_m$ be the connected components of $\tilde{X}_l' \setminus \bigcup_{j \in J_2^l} \overline{T_j^{l-1}}$ it is elementary to see that $m = \# J_2^l + 1$.

Up to a rotation by $\frac{\pi}{2}$ we can assume that each $\overline{A_i}$ intersects the upper and lower boundary of R_l and that $\overline{A_1}$ intersects the left boundary. For convenience we denote the components $(T_j^{l-1})_{j\in J_2^l}$ by $(T_{j_i}^{l-1})_{i=1}^{m-1}$. Suppose $R_l = (0,l_1)\times(0,l_2)$. Let $a_i = \inf_{x\in A_i} x_1$ and $d_i = a_{i+1} - a_i$, where $a_{m+1} = l_1$. Define $T_{j_1}^l = (\overline{T_{j_1}^{l-1}} \cup \overline{(A_1 \cup A_2)})^\circ$ and $T_{j_i}^l = (\overline{T_{j_i}^{l-1}} \cup \overline{A_{i+1}})^\circ$ for $i = 2,\ldots,m-1$. Observe that the sets are pairwise disjoint, connected and that (9.8)(ii) holds for $j_i \in J_2^l$. It is elementary to see that $|T_{j_1}^l|_\infty \le |T_{j_1}^{l-1}|_\infty + d_1 + d_2$ and $|T_{j_i}^l|_\infty \le |T_{j_i}^{l-1}|_\infty + d_{i+1}$ for $i = 2,\ldots,m-1$. Thus, we have

$$\sum_{i=1}^{m-1} |T_{j_i}^l|_\infty \le \sum_{i=1}^{m-1} |T_{j_i}^{l-1}|_\infty + |X_l'|_\infty. \quad (9.11)$$

For $j \notin J^l$ we define $T_j^l = T_j^{l-1}$ and observe that (9.8)(i) holds by construction and the assumptions before (9.8). Together with (9.10) and (9.8)(ii) we then also get

$$|\bigcup_j \partial T_j^l|_{\mathcal{H}} \le |\bigcup_j \partial T_j^{l-1}|_{\mathcal{H}} + |\partial X_l' \setminus (\bigcup_{i=1}^{l-1} \partial X_i' \cup \bigcup_j T_j^0)|_{\mathcal{H}} + \tfrac{1}{2}|\partial X_l' \cap \bigcup_j T_j^0|_{\mathcal{H}}.$$

This in conjunction with (9.9) for W_{l-1} implies that (9.9) holds for W_l. Moreover, by (9.11) it is elementary to see that $\|W_l\|_\infty \le \sum_j |\partial T_j^0|_\infty + \sum_{i=1}^l |\partial X_i'|_\infty$.

Finally, we define $W = W_n$ and observe that W has the desired properties. In fact, by (9.8)(i) we have $|W \triangle W'| = 0$ and thus $|\hat{W} \setminus W| \le ct\|V'\|_*$. Moreover, we clearly get $\|W\|_\infty \le \|U\|_\infty + \|V''\|_\infty \le (1+c\nu)\|V\|_\infty + \|V'\|_\infty$. As each $x \in \mathbb{R}^2$ is contained in at most two different $\partial X_l'$, we find by (9.9)

$$\begin{aligned}\|W\|_{\mathcal{H}} &\le \|V''\|_{\mathcal{H}} + \Big|\bigcup\nolimits_{i=1}^n \partial X_i' \setminus \bigcup\nolimits_j T_j^0\Big|_{\mathcal{H}} + \Big|\bigcup\nolimits_{i=1}^n \partial X_i' \cap \bigcup\nolimits_j T_j^0\Big|_{\mathcal{H}} \\ &= \|V''\|_{\mathcal{H}} + \|U\|_{\mathcal{H}} \le \|V'\|_{\mathcal{H}} + (1+c\nu)\|V\|_{\mathcal{H}},\end{aligned}$$

as desired. Finally, similarly as in Lemma 9.1.1(ii) we obtain $|\pi_i X_j(W)| \le 2t + 4t'$ for $i = 1, 2$ for all j and thus $W \in \mathcal{V}^{s/2}_{t+2t'}$. □

Recall the definition $H(\cdot)$ in (8.32). In addition, for $\lambda > 0$ we define $H^\lambda(W) \supset W$ as the 'variant of W without holes of size smaller than λ': We arrange the sets $(\Gamma_j)_{j=1,\dots,n}$ in the way that $|\Gamma_j|_\infty \le \lambda$ for $j \ge l_\lambda$ and $|\Gamma_j|_\infty > \lambda$ for $j < l_\lambda$. Define

$$H^\lambda(W) = W \cup \bigcup\nolimits_{j=l_\lambda}^n X_j. \tag{9.12}$$

In the following, constants which are much smaller than 1 will frequently appear. For the sake of convenience we introduce one universal parameter. For given $l \ge 1$ and $0 < s, \epsilon, m \le 1$ we let

$$\vartheta = l^9 C_m^2 s^{-1} \epsilon, \tag{9.13}$$

where $C_m = C_1(m, h_*) + C_3(m, h_*) + m^{-4} C_2^{-2}(m, h_*)$ with the constants of Theorem 8.4.2 and Lemma 8.5.7 (for fixed h_*). By Remark 8.4.3, 8.5.8 we find some $z \in \mathbb{N}$ such that $C_m \le C(h_*) m^{-z}$. Moreover, for later let $\hat{m} = C_2(m, h_*)$ and recall that by (8.43) we can assume $\hat{m} \ll m$ as well as $\bar{C}\hat{m} \le m$ for constants $\bar{C} = \bar{C}(h_*)$. Using only one universal parameter the estimates we establish are often not sharp. However, this will not affect our analysis.

Remark 9.1.3. All the constants C in the following may depend on h_* unless they are universal constants indicated as C_{u}. However, to avoid further notation we drop the dependence here. Only at the end of the proof in Section 9.5, when we pass to the limit $h_* \to 0$, we will take the h_* dependence of the constants into account.

9.2 A local rigidity estimate

We now establish a local rigidity estimate on a fine partition of the Lipschitz domain Ω. In the following, ϵ will represent the stored elastic energy. We first construct piecewise constant $SO(2)$-valued mappings approximating the deformation gradient. Afterwards, we employ Theorem 8.4.8 to find a piecewise rigid motion being a good approximation of the deformation.

9.2.1 Estimates for the derivatives

We divide our investigation into two regimes, the 'superatomistic' $k \geq \epsilon$ and the 'subatomisic' $k \leq \epsilon$. Here, we call the ϵ-regime the 'atomistic regime' as in discrete fracture models ϵ is of the same order as the typical interatomic distance (compare e.g. (6.2) with (1.27)). We begin with the superatomistic regime.

Lemma 9.2.1. *Let $k > s \geq \epsilon > 0$ with $1 \ll l := \frac{k}{s}$. Let $m^{-1} \in \mathbb{N}$ and assume that $\frac{km}{s} \in \mathbb{N}$. Then for a constant $C > 0$ we have the following:*
For all $U \in \mathcal{V}^s_k$ with $U \subset \Omega^k$ and for all $y \in H^1(U)$ with $\Delta y = 0$ in U° and

$$\gamma := \| \operatorname{dist}(\nabla y, SO(2))\|^2_{L^2(U)}, \tag{9.14}$$

there is a set $W \in \mathcal{V}^{sm}_{(s,3k)}$ with $W \subset \Omega^{3k}$, $|W \setminus U| = 0$, $|(U \setminus W) \cap \Omega^{3k}| \leq C_u k \|W\|_*$ and

$$\|W\|_* \leq (1 + C_u m)\|U\|_* + C\epsilon^{-1}\gamma. \tag{9.15}$$

Moreover, there are mappings $\hat{R}_i : W^\circ \to SO(2)$, $i = 1, \ldots, 4$, which are constant on the connected components of $Q^k_i(p) \cap W^\circ$, $p \in I^k_i(\Omega^k)$, such that

$$\begin{aligned} &(i) \ \ \|\nabla y - \hat{R}_i\|^2_{L^2(W)} \leq C l^4 \gamma, \\ &(ii) \ \ \|\nabla y - \hat{R}_i\|^4_{L^4(W)} \leq C \vartheta \gamma. \end{aligned} \tag{9.16}$$

Proof. We first construct the set W. Let $J \subset I^k(\Omega^k)$ such that

$$\| \operatorname{dist}(\nabla y, SO(2))\|^2_{L^2(Q^k(p)\cap U)} > \epsilon k \tag{9.17}$$

for all $p \in J$. Define

$$\hat{W} = \Big(U \setminus \bigcup\nolimits_{p\in J} Q^k(p)\Big) \cup \bigcup\nolimits_{p \in J} \partial Q^k(p)$$

and note that $\hat{W} \in \mathcal{V}^s_k$. In particular, the property $\hat{W} \in \mathcal{V}^s_{\mathrm{con}}$ holds since $\max\{|\pi_1 \Gamma_t(U)|, |\pi_2\Gamma_t(U)|\} \leq 2k$. The fact that we add the union of the boundary on the right hand side assures that we do not 'combine' boundary components. Moreover, we derive $\|\hat{W}\|_* \leq \|U\|_* + C\epsilon^{-1}\gamma$. Indeed, $\sum_{p\in J} |\partial Q^k_p|_* \leq 8k \cdot \#J \leq 8k\frac{\gamma}{\epsilon k}$ by (9.14). For all other $\Gamma_t(\hat{W})$ we find a corresponding $\Gamma_t(U)$ (without restriction we use the same index) such that $\Theta_t(\hat{W}) = \Theta_t(U) \setminus \bigcup_{p\in J} \overline{Q^k(p)}$ and thus $|\Theta_t(\hat{W})|_* \leq |\Theta_t(U)|_*$. (Arguments of this form will be used frequently in the following and from now on we will omit the details.) Furthermore, we easily deduce $|U \setminus \hat{W}| \leq C_u k \|\hat{W}\|_*$.

Then we can find a set $W \in \mathcal{V}^{sm}_{2k}$ with $\|W\|_* \leq (1 + C_u m)\|\hat{W}\|_*$, $|U \setminus W| \leq C_u k\|W\|_*$ and $W^\circ \subset \{x \in \Omega^{3k} \cap \hat{W} : \operatorname{dist}_\infty(x, \partial\hat{W}) \leq 2sm\}$, where $\operatorname{dist}_\infty(x, A) := \inf_{y\in A} \max_{i=1,2} |(x - y) \cdot \mathbf{e}_i|$ for $A \subset \mathbb{R}^2$, $x \in \mathbb{R}^2$.

Indeed, let $M(\Gamma_j) \in \mathcal{U}^{sm}$ be the smallest rectangle satisfying $M(\Gamma_j) \supset \{x \in \mathbb{R}^2 : \operatorname{dist}_\infty(x, X_j) \le 2sm\}$, where X_j denotes the component corresponding to $\Gamma_j(\hat{W})$. Clearly, we obtain $|\pi_i \partial M(\Gamma_j)| \le |\pi_i \Gamma_j(\hat{W})| + C_u m |\Gamma_j(\hat{W})|_\infty$ for $i = 1, 2$, $j = 1, \ldots, n$ as $\hat{W} \in \mathcal{V}^s$. Moreover, it is elementary to see that $M(\Gamma_{j_1}) \setminus M(\Gamma_{j_2})$ is connected for $1 \le j_1, j_2 \le n$ since $sm \ll s$. Then by Lemma 9.1.2(i) with $Z_j = M(\Gamma_j)$ we obtain a set $W \in \mathcal{V}^{sm}_{2k}$ which coincides with

$$\Omega^{3k} \cap \big(\hat{W} \setminus \bigcup\nolimits_{j=1}^{n} M(\Gamma_j)\big) = \Omega^{3k} \setminus \bigcup\nolimits_{j=1}^{n} M(\Gamma_j) \tag{9.18}$$

up to a set of negligible measure. Here we used $sm \ll k$. Moreover, we have $|(U \setminus W) \cap \Omega^{3k}| \le C_u k \|W\|_*$ and $\|W\|_* \le (1 + C_u m)\|\hat{W}\|_*$.

Boundary components of W are possibly smaller than $2s$ due to the modification in (9.18). Therefore, we apply Lemma 9.1.1(ii) to get a (not relabeled) set $W \in \mathcal{V}^{sm}_{3k}$ such that (9.15) still holds and (9.3) is satisfied. Now the fact that $U \in \mathcal{V}^s_{(s,k)}$ and (9.3) imply $W \in \mathcal{V}^{sm}_{(s,3k)}$.

Fix $i = 1, \ldots, 4$ and let $F \subset Q^k_i(p) \cap W^\circ$ be a connected component of $Q^k_i(p) \cap W^\circ$. Define $\hat{F} \in \mathcal{U}^s$ as the smallest (connected) set satisfying

$$\hat{F} \supset \{x : \operatorname{dist}_\infty(x, F) < 2sm\}.$$

Due to the construction of W we get $\hat{F} \subset \hat{W}^\circ \subset U$. As $|\hat{F}| \le C_u k^2$, Lemma 7.1.3(i) for $\mu = 2k$ implies that there is a rotation $R \in SO(2)$ such that

$$\|\nabla y - R\|^2_{L^2(\hat{F})} \le C k^4 s^{-4} \|\operatorname{dist}(\nabla y, SO(2))\|^2_{L^2(\hat{F})} = C l^4 \gamma(\hat{F}),$$

where for shorthand we write $\gamma(\hat{F}) = \|\operatorname{dist}(\nabla y, SO(2))\|^2_{L^2(\hat{F})}$. As $\nabla y - R$ is harmonic in $\hat{F}$, the mean value property of harmonic functions for $r = sm$ and Jensen's inequality yield

$$\begin{aligned} |\nabla y(x) - R|^4 &\le \Big| \frac{1}{|B_r(x)|} \int_{B_r(x)} (\nabla y(t) - R)\, dt \Big|^4 \\ &\le C\Big((sm)^{-2} \int_{\hat{F}} |\nabla y - R|^2 \Big)^2 \le C l^8 m^{-4} s^{-4} \gamma(\hat{F})^2 \end{aligned} \tag{9.19}$$

for all $x \in F$. Consequently, as $\hat{F}$ intersects at most nine squares $Q^k(p)$, $p \in I^k(\Omega^k) \setminus J$, by (9.17) and $l = \frac{k}{s}$ we get $\|\nabla y - R\|^2_{L^\infty(F)} \le C l^4 m^{-2} s^{-2} \cdot k\epsilon \le C l^{-4} \vartheta$ as well as

$$\|\nabla y - R\|^4_{L^4(F)} \le C\vartheta l^{-4} \|\nabla y - R\|^2_{L^2(\hat{F})} \le C\vartheta \gamma(\hat{F}).$$

Proceeding in this way for every connected component F of all $Q^k_i(p)$, $p \in I^k_i(\Omega^k)$, and noting that every $Q^s(q)$, $q \in I^s(\Omega^k)$, intersects at most four different associated enlarged sets $\hat{F}$ ($Q^s(q)$ can intersect more than one set if it lies at the boundary of some $Q^k_i(p)$) we obtain a function $\hat{R}_i : W^\circ \to SO(2)$ with the desired properties (9.16). $\square$

We now concern ourselves with the subatomistic regime.

Lemma 9.2.2. *Let $M \geq 0$, $\epsilon > 0$ and $s \leq k \leq \epsilon$. Then for a fixed constant $C = C(M) > 0$ we have the following:*
For all $U \in \mathcal{V}_k^s$ with $U \subset \Omega^k$ and for all $y \in H^1(U)$ with γ as defined in (9.14) and $\|\nabla y\|_\infty \leq M$ there is a set $W \in \mathcal{V}_k^s$ with $W \subset \Omega^{3k}$, $|W \setminus U| = 0$, $|(U \setminus W) \cap \Omega^{3k}| \leq C_u k \|W\|_*$ and

$$\|W\|_* \leq \|U\|_* + C\epsilon^{-1}\gamma \tag{9.20}$$

as well as mappings $\hat{R}_i : \Omega^{3k} \to SO(2)$, $i = 1, \ldots, 4$, which are constant on $Q_i^k(p) \cap W$, $p \in I_i^k(\Omega^k)$, such that

$$\|\nabla y - \hat{R}_i\|^2_{L^2(W)} \leq C\gamma + C\epsilon\|U\|_*. \tag{9.21}$$

Proof. Similarly as in (9.17) we let $J \subset I^k(\Omega^k)$ such that

$$\epsilon\mathcal{H}^1(\partial U \cap Q^k(q)) + \|\operatorname{dist}(\nabla y, SO(2))\|^2_{L^2(Q^k(q)\cap U)} > c_*\epsilon k \tag{9.22}$$

for all $q \in J$. Define $W = \Omega^{3k} \cap \left(\left(U \setminus \bigcup_{p\in J} Q^k(q)\right) \cup \bigcup_{p \in J} \partial Q^k(q)\right)$ and note that the $\|W\|_* \leq \|U\|_* + C\epsilon^{-1}\gamma$ for $c_* = c_*(h_*) > 0$ sufficiently large. Indeed, for the subset $J_1 \subset J$, for which (9.17) holds, we argue as in the previous proof. Then with $J_2 = J \setminus J_1$ we note $\|W\|_\infty \leq \|U\|_\infty + C\epsilon^{-1}\gamma + 2\sqrt{2}k \cdot \#J_2$ and $\|W\|_\mathcal{H} \leq \|U\|_\mathcal{H} + C\epsilon^{-1}\gamma + 8k \cdot \#J_2 - c_*k \cdot \#J_2$. This gives the desired result for c_* large. Moreover, we get $W \in \mathcal{V}_k^s$ and $|(U \setminus W) \cap \Omega^{3k}| \leq C_u k\|W\|_*$.

Consider some $\tilde{Q} := Q_i^k(q)$, $q \in I_i^k(\Omega^k)$. We extend y from $\tilde{Q} \cap W$ to $\tilde{Q}$ by setting $\bar{v} = y$ on $W \cap \tilde{Q}$ and $\bar{v}(x) = x$ for all $x \in \tilde{Q} \setminus W$. Note that $\bar{v} \in SBV(\tilde{Q})$ with $J_{\bar{v}} = \partial W \cap \tilde{Q}$. By Theorem A.1.9 we obtain a function $v \in H^1(\tilde{Q})$ such that by a rescaling argument

$$\|\nabla\bar{v} - \nabla v\|_{L^p(\tilde{Q})} \leq Ck^{\frac{2}{p}-1}\|\nabla\bar{v}\|_\infty\mathcal{H}^1(J_{\bar{v}} \cap \tilde{Q}) \leq CMk^{\frac{2}{p}-1}k^{1-\frac{1}{p}}\beta^{\frac{1}{p}} \leq CM\epsilon^{\frac{1}{p}}\beta^{\frac{1}{p}}$$

for $p < 2$, where $\beta = \mathcal{H}^1(\partial W \cap \tilde{Q})$. In the second step we used $\beta \leq Ck$ by (9.22) and applied $k \leq \epsilon$ in the last step. Consequently, we obtain

$$\|\operatorname{dist}(\nabla v, SO(2))\|^p_{L^p(\tilde{Q})} \leq C\|\operatorname{dist}(\nabla\bar{v}, SO(2))\|^p_{L^p(\tilde{Q})} + C\epsilon\beta.$$

Thus, since $\gamma(\tilde{Q}) := \|\operatorname{dist}(\nabla\bar{v}, SO(2))\|^2_{L^2(\tilde{Q})} = \|\operatorname{dist}(\nabla y, SO(2))\|^2_{L^2(\tilde{Q}\cap W)}$, the rigidity estimate in Theorem B.1 yields a rotation $R \in SO(2)$ such that

$$\begin{aligned}
\|\nabla v - R\|^p_{L^p(\tilde{Q})} &\leq C\|\operatorname{dist}(\nabla v, SO(2))\|^p_{L^p(\tilde{Q})} \leq C|\tilde{Q}|^{1-\frac{p}{2}}\gamma(\tilde{Q})^{\frac{p}{2}} + C\epsilon\beta \\
&\leq C\epsilon^{2-p}\gamma(\tilde{Q})^{\frac{p}{2}-1}\gamma(\tilde{Q}) + C\epsilon\beta \leq C\epsilon^{2-p}\epsilon^{p-2}\gamma(\tilde{Q}) + C\epsilon\beta \\
&\leq C\gamma(\tilde{Q}) + C\epsilon\beta.
\end{aligned}$$

In the second step we used Hölder's inequality and we applied (9.22) in the fourth step. This implies $\|\nabla y - R\|^p_{L^p(W\cap\tilde{Q})} \leq \|\nabla\bar{v} - R\|^p_{L^p(\tilde{Q})} \leq C\gamma(\tilde{Q}) + C\epsilon\beta$

and proceeding in this way for all $Q_i^k(q)$, $q \in I_i^k(\Omega^k)$, we obtain a function $\hat{R}_i : \Omega^{3k} \to SO(2)$ such that for a constant $C = C(h_*)$

$$\|\nabla y - \hat{R}_i\|^p_{L^p(W)} \leq C\gamma + C\epsilon\|U\|_*,$$

where $\hat{R}_i$ is constant on $Q_i^k(p) \cap W$, $p \in I_i^k(\Omega^k)$. Finally, by $\|\nabla y\|_\infty \leq M$ we derive

$$\|\nabla y - \hat{R}_i\|^2_{L^2(W)} \leq (M + \sqrt{2})^{2-p}\|\nabla y - \hat{R}_i\|^p_{L^p(W)} \leq C\gamma + C\epsilon\|U\|_*,$$

as desired. □

We recall (7.2) and similarly as in Section 8.4 we define for shorthand $\alpha_{\hat{R}}(F) = \|\bar{e}_{\hat{R}}(\nabla y)\|^2_{L^2(F)}$ for $F \subset \mathbb{R}^2$ and a function $\hat{R} : F \to SO(2)$. Applying the linearization formula

$$\operatorname{dist}(G, SO(2)) = |\bar{e}_R(G)| + O(|G - R|^2) \tag{9.23}$$

for $R \in SO(2)$ and $G \in \mathbb{R}^{2\times 2}$ we get

$$\alpha_{\hat{R}}(F) = \int_F |\bar{e}_{\hat{R}}(\nabla y)|^2 \leq C_u \int_F \operatorname{dist}^2(\nabla y, SO(2)) + C_u \int_F |\nabla y - \hat{R}|^4. \tag{9.24}$$

Here we already see that it suffices to establish a rigidity estimate of fourth order as in Lemma 9.2.1 in order to bound the symmetric part of the gradient. One of the main ideas in the following will be to choose $l = l(s, \epsilon, m)$ in (9.16) such that $\vartheta \leq 1$ which will imply $\alpha_{\hat{R}}(W) \leq C_u\gamma$.

9.2.2 Estimates in terms of the H^1-norm

We now show that not only the distance of the derivative from a rigid motion can be controlled as derived in (9.16) and (9.21), respectively, but also the distance of the function itself. Once we have such estimates we will be in a position to 'heal' cracks (see Section 9.3 below). After the modification of the deformation $\nu = sd$ will stand for the minimal distance of two different cracks, where d represents the corresponding increase factor. It will turn out that the least crack length will be given by $\lambda = \nu m^{-1}$. Moreover, $k = \lambda m^{-1}$ will denote the size of the cell on which we apply Theorem 8.4.2. Define

$$S_i := \bigcup\nolimits_{p \in I_i^k(\Omega^{3k})} Q_i^{\frac{5}{8}k}(p)$$

and note that $\Omega^{5k} \subset \bigcup_{i=1}^4 S_i$. Recall (A.1), (9.12) and the definition $\hat{m} = C_2(m, h_*)$ (see below (9.13)). We will proceed in two steps similarly as in the proof of Corollary 8.4.7 and Theorem 8.4.8 (see Section 8.4.3). For shorthand we will write $\gamma(F) = \|\operatorname{dist}(\nabla y, SO(2))\|^2_{L^2(F)}$, $\delta_p(F) = \sum_{i=1}^4 \|\nabla y - \hat{R}_i\|^p_{L^p(F)}$ for $p = 2, 4$ and subsets $F \subset W$.

Lemma 9.2.3. *Let $k > s > 0$, $\epsilon > 0$ such that $l := \frac{k}{s} = dm^{-2}$ for $m^{-1}, d \in \mathbb{N}$ with $m^{-1}, d \gg 1$. Let $\lambda = sdm^{-1} = km$. Then for constants $C, c > 0$ we have the following:*
For all $W \in \mathcal{V}^{sm}_{(s,3k)}$ with $W \subset \Omega^{3k}$ and for all $y \in H^1(W)$ with

$$\gamma := \|\operatorname{dist}(\nabla y, SO(2)\|^2_{L^2(W)}, \quad \delta_4 := \sum\nolimits_{i=1}^4 \|\nabla y - \hat{R}_i\|^4_{L^4(W)}$$

for mappings $\hat{R}_i : W^\circ \to SO(2)$, $i = 1, \ldots, 4$, which are constant on the connected components of $Q^k_i(p) \cap W^\circ$, $p \in I^k_i(\Omega^{3k})$, we obtain:
We find sets $U \in \mathcal{V}^{s\hat{m}}_{70k}$, $U_Q \in \mathcal{V}^{s\hat{m}}$ with $U \subset U_Q \subset \Omega^{5k}$, $|U_Q \setminus W| = 0$ and $|(W \setminus U) \cap \Omega^{5k}| \le C_u k \|U\|_*$ such that

$$\|U\|_* \le (1 + C_u m)\|W\|_* + C\epsilon^{-1}(\gamma + \delta_4) \tag{9.25}$$

as well as

$$|Q^\lambda(p) \cap U_Q| \ge cm\lambda^2 \quad \text{for all } p \in J(U_Q), \tag{9.26}$$

where $J(U_Q) := \{p \in I^\lambda(\Omega^{3k}) : Q^\lambda(p) \cap U_Q \neq \emptyset\}$.
Moreover, letting $U_J = \bigcup_{p \in J(U_Q)} \overline{Q^\lambda(p)}$, for $i = 1, \ldots, 4$ we find extensions $\bar{y}_i \in SBV^2(U_J \cap S_i, \mathbb{R}^2)$ with $\bar{y}_i = y$ on $U_Q \cap S_i$ such that for all $\tilde{Q} := Q^{3\lambda}_j(p) \cap U_J$, $p \in I^\lambda_j(\Omega^{3k})$, $j = 1, \ldots, 4$, with $\tilde{Q} \subset S_i$ we have that $R_i = \hat{R}_i|_{W^\circ \cap \tilde{Q}}$ is constant on $W^\circ \cap \tilde{Q}$ and

$$\begin{aligned}(|E(R_i^T \bar{y}_i - \mathbf{id})|(\tilde{Q}))^2 \le Ck^2 C_m \min\Big\{&\epsilon k, \gamma(W \cap Q^{2k}_i(q)) \\ &+ \delta_4(W \cap Q^{2k}_i(q)) + \epsilon|\partial W \cap Q^{2k}_i(q)|_{\mathcal{H}}\Big\},\end{aligned} \tag{9.27}$$

where $q \in I^k_i(\Omega^{3k})$ such that $\tilde{Q} \subset Q^k_i(q)$.

Proof. Similarly as in the previous proof we let $J \subset I^{3k}(\Omega^{3k})$ such that

$$\begin{aligned}\epsilon\mathcal{H}^1(Q^{3k}(p) \cap \partial W) &+ \|\operatorname{dist}(\nabla y, SO(2))\|^2_{L^2(Q^{3k}(p)\cap W)} \\ &+ \sum\nolimits_{i=1}^4 \|\nabla y - \hat{R}_i\|^4_{L^4(Q^{3k}(p)\cap W)} > c_*\epsilon k\end{aligned} \tag{9.28}$$

for all $p \in J$. Define $\hat{W} = \big(W \setminus \bigcup_{p\in J} Q^{3k}(p)\big) \cup \bigcup_{p \in J} \partial Q^{3k}(p)$ and note that choosing c_* sufficiently large and arguing as in the previous proof

$$\|\hat{W}\|_* \le \|W\|_* + C\epsilon^{-1}(\gamma + \delta_4), \tag{9.29}$$

$\hat{W} \in \mathcal{V}^{sm}_{(s,3k)}$ as well as $|(W \setminus \hat{W}) \cap \Omega^{5k}| \le C_u k \|\hat{W}\|_*$. We now subsequently construct sets $\hat{U}_1 \supset \ldots \supset \hat{U}_4$ (the inclusions hold up to sets of negligible measure) by application of Theorem 8.4.2 on connected components of $\hat{W}$ (Step (I)).

Afterwards, since in Theorem 8.4.2 the trace estimate cannot be derived for components near the boundary, we will further modify the sets in a neighborhood of large boundary components (Step (II)). A final modification procedure will then assure property (9.26) (Step (III)).

(I) Begin with $i = 1$ and fix $q \in I_1^k(\Omega^{3k})$. Consider a connected component F of $Q_1^k(q) \cap \hat{W}^\circ$. As $\hat{R}_1 = R$ is constant on F we obtain $\alpha_R(F) \leq C(\gamma(F) + \delta_4(F))$ by (9.24). Define $Q_\mu := Q_1^k(q)$ and recall (9.12). Passing to the closure of F (not relabeled) we can regard F as an element of $\mathcal{V}^{sm}$ with respect to Q_μ (recall (8.1)), where one component is given by $X = Q_\mu \setminus H(F) \in \mathcal{U}^{sm}$. (Observe, however, that $Q_\mu \setminus H(F)$ may intersect several components of $\hat{W}$.) We apply Theorem 8.4.2 on $F \subset Q_\mu$ for $\varepsilon = \epsilon$, $\sigma = m$ to obtain a set $G \in \mathcal{W}^{s\hat{m}}$ with $|G \setminus F| = 0$ and

$$\epsilon \|G\|_* + \alpha_R(G) \leq (1 + C_u m)(\epsilon \|F\|_* + \alpha_R(F)). \tag{9.30}$$

(Recall that the sum in $\|F\|_*$ runs only over the boundary components having empty intersection with ∂Q_μ.) Moreover, similarly as before we have

$$|F \setminus G| \leq C_u k \|G\|_* \tag{9.31}$$

and using (8.41), (8.54) for all $\Gamma_t(G) \in \mathcal{T}(G) := \{\Gamma_t : N^{2\hat{\tau}_t}(\partial R_t) \subset H(G)\}$

$$\int_{\Theta_t(G)} |[\bar{y}_1](x))|^2 \, d\mathcal{H}^1(x) \leq C C_m \epsilon |\Gamma_t(G)|_\infty^2, \tag{9.32}$$

where $\bar{y}_1$ is the extension (cf. (8.40))

$$\bar{y}_1(x) = \begin{cases} y & x \in \hat{W}, \\ R\,(\mathbf{Id} + A_t)\,x + R\,c_t & x \in X_t \ \text{ for } \Gamma_t(G) \in \mathcal{T}(G). \end{cases} \tag{9.33}$$

Here recall that ∂R_t are the rectangles defined in (8.5)(i)(v) as well as $\hat{\tau}_t = C C_2(m, h_*) |\partial R_t|_\infty = C\hat{m} |\partial R_t|_\infty$ for $C = C(h_*)$ (see (8.6), (8.33), (8.43) and Remark 8.3.2).

We proceed in this way for every connected component $(F_j)_j$ of all $Q_1^k(q)$, $q \in I_1^k(\Omega^{3k})$ and define $\hat{U}_1 = (\hat{W} \setminus \bigcup_j F_j) \cup \bigcup_j G_j \in \mathcal{V}^{s\hat{m}}$. (Observe that one may have $H(F_{j_1}) \subset H(F_{j_2})$. In this case the above arguments can be omitted for F_{j_1}.) By $\mathcal{G}$ we denote the set of boundary components $\Gamma(\hat{U}_1)$ which do not coincide with some $\Gamma_t(G_j)$. Note that by (9.24) and (9.29)

$$\begin{aligned} \epsilon \|\hat{U}_1\|_* \leq \epsilon \|\hat{U}_1\|_* + \alpha_{\hat{R}_1}(\hat{U}_1) &\leq (1 + C_u m)(\epsilon \|\hat{W}\|_* + \alpha_{\hat{R}_1}(\hat{W})) \\ &\leq (1 + C_u m)\epsilon \|W\|_* + C(\gamma + \delta_4). \end{aligned} \tag{9.34}$$

The second step follows as by construction for each $\Gamma(\hat{U}_1) \in \mathcal{G}$ there is a $\Gamma(\hat{W}) = \partial X$ such that $\Gamma(\hat{U}_1) \subset \overline{X}$ (recall Remark 8.4.6(i)). By (9.31) we also get $|\hat{W} \setminus \hat{U}_1| \leq$

$C_u k\|\hat{U}_1\|_*$. Moreover, by Remark 8.4.6(ii) we can replace the components of $G_j \in \mathcal{V}^{s\hat{m}}$ by rectangles such that the resulting set G'_j lies in $\mathcal{V}^{s\hat{m}}_{\text{con}}$. Recall that the (rectangular) components of G'_j satisfy $\max_{i=1,2}|\pi_i\Gamma(G'_j)| \le 2k$.

Then we define $\hat{U}''_1 := (\hat{W} \setminus \bigcup_j F_j) \cup \bigcup_j G'_j \in \mathcal{V}^{s\hat{m}}$. We now apply Lemma 9.1.2(ii) for $\nu = 0$, $(Z_j)_j$ the rectangular components of $(G'_j)_j$ and V' the set whose boundary components are given by the elements of $\mathcal{G}$. We obtain a set $\hat{U}'_1 \in \mathcal{V}^{s\hat{m}}_{5k}$ with $\|\hat{U}'_1\|_* \le \|\hat{U}_1\|_*$ and $|\hat{U}''_1 \setminus \hat{U}'_1| \le C_u k\|\hat{U}'_1\|_*$. (Strictly speaking, we need to pass from $\mathcal{V}^{s\hat{m}}$ to $\mathcal{V}^{s\hat{m}/2}$, but do not include it in the notation for convenience.) Likewise we observe $|\hat{U}'_1 \setminus \hat{W}| = 0$ and $|\hat{W} \setminus \hat{U}'_1| \le C_u k\|\hat{U}'_1\|_*$. Additionally, we apply Lemma 9.1.1(ii) and get a (not relabeled) set $\hat{U}'_1 \in \mathcal{V}^{s\hat{m}}_{6k}$ such that (9.3) and (9.34) hold. As in the proof of Lemma 9.2.1 this implies $\hat{U}'_1 \in \mathcal{V}^{s\hat{m}}_{(s,6k)}$ since $\hat{W} \in \mathcal{V}^{sm}_{(s,3k)}$, i.e. the least length of components is bounded from below by s.

In the following, by a slight abuse of notation, we say that a component $\Gamma_t(\hat{U}'_1)$, which coincides with some $\partial X_t = \Gamma_t(G')$ for some component G', satisfies (9.32) if all corresponding $(\Gamma_{t_s}(G))_s$ with $\Gamma_{t_s}(G) \subset \overline{X_t}$ satisfy (9.32). It is not hard to see that (9.32) is satisfied for all boundary components with (recall (9.12))

$$\Gamma_t(\hat{U}'_1) \cap S_1 \neq \emptyset, \quad |\Gamma_t(\hat{U}'_1)|_\infty \le \tfrac{k}{8}, \quad N_*(\Gamma_t(\hat{U}_1)) \subset H^{\frac{k}{8}}(\hat{U}'_1),$$

where $N_*(\Gamma_t(\hat{U}_1)) = \{x : \text{dist}(x, \Gamma_t(\hat{U}'_1)) \le \bar{C}\hat{m}|\Gamma_t(\hat{U}'_1)|_\infty\}$ for some large constant $\bar{C} = \bar{C}(h_*) > 0$. Indeed, assume that there is some $\Gamma_s = \Gamma_{t_s}(G) \subset Q^k_1(q)$ such that for the corresponding rectangle R_s one has $N^{2\hat{\tau}_s}(\partial R_s) \not\subset H(G)$ although the corresponding $\Gamma_t(G') = \partial X_t$ fulfills the above three properties. First, we observe $R_s \subset X_t$ by Remark 8.4.6(ii) and thus $R_s \subset Q^{\frac{3}{4}k}_1(q)$. By (8.6) we get $|\partial R_s|_\infty \le C|\Gamma_s|_\infty$. Consequently, since $2\hat{\tau}_s \ll \frac{1}{C}|\partial R_s|_\infty$ for $\hat{m}$ small enough (recall (8.33)) we have $N^{2\hat{\tau}_s}(\partial R_s) \subset Q^{\frac{7}{8}k}_1(q)$. Since by assumption $N^{2\hat{\tau}_s}(\partial R_s) \not\subset H(G)$, this would imply $|\partial H(G) \cap Q^k_1(q)|_\infty > \frac{k}{8}$.

Consequently, there is a chain of components $(\Gamma_{t_i}(\hat{U}'_1))^n_{i=1} = (\partial X_{t_i}(\hat{U}'_1))^n_{i=1}$ such that $\Gamma_{t_1}(\hat{U}'_1) \cap \partial Q_\mu \neq \emptyset$, $X_{t_n}(\hat{U}'_1) \cap N^{2\hat{\tau}_s}(\partial R_s) \neq \emptyset$ and $\Gamma_{t_{i-1}}(\hat{U}'_1) \cap \Gamma_{t_i}(\hat{U}'_1) \neq \emptyset$. Thus, by (9.3) there is one $\Gamma_*(\hat{U}'_1)$ with $|\Gamma_*(\hat{U}'_1)|_\infty > \frac{k}{8}$ such that $N^{2\hat{\tau}_s}(\partial R_s) \cap X_*(\hat{U}'_1) \neq \emptyset$. Recalling that $R_s \subset X_t$ and $2\hat{\tau}_s < \bar{C}\hat{m}|\Gamma_t(\hat{U}'_1)|_\infty$ for $\bar{C}$ sufficiently large we find $N_*(\Gamma_t(\hat{U}_1)) \cap X_*(U'_1) \neq \emptyset$. This, however, is a contradiction to $N_*(\Gamma_t(\hat{U}_1)) \subset H^{\frac{k}{8}}(\hat{U}'_1)$.

We now iteratively repeat the above construction for $i = 2, 3, 4$ for $\hat{U}'_{i-1}$ instead of $\hat{W}$ and obtain extensions $\bar{y}_2, \bar{y}_3, \bar{y}_4$ as well as $(\hat{U}_i)^4_{i=1}$ and sets $\hat{U}'_4 \subset \ldots \subset \hat{U}'_1 \subset \hat{W}$ (the inclusions hold up to a set of negligible measure) with $\hat{U}'_4 \in \mathcal{V}^{s\hat{m}}_{(s,15k)}$ such that (9.34) holds for a possibly larger constant replacing $\hat{U}_1$ by $\hat{U}_4$. We briefly note that the sets are elements of $\mathcal{V}^{s\hat{m}}$ due to (8.43) and the fact that the least length of components is bounded from below by s. Moreover, for $i = 1, \ldots, 4$, (9.32) is satisfied for $\bar{y}_i$ and all boundary components $\Gamma_t(\hat{U}'_i)$ with $\Gamma_t(\hat{U}'_i) \cap S_i \neq \emptyset$, $|\Gamma_t(\hat{U}'_i)|_\infty \le \frac{k}{8}$ and $N_*(\Gamma_t(\hat{U}'_i)) \subset H^{\frac{k}{8}}(\hat{U}'_i)$.

For later we also observe that due to the local nature of the modification process and (9.30) we get

$$
\begin{aligned}
|\partial\hat{U}_i \cap Q_i^k(q)|_{\mathcal{H}} \le C|\partial\hat{W} \cap Q_i^{2k}(q)|_{\mathcal{H}} \\
+ C\epsilon^{-1}\big(\gamma(\hat{W} \cap Q_i^{2k}(q)) + \delta_4(\hat{W} \cap Q_i^{2k}(q))\big).
\end{aligned}
\tag{9.35}
$$

Although the inclusions for $(\hat{U}_i')_{i=1}^4$ only hold up to segments, we observe that the sets are 'nested' concerning small boundary components in the following sense: Letting $\hat{U}_i^* = \hat{U}_i' \cap \overline{(H^{\frac{k}{8}}(\hat{U}_i'))^\circ}$ we obtain

$$
\hat{U}_4^* \subset \ldots \subset \hat{U}_1^*. \tag{9.36}
$$

Indeed, assume e.g. there was a component $X(\hat{U}_1^*)$ and components $X_1, \ldots, X_n$ of $\hat{U}_2^*$ with $X(\hat{U}_1^*) \subset \bigcup_{j=1}^n \overline{X_j}$ and $\bigcup_{j=1}^n \partial X_j \cap X(\hat{U}_1^*) \neq \emptyset$. Then by construction of the sets we clearly find some X_i with $\partial X_i \cap X(\hat{U}_1^*) \neq \emptyset$, $|X(\hat{U}_1^*) \setminus X_i| > 0$ and $|\partial X_i|_\infty \le \frac{k}{8}$. This, however, together with (9.3) gives a contradiction to $X(\hat{U}_1^*) \subset \bigcup_{j=1}^n \overline{X_j}$. In particular, (9.36) implies $H^{\frac{k}{8}}(\hat{U}_4') \subset \ldots \subset H^{\frac{k}{8}}(\hat{U}_1')$ up to sets of negligible measure and thus for $i = 1, \ldots, 4$, (9.32) is satisfied for $\bar{y}_i$ and all boundary components $\Gamma_t(\hat{U}_i')$ with $\Gamma_t(\hat{U}_i') \cap S_i \neq \emptyset$, $|\Gamma_t(\hat{U}_i')|_\infty \le \frac{k}{8}$, $N_*(\Gamma_t(\hat{U}_i')) \subset H^{\frac{k}{8}}(\hat{U}_4')$. We want to remove the third condition. For that reason, we subtract neighborhoods of large boundary components as follows.

(II) Let $U^* = H^{\frac{k}{8}}(\hat{U}_4')$ and let $\Gamma_1(U^*), \ldots, \Gamma_n(U^*)$ be the boundary components. For $\Gamma_j(U^*)$ let $M(\Gamma_j)$ be the smallest rectangle in $\mathcal{U}^{s\hat{m}}$ satisfying $M(\Gamma_j) \supset \{x \in \mathbb{R}^2 : \mathrm{dist}_\infty(x, X_j) \le \bar{C}k\hat{m}\}$ for the constant $\bar{C} > 0$ introduced above, where X_j denotes the component corresponding component to $\Gamma_j(U^*)$. Clearly, using the fact that $\bar{C}\hat{m} \le m$ (see (9.13)) one has $|\pi_i \partial M(\Gamma_j)| \le |\pi_i \Gamma_j(U^*)| + C_u m|\Gamma_j(U^*)|_\infty \le 31k$ for $i = 1, 2$. As the components $(X_j)_j$ are pairwise disjoint and connected, we obtain $Z(\Gamma_{j_1}) \setminus Z(\Gamma_{j_2})$ is connected for all $1 \le j_1, j_2 \le n$, where $Z(\Gamma_j)$ denotes the smallest rectangle containing X_j. Consequently, since the neighborhoods $M(\Gamma_j) \setminus Z(\Gamma_j)$ all have the same thickness $\sim \bar{C}k\hat{m}$, we get that $M(\Gamma_{j_1}) \setminus M(\Gamma_{j_2})$ is connected for all $1 \le j_1, j_2 \le n$.

Then by Lemma 9.1.2(ii) applied on $V = U^*$, $V' = \Omega^{5k} \setminus \bigcup_{|\Gamma_j(\hat{U}_i')| \le \frac{k}{8}} X_j(\hat{U}_i')$ we obtain sets $\tilde{U}_i$ with $|(\hat{U}_i' \setminus \bigcup_{j=1}^n M(\Gamma_j)) \setminus \tilde{U}_i| \le C_u k \|V'\|_*$. In particular, we set $\tilde{U} = \tilde{U}_4$ and observe that $\tilde{U} \in \mathcal{V}^{s\hat{m}}_{32k}$. Moreover, we obtain $\|\tilde{U}\|_* \le (1 + C_u m)\|V\|_* + \|V'\|_*$. As $\hat{U}_4'$ satisfies (9.3), we derive $(\partial V \cap \partial V') \cap (\Omega^{5k})^\circ = \emptyset$ and therefore $\|\tilde{U}\|_* \le (1 + C_u m)\|\hat{U}_4'\|_*$, i.e. (9.34) holds replacing $\hat{U}_1$ by $\tilde{U}$ (possibly for a larger constant). Applying Lemma 9.1.1(ii) we get (not relabeled) sets $\tilde{U}_i \in \mathcal{V}^{s\hat{m}}_{33k}$ satisfying (9.3). For later we note that $\tilde{U}_4 \subset \ldots \subset \tilde{U}_1$ up to sets of negligible measure. This follows from (9.36) and the fact that in Lemma 9.1.2(ii) the components of V' are replaced by corresponding rectangles. Arguing as in

(9.36) we also find

$$\tilde{U}_4^* \subset \ldots \subset \hat{U}_1^*, \quad \text{where} \quad \tilde{U}_i^* = \tilde{U}_i \cap \overline{(H^{\frac{k}{8}}(\tilde{U}_i))^\circ} \tag{9.37}$$

In particular, this also implies $H^{\frac{k}{8}}(\tilde{U}_4) \subset \ldots \subset H^{\frac{k}{8}}(\tilde{U}_1)$ up to sets of negligible measure.

We now see that for $i = 1, \ldots, 4$, (9.32) holds for $\bar{y}_i$ for all components satisfying

$$\Gamma_t(\tilde{U}_i) \cap S_i \neq \emptyset, \quad |\Gamma_t(\tilde{U}_i)|_\infty \leq \tfrac{1}{8}k. \tag{9.38}$$

(Strictly speaking (9.32) holds for the corresponding components of $\hat{U}_i$.) In fact, since $\bar{C}\hat{m}|\Gamma_t(\hat{U}_i')|_\infty \leq \frac{k}{8}\bar{C}\hat{m}$ for $|\Gamma_t(\hat{U}_i')|_\infty \leq \frac{k}{8}$, due to the construction of $\tilde{U}_i$ components with $|\Gamma_t(\hat{U}_i')|_\infty \leq \frac{k}{8}$ and $N_*(\Gamma_t(\hat{U}_i')) \not\subset H^{\frac{k}{8}}(\hat{U}_4')$ are 'combined' with a boundary component of $\hat{U}_4'$ which is larger than $\frac{k}{8}$.

We apply Lemma 9.1.1(i) to obtain a (not relabeled) set $\tilde{U} \in \mathcal{V}_{33k}^{s\hat{m}}$ satisfying (9.2). For each $\Gamma_t(\tilde{U})$, $t = 1, \ldots, n$, let $N_1(\Gamma_t)$, $N_2(\Gamma_t)$ be the smallest rectangles in $\mathcal{U}^{s\hat{m}}$ satisfying

$$\begin{aligned} N_1(\Gamma_t) &\supset \{x \in \mathbb{R}^2 : \mathrm{dist}_\infty(x, X_t) \leq \min\{Bm|\Gamma_t(\tilde{U})|_\infty, 2\lambda\}\}, \\ N_2(\Gamma_t) &\supset \{x \in \mathbb{R}^2 : \mathrm{dist}_\infty(x, X_t) \leq Bm \min\{|\Gamma_t(\tilde{U})|_\infty, \lambda\}\} \end{aligned}$$

for some $B > 0$ (independent of h_*) and $\lambda = km$, where X_t is the component corresponding to $\Gamma_t(\tilde{U})$. It is not restrictive to assume that

$$\mathcal{H}^1\big(N_2(\Gamma_t) \cap (\partial\tilde{U} \setminus (\Gamma_t(\tilde{U}) \cup \partial H^{\frac{k}{8}}(\tilde{U})))\big) \leq CBm \min\{|\Gamma_t(\tilde{U})|_\infty, \lambda\} \tag{9.39}$$

for all $\Gamma_t(\tilde{U})$ with $|\Gamma_t(\tilde{U})|_\infty \leq \frac{k}{8}$. Indeed, otherwise we replace $\tilde{U}$ by $\tilde{U}' := (\tilde{U} \setminus N_2^*(\Gamma_t)) \cup \partial N_2^*(\Gamma_t)$, where $N_2^*(\Gamma_t) = (N_2(\Gamma_t) \cap H^{\frac{k}{8}}(\tilde{U}))^\circ$, and arguing similarly as in (9.28) and Lemma 8.1.1 we get $\|\tilde{U}'\|_* \leq \|\tilde{U}\|_*$. Let $(X_{t'})_{t'}$, $X_{t'} \neq X_t$, be the components of $\tilde{U}$ having nonempty intersection with $N_2^*(\Gamma_t)$. Clearly, we have $|\partial X_{t'}|_\infty \leq \frac{k}{8}$. We define $T = \overline{N_2^*(\Gamma_t)} \cup \bigcup_{t'} \overline{X_{t'}}$ and modify $\tilde{U}'$ on a set of measure zero by letting $\tilde{U}'' = (\tilde{U}' \setminus T) \cup \partial T$. Arguing similarly as in the proof of Lemma 9.1.1 we find $\tilde{U}'' \in \mathcal{V}_{33k}^{s\hat{m}}$ and $\|\tilde{U}''\|_* \leq \|\tilde{U}\|_*$. Then by Lemma 9.1.1(i) we find a (not relabeled) set $\tilde{U}''$ which additionally satisfies (9.2). We continue with this iterative modification process until (9.39) is satisfied for all components smaller than $\frac{k}{8}$. Finally, by Lemma 9.1.1(ii) we obtain a (not relabeled) set $\tilde{U}'' \in \mathcal{V}_{34k}^{s\hat{m}}$ satisfying (9.3). Noting that during the modification procedure components larger than $\frac{k}{8}$ do not become smaller than $\frac{k}{8}$ we also find $H^{\frac{k}{8}}(\tilde{U}'') \subset H^{\frac{k}{8}}(\tilde{U})$. For convenience the set will still be denoted by $\tilde{U}$ in the following.

(III) We now finally construct the sets U_Q and U. For each $t = 1, \dots, n$ define the rectangle

$$Z_t = \bigcup_{p \in I^\lambda(N_1(\Gamma_t))} \overline{Q^\lambda(p)}. \tag{9.40}$$

We find $Z_t \subset N_1(\Gamma_t)$ and for sufficiently small components one has $Z_t = \emptyset$. Choosing B sufficiently large we get $X_t \subset Z_t$ if $|\partial X_t|_\infty > \frac{k}{8}$. Rearrange the components in a way that $Z_t = \emptyset$ for $t > n'$. This implies

$$\Omega^{5k} \setminus H^{\frac{k}{8}}(\tilde{U}) \subset \bigcup_{t=1}^{n'} Z_t. \tag{9.41}$$

Let $Y_t \in \mathcal{U}^{s\hat{m}}$ be the smallest rectangle containing $Z_t \cup X_t$. By the definition of $N_1(\Gamma_t)$ and Z_t we obtain

$$|\pi_i \partial Y_t| = |\pi_i \partial(\overline{Z_t \cup X_t})| \le |\pi_i \Gamma_t(\tilde{U})| + C_u m |\Gamma_t(\tilde{U})|_\infty, \quad i = 1, 2 \tag{9.42}$$

for some $C_u = C_u(B)$ large enough. As $(X_t)_t$ are pairwise disjoint and connected, it is elementary to see that $Z_{t_1} \setminus Z_{t_2}$ or $Z_{t_2} \setminus Z_{t_1}$ is connected for all $1 \le t_1, t_2 \le n'$. In fact, assume there were $t_1 \neq t_2$ such that $\overline{\pi_1 Z_{t_2}} \subset \pi_1 Z_{t_1}$ and $\overline{\pi_2 Z_{t_1}} \subset \pi_2 Z_{t_2}$. Then due to the definition of the neighborhoods we find $\overline{\pi_1 X_{t_2}} \subset \pi_1 X_{t_1}$ and $\overline{\pi_2 X_{t_1}} \subset \pi_2 X_{t_2}$. This, however, implies $X_{t_1} \cap X_{t_2} \neq \emptyset$ and yields a contradiction. A similar argument yields that $Y_{t_1} \setminus Y_{t_2}$ or $Y_{t_2} \setminus Y_{t_1}$ is connected for all $1 \le t_1, t_2 \le n'$.

Define $U'_Q = \tilde{U} \setminus \bigcup_{j=1}^{n'} Z_j$ and let $\hat{J} \subset I^\lambda(\Omega^{3k})$ such that (cf. also (9.28))

$$\mathcal{H}^1(Q^\lambda(p) \cap \partial U'_Q) > c_* \lambda \tag{9.43}$$

for all $p \in \hat{J}$. Then let $U_Q = (\Omega^{5k} \cap U'_Q) \setminus \bigcup_{p \in \hat{J}} Q^\lambda(p)$. Observe that possibly $U_Q \notin \mathcal{V}^{s\hat{m}}_{\mathrm{con}}$. Therefore, we now define a set $U \subset U_Q$ with connected boundary components.

By Lemma 9.1.2(ii) for $V = \Omega^{5k} \setminus \bigcup_{t=1}^{n'} X_t$, $V' = \Omega^{5k} \setminus \bigcup_{t=n'+1}^{n} X_t$ we obtain a set U' with $|(\tilde{U} \setminus \bigcup_{t=1}^{n'} Y_t) \setminus U'| \le C_u k \|V'\|_*$ such that $U' \in \mathcal{V}^{s\hat{m}}_{\mathrm{con}}$. Moreover, recalling (9.42) as well as $|\partial X_t|_\infty \le \frac{k}{8}$ for $t > n'$, we get $U' \in \mathcal{V}^{s\hat{m}}_{69k}$ for m sufficiently small. Using (9.42) and the fact that $\tilde{U}$ satisfies (9.3) we have $\|U'\|_* \le (1 + C_u m)\|V\|_* + \|V'\|_* \le (1 + C_u m)\|\tilde{U}\|_*$. Finally, again using Lemma 9.1.2(ii) we find a set $U \in \mathcal{V}^{s\hat{m}}_{70k}$ with

$$\Big|\Big(\Omega^{5k} \setminus \big(\bigcup_{t=1}^{n'} Y_t \cup \bigcup_{t=n'+1}^{n} X_t \cup \bigcup_{p \in \hat{J}} Q^\lambda(p)\big)\Big) \setminus U\Big| \le C_u k \|U\|_* \tag{9.44}$$

Arguing similarly as in (9.22), (9.28) we find $\|U\|_* \le \|U'\|_* \le (1 + C_u m)\|\tilde{U}\|_*$. This implies (9.25) since $\tilde{U}$ satisfies (9.34). Moreover, we derive $|(W \setminus U) \cap \Omega^{5k}| \le C_u k \|U\|_*$.

Define U_J as in the assertion of Lemma 9.2.3. We see that all $\Gamma_t(\tilde{U}_i) = \partial X_t$ with $\Gamma_t(\tilde{U}_i) \cap U_J^\circ \neq \emptyset$ satisfy $|\Gamma_t(\tilde{U}_i)|_\infty \le \frac{k}{8}$. In fact, if $|\Gamma_t(\tilde{U}_i)|_\infty > \frac{k}{8}$, we would have $X_t \subset \Omega^{5k} \setminus (H^{\frac{k}{8}}(\tilde{U}_i))^\circ$ and thus $X_t \subset \Omega^{5k} \setminus (H^{\frac{k}{8}}(\tilde{U}))^\circ$, where we used $H^{\frac{k}{8}}(\tilde{U}'') \subset H^{\frac{k}{8}}(\tilde{U}) \subset H^{\frac{k}{8}}(\tilde{U}_4) \subset H^{\frac{k}{8}}(\tilde{U}_i)$ up to a set of negligible measure (see (9.37)). (Recall that the set $\tilde{U}''$ given by the modification described below (9.39) is also denoted by $\tilde{U}$ for convenience.) Therefore, by (9.41) we get $\Gamma_t(\tilde{U}_i) \subset \overline{X_t} \subset \bigcup_{j=1}^{n'} \overline{Z_j}$ and thus $\Gamma_t(\tilde{U}_i) \cap U_J^\circ = \emptyset$ giving a contradiction. Consequently, by (9.38)

$$(9.32) \text{ holds for } \bar{y}_i \text{ for all } \Gamma_t(\tilde{U}_i) \text{ with } \Gamma_t(\tilde{U}_i) \cap U_J^\circ \cap S_i \neq \emptyset. \tag{9.45}$$

For later we recall that the corresponding components $(\Gamma_{t_s}(G))_s$ with $\Gamma_{t_s}(G) \subset X_t(\tilde{U}_i)$ (which satisfy (9.32)) also satisfy (8.5) since $G \in \mathcal{W}^{s\hat{m}}$. Consider $\tilde{Q} := Q_j^{3\lambda}(p) \cap U_J \subset S_i$. We observe that $\tilde{Q}$ consists of a bounded number of squares and that $\tilde{Q} \cap U_Q$ is contained in a connected component F of $Q_i^k(q) \cap \hat{W}^\circ$. Indeed, this follows from the fact that due to the construction of U_Q, in particular (9.40), two connected components $F_1 \neq F_2$, $F_t \cap S_i \neq \emptyset$ for $t = 1, 2$, for which $H(F_t)$ is not completely contained in another component $H(F_{t'})$, fulfill $\operatorname{dist}(F_1 \cap U_J, F_2 \cap U_J) \geq 2\lambda$. This observation also implies that $\tilde{Q}^\circ$ is connected, i.e. each $Q \subset \tilde{Q}$ shares at least one face with the rest of $\tilde{Q}$. Consequently, Corollary 8.4.7 together with (9.32) yield

$$(|E(R_i^T \bar{y}_i - \mathbf{id})|(\tilde{Q}))^2 \leq C\lambda^2 \alpha_{R_i}(U_Q \cap \tilde{Q}) + CkC_m\epsilon|\partial\hat{U}_i \cap Q_i^k(q)|^2_{\mathcal{H}},$$

where R_i is the value of the constant function $\hat{R}_i|_F$. Then (9.35) and (9.28) imply $|\partial\hat{U}_i \cap Q_i^k(q)|_{\mathcal{H}} \leq Ck$ which together with (9.24) yields (9.27). For later we note that Corollary 8.4.7 also yields

$$(|D^j(\bar{y}_i - R_i\,\mathbf{id})|(\tilde{Q}))^2 \leq C\lambda^2 \alpha_{R_i}(U_Q \cap \tilde{Q}) + CkC_m\epsilon|\partial\hat{U}_i \cap Q_i^k(q)|^2_{\mathcal{H}}. \tag{9.46}$$

It remains to show (9.26). Consider $\hat{Q} = Q^\lambda(p)$ with $\hat{Q} \cap U_Q \neq \emptyset$ and show that $|\hat{Q} \cap U_Q| \geq cm\lambda^2$. First note that $\hat{Q} \cap U_Q = \hat{Q} \cap \tilde{U}$. Let $\Gamma = \Gamma(\tilde{U}) = \partial X$ be the boundary component maximizing $|X \cap \hat{Q}|_\infty$. If $|\Gamma|_\infty \geq \frac{k}{8}$ we get a contradiction for B large enough as then $\hat{Q} \cap U_J = \emptyset$. Assume $|X \cap \hat{Q}|_\infty \ll \lambda$. Then (9.43) and the isoperimetric inequality imply $|\hat{Q} \setminus U_Q| \leq C_u \sum_t |X_t(\tilde{U}) \cap \hat{Q}|^2_\infty \ll C_u\lambda \sum_t |X_t(\tilde{U}) \cap \hat{Q}|_\infty \leq C_u\lambda^2$ and thus $|\hat{Q} \cap U_Q| \geq cm\lambda^2$ for m small enough. Therefore, we may assume that

$$\tfrac{1}{8}k = \tfrac{1}{8}m^{-1}\lambda \geq |\Gamma|_\infty \geq |X \cap \hat{Q}|_\infty \geq \bar{c}\lambda \tag{9.47}$$

for $\bar{c} > 0$ small enough. It is not hard to see that $|(N_2(\Gamma) \setminus X) \cap \hat{Q}| \geq CBm\bar{c}^2\lambda^2$. Indeed, an elementary argument yields $|N_2(\Gamma) \cap \hat{Q}| \geq CBm\bar{c}^2\lambda^2$. Moreover, if we had $|\hat{Q} \setminus X| \ll Bm\bar{c}^2\lambda^2$, we would get $\hat{Q} \subset N_1(\Gamma)$ and thus $\hat{Q} \cap U_Q = \emptyset$ by the construction of U_Q. We can assume that $N_2(\Gamma) \cap \partial H^{\frac{k}{8}}(\tilde{U}) = \emptyset$ since otherwise a component larger than $\frac{k}{8}$ intersects $\hat{Q}$ and we derive $\hat{Q} \cap U_J = \emptyset$ as before. By (9.2) this also implies that all components $X_j(\tilde{U})$ with $X_j(\tilde{U}) \cap N_2(\Gamma) \neq \emptyset$

satisfy $\overline{X_j(\tilde{U})} \cap \Gamma = \emptyset$. Thus by the isoperimetric inequality and by (9.39) we get $|N_2(\Gamma) \cap \hat{Q} \cap \tilde{U}| \geq |(N_2(\Gamma) \setminus X) \cap \hat{Q}| - C(B\lambda m)^2$. This implies

$$\begin{aligned}|\hat{Q} \cap U_Q| = |\hat{Q} \cap \tilde{U}| \geq |\hat{Q} \cap \tilde{U} \cap N_2(\Gamma)| \geq |(N_2(\Gamma) \setminus X) \cap \hat{Q}| - C(B\lambda m)^2 \\ \geq -CB^2\lambda^2 m^2 + C\bar{c}^2 Bm\lambda^2 \geq cm\lambda^2\end{aligned}$$

for m sufficiently small. □

Remark 9.2.4. (i) For later we observe that there is a set $U^H \in \mathcal{V}^\lambda_{35k}$ with

$$(i)\ \|U^H\|_* \leq (1 + C_u m)\|W\|_* + C\epsilon^{-1}(\gamma + \delta_4), \quad (ii)\ \|U^H\|_{\mathcal{H}} \leq C_u\|U^H\|_* \quad (9.48)$$

which coincides with the set U_J considered in the previous lemma up to a set of negligible measure. In fact, we apply Lemma 9.1.2(i) on the rectangles $(Z_t)_{t=1}^{n'}$ considered in (9.40) and find pairwise disjoint $(Z'_t)_{t=1}^{n'}$ with $\bigcup_{j=1}^{n'} \overline{Z_j} = \bigcup_{j=1}^{n'} \overline{Z'_j}$. We define

$$U^H := \Omega^{5k} \setminus \Big(\bigcup\nolimits_{j=1}^{n'} Z'_j \cup \bigcup\nolimits_{p \in \hat{J}} Q^\lambda(p)\Big),$$

where $\hat{J}$ as in (9.44). By Lemma 9.1.2(i) we get (i) and $U^H \in \mathcal{V}^\lambda_{35k}$ since $|\pi_i \partial Z_t| \leq 2 \cdot 34k + C_u mk \leq 70k$ for $i = 1, 2$. Moreover, (9.48)(ii) is a consequence of Lemma 8.1.1 and the fact that $(Z_t)_t$, $(Q^\lambda(p))_{p \in J}$ are rectangles.

Clearly $U_J \subset U^H$. Moreover, we see that $|U^H \setminus U_J| > 0$ can only happen if there is a square $Q^\lambda(p) \subset U^H$ and components $(X_t(\tilde{U}))_t$ of $\tilde{U}$ such that $Q^\lambda(p) \subset \bigcup_t \overline{X_t(\tilde{U})}$. Since we can suppose $|\partial X_t(\tilde{U})|_\infty \leq \frac{k}{8}$ (otherwise the components are contained in some rectangle Z_t), this yields a contradiction to (9.2).

(ii) For $i = 1, \ldots, 4$ we have

$$|\partial \hat{U}_i \cap U_J^\circ|_{\mathcal{H}} \leq C_u(\|W\|_* + C\epsilon^{-1}(\gamma + \delta_4)).$$

In fact, recalling (9.45) we get that all $\Gamma_t(\hat{U}_i)$ with $\Gamma_t(\hat{U}_i) \cap U_J^\circ \neq \emptyset$ fulfill (9.32) and (8.5). Thus, we obtain $|\Theta_t(\hat{U}_i)|_{\mathcal{H}} \leq C_u|\Theta_t(\hat{U}_i)|_*$ and the claim follows from (9.34) replacing $\hat{U}_1$ by $\hat{U}_i$.

We are now in a position to prove the main result of this section. Recall the definition $\lambda = sdm^{-1} = km$ and (9.13).

Lemma 9.2.5. *Let $k > s, \epsilon > 0$ such that $l := \frac{k}{s} = dm^{-2}$ for $m^{-1}, d \in \mathbb{N}$ with $m^{-1}, d \gg 1$. Then for a fixed constant $C > 0$ we have the following:*
For all $W \in \mathcal{V}^{sm}_{(s,3k)}$ with $W \subset \Omega^{3k}$ and for all $y \in H^1(W)$ with $\|\nabla y\|_\infty \leq C$, γ as defined in (9.14) and

$$\delta_4 := \sum\nolimits_{i=1}^4 \|\nabla y - \hat{R}_i\|^4_{L^4(W)}, \quad \delta_2 := \sum\nolimits_{i=1}^4 \|\nabla y - \hat{R}_i\|^2_{L^2(W)} \quad (9.49)$$

for mappings $\hat{R}_i : W^\circ \to SO(2)$, $i = 1, \ldots, 4$, which are constant on the connected components of $Q_i^k(p) \cap W^\circ$, $p \in I_i^k(\Omega^{3k})$, we obtain:

We find sets $V \in \mathcal{V}^{s\hat{m}^2}_{71k}$, $U_J \in \mathcal{V}^\lambda$ *with* $V \subset U_J$ *and* $V \subset \Omega^{6k}$, $|V \setminus W| = 0$, $|(W \setminus V) \cap \Omega^{6k}| \le C_u k \|V\|_*$ *such that*

$$\|V\|_* \le (1 + C_u m)\|W\|_* + C\epsilon^{-1}(\gamma + \delta_4) \tag{9.50}$$

as well as mappings $\bar{R}_j : U_J \to SO(2)$ *and* $\bar{c}_j : U_J \to \mathbb{R}^2$, *which are constant on* $Q^\lambda_j(p)$, $p \in I^\lambda_j(\Omega^{3k})$, *such that*

$$\begin{aligned}
&(i) && \|y - (\bar{R}_j \,\cdot + \bar{c}_j)\|^2_{L^2(V)} \le CC^2_m \lambda^2 \min\nolimits_{p=2,4}(1 + \vartheta_p)(\gamma + \delta_p + \epsilon\|W\|_*),\\
&(ii) && \|\nabla y - \bar{R}_j\|^p_{L^p(V)} \le CC^2_m\big(\delta_p + \vartheta_p(\gamma + \delta_4 + \epsilon\|W\|_*)\big),\ p = 2, 4,\\
&(iii) && \|\bar{R}_{j_1} - \bar{R}_{j_2}\|^p_{L^p(U_J)} \le CC^2_m\big(\delta_p + \vartheta_p(\gamma + \delta_4 + \epsilon\|W\|_*)\big),\ p = 2, 4,\\
&(iv) && \|(\bar{R}_{j_1} \,\cdot + \bar{c}_{j_1}) - (\bar{R}_{j_2} \,\cdot + \bar{c}_{j_2})\|^2_{L^2(U_J)} \le CC^2_m \lambda^2 \min_{p=2,4}(1 + \vartheta_p)(\gamma + \delta_p + \epsilon\|W\|_*)
\end{aligned} \tag{9.51}$$

for $j_1, j_2 = 1, \ldots, 4$, $j = 1, \ldots, 4$, *where* $\vartheta_4 = \vartheta$ *and* $\vartheta_2 = 1$. *Moreover, we have*

$$\lambda^{-2}\|(\bar{R}_{j_1} \,\cdot + \bar{c}_{j_1}) - (\bar{R}_{j_2} \,\cdot + \bar{c}_{j_2})\|^2_{L^\infty(U_J)} + \|\bar{R}_{j_1} - \bar{R}_{j_2}\|^4_{L^\infty(U_J)} \le C\bar{\vartheta} \tag{9.52}$$

for $\bar{\vartheta} = \min\{\vartheta(1 + \vartheta), C^3_m\}$ *and under the additional assumption that* $\Delta y = 0$ *in* W° *we obtain*

$$\lambda^{-2}\|y - (\bar{R}_j \,\cdot + \bar{c}_j)\|^2_{L^\infty(V)} \le C\vartheta(1 + \vartheta). \tag{9.53}$$

Proof. Apply Lemma 9.2.3 to obtain $U \in \mathcal{V}^{s\hat{m}}_{70k}$, $U_Q \in \mathcal{V}^{s\hat{m}}$ with $|U_Q \setminus W| = 0$, U_J and extensions $\bar{y}_i : S_i \cap U_J \to \mathbb{R}^2$ such that (9.25), (9.26) and (9.27) hold. Consider $Q = Q^\lambda_j(p)$, $p \in I^\lambda_j(\Omega^{3k})$, $j = 1, \ldots, 4$, with $Q \cap U_J \neq \emptyset$. Moreover, let $\tilde{Q} = Q^{3\lambda}_j(p) \cap U_J$. As $6\lambda < \frac{k}{4}$ by $m \ll 1$, we find some $Q^k_i(q)$ for some $i = 1, \ldots, 4$ with $\tilde{Q} \subset Q^{\frac{5}{8}k}_i(q) \subset S_i$ and therefore we can apply (9.27). Recall that $\hat{R} := \hat{R}_i|_{W^\circ \cap \tilde{Q}}$ is constant due to the construction in Lemma 9.2.3 (see below (9.45)). By Theorem B.4 we find $A \in \mathbb{R}^{2\times 2}_{\mathrm{skew}}$ and $c \in \mathbb{R}^2$ such that

$$\begin{aligned}
\|\bar{y}_i - \hat{R}\,(\mathbf{Id} + A)\,\cdot - \hat{R}\,c\|^2_{L^2(\tilde{Q})} &= \|\hat{R}^T\bar{y}_i - \cdot - (A\,\cdot + c)\|^2_{L^2(\tilde{Q})}\\
&\le C(|E(\hat{R}^T\bar{y}_i - \mathbf{id})|(\tilde{Q}))^2 \le Ck^2 G,
\end{aligned} \tag{9.54}$$

where

$$G := C_m \min\Big\{\epsilon k, \gamma(W \cap Q^{2k}_i(q)) + \delta_4(W \cap Q^{2k}_i(q)) + \epsilon|\partial W \cap Q^{2k}_i(q)|_{\mathcal{H}}\Big\}.$$

The constant C is independent of $\tilde{Q}$ as there are (up to rescaling) only a finite number of different shapes of $\tilde{Q}$. (Also recall that each $Q \subset \tilde{Q}$ shares at least one face with the rest of $\tilde{Q}$.)

In the proof of Lemma 9.2.3 we have seen that all $\Gamma_t = \Gamma_t(\tilde{U}_i)$ with $\tilde{Q} \cap \Gamma_t \neq \emptyset$ satisfy (9.32) for $\bar{y}_i$ and $|\Gamma_t|_\infty \le \frac{k}{8}$ as well as $N^{\hat{\tau}_l}(\partial R_t) \subset Q_i^k(q)$ (cf. (9.45)). Thus, by Lemma 8.5.7 for $V = Q_i^k(q)$ we get

$$\begin{aligned}\|\nabla \bar{y}_i - \hat{R}\|^p_{L^p(\tilde{Q})} &\le C\|\nabla y - \hat{R}\|^p_{L^p(\tilde{Q}\cap\hat{W})} + C\sum\nolimits_{\Gamma_t \in \mathcal{F}(Q_i^k(q))} |X_t|^2_\infty |A_t|^p \\ &\le CC_m\delta_p(Q_i^k(q)\cap\hat{W}) + CC_m(\epsilon s^{-1})^{\frac{p}{2}-1}\epsilon|\partial\hat{U}_i \cap Q_i^k(q)|_{\mathcal{H}}\end{aligned} \tag{9.55}$$

for $p = 2, 4$, where $\hat{W}, \hat{U}_i$ as defined in the previous proof and X_t, A_t as in (9.33). Recall that the factor s^{-1} appearing in the estimate is related to the fact that the least length of boundary components of $\hat{U}_i$ is s. Thus, recalling that $\hat{U}_i$ fulfills (9.35) we obtain by the definition of G

$$\|\nabla \bar{y}_i - \hat{R}\|^p_{L^p(\tilde{Q})} \le CC_m\delta_p(Q_i^k(q)\cap\hat{W}) + C(\epsilon s^{-1})^{\frac{p}{2}-1}G =: H_p. \tag{9.56}$$

We repeat the estimate (9.54) with Theorem B.3 instead of Theorem B.4 and obtain by (9.46) and Hölder's inequality

$$\begin{aligned}\|\bar{y}_i - \hat{R}\ \cdot - \tilde{c}\|^2_{L^2(\tilde{Q})} &\le C\|\nabla \bar{y}_i - \hat{R}\|^2_{L^1(\tilde{Q})} + C(|D^j(\bar{y}_i - \hat{R}\,\mathbf{id})|(\tilde{Q}))^2 \\ &\le C\lambda^{4(1-\frac{1}{p})}H_p^{\frac{2}{p}} + Ck^2G,\end{aligned}$$

for $\tilde{c} \in \mathbb{R}^2$ for $p = 2, 4$. This together with (9.54) and an argumentation similar to (7.10) yields $\lambda^4|A|^2 \le C\lambda^{4-4/p}H_p^{2/p} + Ck^2G$ and therefore by (9.56)

$$\begin{aligned}\lambda^2|A|^2 &\le CH_2 + Cm^{-2}G \le CC_m\delta_2(Q_i^k(q)\cap\hat{W}) + Cm^{-2}G =: \hat{H}_2, \\ \lambda^2|A|^4 &\le CH_4 + C\lambda^{-2}m^{-4}G^2 \le CH_4 + C\lambda^{-1}m^{-5}C_m\epsilon G \\ &\le CC_m\delta_4(Q_i^k(q)\cap\hat{W}) + C\vartheta G =: \hat{H}_4.\end{aligned} \tag{9.57}$$

Observe that $\hat{H}_4 \le C(1+\vartheta)G$. By (9.23) there is a rotation $\bar{R} \in SO(2)$ such that

$$\begin{aligned}|\bar{R} - \hat{R}(\mathbf{Id} + A)|^2 &= \mathrm{dist}^2(\hat{R}(\mathbf{Id}+A), SO(2)) \\ &\le 0 + C|\hat{R}(\mathbf{Id}+A) - \hat{R}|^4 = C|A|^4 \le C\lambda^{-2}\hat{H}_4,\end{aligned} \tag{9.58}$$

as $\bar{e}_{\hat{R}}(\hat{R}(\mathbf{Id}+A)) = 0$. Likewise, as $|A| \le C$ by $\|\nabla y\|_\infty \le C$ we get $|\bar{R} - \hat{R}(\mathbf{Id}+A)|^2 \le C|A|^2 \le C\lambda^{-2}\hat{H}_2$. Consequently, the Poincaré inequality, (9.54) and (9.57) yield

$$\|\bar{y}_i - (\bar{R}\ \cdot + \bar{c})\|^2_{L^2(\tilde{Q})} \le Ck^2G + C\lambda^4|A|^4 \le Ck^2G + Ck^2\min\nolimits_{p=2,4}\hat{H}_p \tag{9.59}$$

for some possibly different $\bar{c} \in \mathbb{R}^2$. Moreover, we get

$$\begin{aligned}\lambda^2|\hat{R} - \bar{R}|^4 &\le C\lambda^2|\bar{R} - \hat{R}(\mathbf{Id}+A)|^4 + C\lambda^2|A|^4 \\ &\le C\lambda^2|\bar{R} - \hat{R}(\mathbf{Id}+A)|^2 + C\lambda^2|A|^4 \le C\hat{H}_4.\end{aligned} \tag{9.60}$$

and likewise

$$\lambda^2|\hat{R}-\bar{R}|^2 \le C\hat{H}_2. \tag{9.61}$$

For fixed $j=1,\ldots,4$ we proceed in this way on each $Q_t = Q_j^\lambda(p)$, $p \in I_j^\lambda(\Omega^{3k})$, with $Q_t \cap U_J \neq \emptyset$ and for the corresponding $\tilde{Q}_t = Q_j^{3\lambda}(p) \cap U_J$ we obtain constants $\hat{R}_t, \bar{R}_t \in SO(2)$ and $\bar{c}_t \in \mathbb{R}^2$ as given in (9.59)-(9.61). Consequently, we find mappings $\bar{R}_j : U_J \to SO(2)$ and $\bar{c}_j : U_J \to \mathbb{R}^2$ being constant on each Q_t, where on each $Q_t \subset \tilde{Q}_t$ we choose $\bar{R}_j = \bar{R}_t$ and $\bar{c}_j = \bar{c}_t$. By (9.59) and the observation that every $Q_i^{2k}(q)$ is intersected only by $\sim m^{-2}$ squares $\tilde{Q}_t$ we obtain

$$\begin{aligned} \|y-(\bar{R}_j\ \cdot+\bar{c}_j)\|^2_{L^2(U)} &\le Ck^2 \min_{p=2,4}(1+\vartheta_p)m^{-2}C_m m^{-2}(\gamma+\delta_p+\epsilon\|W\|_*) \\ &\le C\lambda^2 \min\nolimits_{p=2,4}(1+\vartheta_p)m^2C_m^2(\gamma+\delta_p+\epsilon\|W\|_*) \end{aligned} \tag{9.62}$$

where $\vartheta_2 = 1$ and $\vartheta_4 = \vartheta$. Here we used that $\delta_4 \le C\delta_2$. Likewise, applying (9.49), (9.57), (9.60), (9.61) as well as the triangle inequality we get

$$\begin{aligned} \|\nabla y - \bar{R}_j\|^p_{L^p(U)} &\le Cm^{-2}C_m\big(\delta_p + m^{-2}\vartheta_p(\gamma+\delta_4+\epsilon\|W\|_*)\big) \\ &\le CmC_m^2\big(\delta_p+\vartheta_p(\gamma+\delta_4+\epsilon\|W\|_*)\big) \end{aligned} \tag{9.63}$$

for $p=2,4$. We now consider $Q_1 := Q_{j_1}^\lambda(p_1)$, $Q_2 := Q_{j_2}^\lambda(p_2)$ with $Q_1 \cap Q_2 \neq \emptyset$ and $Q_1, Q_2 \cap U_J \neq \emptyset$. Moreover, let $\tilde{Q}_i = Q_{j_i}^{3\lambda}(p_i) \cap U_J$ be the corresponding enlarged sets. It is not hard to see that there is some $Q^\lambda(p)$, $p \in J(U_Q)$, with $Q^\lambda(p) \subset \tilde{Q}_1, \tilde{Q}_2$ and therefore by the definition of U_J, in particular (9.26), we derive $|\tilde{Q}_1 \cap \tilde{Q}_2 \cap U_Q| \ge cm\lambda^2$. Let $\bar{R}_{j_i} \in SO(2)$, $\bar{c}_{j_i} \in \mathbb{R}^2$, $i=1,2$, be the constants constructed above. We compute

$$\begin{aligned} \lambda^2\|\bar{R}_{j_1}-\bar{R}_{j_2}\|^p_{L^\infty(Q_1\cap Q_2)} &\le Cm^{-1}\|\bar{R}_{j_1}-\bar{R}_{j_2}\|^p_{L^p(\tilde{Q}_1\cap\tilde{Q}_2\cap U_Q)} \\ &\le Cm^{-1}\sum\nolimits_{j=1}^4 \|\nabla y - \bar{R}_j\|^p_{L^p(\tilde{Q}_1\cap\tilde{Q}_2\cap U_Q)} \end{aligned} \tag{9.64}$$

and summing over all squares we get by (9.63)

$$\|\bar{R}_{j_1}-\bar{R}_{j_2}\|^p_{L^p(U_J)} \le CC_m^2\big(\delta_p+\vartheta_p(\gamma+\delta_4+\epsilon\|W\|_*)\big) \tag{9.65}$$

for $1 \le j_1, j_2 \le 4$ and $p=2,4$. Here we used that each $Q_j^{3\lambda}(p) \cap U_J$ only appears in a finite number of addends. Note that $\frac{|\pi_1(Q_1\cap Q_2)|+|\pi_2(Q_1\cap Q_2)|}{\max_{i=1,2}|\pi_i(\tilde{Q}_1\cap\tilde{Q}_2\cap U_Q)|} \le Cm^{-1/2}$ and $\frac{|Q_1\cap Q_2|}{|\tilde{Q}_1\cap\tilde{Q}_2\cap U_Q|} \le Cm^{-1}$. Consequently, arguing similarly as in (7.12) we find

$$\begin{aligned} &\lambda^2\|(\bar{R}_{j_1}\ \cdot+\bar{c}_{j_1})-(\bar{R}_{j_2}\ \cdot+\bar{c}_{j_2})\|^2_{L^\infty(Q_1\cap Q_2)} \\ &\quad\le C(m^{-\frac{1}{2}})^2m^{-1}\|(\bar{R}_{j_1}\ \cdot+\bar{c}_{j_1})-(\bar{R}_{j_2}\ \cdot+\bar{c}_{j_2})\|^2_{L^2(\tilde{Q}_1\cap\tilde{Q}_2\cap U_Q)}. \end{aligned} \tag{9.66}$$

($\mathbb{R}^{2\times 2}_{\text{skew}}$ may be replaced by $SO(2)$, see Remark 7.1.4(iii).) Replacing (9.63) by (9.62) in the above argument we then get

$$\begin{aligned}\|(\bar{R}_{j_1}\ \cdot + \bar{c}_{j_1}) - (\bar{R}_{j_2}\ \cdot + \bar{c}_{j_2})\|^2_{L^2(U_J)} &\le Cm^{-2}\sum\nolimits_{j=1}^{4}\|y - (\bar{R}_j\ \cdot + \bar{c}_j)\|^2_{L^2(U_Q)}\\ &\le CC_m^2\lambda^2 \min\nolimits_{p=2,4}(1+\vartheta_p)(\gamma + \delta_p + \epsilon\|W\|_*).\end{aligned} \tag{9.67}$$

Similarly as in the proof of Lemma 9.2.1 (see the construction in (9.18)) we can define $V \in \mathcal{V}^{s\hat{m}^2}_{71k}$ with $|V \setminus U| = 0$, $V^\circ \subset \{x \in U \cap \Omega^{6k} : \text{dist}_\infty(x, \partial U) \ge 2s\hat{m}m\}$, $\|V\|_* \le (1 + C_u m)\|U\|_*$ and $|(W \setminus V) \cap \Omega^{6k}| \le C_u k\|V\|_*$. By (9.25) this implies (9.50). We note that in this case for components $\Gamma_j = \partial X_j$ with $X_j \subset U_J$ it suffices to consider a corresponding rectangle $M(\Gamma_j)$ with $M(\Gamma_j) \subset U_J$. For later we observe that this construction yields

$$V \subset U_J, \qquad \Big|\big(\Omega^{6k} \setminus \bigcup M(\Gamma_j)\big)\triangle V\Big| = 0. \tag{9.68}$$

We now see that (9.51) follows directly from (9.62)-(9.67).

It remains to show (9.52) and (9.53). By (9.57), (9.60) and (9.64) we find $\|\bar{R}_{j_1} - \bar{R}_{j_2}\|^4_{L^\infty(Q_1\cap Q_2)} \le C\lambda^{-2}(1+\vartheta)G + C\lambda^{-2}m^{-1}G$ for sets $Q_1, Q_2 \subset U_J$ as considered above. Recalling the definition of G we then get

$$\|\bar{R}_{j_1} - \bar{R}_{j_2}\|^4_{L^\infty(Q_1\cap Q_2)} \le C(1+\vartheta)\lambda^{-2}m^{-1}C_m\epsilon k \le Cs^{-1}(1+\vartheta)C_m^2\epsilon \le C(1+\vartheta)\vartheta$$

Likewise, we derive $\lambda^{-2}\|(\bar{R}_{j_1}\ \cdot + \bar{c}_{j_1}) - (\bar{R}_{j_2}\ \cdot + \bar{c}_{j_2})\|^2_{L^\infty(Q_1\cap Q_2)} \le C(1+\vartheta)\vartheta$ recalling the definition of G and taking (9.66), (9.59) (for $p = 4$) and the triangle inequality into account. Similarly, by (9.59) for $p = 2$ and the observation that $\delta_2(Q_i^k(q) \cap \hat{W}) \le Ck^2$ as $\|\nabla y\|_\infty \le C$ we find using $\epsilon \le k$

$$\begin{aligned}\lambda^{-2}\|(\bar{R}_{j_1}\ \cdot + \bar{c}_{j_1}) - (\bar{R}_{j_2}\ \cdot + \bar{c}_{j_2})\|^2_{L^\infty(Q_1\cap Q_2)} &\le C\lambda^{-4}m^{-2}k^2(G + \hat{H}_2)\\ &\le C\lambda^{-2}m^{-4}(m^{-2}G + C_m k^2) \le C\lambda^{-2}C_m^2 k^2 \le CC_m^3.\end{aligned}$$

This finishes the proof of (9.52).

Finally, to see (9.53), we repeat the argument in (9.19): Let $x \in Q \cap V \subset \tilde{Q}$ for $Q = Q_j^\lambda(p)$, $\tilde{Q} = Q_j^{3\lambda}(p) \cap U_J$ as considered above and let $\bar{R}\ \cdot + \bar{c}$ be the corresponding rigid motion as given in (9.59). Since y is assumed to be harmonic in U° the mean value property of harmonic function for $r \le s\hat{m}m$ and Jensen's inequality yield

$$\begin{aligned}|y(x) - (\bar{R}\,x + \bar{c})|^2 &\le \Big|\frac{1}{|B_r(x)|}\int_{B_r(x)} (y(t) - (\bar{R}\,t + \bar{c}))\,dt\Big|^2\\ &\le C|B_r(x)|^{-1}(1+\vartheta)k^2 G \le C(1+\vartheta)m^{-2}\hat{m}^{-2}s^{-2}k^2 G\\ &\le C(1+\vartheta)C_m m^{-4}\hat{m}^{-2}l\epsilon s^{-1}\lambda^2 \le C(1+\vartheta)\vartheta\lambda^2.\end{aligned}$$

Here we used (9.59) and the fact that $B_r(x) \subset U^\circ \cap \tilde{Q}$ for all $x \in Q \cap V$. □

9.2.3 Local rigidity for an extended function

We now state a version of Lemma 9.2.5 for an extension of the function y.

Corollary 9.2.6. *Let be given the assumptions of Lemma 9.2.3, Lemma 9.2.5 and let $U \in \mathcal{V}^{s\hat{m}}_{70k}$, $U^H \in \mathcal{V}^{\lambda}_{35k}$ be the sets provided by Lemma 9.2.3, Remark 9.2.4, respectively. Moreover, assume that $\vartheta \le 1$. Then the estimates* (9.51)*(iii),(iv) hold on U^H for functions $\bar{R}_j$, $\bar{c}_j$, $j = 1, \ldots, 4$. Moreover, we find an extension $\hat{y} \in SBV^2(U^H, \mathbb{R}^2)$ with $\hat{y} = y$ on U and $\nabla\hat{y} \in SO(2)$ on $U^H \setminus W$ a.e. such that for every $Q = Q^{\lambda}_j(p)$, $p \in I^{\lambda}_j(\Omega^{3k})$, with $Q \cap U^H \neq \emptyset$ we have*

$$\begin{aligned} &(i) \ \ \|\nabla\hat{y} - \bar{R}_j\|^p_{L^p(Q)} \le CC^2_m\,(\bar{G}(N) + \delta_p(N)), \ \ p = 2, 4 \\ &(ii) \ \ \|\hat{y} - (\bar{R}_j\,\cdot + \bar{c}_j)\|^2_{L^2(Q)} \le C\lambda^2 C^2_m\, \min\{\epsilon k, \bar{G}(N)\}, \\ &(iii) \ \ \|\hat{y} - (\bar{R}_j\,\cdot + \bar{c}_j)\|^2_{L^1(\partial Q)} \le C\lambda^2 C^2_m\, \min\{\epsilon k, \bar{G}(N)\}, \end{aligned} \tag{9.69}$$

where $N = N(Q) = \{x \in W : \operatorname{dist}(x, Q) \le Ck\}$ and for shorthand $\bar{G}(N) = \gamma(N) + \delta_4(N) + \epsilon\mathcal{H}^1(N \cap \partial W)$. Furthermore, we have

$$\mathcal{H}^1(J_{\hat{y}}) \le C_u(\|W\|_* + C\epsilon^{-1}(\gamma + \delta_4)). \tag{9.70}$$

Proof. Recall the definition of U in (9.44) and that U_J and U^H coincide up to a set of measure zero by Remark 9.2.4. In Lemma 9.2.3 we have defined sets $(\tilde{U}_j)^4_{j=1}$, $\tilde{U}^*_4 \subset \ldots \subset \tilde{U}^*_1$ (see (9.37)) and corresponding extensions $\bar{y}_i|_{U_J \cap S_i}$. Moreover, in (9.45) have seen that all $\Gamma_t(\tilde{U}_i)$ with $\Gamma_t(\tilde{U}_i) \cap U^\circ_J \cap S_i \neq \emptyset$ satisfy (9.32) for $\bar{y}_i$ and $|\Gamma_t(\tilde{U}_i)|_\infty \le \frac{k}{8}$. By Lemma 9.2.5 we get that (9.51)(iii),(iv) hold.

The goal is to provide one single extension $\hat{y} : U^H \to \mathbb{R}^2$ and to confirm (9.69). Define

$$\hat{S}_i := \bigcup\nolimits_{p \in I^k_i(\Omega^{3k})} \overline{Q_i^{\frac{9}{16}k}(p)} \subset S_i$$

and let $D_i = (\tilde{U}_i \cap U^\circ_J) \cup \bigcup_{\Gamma_t(\tilde{U}_i) \subset \hat{S}_i} X_t(\tilde{U}_i)$, where $X_t(\tilde{U}_i)$ is the component corresponding to $\Gamma_t(\tilde{U}_i)$. We now show that $U^\circ_J \subset \bigcup^4_{i=1} D_i$. To see this, it suffices to prove

$$S_i \cap U^\circ_J \subset \bigcup\nolimits^4_{n=1} D_n, \ \ i = 1, \ldots, 4. \tag{9.71}$$

Fix i and assume that (9.71) has already be established for $j > i$. As $S_i \cap U^\circ_J \subset \Omega^{5k} \subset H(\tilde{U}_i) = \tilde{U}_i \cup \bigcup_{\Gamma_t(\tilde{U}_i)} X_t(\tilde{U}_i)$ by the definition of U_J, we find $(S_i \cap U^\circ_J) \setminus D_i \subset (S_i \cap U^\circ_J) \cap \bigcup_{\Gamma_t(\tilde{U}_i) \not\subset \hat{S}_i} X_t(\tilde{U}_i)$. To see (9.71) for i, it now suffices to show that each $\Gamma_t(\tilde{U}_i)$ with $\Gamma_t(\tilde{U}_i) \cap U^\circ_J \cap S_i \neq \emptyset$ satisfies $U^\circ_J \cap X_t(\tilde{U}_i) \subset \bigcup^4_{n=1} D_n$. Since $|\Gamma_t(\tilde{U}_i)|_\infty \le \frac{k}{8}$ for all such components, we derive $X_t(\tilde{U}_i) \subset \hat{S}_j$ for some $j = 1, \ldots, 4$. If $j < i$, by the construction of the sets $\tilde{U}^*_1 \supset \ldots \supset \tilde{U}^*_4$ we find $(X_{t_s}(\tilde{U}_j))_s$ such that

$$X_t(\tilde{U}_i) = (\tilde{U}_j \cap X_t(\tilde{U}_i)) \cup \bigcup\nolimits_s X_{t_s}(\tilde{U}_j).$$

As $X_{t_s}(\tilde{U}_j) \subset \hat{S}_j$, this implies $X_t(\tilde{U}_i) \cap U_J^\circ \subset D_j$. The case $j = i$ is clear. If $j > i$, we obtain $X_t(\tilde{U}_i) \cap U_J^\circ \subset S_j \cap U_J^\circ \subset \bigcup_{n=1}^4 D_n$ by (9.71). This yields the claim.

Set $\bar{y} = \bar{y}_4$ on $D_4 \cap U_J$, $\bar{y} = \bar{y}_j$ on $(D_j \setminus D_{j+1}) \cap U_J$ for $j = 3, 2, 1$. It is not hard to see that $\bar{y}$ is defined on U^H (as $|U^H \setminus U_J^\circ| = 0$) and $\bar{y} = y$ on U. Moreover, by construction there is a set of components $(X_t)_t$ consisting of components of $(\hat{U}_i)_i$ such that

$$J_{\bar{y}} \subset \bigcup\nolimits_t \partial X_t \subset \bigcup\nolimits_{i=1}^4 \bigcup\nolimits_t \Gamma_t(\hat{U}_i).$$

By (9.33) we have $\bar{y}(x) = \bar{y}_{i_t}(x) = R_t\,(\mathbf{Id} + A_t)\,x + R_t\,c_t$ for $x \in X_t$, where $R_t \in SO(2)$, $A_t \in \mathbb{R}^{2\times 2}_{\rm skew}$, $c_t \in \mathbb{R}^2$ and $1 \le i_t \le 4$ appropriately. Note that due the the definition of the extensions in (9.33) the components X_t are associated to the sets $(\hat{U}_i)_i$, not to $(\tilde{U}_i)_i$. By Remark 9.2.4(ii) this yields (9.70) for $\bar{y}$.

Consider $Q = Q_j^\lambda(p)$ with $Q \cap U_J \neq \emptyset$. Let $\tilde{Q} = Q_j^{3\lambda}(p) \cap U_J$ and observe $|\tilde{Q} \cap U_J| \sim \lambda^2$. Let $\mathcal{I} \subset \{1, \ldots, 4\}$ such that for each $\iota \in \mathcal{I}$ we can select some $Q_\iota^k(q_\iota)$ such that $\tilde{Q} \subset Q_\iota^{\frac{5}{8}k}(q_\iota)$. Note that $\#\mathcal{I} > 1$ is possible. It is not hard to see that for all X_t with $X_t \cap Q \neq \emptyset$ we get $i_t \in \mathcal{I}$. This follows from the construction of the sets $(D_i)_i$ and the fact that $\tilde{Q} \not\subset S_\iota$ implies $\tilde{Q} \cap \hat{S}_\iota = \emptyset$ as $\lambda \ll k$. Following the lines of (9.56), (9.59)-(9.61) and using $\hat{H}_4 \le CG$ we find $\bar{R}^\iota \in SO(2)$, $\bar{c}^\iota \in \mathbb{R}^2$ such that

$$\|\bar{y}_\iota - (\bar{R}^\iota\,\cdot + \bar{c}^\iota)\|^2_{L^2(\tilde{Q})} \le Ck^2 G, \qquad \|\nabla \bar{y}_\iota - \bar{R}^\iota\|^p_{L^p(\tilde{Q})} \le C\hat{H}_p \tag{9.72}$$

for $\iota \in \mathcal{I}$. Note that for a special choice of $\iota \in \mathcal{I}$ (for $\iota = i$ with i as considered in (9.54)ff.) we obtain the rigid motion $\bar{R}_j\,x + \bar{c}_j$ which we defined in Lemma 9.2.5. Then arguing as in (9.64) and (9.66), in particular employing the triangle inequality and using (9.72), we derive

$$\begin{aligned} &\|(\bar{R}_j\,\cdot + \bar{c}_j) - (\bar{R}^\iota\,\cdot + \bar{c}^\iota)\|^2_{L^2(\tilde{Q})} \le Cm^{-2}k^2 G, \\ &\|\bar{R}_j - \bar{R}^\iota\|^p_{L^p(\tilde{Q})} \le Cm^{-1}\hat{H}_p \end{aligned} \tag{9.73}$$

for $\iota \in \mathcal{I}$. Likewise we obtain by (9.32)

$$\int_{J_{\bar{y}} \cap \overline{Q}} |[\bar{y}]|^2\, d\mathcal{H}^1 \le C \sum_{\iota \in \mathcal{I}} \int_{J_{\bar{y}_\iota} \cap \overline{Q}} |[\bar{y}_\iota]|^2\, d\mathcal{H}^1 \le C \sum_{\iota \in \mathcal{I}} kC_m \epsilon |\partial \hat{U}_\iota \cap Q_\iota^k(q_\iota)|_{\mathcal{H}}. \tag{9.74}$$

Here we used that all X_t with $\overline{Q} \cap X_t \neq \emptyset$ satisfy $|\partial X_t|_\infty \le \frac{k}{8}$ and thus $X_t \subset Q_\iota^k(q)$. Now we obtain

$$\begin{aligned} \|\nabla \bar{y} - \bar{R}_j\|^p_{L^p(Q)} &\le \sum\nolimits_{\iota \in \mathcal{I}} \|\nabla \bar{y}_\iota - \bar{R}_j\|^p_{L^p(Q)} \\ &\le C \sum\nolimits_{\iota \in \mathcal{I}} \left(\|\nabla \bar{y}_\iota - \bar{R}^\iota\|^p_{L^p(Q)} + \|\bar{R}^\iota - \bar{R}_j\|^p_{L^p(Q)} \right) \end{aligned}$$

for $p = 2, 4$. Choosing the constant in the definition of N sufficiently large and recalling the definition of G and $\hat{H}_p$ (see (9.57)) we obtain by (9.72) and (9.73)

$$\|\nabla \bar{y} - \bar{R}_j\|^p_{L^p(Q)} \le CC_m^2(\gamma(N) + \delta_p(N) + \epsilon|\partial W \cap N|_{\mathcal{H}}).$$

Similarly, recalling $\lambda = mk$ we derive

$$\|\bar{y} - (\bar{R}_j \, \cdot + \bar{c}_j)\|^2_{L^2(Q)} \le C\lambda^2 C_m^2 \min\{(\gamma(N) + \delta_4(N) + \epsilon|\partial W \cap N|_{\mathcal{H}}), \epsilon k\}.$$

Consequently, (9.69)(i),(ii) hold for $\bar{y}$.

For later purposes, it is convenient to have an extension satisfying $\nabla\hat{y}(x) \in SO(2)$ for a.e. $x \in U^H \setminus W$. Arguing as in (9.58) for all components X_t we find $\tilde{R}_t \in SO(2)$ such that $|\tilde{R}_t - (R_t + R_t A_t)|^2 \le C|A_t|^4$. Therefore, by Poincaré's inequality we find for some possibly different $\tilde{c}_t \in \mathbb{R}^2$

$$\|\tilde{R}_t \, \cdot + \tilde{c}_t - (R_t \, (\mathbf{Id} + A_t) \, \cdot + R_t \, c_t)\|^2_{L^2(X_t)} \le C|\partial X_t|^2_\infty \, |X_t||A_t|^4 \tag{9.75}$$

for all X_t and likewise passing to the trace (e.g., argue as in (7.10)ff.) we get

$$\|\tilde{R}_t \, \cdot + \tilde{c}_t - (R_t \, (\mathbf{Id} + A_t) \, \cdot + R_t \, c_t)\|^2_{L^2(\partial X_t)} \le C|\partial X_t|^2_\infty \, |\partial X_t|_{\mathcal{H}}|A_t|^4.$$

In particular, note the the constants above do not depend on the shape of X_t as the involved functions are affine. We set $\hat{y} : U^H \to \mathbb{R}^2$ by $\hat{y}(x) = \tilde{R}_t \, x + \tilde{c}_t$ for $x \in X_t$ and $\hat{y} = y$ else. First, we see that (9.70) holds since $\mathcal{H}^1(J_{\bar{y}}) = \mathcal{H}^1(J_{\hat{y}})$. The definition together with (9.74) yields

$$\begin{aligned}
\int_{J_{\hat{y}} \cap \overline{Q}} |[\hat{y}]|^2 \, d\mathcal{H}^1 &\le \int_{J_{\bar{y}} \cap \overline{Q}} |[\bar{y}]|^2 \, d\mathcal{H}^1 + C \sum\nolimits_{X_t \cap \overline{Q} \neq \emptyset} |\partial X_t|^2_\infty |\partial X_t|_{\mathcal{H}} |A_t|^4 \\
&\le \int_{J_{\bar{y}} \cap \overline{Q}} |[\bar{y}]|^2 \, d\mathcal{H}^1 + Ck \sum\nolimits_{X_t \cap \overline{Q} \neq \emptyset} |\partial X_t|^2_\infty |A_t|^4 \\
&\le CC_m k \sum\nolimits_{\iota=1}^4 \epsilon |\partial \hat{U}_\iota \cap Q^k_\iota(q_\iota)|_{\mathcal{H}}.
\end{aligned}$$

In the second step we used $|\partial X_t|_{\mathcal{H}} \le Ck$ which follows from (9.35) and (9.28). In the last step we used Lemma 8.5.7 similarly as in the derivation of (9.55) and employed $s \ge \epsilon$. Using once more that $|J_{\bar{y}} \cap \overline{Q}|_{\mathcal{H}} \le \sum_{\iota=1}^4 |\partial\hat{U}_\iota \cap Q^k_\iota(q_\iota)|_{\mathcal{H}} \le Ck$, Hölder's inequality and (9.35) yield

$$\begin{aligned}
\Big(\int_{J_{\hat{y}} \cap \overline{Q}} |[\hat{y}]| \, d\mathcal{H}^1 \Big)^2 &\le |J_{\bar{y}} \cap \overline{Q}|_{\mathcal{H}} \cdot \int_{J_{\hat{y}} \cap \overline{Q}} |[\hat{y}]|^2 \, d\mathcal{H}^1 \\
&\le CC_m k^2 \sum\nolimits_{\iota=1}^4 \epsilon |\partial \hat{U}_\iota \cap Q^k_\iota(q_\iota)|_{\mathcal{H}}. \\
&\le CC_m^2 \lambda^2 \min\big\{ \big(\gamma(N) + \delta_4(N) + \epsilon|\partial W \cap N|_{\mathcal{H}}\big), \epsilon k \big\}.
\end{aligned} \tag{9.76}$$

Recalling $|\tilde{R}_t - (R_t + R_t A_t)|^2 \le C|A_t|^4$, $|A_t| \le C$ and again using (9.55), (9.35) we obtain

$$\|\nabla\bar{y} - \nabla\hat{y}\|^p_{L^p(Q)} \le C \sum_{X_t \cap Q \neq \emptyset} |\partial X_t|^2_\infty |A_t|^4 \le CC_m(\gamma(N) + \delta_4(N) + \epsilon|\partial W \cap N|_{\mathcal{H}})$$

for $p = 2, 4$, and analogously by (9.75) we get

$$\|\bar{y} - \hat{y}\|_{L^2(Q)}^2 \le C \sum_{X_t \cap Q \neq \emptyset} |\partial X_t|_\infty^4 |A_t|^4 \le CC_m^2 \lambda^2 (\gamma(N) + \delta_4(N) + \epsilon |\partial W \cap N|_{\mathcal{H}}),$$

where we employed $|\partial X_t|_\infty \le Ck = C\lambda m^{-1}$. Likewise we derive $\|\bar{y} - \hat{y}\|_{L^2(Q)}^2 \le CC_m^2 \lambda^2 \epsilon k$. Together with the estimates for $\bar{y}$ this shows (9.69)(i),(ii). It remains to prove (9.69)(iii). By (9.69)(i) for $p = 4$, (9.23) and the fact that $\nabla \hat{y}(x) \in SO(2)$ for a.e. $x \in U^H \setminus W$ we find $\|\bar{e}_{\bar{R}_j}(\nabla \hat{y})\|_{L^2(Q)}^2 \le CC_m^2(\gamma(N) + \delta_4(N) + \epsilon |N \cap \partial W|_{\mathcal{H}})$. This together with (9.76), $|Q| \le C\lambda^2$ and Hölder's inequality yields

$$(|E(\bar{R}_j^T \hat{y} - \mathbf{id})|(Q))^2 \le CC_m^2 \lambda^2 (\gamma(N) + \delta_4(N) + \epsilon |\partial W \cap N|_{\mathcal{H}}).$$

Then Theorem B.5 and a rescaling argument show

$$\begin{aligned} \|\hat{y} - (\bar{R}_j \,\cdot + \bar{c}_j)\|_{L^1(\partial Q)}^2 &\le C\lambda^{-2} \|\hat{y} - (\bar{R}_j \,\cdot + \bar{c}_j)\|_{L^1(Q)}^2 + C(|E(\bar{R}_j^T \hat{y} - \mathbf{id})|(Q))^2 \\ &\le C\lambda^2 C_m^2 (\gamma(N) + \delta_4(N) + \epsilon |\partial W \cap N|_{\mathcal{H}}). \end{aligned}$$

In the last step we have used Hölder's inequality and (9.69)(ii). Similarly as before we also derive $\|\hat{y} - (\bar{R}_j \,\cdot + \bar{c}_j)\|_{L^1(\partial Q)}^2 \le CC_m^2 \lambda^2 \epsilon k$. □

9.3 Modification of the deformation

The goal of the section is to replace the deformation by an H^1-function on U_J. In particular, we modify the deformation in such a way that the least crack length is increased. Recall $\nu = sd = \lambda m$.

Lemma 9.3.1. *Let $k > s, \epsilon > 0$ such that $l := \frac{k}{s} = dm^{-2}$ for $m^{-1}, d \in \mathbb{N}$ with $m^{-1}, d \gg 1$. Then there is a constant $C > 0$ such that for all $W \in \mathcal{V}^{sm}_{(s,3k)}$ with $W \subset \Omega^{3k}$ and for all $y \in H^1(W)$ with $\|\nabla y\|_\infty \le C$, γ as defined in (9.14) and δ_2, δ_4 as given in (9.49) we have the following:*
There are sets $U \in \mathcal{V}^{s\hat{m}^2}_{71k}$ and $U^H \in \mathcal{V}^{\nu}_{72k}$ with $U, U^H \subset \Omega^{6k}$, $|U \setminus W| = 0$, $|U^H \setminus H^\lambda(U)| = 0$, $|(W \setminus U) \cap \Omega^{6k}| + |U \setminus U^H| \le C_u k \|U\|_*$ and

$$\|U\|_* \le (1 + C_u m)\|W\|_* + C\epsilon^{-1}(\gamma + \delta_4) \tag{9.77}$$

as well as a function $\tilde{y} \in H^1(U^H)$ such that

$$\begin{aligned} &(i) \quad \|\operatorname{dist}(\nabla \tilde{y}, SO(2))\|_{L^2(U^H)}^2 \le C \min_{p=2,4} (1 + \vartheta_p^3) C_m^2 (\gamma + \delta_p + \epsilon \|W\|_*), \\ &(ii) \quad \|\operatorname{dist}(\nabla \tilde{y}, SO(2))\|_{L^\infty(U^H)}^2 \le C\bar{\vartheta}(1 + \bar{\vartheta}), \\ &(iii) \quad \|\nabla y - \nabla \tilde{y}\|_{L^2(U)}^2 \le CC_m^2 (\gamma + \delta_2 + \epsilon \|W\|_*), \\ &(iv) \quad \|\tilde{y} - y\|_{L^2(U)}^2 \le CC_m^2 (1 + \vartheta) \lambda^2 (\gamma + \delta_4 + \epsilon \|W\|_*), \end{aligned} \tag{9.78}$$

where $\bar{\vartheta} = \min\{\vartheta(1+\vartheta), C_m^3\}$ *and* $\vartheta_2 = 1$, $\vartheta_4 = \vartheta$. *Under the additional assumption that* $\Delta y = 0$ *in* W° *we get*

$$\|\nabla y - \nabla \tilde{y}\|^4_{L^4(U)} \le CC_m^2 \delta_4 + CC_m^2 \vartheta(1+\vartheta)^2(\gamma + \delta_4 + \epsilon\|W\|_*). \tag{9.79}$$

Proof. Apply Lemma 9.2.5 to obtain sets $V \in \mathcal{V}^{s\hat{m}^2}_{71k}$, $U_J \in \mathcal{V}^\lambda$ satisfying (9.50) and (9.51) for mappings $\bar{R}_j : U_J \to SO(2)$ and $\bar{c}_j : U_J \to \mathbb{R}^2$, $j = 1, \ldots, 4$. We first define $U = V$ and see that the estimate in (9.77). Moreover, we recall that $\Omega^{6k} \setminus U$ is the union of rectangular components (see (9.68)). For the components $\Gamma_1(H^\lambda(V)), \ldots, \Gamma_n(H^\lambda(V))$ we let $N(\Gamma_j) \in \mathcal{U}^\nu$ denote the smallest rectangle with $N(\Gamma_j) \supset X_j$, where as before X_j denotes the component corresponding to $\Gamma_j(H^\lambda(V))$.

As $\frac{\nu}{\lambda} = m$, we find $|\pi_i \partial N(\Gamma_j)| \le |\pi_i \Gamma_j(H^\lambda(V))| + C_u m |\Gamma_j(H^\lambda(V))|_\infty$ for $i = 1, 2$. Arguing similarly as in the construction of (9.18) we have that $N(\Gamma_{j_1}) \setminus N(\Gamma_{j_2})$ is connected for $1 \le j_1, j_2 \le n$. We apply Lemma 9.1.2(i) to obtain pairwise disjoint, connected sets $(X_j')_{j=1}^n$ such that $\bigcup_{j=1}^n \overline{N(\Gamma_j)} = \bigcup_{j=1}^n \overline{X_j'}$ and define

$$U^H = \Omega^{6k} \setminus \bigcup\nolimits_{j=1}^n X_j'.$$

It is not hard to see that $U^H \in \mathcal{V}^\nu_{72k}$. Moreover, we find $U^H \subset H^\lambda(U)$ up to a set of negligible measure and recalling (9.68) we obtain $(U^H)^\circ \subset U_J$. For later we also observe that

$$\|U^H\|_* \le (1 + C_u m)\|H^\lambda(U)\|_*. \tag{9.80}$$

This also implies $|U \setminus U^H| \le C_u k \|U\|_*$.

Let $T_j = \bigcup_{p \in I_j^\lambda(\Omega^{3k})} Q_j^{\frac{3}{4}\lambda}(p)$ and define a partition of unity $(\eta_j)_{j=1}^4$ with $\eta_j \in C^\infty(U_J, [0,1])$, $\mathrm{supp}(\eta_j) \subset T_j$ and $\|\nabla \eta_j\|_\infty \le \frac{C}{\lambda}$. Define $\tilde{y} : U_J \to \mathbb{R}^2$ by

$$\tilde{y}(x) = \sum\nolimits_{j=1}^4 \eta_j(x)(\bar{R}_j\, x + \bar{c}_j)$$

and observe that $\tilde{y} \in H^1(U_J)$ as the functions $\bar{R}_j, \bar{c}_j$ are constant on each $Q_j^\lambda(p)$, $p \in I_j^\lambda(U_J)$. The derivative reads as

$$\nabla \tilde{y}(x) = \sum\nolimits_{j=1}^4 \big(\eta_j(x)\bar{R}_j + (\bar{R}_j\, x + \bar{c}_j) \otimes \nabla \eta_j(x)\big). \tag{9.81}$$

Since $\sum_{j=1}^4 \nabla \eta_j = 0$ we find

$$\nabla \tilde{y}(x) = \bar{R}_1 + \sum\nolimits_{j=2}^4 \big(\eta_j(x)(\bar{R}_j - \bar{R}_1) + (\bar{R}_j\, x + \bar{c}_j - (\bar{R}_1\, x + \bar{c}_1)) \otimes \nabla \eta_j(x)\big).$$

First, we compute by (9.52)

$$\|\nabla\tilde{y} - \bar{R}_1\|^4_{L^4(U_J)} \le C\sum_{j=2}^4 \Big(\|\bar{R}_j - \bar{R}_1\|^4_{L^4(U_J)} + \frac{1}{\lambda^4}\|\bar{R}_j \ \cdot + \bar{c}_j - (\bar{R}_1 \ \cdot + \bar{c}_1)\|^4_{L^4(U_J)}\Big)$$

$$\le C\sum_{j=2}^4 \Big(\|\bar{R}_j - \bar{R}_1\|^4_{L^4(U_J)} + \frac{\bar{\vartheta}}{\lambda^2}\|\bar{R}_j \ \cdot + \bar{c}_j - (\bar{R}_1 \ \cdot + \bar{c}_1)\|^2_{L^2(U_J)}\Big),$$

where $\bar{\vartheta} = \min\{\vartheta(1+\vartheta), C_m^3\}$. By (9.23) we find $\bar{e}_{\bar{R}_1}(\bar{R}_j) \le C|\bar{R}_j - \bar{R}_1|^2$ and thus

$$\|\bar{e}_{\bar{R}_1}(\nabla\tilde{y})\|^2_{L^2(U_J)} \le C\sum_{j=2}^4 \Big(\|\bar{e}_{\bar{R}_1}(\bar{R}_j)\|^2_{L^2(U_J)} + \frac{1}{\lambda^2}\|\bar{R}_j \ \cdot + \bar{c}_j - (\bar{R}_1 \ \cdot + \bar{c}_1)\|^2_{L^2(U_J)}\Big)$$

$$\le C\sum_{j=2}^4 \Big(\|\bar{R}_j - \bar{R}_1\|^4_{L^4(U_J)} + \frac{1}{\lambda^2}\|\bar{R}_j \ \cdot + \bar{c}_j - (\bar{R}_1 \ \cdot + \bar{c}_1)\|^2_{L^2(U_J)}\Big).$$

Again using (9.23) and (9.51)(iii),(iv) we derive

$$\|\operatorname{dist}(\nabla\tilde{y}, SO(2))\|^2_{L^2(U_J)} \le C(1+\vartheta^3)C_m^2(\gamma + \delta_4 + \epsilon\|W\|_*).$$

Similarly, we get

$$\|\nabla\tilde{y} - \bar{R}_1\|^2_{L^2(U_J)} \le C\sum_{j=2}^4 \Big(\|\bar{R}_j - \bar{R}_1\|^2_{L^2(U_J)} + \frac{1}{\lambda^2}\|\bar{R}_j \ \cdot + \bar{c}_j - (\bar{R}_1 \ \cdot + \bar{c}_1)\|^2_{L^2(U_J)}\Big)$$

and thus we find by (9.51)(iii),(iv)

$$\|\operatorname{dist}(\nabla\tilde{y}, SO(2))\|^2_{L^2(U_J)} \le CC_m^2(\gamma + \delta_2 + \epsilon\|W\|_*),$$

where we used that $\delta_4 \le C\delta_2$. This gives (9.78)(i) as $(U^H)^\circ \subset U_J$. Likewise, we may replace the L^2, L^4-norms in the above estimates by the L^∞-norm. Consequently, by (9.52) we obtain $\|\nabla\tilde{y} - \bar{R}_1\|^4_{L^\infty(U_J)} \le C\bar{\vartheta}(1+\bar{\vartheta})$ and $\|\bar{e}_{\bar{R}_1}(\nabla\tilde{y})\|^2_{L^\infty(U_J)} \le C\bar{\vartheta}$ which then implies $\|\operatorname{dist}(\nabla\tilde{y}, SO(2))\|^2_{L^\infty(U_J)} \le C\bar{\vartheta}(1+\bar{\vartheta})$. It remains to show (9.78)(iii),(iv) and (9.79). By (9.51)(i) and the fact that $U = V$ we obtain

$$\|\tilde{y} - y\|^2_{L^2(U)} \le \sum\nolimits_{j=1}^4 C\|y - (\bar{R}_j \ \cdot + \bar{c}_j)\|^2_{L^2(U)} \le CC_m^2\lambda^2(1+\vartheta)(\gamma + \delta_4 + \epsilon\|W\|_*).$$

By (9.81) and the fact that $\sum_{j=1}^4 \eta_j = 1$, $\sum_{j=1}^4 \nabla\eta_j = 0$ we derive

$$\nabla y(x) - \nabla\tilde{y}(x) = \sum\nolimits_{j=1}^4 \big(\eta_j(x)(\nabla y(x) - \bar{R}_j) + (y(x) - (\bar{R}_j\, x + \bar{c}_j)) \otimes \nabla\eta_j(x)\big).$$

Therefore, by (9.51)(i)(ii) for $p = 2$ we get

$$\|\nabla\tilde{y} - \nabla y\|^2_{L^2(U)} \le C\sum\nolimits_{j=1}^4 \Big(\|\nabla y - \bar{R}_j\|^2_{L^2(U)} + \frac{1}{\lambda^2}\|y - (\bar{R}_j \ \cdot + \bar{c}_j)\|^2_{L^2(U)}\Big)$$
$$\le CC_m^2(\gamma + \delta_2 + \epsilon\|W\|_*),$$

where we used that $\delta_4 \le C\delta_2$. Finally, in the case that $\Delta y = 0$ in W° we obtain by (9.51)(i)(ii) for $p = 4$ and (9.53)

$$\begin{aligned}\|\nabla \tilde{y} - \nabla y\|^4_{L^4(U)} &\le C\sum\nolimits_{j=1}^4 \Big(\|\nabla y - \bar{R}_j\|^4_{L^4(U)} + \frac{\vartheta(1+\vartheta)}{\lambda^2}\|y - (\bar{R}_j\,\cdot + \bar{c}_j)\|^2_{L^2(U)}\Big)\\ &\le CC_m^2\delta_4 + CC_m^2\vartheta(1+\vartheta)^2(\gamma + \delta_4 + \epsilon\|W\|_*).\end{aligned}$$

□

9.4 SBD-rigidity up to small sets

In this section we prove a slightly weaker version of the rigidity estimate given in Theorem 6.1.1 and postpone the proof of the general version to the next section. Recall definition (6.1).

Theorem 9.4.1. *Let $\Omega \subset \mathbb{R}^2$ open, bounded with Lipschitz boundary. Let $M > 0$ and $0 < \eta, \rho, h_* \ll 1$. Let $q \in \mathbb{N}$ sufficiently large. Then there are constants $C_1 = C_1(\Omega, M, \eta)$, $C_2 = C_2(\Omega, M, \eta, \rho, h_*, q)$ and a universal constant $c > 0$ such that the following holds for $\varepsilon > 0$ small enough:*
For each $y \in SBV_M(\Omega)$ with $\mathcal{H}^1(J_y) \le M$ and $\int_\Omega \mathrm{dist}^2(\nabla y, SO(2)) \le M\varepsilon$, there is a set $\Omega_y \in \mathcal{V}^{\hat{s}}_{c\rho^{q-1}}$, $\hat{s} > 0$, with $\Omega_y \subset \Omega$, $|\Omega \setminus \Omega_y| \le C_1\rho$, a modification $\tilde{y} \in H^1(\Omega_y) \cap SBV_{cM}(\Omega_y)$ with $\|y - \tilde{y}\|^2_{L^2(\Omega_y)} + \|\nabla y - \nabla\tilde{y}\|^2_{L^2(\Omega_y)} \le C_1\varepsilon\rho$, a partition $(P_i)_i$ of Ω_y and for each P_i a corresponding rigid motion $R_i\,x + c_i$, $R_i \in SO(2)$ and $c_i \in \mathbb{R}^2$, such that the function $u : \Omega \to \mathbb{R}^2$ defined by

$$u(x) := \begin{cases} \tilde{y}(x) - (R_i\,x + c_i) & \text{for } x \in P_i \\ 0 & \text{else} \end{cases} \tag{9.82}$$

satisfies

$$\begin{aligned} &(i)\ \|\Omega_y\|_* \le (1 + C_1h_*)\mathcal{H}^1(J_y) + C_1\rho, \quad &&(ii)\ \|u\|^2_{L^2(\Omega_y)} \le C_2\varepsilon,\\ &(iii)\ \sum\nolimits_i \|e(R_i^T\nabla u)\|^2_{L^2(P_i)} \le C_2\varepsilon, &&(iv)\ \|\nabla u\|^2_{L^2(\Omega_y)} \le C_2\varepsilon^{1-\eta}. \end{aligned} \tag{9.83}$$

We divide the proof into three steps. We begin with a version where the least crack length is almost of macroscopic size. Afterwards, we assume that the jump set consists only of a finite number of cracks of arbitrary size. Finally, we treat the general case applying a suitable approximation argument.

In what follows, constants indicated by C_1 only depend on M, η, Ω. Generic constants C may additionally depend on h_*. All constants do not depend on ρ and q unless stated otherwise. As we will eventually let $h_* \sim \rho$ in Section 9.5, it is essential that the constant in (9.83)(i) does not depend on h_*.

9.4.1 Step 1: Deformations with least crack length

We first treat the case that the least crack length is almost of macroscopic size.

Theorem 9.4.2. *Theorem 9.4.1 holds under the additional assumption that there is an $\tilde{\Omega}_y \subset \Omega^s$, $\tilde{\Omega}_y \in \mathcal{V}^s_{\rho^{q-1}}$ for some $s \ge \rho^{q-1}\varepsilon^{\frac{\eta}{8}}$ such that $y \in H^1(\tilde{\Omega}_y)$, $\|\tilde{\Omega}_y\|_* \le (1+C_1 h_*)\mathcal{H}^1(J_y) + C_1\rho$ and $|\Omega \setminus \tilde{\Omega}_y| \le C_1\rho$ for a constant $C_1 = C_1(\Omega, M, \eta)$.*

Proof. Let $y \in H^1(\tilde{\Omega}_y)$ be given. Let ρ and define $\varrho = \rho^q$ for some $q \in \mathbb{N}$, $q \ge 2$ large enough to be specified in the proof of Theorem 6.1.1 (see Section 9.5). Assume without restriction $\rho^{-1} \in \mathbb{N}$ large. We apply Theorem B.2 and consider the harmonic part w of y satisfying

$$\begin{aligned} \|\nabla y - \nabla w\|^2_{L^2(\tilde{\Omega}_y)} &\le C\|\operatorname{dist}(\nabla y, SO(2))\|^2_{L^2(\tilde{\Omega}_y)} \le C\varepsilon, \\ \|\nabla y - \nabla w\|^4_{L^4(\tilde{\Omega}_y)} &\le C\|\operatorname{dist}(\nabla y, SO(2))\|^4_{L^4(\tilde{\Omega}_y)} \le C\varepsilon. \end{aligned} \tag{9.84}$$

In the last inequality we used $\|\nabla y\|_\infty \le M$. Let $k = \varrho\rho^{-1} = \rho^{q-1}$. Apply Lemma 9.2.1 on $\tilde{\Omega}_y \cap \Omega^k$ for the function w and $\epsilon = \hat{c}\rho^{-1}\varepsilon$, $m = \rho$, where $\hat{c} > 0$ is sufficiently large. (Possibly passing to a smaller s we can assume that $k\varepsilon^{\frac{\eta}{8}} \le s \ll k = \rho^{q-1}$.) We find a set $W \subset \Omega^{3k}$, $W \in \mathcal{V}^{sm}_{(s,3k)}$ such that

$$\|W\|_* \le (1 + C_1\rho)\|\tilde{\Omega}_y\|_* + C\epsilon^{-1}\varepsilon \le (1 + C_1\rho)\|\tilde{\Omega}_y\|_* + \rho \tag{9.85}$$

by (9.15) and $|(\tilde{\Omega}_y \setminus W) \cap \Omega^{3k}| \le C_1 k \le C_1\rho$. (Here and in the following we choose the constant $\hat{c} = \hat{c}(h_*)$ always larger then the constant C.) Moreover, there are mappings $\hat{R}_i : W^\circ \to SO(2)$, $i = 1, \ldots, 4$, which are constant on the connected components of $Q^k_i(p) \cap W^\circ$, $p \in I^k_i(\Omega)$, such that by (9.16)(i) for $i = 1, \ldots, 4$

$$\|\nabla y - \hat{R}_i\|^2_{L^2(W)} \le C\varepsilon + C\|\nabla w - \hat{R}_i\|^2_{L^2(W)} \le Cl^4\varepsilon \le C\varepsilon^{1-\eta}, \tag{9.86}$$

where $l = ks^{-1} \le C\varepsilon^{-\frac{\eta}{8}}$. Moreover, as $\vartheta = l^9 C^2_m s^{-1}\varepsilon \le C(\rho)s^{-10}\varepsilon \le C(\rho)\varepsilon^{1-\frac{5}{4}\eta} \le 1$ for η, ε small enough (recall (9.13)) we also get

$$\|\nabla y - \hat{R}_i\|^4_{L^4(W)} \le C\varepsilon + C\|\nabla w - \hat{R}_i\|^4_{L^4(W)} \le C\varepsilon \tag{9.87}$$

by (9.16)(ii). Now we apply Corollary 9.2.6 on $W \subset \Omega^{3k}$ for $k = \rho^{q-1}$, $\lambda = 3\varrho$, $m = 3\rho$ and $\epsilon = \hat{c}\rho^{-1}\varepsilon$. We obtain a set $\Omega_y \in \mathcal{V}^{s\hat{m}}_{9k}$ with $\Omega_y \subset \Omega^{5k}$, $|\Omega_y \setminus \tilde{\Omega}_y| = 0$ such that by (9.25), (9.85) and (9.87) we find

$$\|\Omega_y\|_* \le (1 + C_1\rho)\|W\|_* + C\epsilon^{-1}\varepsilon \le (1 + C_1 h_*)\mathcal{H}^1(J_y) + C_1\rho \tag{9.88}$$

and $|(\tilde{\Omega}_y \setminus \Omega_y) \cap \Omega^{5k}| \le C_1 k$. This together with the assumption $|\Omega \setminus \tilde{\Omega}_y| \le C_1\rho$ and the fact that $|\Omega \setminus \Omega^{5k}| \le C(\Omega)k$ yields $|\Omega \setminus \Omega_y| \le C_1\rho$. Moreover, there is a set $\Omega^H_y \in \mathcal{V}^\lambda$ with $H^\lambda(\Omega_y) \subset \Omega^H_y$ and mappings $\bar{R}_j : \Omega^H_y \to SO(2)$, $\bar{c}_j : \Omega^H_y \to \mathbb{R}^2$

being constant on $Q_j^{3\varrho}(p)$, $p \in I_j^{3\varrho}(\Omega^{3k})$, and an extension $\hat{y} \in SBV_M(\Omega_y^H, \mathbb{R}^2)$ such that by (9.69)(ii) we derive

$$\|\hat{y} - (\bar{R}_j \ \cdot + \bar{c}_j)\|^2_{L^2(\Omega_y^H)} \le C\varrho^2\rho^{-2}C_\rho^4(\varepsilon + \epsilon\|W\|_*) \le C\rho^{2q-3}C_\rho^4\varepsilon \tag{9.89}$$

where $C_\rho = C_{\frac{m}{3}}$ is the constant defined in (9.13). Here we used that each $x \in W$ is contained in at most $\sim \rho^{-2}$ different neighborhoods $N(Q)$ considered in Corollary 9.2.6. Moreover, the constant $\hat{c}$ was absorbed in C. Similarly, recalling $\vartheta \le 1$ we get by (9.51)(iii),(iv), (9.69)(i) and (9.86), (9.87)

$$\begin{aligned}
&\|\nabla\hat{y} - \bar{R}_j\|^2_{L^2(\Omega_y^H)} + \|\bar{R}_{j_1} - \bar{R}_{j_2}\|^2_{L^2(\Omega_y^H)} \le C\rho^{-3}C_\rho^2\varepsilon^{1-\eta},\\
&\|\nabla\hat{y} - \bar{R}_j\|^4_{L^4(\Omega_y^H)} + \|\bar{R}_{j_1} - \bar{R}_{j_2}\|^4_{L^4(\Omega_y^H)} \le C\rho^{-3}C_\rho^2\varepsilon,\\
&\|(\bar{R}_{j_1} \ \cdot + \bar{c}_{j_1}) - (\bar{R}_{j_2} \ \cdot + \bar{c}_{j_2})\|^2_{L^2(\Omega_y^H)} \le C\rho^{2q-3}C_\rho^2\varepsilon,
\end{aligned} \tag{9.90}$$

for $j = 1, \dots, 4$ and $1 \le j_1, j_2 \le 4$.

Denote the connected components of $(\Omega_y^H)^\circ \in \mathcal{U}^{3\varrho}$ by $(P_i^H)_i$ and define $P_i = P_i^H \cap \Omega_y$. Let $J_i \subset I^\varrho(\Omega)$ be the index set such that $Q^\varrho(p) \subset P_i^H$ for all $p \in J_i$. We now estimate the variation of the rigid motions defined on these squares. Let $Q_1 = Q^\varrho(p_1)$, $Q_2 = Q^\varrho(p_2)$ for $p_1, p_2 \in J_i$ such that $\overline{Q_1} \cap \overline{Q_2} \neq \emptyset$. Let $R_t = \bar{R}_4|_{Q_t}$ and $c_t = \bar{c}_4|_{Q_t}$ for $t = 1, 2$. Then we find some $j = 1, \dots, 4$ such that $\bar{R}_j$ is constant on $Q_1 \cup Q_2$ and thus $\varrho^2|R_1 - R_2|^p \le C\sum_{t=1,2}\|\bar{R}_j - R_t\|^p_{L^p(Q_1\cup Q_2)}$ for $p = 2, 4$. Using the arguments in (7.10), (7.12), and Remark 7.1.4(iii) we get

$$\begin{aligned}
\varrho^4|R_1 - R_2|^2 + &\|(R_1 - R_2) \ \cdot + c_1 - c_2\|^2_{L^2(Q_1\cup Q_2)}\\
&\le C\sum\nolimits_{t=1,2}\|(\bar{R}_j \ \cdot + \bar{c}_j) - (R_t \ \cdot + c_t)\|^2_{L^2(Q_1\cup Q_2)}.
\end{aligned} \tag{9.91}$$

Consequently, considering chains as in (7.6) and (7.18), respectively, following the arguments in the proof of Lemma 7.1.3(i),(ii) and recalling Remark 7.1.4(iv), we obtain $R_i \in SO(2)$, $c_i \in \mathbb{R}^2$ such that

$$\begin{aligned}
\|\hat{y} - (R_i \ \cdot + c_i)\|^2_{L^2(P_i^H)} &\le C\|\hat{y} - (\bar{R}_4 \ \cdot + \bar{c}_4)\|^2_{L^2(P_i^H)}\\
&\quad + C\varrho^{-8}\sum\nolimits_{1\le j_1,j_2\le 4}\|(\bar{R}_{j_1} \ \cdot + \bar{c}_{j_1}) - (\bar{R}_{j_2} \ \cdot + \bar{c}_{j_2})\|^2_{L^2(P_i^H)},\\
\|\nabla\hat{y} - R_i\|^p_{L^p(P_i^H)} &\le C\|\nabla\hat{y} - \bar{R}_4\|^p_{L^p(P_i^H)}\\
&\quad + C\varrho^{-2p}\sum\nolimits_{1\le j_1,j_2\le 4}\|\bar{R}_{j_1} - \bar{R}_{j_2}\|^p_{L^p(P_i^H)}, \quad p = 2, 4.
\end{aligned}$$

In the first estimate we used Hölder's inequality (cf. (7.19)). Summing over all connected components, (9.89) and (9.90) implies

$$\begin{aligned}
&\sum\nolimits_i\|\hat{y} - (R_i \ \cdot + c_i)\|^2_{L^2(P_i^H)} \le C(\rho, q)\varepsilon,\\
&\sum\nolimits_i\|\nabla\hat{y} - R_i\|^4_{L^4(P_i^H)} \le C(\rho, q)\varepsilon, \ \sum\nolimits_j\|\nabla\hat{y} - R_i\|^2_{L^2(P_i^H)} \le C(\rho, q)\varepsilon^{1-\eta}
\end{aligned} \tag{9.92}$$

for $C(\rho, q)$ large enough. Defining u as in (9.82) (for $\tilde{y} = y$) and taking also (9.88) into account, we immediately get (9.83)(i)(ii),(iv). Finally, (9.83)(iii) is a consequence of the linearization formula (9.24) and (9.92). □

9.4.2 Step 2: Deformations with a finite number of cracks

We now prove a version where the crack set consists of a finite number of components. We first assume that each crack is at least of atomistic size. The strategy will be to establish an estimate of the form (9.86) and (9.87) by iterative modification of y according to Lemma 9.3.1.

First, we introduce some notation and derive preliminary estimates. Let $\rho > 0$, set $\varrho = \rho^q$ and assume without restriction $\rho^{-1} \in \mathbb{N}$ large. As before we assume $\|\operatorname{dist}(\nabla y, SO(2))\|^2_{L^2(\Omega)} \le C\varepsilon$. Choose $t^{-1} \in \mathbb{N}$ such that $t \le \rho$ and set $t_j = t^{j+1}$. By Remark 8.4.3, 8.5.8 we can assume that $T := t^{z+18} \le C_t^{-2} t^{18}$ for $z \in \mathbb{N}$ sufficiently large (recall (9.13) for the definition of C_t). Moreover, set $T_j = T^{j+1}$. Let $\tilde{\Omega}_y \subset \Omega^s$ for some $s > 0$ be given. Let

$$B_j = \left(\|\tilde{\Omega}_y\|_* + C_*\rho\right) \cdot \textstyle\sum_{i=0}^{j-1} t^i \cdot \Pi_{i=0}^{j-1}(1 + C_* t^{i+1}) \tag{9.93}$$

and $B = \lim_{j\to\infty} B_j$ for a constant $C_* = C_*(M, \eta, \Omega) \ge 1$ to be specified below. Furthermore, let $P = \hat{c}^2(1 + \rho^{-1}B)$ for $\hat{c} = \hat{c}(h_*)$ sufficiently large. Set $s_0 = \kappa\varepsilon$ for κ sufficiently large, let $\epsilon_0 = \hat{c}^2\rho^{-1}\varepsilon$ and subsequently define $\epsilon_{j+1} = PT_j^{-1}\epsilon_j$. We set $r = \frac{1}{18}$, $\omega = \frac{\eta}{36}$ for notational convenience and for $j \ge 0$ we define

$$d_j = \left\lfloor \min\left\{ \left(\frac{s_j}{\epsilon_j}\right)^r, \varepsilon^{-\omega} \right\} \right\rfloor, \tag{9.94}$$

where $s_j = s_0 \Pi_{i=0}^{j-1} d_i$. In accordance with Sections 9.2, 9.3 we also define

$$l_j = d_j t_j^{-2}, \qquad \lambda_j = s_j d_j t_j^{-1}, \qquad k_j = s_j l_j. \tag{9.95}$$

As noted before, d_j describes the increase of the minimal distance of different cracks and PT_j^{-1} will be the factor of energy increase. Below we will show that indeed $d_j \gg 1$ for all $0 \le j \le J^*$, where

$$J^* = \lceil \log_{1+r}(\log_T \varepsilon^\omega)) + \tfrac{1}{\omega} \rceil.$$

One of the main reasons why the iterative application of Lemma 9.3.1 works is the fact that d_j increases much faster than PT_j^{-1}. We define the quotient $q_j := \frac{d_j}{PT_j^{-1}}$ and observe $q_0 = \frac{d_0 T_0}{P} = TP^{-1}(s_0\epsilon_0^{-1})^r$ for ε sufficiently small. Recalling (9.93) and the definition $s_0 = \kappa\varepsilon$, $\epsilon_0 = \hat{c}^2\rho^{-1}\varepsilon$ we can first choose $T = T(\rho, h_*)$ so small and then $\kappa = \kappa(T, \rho, h_*, \bar{z})$ so large that

$$q_0 T^{1/r} \ge T^{-\bar{z}} \ge T^{-1} \ge \hat{c}^4 P^2 > 1 \tag{9.96}$$

for $\bar{z} \in \mathbb{N}$ to be specified below. For the third inequality we used the fact that $P \le C$ for some $C = C(C_*, \rho, h_*, M)$ independent of T. We find

$$q_j = T^{-1/r}(q_0 T^{1/r})^{(1+r)^j} \tag{9.97}$$

for $j \le \hat{J}$, where $\hat{J} \in \mathbb{N}$ is the largest index such that $\frac{s_j}{\epsilon_j} \le \varepsilon^{-\frac{\eta}{2}}$ for all $j \le \hat{J}$. Indeed, we first note that the formula is trivial for $j = 0$. Assume (9.97) holds for $j \le \hat{J} - 1$, then we compute

$$q_{j+1} = \frac{T_{j+1}}{P}\Big(\frac{s_{j+1}}{\epsilon_{j+1}}\Big)^r = \frac{T_{j+1}}{P}\Big(\frac{s_j d_j}{PT_j^{-1}\epsilon_j}\Big)^r = \frac{q_j^r T_{j+1}}{P}\Big(\frac{s_j}{\epsilon_j}\Big)^r = \frac{q_j^r d_j T_j T}{P} = Tq_j^{1+r}$$

which gives (9.97) for $j+1$, as desired. In particular, taking (9.96) into account, (9.97) implies $q_j > 1$ and thus $d_j = q_j P T_j^{-1} \gg 1$ for all $j \le \hat{J}$. For $\hat{J} < j \le J^*$ we get $d_j = \varepsilon^{-\omega}$. In fact, using (9.96) and $\epsilon_0 \le \hat{c}^2 t^{-1}\varepsilon$ we observe for C sufficiently large

$$\begin{aligned} \epsilon_j &= \epsilon_0 \Pi_{i=0}^{j-1}(PT_i^{-1}) \le \hat{c}^{-2}\epsilon_0 \Pi_{i=0}^{j-1}(T^{-(i+1)}T^{-\frac{1}{2}}) \le \hat{c}^{-2}\epsilon_0 T^{-\frac{1}{2}(j+1)^2} \\ &\le \varepsilon T^{-C-[\log_{1+r}(\log_T \varepsilon^{\omega})]^2} = \varepsilon o\big(T^{-\log_{T^{-1}}\varepsilon^{-\omega}}\big) = \varepsilon \cdot o(\varepsilon^{-\omega}) \end{aligned} \tag{9.98}$$

for $\varepsilon \to 0$ for all $1 \le j \le J^*$. Consequently, if $\frac{s_j}{\epsilon_j} \ge \varepsilon^{-\frac{\omega}{r}} = \varepsilon^{-\frac{\eta}{2}}$, then $d_j = \varepsilon^{-\omega}$, $PT_j^{-1} = o(\varepsilon^{-\omega})$ (see (9.98)) and thus $\frac{s_{j+1}}{\epsilon_{j+1}} = \frac{d_j s_j}{PT_j^{-1}\epsilon_j} \ge \varepsilon^{-\frac{\omega}{r}}$. This then implies $d_j = \varepsilon^{-\omega}$ for all $\hat{J} < j \le J^*$.

We introduce $\vartheta_j = s_j^{-1}\epsilon_j l_j^9 C_{t_j}^2$ (recall definition (9.13) and $l_j = d_j t_j^{-2}$) and close the preparations by showing that

$$\vartheta_j \le \frac{\epsilon_0}{\hat{c}^2 \epsilon_{j+1}} T_j \quad \text{for} \quad 0 \le j \le J^*. \tag{9.99}$$

This particularly implies $\vartheta_j \le 1$ for all j as $\epsilon_j \ge \epsilon_0$ for all j. By (9.94)-(9.97) we obtain

$$s_j \ge \epsilon_j \varepsilon^{-\frac{\eta}{2}} \quad \text{or} \quad s_j = \epsilon_j d_j^{1/r} \ge \epsilon_j q_j^{1/r} \ge \epsilon_j T^{-\frac{\bar{z}}{r}(1+r)^j} \ge \epsilon_j T^{-9(j+1)^2}. \tag{9.100}$$

for all $0 \le j \le J^*$. The last step holds for $\bar{z} \in \mathbb{N}$ sufficiently large as $\lim_{j\to\infty} \frac{1}{r}(1+r)^j(9(j+1)^2)^{-1} = \infty$. Similarly as in (9.98) we see that $T^{-9(j+1)^2} = o(\varepsilon^{-\omega})$ for $j \le J^*$ as $\varepsilon \to 0$. Since $\varepsilon^{-\omega} = o(\varepsilon^{-\frac{\eta}{2}})$, we find $s_j \ge \epsilon_j T^{-9(j+1)^2}$ for all $0 \le j \le J^*$. Therefore, we derive by (9.94), (9.96), the first line of (9.98) and $r = \frac{1}{18}$

$$\vartheta_j \epsilon_{j+1} = s_j^{-1}\epsilon_j d_j^9 t_j^{-18} C_{t_j}^2 \; PT_j^{-1}\epsilon_j \le s_j^{-\frac{1}{2}}\epsilon_j^{\frac{3}{2}}\hat{c}^{-2}T_j^{-3} \le \hat{c}^{-2}\epsilon_0 T^{4(j+1)^2} T_j^{-3} \le \hat{c}^{-2}\epsilon_0 T_j$$

for all $0 \le j \le J^*$, as desired. In the second step we used $C_{t_j}^2 t_j^{-18} \le T_j^{-1}$ and $P \le T_j^{-1}$. Recall the definition of κ and k_0 above (see (9.95) and (9.96)).

Theorem 9.4.3. *Theorem 9.4.1 holds under the additional assumption that there is an $\tilde{\Omega}_y \subset \Omega^s$, $\tilde{\Omega}_y \in \mathcal{V}^s_{k_0}$ for some $s \ge \kappa\varepsilon$, such that $y \in H^1(\tilde{\Omega}_y)$, $\|\tilde{\Omega}_y\|_* \le (1 + C_1 h_*)\mathcal{H}^1(J_y) + C_1\rho$ and $|\Omega \setminus \tilde{\Omega}_y| \le C_1\rho$ for a constant $C_1 = C_1(\Omega, M, \eta)$.*

Proof. Let $y \in H^1(\tilde{\Omega}_y)$ be given. If $s \geq \varepsilon^{\frac{\eta}{8}}$ we can apply Theorem 9.4.2, so it suffices to consider $s \leq \varepsilon^{\frac{\eta}{8}}$. Recall $s_0 = \kappa\varepsilon$ for some $\kappa = \kappa(T, \rho, h_*, \bar{z}) \gg 1$ and assume $s \geq s_0$. The strategy is to apply Lemma 9.3.1 iteratively. Set $W_0 = W^H_{-1} = W^H_0 = \tilde{\Omega}_y \in \mathcal{V}^s_{k_0}$ and $y_0 = y$. Recall $\epsilon_0 = \hat{c}^2\rho^{-1}\varepsilon$ and define

$$\gamma_0 := \| \operatorname{dist}(\nabla y_0, SO(2))\|^2_{L^2(\tilde{\Omega}_y)} \leq C\frac{\rho\epsilon_0}{\hat{c}^2}, \ \alpha_0 := \| \operatorname{dist}(\nabla y_0, SO(2))\|^4_{L^4(\tilde{\Omega}_y)} \leq C\frac{\rho\epsilon_0}{\hat{c}^2}.$$

In the last inequality we used $\|\nabla y\|_\infty \leq M$. Recall (9.95). Set $\hat{s}_j = s_j\hat{t}_j^2$ for $j \geq 0$ and $\hat{s}_{-1} = s$, where $\hat{t}_j = C_2(t_j, h_*)$ (see (9.13)). Assume $W_j \in \mathcal{V}^{\hat{s}_{j-1}}_{k_j}$, $W^H_j \in \mathcal{V}^{s_j}_{k_j}$ are given with $W_j, W^H_j \subset \Omega^{6k_{j-1}}$, $|W_j \setminus W^H_{j-1}| = 0$ and $|\tilde{\Omega}_y \setminus W_j| \leq C_1 \sum_{i=0}^{j-1} k_i$, where we set $k_{-1} = s$. Recall that $|W_j \setminus W^H_j| \leq C_1 k_{j-1}$ and $|W^H_j \setminus H^{\lambda_{j-1}}(W_j)| = 0$, where $\lambda_{-1} = 0$. Set $\beta_j = \|H^{\lambda_{j-1}}(W_j)\|_*$ and $\beta^d_j = \|W_j\|_* - \|H^{\lambda_{j-1}}(W_j)\|_*$. Moreover, suppose there is a function $y_j \in H^1(W^H_j)$ with

$$\gamma_j := \| \operatorname{dist}(\nabla y_j, SO(2))\|^2_{L^2(W^H_j)}, \quad \alpha_j := \| \operatorname{dist}(\nabla y_j, SO(2))\|^4_{L^4(W^H_j)}$$

such that for $j \geq 1$

$$\begin{aligned}
&(i) \ \ \beta_j + \beta^d_j \leq (1 + C_1 t_{j-1})\beta_{j-1} + C\epsilon^{-1}_{j-1}\gamma_{j-1} \leq B_j,\\
&(ii) \ \ \gamma_j \leq CT^{-1}_{j-1} t_{j-1}(\gamma_{j-1} + \epsilon_{j-1}\beta_{j-1}) \leq \hat{c}^{-1} t_{j-1}\rho\epsilon_j,\\
&(iii) \ \ \alpha_j \leq C\vartheta_{j-1}\gamma_j \leq C\varepsilon T_{j-1},\\
&(iv) \ \ \| \operatorname{dist}(\nabla y_j, SO(2))\|^2_{L^\infty(W^H_j)} \leq C\vartheta_{j-1},\\
&(v) \ \ \|\nabla y_j - \nabla y_{j-1}\|^4_{L^4(W_j)} \leq C\varepsilon T_{j-1},\\
&(vi) \ \ \|\nabla y_j - \nabla y_{j-1}\|^2_{L^2(W_j)} \leq CT^{-1}_{j-1}(l^4_{j-1}\gamma_{j-1} + \epsilon_{j-1}\beta_{j-1}) \leq Cl^4_{j-1}\epsilon_j.
\end{aligned} \tag{9.101}$$

Setting $\vartheta_{-1} = 1$ and $t_{-1} = 1$, we note that, provided $\hat{c}$ is sufficiently large, in the case $j = 0$ (iii),(iv) are clearly satisfied for $y_0 = y$ and (i),(ii) hold neglecting the second terms. We now construct y_{j+1}, $W_{j+1} \in \mathcal{V}^{\hat{s}_j}_{k_{j+1}}$ with $W_{j+1} \subset \Omega^{6k_j}$, $|W_{j+1} \setminus W^H_j| = 0$ and $|\tilde{\Omega}_y \setminus W_{j+1}| \leq C_1 \sum_{i=0}^{j} k_i$ as well as $W^H_{j+1} \in \mathcal{V}^{s_{j+1}}_{k_{j+1}}$.

First we apply Theorem B.2 and let $w_j \in H^1(W^H_j)$ be the harmonic part of y_j such that similarly as in (9.84)

$$\|\nabla y_j - \nabla w_j\|^2_{L^2(W^H_j)} \leq C\gamma_j, \quad \|\nabla y_j - \nabla w_j\|^4_{L^4(W^H_j)} \leq C\alpha_j \tag{9.102}$$

and so in particular $\| \operatorname{dist}(\nabla w_j, SO(2))\|^2_{L^2(W^H_j)} \leq C\gamma_j$. Recall $W^H_j \in \mathcal{V}^{s_j}_{k_j}$, $W_j \subset \Omega^{6k_{j-1}}$ and note $\Omega^{k_j} \subset \Omega^{6k_{j-1}}$. Then apply Lemma 9.2.1 with $s = s_j$, $k = k_j = s_j l_j$, $m = t_j = t^{j+1}$, $\epsilon = \epsilon_j$, $U = W^H_j \cap \Omega^{k_j}$, $y = w_j$ and obtain a set $\tilde{W}^H_j \in \mathcal{V}^{s_j t_j}_{(s_j, 3k_j)}$ such that

$$\delta_4 := \sum\nolimits_{i=1}^4 \|\nabla w_j - \hat{R}_i\|^4_{L^4(\tilde{W}^H_j)} \leq C\vartheta_j\gamma_j, \ \delta_2 := \sum\nolimits_{i=1}^4 \|\nabla w_j - \hat{R}_i\|^2_{L^2(\tilde{W}^H_j)} \leq Cl^4_j\gamma_j$$

for mappings $\hat{R}_i : (\tilde{W}_j^H)^\circ \to SO(2)$, $i = 1, \dots, 4$, which are constant on the connected components of $Q_i^k(p) \cap (\tilde{W}_j^H)^\circ$, $p \in I_i^k(\Omega^k)$. We now use Lemma 9.3.1 with $m = t_j$, $s = s_j$, $\epsilon = \epsilon_j$, $d = d_j$, $W = \tilde{W}_j^H$, $y = w_j$ and show (9.101) for $j+1$. First, we obtain $W_{j+1} \in \mathcal{V}^{\hat{s}_j}_{71k_j} \subset \mathcal{V}^{\hat{s}_j}_{k_{j+1}}$, with $W_{j+1} \subset \Omega^{6k_j}$, $|W_{j+1} \setminus W_j^H| = 0$, $|(W_j^H \setminus W_{j+1}) \cap \Omega^{6k_j}| \le Ck_j \|W_{j+1}\|_*$ and $W_{j+1}^H \in \mathcal{V}^{s_{j+1}}_{72k_j} \subset \mathcal{V}^{s_{j+1}}_{k_{j+1}}$ with $|W_{j+1}^H \setminus H^{\lambda_j}(W_{j+1})| = 0$ and $|W_{j+1} \setminus W_{j+1}^H| \le C_1 k_j$. Recall $\|W_j^H\|_* \le (1 + C_1 t_j)\beta_j$ by (9.80). Thus, we have

$$\|W_{j+1}\|_* \le (1 + C_1 t_j)\|W_j^H\|_* + C\epsilon_j^{-1}(\gamma_j + \vartheta_j \gamma_j) \le (1 + C_1 t_j)\beta_j + C\epsilon_j^{-1}\gamma_j \tag{9.103}$$

by (9.15), (9.77) and the fact that $\vartheta_j \le 1$ (see (9.99)). Moreover, we get a function $y_{j+1} \in H^1(W_{j+1}^H)$ with (see (9.78), (9.79))

$$\begin{aligned}
&(i) \quad \|\operatorname{dist}(\nabla y_{j+1}, SO(2))\|^2_{L^2(W_{j+1}^H)} \le CC_{t_j}^2(\gamma_j + \epsilon_j \beta_j),\\
&(ii) \quad \|\nabla w_j - \nabla y_{j+1}\|^2_{L^2(W_{j+1})} \le CC_{t_j}^2(\gamma_j + l_j^4 \gamma_j + \epsilon_j \beta_j),\\
&(iii) \quad \|\nabla w_j - \nabla y_{j+1}\|^4_{L^4(W_{j+1})} \le CC_{t_j}^2 \vartheta_j(\gamma_j + \epsilon_j \beta_j),\\
&(iv) \quad \|\operatorname{dist}(\nabla y_{j+1}, SO(2))\|^2_{L^\infty(W_{j+1}^H)} \le C\vartheta_j,
\end{aligned} \tag{9.104}$$

where we again used that $\vartheta_j \le 1$. The first inequality in (9.101)(ii) follows directly noting that $T_j^{-1} t_j \ge C_{t_j}^2$ and for the second inequality we use (9.101)(i),(ii) for iteration step j as well as (9.93) to see

$$CT_j^{-1}(\gamma_j + \epsilon_j \beta_j) \le CT_j^{-1}\rho\epsilon_j(1 + \rho^{-1}B_j) \le \rho\hat{c}^{-1}PT_j^{-1}\epsilon_j = \hat{c}^{-1}\rho\epsilon_{j+1}, \tag{9.105}$$

where we choose $\hat{c}$ sufficiently large. Likewise, (9.101)(i) follows by (9.103), the fact that $\|W_{j+1}\|_* = \beta_{j+1} + \beta_{j+1}^d$ and

$$\begin{aligned}
\beta_{j+1} + \beta_{j+1}^d &\le (1 + C_1 t_j)B_j + \rho t_{j-1}\\
&\le \big(\|\tilde{\Omega}_y\|_* + C_*\rho\big) \cdot \textstyle\sum_{i=0}^{j-1} t^i \cdot \pi_{t=0}^j (1 + C_* t^{i+1}) + \rho t^j\\
&\le \big(\|\tilde{\Omega}_y\|_* + C_*\rho\big) \cdot \textstyle\sum_{i=0}^{j} t^i \cdot \pi_{t=0}^j (1 + C_* t^{i+1}) = B_{j+1}.
\end{aligned}$$

Here we have again chosen $\hat{c}$ and C_* large enough (with respect to C and C_1, respectively). This also implies $|(W_j \setminus W_{j+1}) \cap \Omega^{6k_j}| \le Ck_j$ by (9.101)(i) and thus $|(\tilde{\Omega}_y \setminus W_{j+1})| \le C\sum_{i=0}^j k_i + |\Omega \setminus \Omega^{6k_j}| \le C\sum_{i=0}^j k_i$.

Estimate (9.101)(iv) follows from (9.104)(iv). The first inequality in (9.101)(iii) is a consequence of (9.101)(iv), the second inequality is implied by the fact that $\varepsilon = \hat{c}^{-2}\rho\epsilon_0$, (9.101)(ii) and (9.99). Moreover, (9.101)(v) follows from (9.101)(iii), (9.102), (9.104)(iii) and the fact that $\vartheta_j C_{t_j}^2(\gamma_j + \epsilon_j\beta_j) \le \vartheta_j \rho\epsilon_{j+1} \le C\varepsilon T_j$ by (9.99) and (9.105). Similarly, (9.101)(vi) follows from (9.104)(ii), (9.102) and (9.105).

We now choose $j^* \in \mathbb{N}$ such that

$$\varepsilon^{3\omega} \ge s_{j^*} \ge \varepsilon^{4\omega}, \qquad \epsilon_{j^*} \le C\varepsilon^{1-\omega} T_{j^*}^2 \tag{9.106}$$

holds for ε sufficiently small. The first inequality is possible by (9.94) and we obtain $j^* \le J^* = \lceil \log_{1+r}(\log_T \varepsilon^\omega)) + \frac{1}{\omega}\rceil$. Indeed, by (9.100) and the fact that $\bar{z} \ge 1$ we get $s_j \ge \varepsilon^{-\frac{\omega}{r}}\epsilon_j = \varepsilon^{-\frac{\eta}{2}}\epsilon_j$ for $j > \lceil \log_{1+r}(\log_T \varepsilon^\omega))\rceil$ and therefore $\hat{J} \le \lceil \log_{1+r}(\log_T \varepsilon^\omega))\rceil$. The second inequality can be derived arguing as in (9.98). Similarly, proceeding as in (9.98) we have $t_{j^*}^{-2} = o(\varepsilon^{-\omega})$ for $\varepsilon \to 0$ and thus $k_{j^*} = s_{j^*}d_{j^*}t_{j^*}^{-2} = o(\varepsilon^\omega)$. This implies $\Omega^{6k_{j^*}} \supset \Omega^\varrho$ for ε small enough. We let

$$y_* = y_{j^*}, \quad W_*^H = W_{j^*}^H \cap \Omega^\varrho, \quad W_* = \bigcap\nolimits_{i=0}^{j^*} W_i \cap \Omega^\varrho.$$

It is not hard to see that $|\tilde{\Omega}_y \setminus W_*| \le C_1 \sum_{i=0}^{j^*} k_i \le C\varrho$. As $\hat{s}_j = s_j\hat{t}_j^2$ is increasing in j (note that $d_j \ge \hat{t}_j^{-2}$ for all j, see e.g. (9.100)), we find $W_* \in \mathcal{V}^{\hat{s}_0}$.

The strategy is now to establish an estimate of the form (9.86) and (9.87). Observe that $s_{j^*} \ge \varepsilon^{\frac{\eta}{8}}$, i.e. for the function $y_* \in H^1(W_*^H)$ we may proceed as in Theorem 9.4.2 (replacing s by s_{j^*}). Similarly as in (9.84), we apply Theorem B.2 and let w_* be the harmonic part of y_* with

$$\|\nabla w_* - \nabla y_*\|^2_{L^2(W_*^H)} \le C\varepsilon^{1-\frac{\eta}{2}}, \quad \|\nabla w_* - \nabla y_*\|^4_{L^4(W_*^H)} \le C\varepsilon T^{j^*}. \tag{9.107}$$

by (9.101), (9.106) and $\omega \le \frac{\eta}{2}$. Apply Lemma 9.2.1 on $W_*^H \subset \Omega^\varrho$ for the function w_* and $k = \rho^{q-1} = \varrho\rho^{-1}$, $s = \varepsilon^{4\omega}$, $\epsilon = \hat{c}\rho^{-1}\varepsilon^{1-\frac{\eta}{2}}$, $m = \rho$. (Without restriction we can assume $s^{-1} \in \mathbb{N}$.) We find a set $W^H \subset \Omega^{3k}$, $W^H \in \mathcal{V}_{3k}^{s_{j^*}m}$ such that

$$\|W^H\|_* \le (1 + C_1\rho)\|W_*^H\|_* + C\hat{c}^{-1}\rho\varepsilon^{\frac{\eta}{2}-1}\varepsilon^{1-\frac{\eta}{2}} \le \|W_*^H\|_* + C_1\rho \tag{9.108}$$

by (9.15) as well as $|W_*^H \setminus W^H| \le |(W_*^H \setminus W^H) \cap \Omega^{3k}| + C_1 k \le C_1 k \le C_1\rho$. Moreover, there are mappings $\hat{R}_i : (W^H)^\circ \to SO(2)$, $i = 1, \ldots, 4$, which are constant on the connected components of $Q_i^k(p) \cap (W^H)^\circ$, $p \in I_i^k(\Omega)$, such that by (9.16)(i) and (9.107)

$$\|\nabla y_* - \hat{R}_i\|^4_{L^4(W^H)} \le C\|\nabla w_* - \hat{R}_i\|^4_{L^4(W^H)} + C\varepsilon T^{j^*} \le C\vartheta\varepsilon^{1-\frac{\eta}{2}} + C\varepsilon \le C\varepsilon,$$

where similarly as before equation (9.87) we compute (recall (9.106) and $\omega = \frac{\eta}{36}$) $\vartheta \le C(\rho,q)s^{-10}\epsilon \le C(\rho,q)\varepsilon^{-40\omega}\varepsilon^{1-\omega} = C(\rho,q)\varepsilon^{1-\frac{41}{36}\eta} \le \varepsilon^{\frac{\eta}{2}}$ for ε, η small enough. Likewise, we derive

$$\|\nabla y_* - \hat{R}_i\|^2_{L^2(W^H)} \le C\|\nabla w_* - \hat{R}_i\|^2_{L^2(W^H)} + C\varepsilon^{1-\frac{\eta}{2}} \le C(1+l^4)\varepsilon^{1-\frac{\eta}{2}} \le C\varepsilon^{1-\eta}$$

as $l = \frac{k}{s} \le C\varepsilon^{-4\omega} \le \varepsilon^{-\frac{\eta}{8}}$.

We now will construct a set $W \in \mathcal{V}_{143k}^{\hat{s}_0}$ which is contained in $W^H \cap W_* \cap \Omega^{3k} \in \mathcal{V}^{\hat{s}_0}$, where the two sets coincide up to a set of measure smaller than $C_1\rho$. (Similarly as before the difference of the sets is related to the definition of the boundary components.) Before we give the exact definition of W and establish an estimate of the form (9.85), we first observe $|\tilde{\Omega}_y \setminus W| \le C_1\rho$ arguing as before and derive estimates similar to (9.86) and (9.87).

We iteratively apply (9.101)(v) and derive for $i = 1, \dots, 4$

$$\|\nabla y - \hat{R}_i\|^4_{L^4(W)} \le C\Big(\sum_{\iota=1}^{j^*} (\varepsilon T_{\iota-1})^{\frac{1}{4}}\Big)^4 + C\|\nabla y_* - \hat{R}_i\|^4_{L^4(W)} \le C\varepsilon. \quad (9.109)$$

Likewise, observe that by (9.94), (9.95) and (9.106) we have $l^4_{j-1}\epsilon_j \le l^4_j \epsilon_j = d^4_j t^{-8(j+1)}\epsilon_j \le \varepsilon^{-4\omega}\varepsilon^{1-\omega}T_j \le \varepsilon^{1-\eta}T_j$. We derive by (9.101)(vi)

$$\|\nabla y - \hat{R}_i\|^2_{L^2(W)} \le C\varepsilon^{1-\eta}\Big(\sum_{\iota=1}^{j^*} T_\iota^{\frac{1}{2}}\Big)^2 + C\|\nabla y_* - \hat{R}_i\|^2_{L^2(W)} \le C\varepsilon^{1-\eta}$$

for $i = 1, \dots, 4$.

It remains to give the exact definition of $W \in \mathcal{V}^{\hat{s}_0}_{143k}$ and to establish $\|W\|_* \le (1 + Ch_*)\mathcal{H}^1(J_y) + C\rho$. Recall $W_0 = \tilde{\Omega}_y$ and define $W_{j^*+1} := W^H$ for notational convenience. We now define W inductively.

Let $Y_0 = Y_0' = Y_0'' = W_0$. Assume $Y_j \in \mathcal{V}^{\hat{s}_0}$ and $Y_j' \in \mathcal{V}^{\hat{s}_0}_{k_j}$, $Y_j'' \in \mathcal{V}^{\hat{s}_0}$ are given with $|Y_j' \setminus Y_j| + |Y_j' \triangle Y_j''| = 0$, $|Y_j \setminus Y_j'| \le C_1 k_{j-1}$ and

$$\max\{\|Y_j'\|_*, \|Y_j''\|_*\} \le \|Y_j\|_* \le \|W_j\|_* + \sum_{i=1}^{j-1} \beta_i^d,$$

where Y_j'' has the property that all components not intersecting $\partial H^{\lambda_{j-1}}(W_j)$ coincide with components of Y_j' and the set $\big(X_t(H^{\lambda_{j-1}}(W_j))\big)_t$ of components of $H^{\lambda_{j-1}}(W_j)$ is a subset of the components of Y_j''. Moreover, suppose that $|Y_j' \setminus \bigcap_{i=0}^j W_i| = 0$ and $|\bigcap_{i=0}^j W_i \setminus Y_j'| \le \sum_{i=0}^{j-1} k_i$.

We now pass to step $j+1$. Let $X_1(W_{j+1}), \dots, X_{n_{j+1}}(W_{j+1})$ be the components of W_{j+1} and define

$$Y_{j+1} = \big(Y_j'' \setminus \bigcup_{t=1}^{n_{j+1}} X_t(W_{j+1})\big) \cup \bigcup_{t=1}^{n_{j+1}} \partial X_t(W_{j+1}) \in \mathcal{V}^{\hat{s}_0}.$$

First observe that Y_{j+1} satisfies $|Y_{j+1} \setminus \bigcap_{i=0}^{j+1} W_i| = 0$ and $|\bigcap_{i=0}^{j+1} W_i \setminus Y_{j+1}| \le \sum_{i=0}^{j-1} k_i$. As $|W_{j+1} \setminus W_j^H| = 0$, we obtain $\bigcup_{t=1}^{n_{j+1}} \overline{X_t(W_{j+1})} \supset \bigcup_t \overline{X_t(W_j^H)}$ and then by the fact that $|W_j^H \setminus H^{\lambda_{j-1}}(W_j)| = 0$ we get $\bigcup_{t=1}^{n_{j+1}} \overline{X_t(W_{j+1})} \supset \bigcup_t \overline{X_t(H^{\lambda_{j-1}}(W_j))}$. As by hypothesis the components of $H^{\lambda_{j-1}}(W_j)$ are also components of Y_j'', we derive recalling $\beta_i^d = \|W_i\|_* - \|H^{\lambda_{j-1}}(W_j)\|_*$ and $\beta_0^d = 0$

$$\|Y_{j+1}\|_* \le \|Y_j''\|_* + \|W_{j+1}\|_* - \|H^{\lambda_{j-1}}(W_j)\|_* = \|W_{j+1}\|_* + \sum_{i=1}^{j} \beta_i^d.$$

Observe that possibly $Y_{j+1} \notin \mathcal{V}^{\hat{s}_0}_{\mathrm{con}}$. However, by Lemma 9.1.2(ii) we find a set $Y_{j+1}' \in \mathcal{V}^{\hat{s}_0}$ with $|Y_{j+1} \setminus Y_{j+1}'| \le C_1 k_j$ and $\|Y_{j+1}'\|_* \le \|Y_{j+1}\|_*$. Here we essentially used the rectangular shape of the boundary components given by (9.68) and (9.18), respectively. Then it is elementary to see that $Y_{j+1}' \in \mathcal{V}^{\hat{s}_0}_{143k_j} \subset \mathcal{V}^{\hat{s}_0}_{k_{j+1}}$ and $|\bigcap_{i=0}^{j+1} W_i \setminus Y_{j+1}'| \le \sum_{i=0}^{j} k_i$. Moreover, if $j + 1 \le j^*$, we let $Y_{j+1}'' = (Y_{j+1}' \cap H^{\lambda_j}(W_{j+1})) \cup \partial H^{\lambda_j}(W_{j+1})$ and observe that Y_{j+1}'' has the desired properties. In fact, $\|Y_{j+1}''\|_* \le \|Y_{j+1}\|_*$ follows as before. Components not intersecting

$\partial H^{\lambda_j}(W_{j+1})$ are clearly components of Y'_{j+1}. Finally, by definition components of $H^{\lambda_j}(W_{j+1})$ are also components of Y''_{j+1}.

We finally define $W = Y'_{j^*+1} \cap \Omega^{3k} \in \mathcal{V}^{\hat{s}_0}_{143k}$. By (9.93) and (9.101)(i),(ii) we have

$$\beta_i^d \le \beta_{i-1} - \beta_i + C_1 t^i \beta_{i-1} + C\epsilon_{i-1}^{-1}\gamma_{i-1} \le \beta_{i-1} - \beta_i + C_1 t^i B + \rho t^{i-1}$$

for $i = 1, \ldots, j^*$. Recalling $\beta_0 = \|\tilde{\Omega}_y\|_*$, $\|W_*^H\|_* \le (1 + C_1 t_{j^*})\beta_{j^*}$ and using (9.93), (9.108) as well as $t \le \rho$ we conclude

$$\begin{aligned}\|W\|_* &\le \|Y'_{j^*+1}\|_* \le \|W^H\|_* + \sum\nolimits_{i=1}^{j^*} (\beta_{i-1} - \beta_i + C_1 t^i B + \rho t^{i-1}) \\ &\le \|W^H\|_* - \beta_{j^*} + \beta_0 + C_1\rho B + C_1\rho \le C_1\rho + \|\tilde{\Omega}_y\|_* + C_1\rho B + C_1\rho \\ &\le (1 + C_1\rho)\|\tilde{\Omega}_y\|_* + C_1\rho \le (1 + C_1 h_*)\mathcal{H}^1(J_y) + C_1\rho,\end{aligned}$$

as derided.

We now proceed as in the proof of Theorem 9.4.2 after equation (9.87) with the only difference that we take $\hat{s}_0$ instead of $s \sim \varepsilon^{\frac{\eta}{8}}$ in the application of Corollary 9.2.6. However, this does not change the analysis. This leads to a set $\Omega_y \in \mathcal{V}^{\hat{s}_0\hat{m}}_{ck}$ with $\Omega_y \subset \Omega^{5k}$ and $|\Omega \setminus \Omega_y| \le C_1\rho$ for $k = \rho^{q-1}, m = 3\rho$ for which (9.83) can be established. □

We now additionally treat the subatomistic regime by dropping the assumption $s \ge \kappa\varepsilon$.

Theorem 9.4.4. *Theorem 9.4.1 holds under the additional assumption that there is an $\tilde{\Omega}_y \subset \Omega^s$, $\tilde{\Omega}_y \in \mathcal{V}^s_\varepsilon$ for some $0 < s \ll \varepsilon$ such that $y \in H^1(\tilde{\Omega}_y)$, $\|\tilde{\Omega}_y\|_* \le (1 + C_1 h_*)\mathcal{H}^1(J_y) + C_1\rho$ and $|\Omega \setminus \tilde{\Omega}_y| \le C_1\rho$ for a constant $C_1 = C_1(\Omega, M, \eta)$.*

Proof. Let again $\rho^{-1} \in \mathbb{N}$, $s_0 = \kappa\varepsilon$ and recall $\|\operatorname{dist}(\nabla y, SO(2))\|^2_{L^2(\Omega)} \le C\varepsilon$. As $\kappa \gg 1$ was chosen in dependence of T and $T = T(\rho, h_*)$ (see (9.96)), we can suppose $\kappa = \kappa(\rho, h_*)$. Applying Lemma 9.2.2 for s, $k = \rho^{-2}\kappa\varepsilon$, $m = \rho$ and $\epsilon = \rho^{-2}\kappa\varepsilon$, $U = \tilde{\Omega}_y \cap \Omega^k$ there is a set $W \subset \Omega^{3k}$ with $W \in \mathcal{V}^s_k$, $|\tilde{\Omega}_y \setminus W| \le C_1 k \le C_1\rho$ for ε small enough and

$$\|W\|_* \le \|\tilde{\Omega}_y\|_* + C\epsilon^{-1}\varepsilon \le \|\tilde{\Omega}_y\|_* + \rho.$$

The last inequality holds by choosing κ larger than C. Moreover, there are mappings $\hat{R}_i : \Omega^{3k} \to SO(2)$, $i = 1, \ldots, 4$, which are constant on $Q_i^k(q) \cap W$, $q \in I_i^k(\Omega^k)$, such that

$$\|\nabla y - \hat{R}_i\|^2_{L^2(W)} \le C\varepsilon + C\varepsilon\rho^{-2}\kappa\|\tilde{\Omega}_y\|_* \le C\rho^{-2}\kappa\varepsilon.$$

Clearly, we also get $\|\nabla y - \hat{R}_i\|^4_{L^4(W)} \le C\rho^{-2}\kappa\varepsilon$ as $\|\nabla y\|_\infty \le M$. Then we apply Lemma 9.3.1 for $k = \rho^{-2}s_0$, $\nu = s_0$, $m = \rho$ and $\epsilon = \hat{c}\rho^{-3}\kappa\varepsilon$ to get sets $U \in \mathcal{V}^{s\hat{m}^2}_{71k}$ and $U^H \in \mathcal{V}^\nu_{72k}$ with $U, U^H \subset \Omega^{6k}$, $|U \setminus W| = 0$, $|U^H \setminus H^{\frac{\nu}{m}}(U)| = 0$ and

$$\|U\|_* \le (1 + C_1\rho)\|W\|_* + C\epsilon^{-1}\rho^{-2}\kappa\varepsilon \le \|\tilde{\Omega}_y\|_* + C_1\rho$$

as well as $|W \setminus U| \le C_1 k \le C_1 \rho$ for ε small enough. Moreover, we find a function $\hat{y} \in H^1(U^H)$ such that by (9.78)

(i) $\| \operatorname{dist}(\nabla \hat{y}, SO(2))\|^2_{L^2(U^H)} \le CC_\rho^2(\rho^{-2}\kappa\varepsilon + \rho^{-3}\kappa\varepsilon\|W\|_*) \le CC_\rho^2\rho^{-3}\kappa\varepsilon,$

(ii) $\| \operatorname{dist}(\nabla \hat{y}, SO(2))\|^2_{L^\infty(U^H)} \le CC_\rho^6,$

(iii) $\|\nabla y - \nabla \hat{y}\|^2_{L^2(U')} \le CC_\rho^2\rho^{-3}\kappa\varepsilon, \quad \|\nabla y - \nabla \hat{y}\|^4_{L^4(U')} \le CC_\rho^8\rho^{-3}\kappa\varepsilon,$

where the second part of (iii) follows from (ii). Note that this also implies $\| \operatorname{dist}(\nabla \hat{y}, SO(2))\|^4_{L^4(U^H)} \le CC_\rho^8\rho^{-3}\kappa\varepsilon$. Setting $W_1 = U$, $W_1^H = U^H$, $y_1 = \hat{y}$ we can now follow the proof of Theorem 9.4.3 beginning with (9.101) with the essential difference that we have to replace ε by $CC_\rho^8\rho^{-3}\kappa\varepsilon$. We then obtain the desired result for a possibly larger constant C_2 in (9.83). □

9.4.3 Step 3: General case

We are now in a position to prove the general version of Theorem 9.4.1.

Proof of Theorem 9.4.1. Let $y \in SBV_M(\Omega)$ be given and let $\rho > 0$. It suffices to find a set $\tilde{\Omega} \in \mathcal{V}^s_\varepsilon$, $s > 0$, and a function $\tilde{y} \in H^1(\tilde{\Omega})$ with $\|\tilde{y}\|_{L^\infty(\tilde{\Omega})} + \|\nabla \tilde{y}\|_{L^\infty(\tilde{\Omega})} \le cM$ for a universal constant $c > 0$ such that

$$\begin{aligned} &|\Omega \setminus \tilde{\Omega}| \le C_1\rho, \quad \|\tilde{\Omega}\|_* \le (1 + C_1 h_*)\mathcal{H}^1(J_y) + C_1\rho, \\ &\|y - \tilde{y}\|^2_{L^2(\tilde{\Omega})} + \|\nabla y - \nabla \tilde{y}\|^2_{L^2(\tilde{\Omega})} \le C_1\varepsilon\rho. \end{aligned} \tag{9.110}$$

Then the result follows from Theorem 9.4.4 applied on the function $\tilde{y}$. (Accordingly, replace M by cM in all estimates.) Note that we cannot just apply density results for SBV functions (see Theorem A.1.6) since in general such approximations do not preserve an L^∞ bound for the derivative. The problem may be bypassed by construction of a different approximation at the cost of a non exact approximation of the jump set which, however, suffices for our purposes.

Let $\mu = \varepsilon\rho$. Recall that J_y is rectifiable (see [6, Section 2.9]), i.e. there is a countable union of C^1 curves $(\Gamma_i)_{i\in\mathbb{N}}$ such that $\mathcal{H}^1(J_y \setminus \bigcup_i \Gamma_i) = 0$. Covering J_y with small balls and applying Besicovitch's covering theorem (see [40, Corollary 1, p. 35]) we find finitely many closed, pairwise disjoint balls $\overline{B_j} = \overline{B_{r_j}(x_j)}$, $j = 1, \ldots, n$ with $r_j \le \mu$ such that $\mathcal{H}^1(J_y \setminus \bigcup_{j=1}^n B_j) \le \mu$. Moreover, we get $\mathcal{H}^1(J_y \cap \overline{B_j}) \ge 2(1-\mu)r_j$ and for each B_j we find a C^1 curve Γ_{i_j} such that $\Gamma_{i_j} \cap \overline{B_j}$ is connected and $\mathcal{H}^1((\Gamma_{i_j} \triangle J_y) \cap \overline{B_j}) \le 2\mu r_j \le \frac{\mu}{1-\mu}\mathcal{H}^1(J_y \cap \overline{B_j})$. For a detailed proof we refer to [26, Theorem 2].

We choose rectangles R_j with $|\partial R_j|_\infty \le 2\sqrt{2}r_j$ such that $\mathcal{H}^1(\Gamma_{i_j} \cap (B_j \setminus R_j)) = 0$ and $|\partial R_j|_\infty \le \mathcal{H}^1(\Gamma_{i_j} \cap \overline{B_j})$. We then obtain

$$\begin{aligned} \sum\nolimits_j |\partial R_j|_\infty &\le \sum\nolimits_j \mathcal{H}^1(\Gamma_{i_j} \cap \overline{B_j}) \\ &\le \Big(1 + \frac{\mu}{1-\mu}\Big) \sum\nolimits_j \mathcal{H}^1(J_y \cap \overline{B_j}) \le (1 + C_1\mu)\mathcal{H}^1(J_y) \end{aligned}$$

and likewise $\sum_j |\partial R_j|_{\mathcal{H}} \le C_1 \mathcal{H}^1(J_y)$. Choose rectangles Q_j with $R_j \subset\subset Q_j$ such that $|\partial Q_j|_* \le (1+\mu)|\partial R_j|_*$ and

$$\mathcal{H}^1\Big(\bigcup\nolimits_j \partial Q_j \cap J_y\Big) = 0. \tag{9.111}$$

As before it is not hard to see that $R_{j_1} \setminus R_{j_2}$ is connected for $1 \le j_1, j_2 \le n$. The rectangles $(Q_j)_j$ can be chosen in a way such that they fulfill the same property. Possibly replacing the rectangles by infinitesimally larger rectangles we can assume that there is some $s > 0$ such that $R_j, Q_j \in \mathcal{U}^s$ for $j = 1, \dots, n$. Then by Lemma 9.1.2(i) we find sets $W, V \in \mathcal{V}^s_\varepsilon$ with $|V \triangle (\Omega^\rho \setminus \bigcup_j R_j)| = 0$ and $|W \triangle (\Omega^\rho \setminus \bigcup_j Q_j)| = 0$. Note that $W^\circ \subset\subset V^\circ$ and $|\Omega \setminus W| \le C_1\rho$. It is not restrictive to assume that corners of R_j, Q_j do not coincide and thus W°, V° are Lipschitz domains. We get (recall Lemma 8.1.1)

$$\|W\|_* \le (1+\mu)\sum\nolimits_j |\partial R_j|_* \le (1 + C_1\rho + C_1 h_*)\mathcal{H}^1(J_y). \tag{9.112}$$

Moreover, as $\mathcal{H}^1(J_y \setminus \bigcup_{j=1}^n B_j) \le \mu$ we get

$$\begin{aligned}
\mathcal{H}^1(J_y \setminus \bigcup\nolimits_{j=1}^n R_j) &\le \mu + \mathcal{H}^1\Big(\bigcup\nolimits_{j=1}^n J_y \cap (B_j \setminus R_j)\Big) \\
&\le \mu + \mathcal{H}^1\Big(\bigcup\nolimits_{j=1}^n \Gamma_{i_j} \cap (B_j \setminus R_j)\Big) + \mathcal{H}^1\Big(\bigcup\nolimits_{j=1}^n (\Gamma_{i_j} \triangle J_y) \cap \overline{B_j}\Big) \\
&\le \mu + \frac{\mu}{1-\mu}\mathcal{H}^1(J_y) \le C_1\mu,
\end{aligned} \tag{9.113}$$

where in the last step we have used $\mathcal{H}^1(\Gamma_{i_j} \cap (B_j \setminus R_j)) = 0$. We now show that there is a function $\hat{y} \in SBV(W^\circ)$ with $\|y - \hat{y}\|^2_{L^2(W)} + \|\nabla y - \nabla\hat{y}\|^2_{L^2(W)} \le C_1\varepsilon\rho$ such that $\|\nabla\hat{y}\|_\infty \le cM$ and $J_{\hat{y}}$ is a finite union of closed segments satisfying $\mathcal{H}^1(J_{\hat{y}}) \le C_1\mu \le C_1\rho$. We apply a result by Chambolle obtained in [26] in an SBD-setting and rather cite the result as repeating the arguments. Therefore, we first obtain a control only over the symmetric part of the gradient. To derive the desired result we repeat the arguments for the function $v = (y^2, y^1)$ instead of $y = (y^1, y^2)$ to control also the skew part.

We define

$$E(y, W^\circ) = \int_{W^\circ} V(e(\nabla y)) + \mathcal{H}^1(J_y \cap W^\circ)$$

and $E_c(y, W^\circ) = E(y, W^\circ) + c\mathcal{H}^1(J_y \cap W^\circ)$, where $V(A) := \frac{1}{2\pi}\int_{S^1} (\xi^T A \xi)^2\,d\xi$ for $A \in \mathbb{R}^{2\times 2}$. As $y \in SBV_M(W^\circ) \cap L^2(W^\circ)$ with $E(y, W^\circ) < +\infty$ and W° has Lipschitz boundary, by [26, Theorem 1] we find a sequence $y_n \in SBD(W^\circ) \cap L^2(W^\circ)$ with $\|y_n - y\|_{L^2(W^\circ)} \to 0$ such that $\overline{J_{y_n}}$ is a finite union of closed segments and

$$\begin{aligned}
\limsup_{n\to\infty} E(y_n, W^\circ) &\le E_c(y, W^\circ) \le E(y, W^\circ) + C_1\mu \\
&\le \int_{W^\circ} V(e(\nabla y)) + C_1\mu.
\end{aligned} \tag{9.114}$$

In the second and third step we used (9.113). The proof is based on a discretization argument. Consequently, as a preparation an extension y' to some set $W' \supset\supset W^\circ$ with $E(y', W') \le E(y, W^\circ) + \delta$ for arbitrary $\delta > 0$ had to be constructed (see [26, Lemma 3.2]). In our framework we can choose $y' = y$ due to $W^\circ \subset\subset V^\circ$ and (9.111). Moreover, $\|y_n\|_\infty \le \|y\|_\infty$ holds. Although not stated explicitly in the theorem, the approximations satisfy $\|\nabla y_n\|_{L^\infty(W^\circ)} \le c\|\nabla y'\|_{L^\infty(W')} \le c\|\nabla y\|_{L^\infty(V)} \le cM$. (For a precise argument see the proof of [25, Theorem 3.1], where a similar construction is used.) Strictly speaking, the theorem only states that J_{y_n} is essentially closed and contained in a finite union of closed segments. However, the proof shows that up to an infinitesimal perturbation of y_n (do not set $y_n = 0$ on a 'jump square', but $y_n = \tilde{c}$ for $\tilde{c} \approx 0$) the desired property can be achieved.

By [26, Lemma 5.1] we obtain weak convergence $e(\nabla y_n) \rightharpoonup e(\nabla y)$ in $L^2(W^\circ)$ up to a not relabeled subsequence. Together with the lower semicontinuity results $\int_{W^\circ} V(e(\nabla y)) \le \liminf_{n\to\infty} \int_{W^\circ} V(e(\nabla y_n))$ and $\mathcal{H}^1(J_y) \le \liminf_{n\to\infty} \mathcal{H}^1(J_{y_n})$ (see [26, Lemma 5.1]) we find by (9.114)

$$\int_{W^\circ} V(e(\nabla y)) \le \limsup_{n\to\infty} \int_{W^\circ} V(e(\nabla y_n)) \le \int_{W^\circ} V(e(\nabla y)) + C_1\mu.$$

Consequently, by weak convergence we obtain

$$\begin{aligned}\limsup_{n\to\infty} \|e(\nabla y_n) - e(\nabla y)\|^2_{L^2(W^\circ)} &\le c \limsup_{n\to\infty} \int_{W^\circ} V(e(\nabla y_n - \nabla y)) \\ &\le c \limsup_{n\to\infty} \Big(\int_{W^\circ} V(e(\nabla y_n)) - \int_{W^\circ} V(e(\nabla y)) \Big) \\ &\le C_1\mu = C_1\varepsilon\rho.\end{aligned}$$

Then by (9.114) we also get $\limsup_{n\to\infty} \mathcal{H}^1(J_{y_n}) \le C_1\mu \le C_1\rho$. We now repeat the argument for $v = (y^2, y^1)$ instead of y and observe that by construction the approximations can be chosen as $v_n = (y_n^2, y_n^1)$. We find that $y_n \in SBV(W^\circ)$ and $\limsup_{n\to\infty} \|\nabla y_n - \nabla y\|^2_{L^2(W^\circ)} \le C_1\varepsilon\rho$. Now choose n large enough such that $\hat{y} := y_n$ satisfies

$$\|y - \hat{y}\|^2_{L^2(W^\circ)} + \|\nabla y - \nabla \hat{y}\|^2_{L^2(W^\circ)} \le C_1\varepsilon\rho, \quad \mathcal{H}^1(J_{\hat{y}}) \le C_1\rho$$

for $C_1 > 0$ large enough. Choose a finite number of closed segments $(S_i)_i^m$ such that $\overline{J_{\hat{y}}} \cap W^\circ \subset \bigcup_i S_i$ and $\mathcal{H}^1(\bigcup_i S_i) \le C_1\rho$. For $s > 0$ small choose $T_i \in \mathcal{U}^s$ as the smallest rectangle with $S_i \subset T_i$. Then by Lemma 9.1.2(i) we obtain a set $\tilde{\Omega} \in \mathcal{V}^s_\varepsilon$ with

$$\big|\tilde{\Omega} \triangle (W \setminus \bigcup\nolimits_{j=1}^m T_m)\big| = 0.$$

Observe that for s sufficiently small we obtain $\|\tilde{\Omega}\|_* \le \|W\|_* + C_1\rho$ and $|W \setminus \tilde{\Omega}| \le C_1\rho$. This together with (9.112) gives the two first parts of (9.110). Finally, define the function $\tilde{y} \in H^1(\tilde{\Omega})$ by $\tilde{y} = \hat{y}|_{\tilde{\Omega}}$ and observe that $\tilde{y}$ satisfies (9.110). □

9.5 Proof of the main SBD-rigidity result

This Section is devoted to the proof of the main SBD-rigidity result. We start with some preparations and then split up the proof into two steps concerning a suitable construction of the jump set and the definition of an extension. As before constants indicated by C_1 only depend on M, η, Ω and all constants do not depend on ρ and q unless stated otherwise.

Let $y \in SBV_M(\Omega)$ be given and let $\rho > 0$, $\varrho = \rho^q$ for $q \in \mathbb{N}$ to be specified below. Set $k = \rho^{q-1}$ and $m = \rho$. Recall the definition $\Omega_\rho = \{x \in \Omega : \mathrm{dist}(x, \partial\Omega) > C\rho\}$. We apply Theorem 9.4.1 and obtain a set $\Omega_y \subset \Omega_\rho$ with $\Omega_y \in \mathcal{V}^s_{ck}$ for s sufficiently small and $|\Omega \setminus \Omega_y| \le C_1\rho$ such that (9.83) holds for a modification $\tilde{y} \in H^1(\Omega_y) \cap SBV_{cM}(\Omega_y)$ with $\|y - \tilde{y}\|^2_{L^2(\Omega_y)} + \|\nabla y - \nabla\tilde{y}\|^2_{L^2(\Omega_y)} \le C_1\varepsilon\rho$. Recall from the proof of Theorem 9.4.2 and Corollary 9.2.6 that there is a set $\Omega^H_y \in \mathcal{V}^{3\varrho}_{ck}$ with $\Omega^\circ_y \subset \Omega^H_y$ and an extension $\hat{y} : \Omega^H_y \to \mathbb{R}^2$ of $\tilde{y}$ satisfying (9.70) and estimates of the form (9.69).

We first construct a modification of Ω^H_y and appropriate Jordan curves which separate the connected components. For a (closed) Jordan curve γ we denote by $\mathrm{int}(\gamma)$ the interior of the curve. As connected components may be not simply connected we further introduce a generalization: We say a curve $\gamma = \gamma_0 \cup \bigcup_{j=1}^m \gamma_j$ is a *generalized Jordan curve* if it consists of pairwise disjoint Jordan curves $\gamma_0, \dots, \gamma_m$ with $\gamma_j \in \mathrm{int}(\gamma_0)$ for $j = 1, \dots, m$. We define the interior of γ by $\mathrm{int}(\gamma) = \mathrm{int}(\gamma_0) \setminus \bigcup_{j=1}^m \mathrm{int}(\gamma_j)$.

Lemma 9.5.1. *Let $\rho > 0$, $M > 0$ and $q \in \mathbb{N}$. There is a constant $C_1 = C_1(M) > 0$ such that for all $\Omega^H_y \in \mathcal{V}^{3\varrho}_{ck}$ as given above we find $\hat{\Omega} \subset \Omega_\rho$ with $\mathcal{H}^1(\partial\hat{\Omega}) \le C_1$, $|\Omega^H_y \setminus \hat{\Omega}| \le C_1\rho$ and a set $S \subset \Omega_\rho \setminus \hat{\Omega}$ such that*

(i) $\mathcal{H}^1(S) \le \|\Omega^H_y\|_* + C_1\rho$,

(ii) for all $\hat{P}_i$ there is a generalized Jordan curve γ in $S \cup \partial\Omega_\rho$ such that $\mathrm{int}(\gamma) \cap \hat{\Omega} = \hat{P}_i$, where $(\hat{P}_i)_i$ denote the connected components of $\hat{\Omega}$,

(iii) $\mathrm{int}(\gamma) \cap \hat{\Omega} \neq \emptyset$ *for all Jordan curves γ in $S \cup \partial\Omega_\rho$,*

(iv) $\mathrm{dist}(x, S) \le C_1\rho^{q-2}$ *for all $x \in \Omega_\rho \setminus \hat{\Omega}$,*

(v) $(S \cup \partial\Omega_\rho) \cap X_t(\hat{\Omega})$ *is connected for all components $X_t(\hat{\Omega})$ of $\Omega_\rho \setminus \hat{\Omega}$.*

Proof. In contrast to the previous sections, where it was essential to avoid the combination of different cracks, we now combine boundary components: Choose a set $\hat{\Omega}^H_y \in \mathcal{V}^{3\varrho}$ satisfying $\hat{\Omega}^H_y \subset \Omega^H_y$, $|\Omega^H_y \setminus \hat{\Omega}^H_y| = 0$ and $|\Gamma_j(\hat{\Omega}^H_y) \cap \Gamma_l(\hat{\Omega}^H_y)|_\mathcal{H} = 0$ for $j \neq l$. Clearly, by (9.88) and (9.48) we have $\mathcal{H}^1(\hat{\Omega}^H_y) \le \mathcal{H}^1(\Omega^H_y) \le C_1$.

Letting $Y_1, \dots, Y_m$ be the connected components of $\hat{\Omega}^H_y$ satisfying $|\partial Y_j|_\infty \le \rho^{q-2}$ for $j = 1, \dots, m$ we define $\tilde{\Omega} = \hat{\Omega}^H_y \setminus \bigcup_{j=1}^m Y_j$. As $|\partial Y_j|_\infty \le \rho^{q-2}$ for $j =$

$1, \ldots, m$, the isoperimetric inequality implies $|\bigcup_{j=1}^m Y_j| \le C_1\rho^{q-2}\|\hat{\Omega}_y^H\|_{\mathcal{H}} \le C_1\rho$ and thus $|\Omega_y^H \setminus \tilde{\Omega}| \le C_1\rho$.

Let $Z \subset \Omega_\rho \setminus \tilde{\Omega}$ be the largest set in $\mathcal{U}^{\rho^{q-2}}$ such that $\mathrm{dist}_\infty(x, \partial\tilde{\Omega} \setminus \partial\Omega_\rho) \ge \rho^{q-2}$ for all $x \in Z$ and define $\hat{\Omega} = \tilde{\Omega} \cup \overline{Z}$. (Observe that Z is typically not connected.) It is not hard to see that

$$\mathrm{dist}(x, \partial\hat{\Omega} \setminus \partial\Omega_\rho) \le C_1\rho^{q-2} \quad \text{for all} \;\; x \in \Omega_\rho \setminus \hat{\Omega}. \tag{9.115}$$

Moreover, we get $|\Omega_y^H \setminus \hat{\Omega}| \le C_1\rho$ and $\mathcal{H}^1(\hat{\Omega}) \le C_1$. In fact, for each connected component Z^i of $\overline{Z}$ we find boundary components $(X_j^i = X_j^i(\Omega_y^H))_j$ and $(Y_j^i)_j$ such that $\partial Z^i \subset \bigcup_j \overline{X_j^i} \cup \bigcup_j \overline{Y_j^i}$ and thus by $|\partial X_j^i|_\infty \le c\rho^{q-1}$, $|\partial Y_j^i|_\infty \le \rho^{q-2}$ we obtain $|\partial Z^i|_{\mathcal{H}} \le C_1(\sum_j |\partial X_j^i|_{\mathcal{H}} + \sum_j |\partial Y_j^i|_{\mathcal{H}})$. We recall $\mathcal{H}^1(\Omega_y^H) \le C_1$ and observe that for different components Z^{i_1}, Z^{i_2} one has $(\bigcup_j \overline{X_j^{i_1}} \cup \bigcup_j \overline{Y_j^{i_1}}) \cap (\bigcup_j \overline{X_j^{i_2}} \cup \bigcup_j \overline{Y_j^{i_2}}) = \emptyset$.

Let $\hat{P}_1, \ldots, \hat{P}_n$ be the connected components of $\hat{\Omega}$ and define $\mathcal{F}(\hat{P}_i) = \{X_j = X_j(\Omega_y^H) : \overline{X_j} \cap \hat{P}_i \ne \emptyset\}$. (Here it is essential that we take the components of Ω_y^H.) By $Z_j \in \mathcal{U}^{3\varrho}$ we denote the smallest rectangle containing X_j.

(I) As a preparation we consider the special case that there is only one connected component $\hat{P}_1$. Moreover, we first suppose that $\Omega_\rho \setminus \hat{\Omega}$ consists of one connected component only. We can choose a set S in $\bigcup_{Z_j \in \mathcal{F}(\hat{P}_1)} \overline{Z_j}$ consisting of segments such that $S \cup (\partial\Omega_\rho \setminus \hat{\Omega})$ is connected,

$$\mathcal{H}^1(S) \le (1 + C_1\rho) \sum\nolimits_{X_j \in \mathcal{F}(\hat{P}_1)} |\Gamma_j|_\infty \le (1 + C_1\rho) \sum\nolimits_{X_j \in \mathcal{F}(\hat{P}_1)} |\Gamma_j|_* \tag{9.116}$$

and $\mathrm{dist}(x, S) \le C_1\rho^{q-2}$ for all $x \in \partial\hat{P}_1 \setminus \partial\Omega_\rho$ for a sufficiently large constant. Indeed, a set with the desired properties can be constructed in the following way. By the definition of $|\cdot|_\infty$ we first see that we can choose a piecewise affine Jordan curve γ in $\bigcup_{X_j \in \mathcal{F}(\hat{P}_1)} \overline{Z_j} \cup \partial\Omega_\rho$ such that $\hat{P}_1 \subset \mathrm{int}(\gamma)$ and $S_0 := \gamma \cap \Omega_\rho^\circ$ satisfies $\mathcal{H}^1(S_0) \le \sum_{X_j \in \mathcal{G}(S_0)} |\Gamma_j|_\infty$, where $\mathcal{G}(S_0) = \{X_j : X_j \cap S_0 \ne \emptyset\}$. (If $\gamma \cap \Omega_\rho^\circ = \emptyset$, we let $S_0 = \{p_0\}$ for some point $p_0 \in \Omega_\rho \setminus \hat{\Omega}$.) Assume a connected set S_l consisting of segments has been constructed such that

$$\mathcal{H}^1(S_l) \le \sum\nolimits_{X_j \in \mathcal{G}(S_l)} |\Gamma_j|_\infty + C_1 l \rho^{q-1}. \tag{9.117}$$

If $\mathrm{dist}(x, S_l) \le C_1\rho^{q-2}$ for all $x \in \partial\hat{P}_1 \setminus \partial\Omega_\rho$, we stop. Otherwise, there is some $y \in \partial\hat{P}_1 \setminus \partial\Omega_\rho$ such that $\mathrm{dist}(y, S_l) > C_1\rho^{q-2}$. By the definition of $|\cdot|_\infty$ it is elementary to see that we can find a piecewise affine, continuous curve T_{l+1} with $T_{l+1} \cap S_l \ne \emptyset$, $y \in T_{l+1}$, $\#(\mathcal{G}(T_{l+1}) \cap \mathcal{G}(S_l)) = 1$ such that $\mathcal{H}^1(T_{l+1}) \le \sum_{X_j \in \mathcal{G}(T_{l+1})} |\Gamma_j|_\infty$. Then using that $|\Gamma(\Omega_y^H)|_\infty \le 2\sqrt{2} \cdot ck \le C_1\rho^{q-1}$ and $\#(\mathcal{G}(T_{l+1}) \cap \mathcal{G}(S_l)) = 1$ we find that (9.117) is satisfied for $S_{l+1} := S_l \cup T_{l+1}$.

After a finite number of iterations $n \in \mathbb{N}$ we find that $\mathrm{dist}(y, S_n) \le C_1\rho^{q-2}$ for all $y \in \partial\hat{P}_1 \setminus \partial\Omega_\rho$ and set $S_* = S_n$. Indeed, this follows from the fact that in each iteration $\mathcal{G}(S_l)$ increases and the assertion clearly holds if S_l intersects all boundary components since $\max_j |\Gamma_j(\Omega_y^H)|_\infty \le C_1\rho^{q-1}$. As $\mathcal{H}^1(T_l) > C_1\rho^{q-2}$, it is not hard to see that $n \le C_1\rho^{2-q}\sum_{X_j\in\mathcal{F}(\hat{P}_1)}|\Gamma_j|_\infty$ and thus (9.116) holds replacing S by S_*.

Observe that possibly $S_*\cup(\partial\Omega_\rho\setminus\hat{\Omega})$ is not connected. Therefore, we choose some point y in each connected component of $\partial\Omega_\rho\setminus\hat{\Omega}$ (which may be several if Ω_ρ is not simply connected) and repeat the construction below (9.117) for each y. We obtain a set S such that (9.116) still holds and $S\cup(\partial\Omega_\rho\setminus\hat{\Omega})$ is connected.

If $\Omega_\rho\setminus\hat{\Omega}$ consists of several connected components $X_t(\hat{\Omega})$, we repeat the arguments on each component separately possibly starting with $S_0=\{p_0\}$ for some $p_0\in X_t(\hat{\Omega})$.

We see that (i),(v) are satisfied, (ii) holds with γ and (iii) follows from the fact that in the construction of the sets T_l above we do not obtain additional 'loops'. Moreover, (iv) follows from the fact that each $x\in\Omega_\rho\setminus\hat{\Omega}$ satisfies $\mathrm{dist}(x,\partial\hat{P}_1\setminus\partial\Omega_\rho)\le C_1\rho^{q-2}$ by (9.115).

(II) We now consider an arbitrary number of connected components. Choose Jordan curves γ^i in $\bigcup_{X_j\in\mathcal{F}(\hat{P}_i)}\overline{Z_j}\cup\partial\Omega_\rho$ such that $\hat{P}_i\subset\mathrm{int}(\gamma^i)\cap\hat{\Omega}$ and $\mathcal{H}^1(\gamma^i\cap\Omega_\rho^\circ)\le\sum_{X_j\in\mathcal{G}(\gamma^i)}|\Gamma_j|_\infty$. We first assume that $\hat{P}_i=\mathrm{int}(\gamma^i)\cap\hat{\Omega}$, i.e. $\mathrm{int}(\gamma^i)$ does not contain other components of $\hat{\Omega}$, and treat the general case in (III). As the sets $(\mathcal{F}(\hat{P}_i))_{i=1}^n$ might be not disjoint, we have to combine the different curves in a suitable way. Define $G_i=\bigcup_{X_j\in\mathcal{G}(\gamma^i)}\overline{Z_j}$. It is not restrictive to assume that $\bigcup_{1\le i\le n}G_i$ is connected as otherwise we apply the following arguments on each component separately. For $B\subset\mathbb{R}^2$ we define

$$\mathrm{Int}(B)=\{x\in\mathbb{R}^2:\exists\text{ Jordan curve }\gamma^i\text{ in }B: x\in\mathrm{int}(\gamma^i)\}.$$

It is not hard to see that we can order the sets $(\hat{P}_i)_i$ in a way such that for all $1\le l\le n$ we have $\bigcup_{1\le i\le l}G_i$ is connected and $\mathrm{Int}(\bigcup_{1\le i\le l}G_i)\cap\hat{P}_j=\emptyset$ for all $j>l$. In fact, to see the second property, assume the first l sets $G_1,\ldots,G_l$ have already been chosen. Select some other component $\hat{P}_k$, $k>l$, with corresponding G_k. If the desired property is satisfied, we reorder and set $G_{l+1}=G_k$, otherwise we find some $\hat{P}_{k'}$, $k'>l, k'\neq k$, with corresponding $G_{k'}$ such that $\hat{P}_{k'}\subset\mathrm{Int}(\bigcup_{1\le i\le l}G_i\cup G_k)$. Possibly repeating this procedure we finally find a set G_{l+1} such that $\mathrm{Int}(\bigcup_{1\le i\le l+1}G_i)\cap\hat{P}_j=\emptyset$ for all $j>l+1$.

We now proceed iteratively. Set $S_0=\emptyset$ and assume a connected set S_l has been constructed with

$$\begin{aligned}
&(a)\;\; \mathcal{H}^1(S_l\cap\Omega_\rho)\le(1+C_1\rho)\sum\nolimits_{X_j\in\bigcup_{1\le i\le l}\mathcal{G}(\mathrm{int}(\gamma^i))}|\Gamma_j|_*+C_1(l-1)\rho^{q-1},\\
&(b)\;\; \text{for all } 1\le i\le l \text{ there is a Jordan curve } \gamma \text{ in } S_l \text{ such that } \mathrm{int}(\gamma)\cap\hat{\Omega}=\hat{P}_i,\\
&(c)\;\; \mathrm{dist}(x,S_l)\le C_1\rho^{q-2} \text{ for all } x\in\bigcup\nolimits_{i=1}^{l}\partial\hat{P}_i\setminus\partial\Omega_\rho. \qquad (9.118)
\end{aligned}$$

Let T_{l+1} be the (unique) connected component of $\gamma^{l+1}\setminus\bigcup_{1\le i\le l}G_i$ such that $\hat{P}_{l+1}\subset \mathrm{Int}(\bigcup_{1\le i\le l}G_i\cup T_{l+1})$. Now choose two segments T^j_{l+1}, $j=1,2$, with $\mathcal{H}^1(T^j_{l+1})\le C_1\rho^{q-1}$, $T^j_{l+1}\cap S_l\neq\emptyset$, $T^j_{l+1}\cap T_{l+1}\neq\emptyset$ such that $\hat{S}_{l+1}:=S_l\cup T_{l+1}\cup\bigcup_{j=1,2}T^j_{l+1}$ satisfies $\hat{P}_{l+1}\subset\mathrm{Int}(\hat{S}_{l+1})$ and

$$\mathcal{H}^1(\hat{S}_{l+1}\cap\Omega_\rho)\le(1+C_1\rho)\sum\nolimits_{X_j\in\bigcup_{1\le i\le l}\mathcal{G}(\mathrm{int}(\gamma^i))\cup\mathcal{G}(T_{l+1})}|\Gamma_j|_*+C_1l\rho^{q-1}.$$

By the order of the sets $(\hat{P}_i)_i$ it is not hard to see that there is a Jordan curve γ in $\hat{S}_{l+1}$ with $\mathrm{int}(\gamma)\cap\hat{\Omega}=\hat{P}_{l+1}$. Observe that $\mathrm{dist}(x,\gamma^{l+1})\le C_1\rho^{q-2}$ for all $x\in\partial\hat{P}_{l+1}\setminus\partial\Omega_\rho$ might not hold. Therefore, following the lines of (I) we choose a (possibly not connected) set $R_{l+1}\subset\mathrm{int}(\gamma^{l+1})$ such that such that $S_{l+1}:=\hat{S}_{l+1}\cup R_{l+1}$ is connected in each component of $\Omega_\rho\setminus\hat{\Omega}$, $\mathrm{dist}(x,S_{l+1})\le C_1\rho^{q-2}$ for all $x\in\partial\hat{P}_{l+1}\setminus\partial\Omega_\rho$ and

$$\mathcal{H}^1(R_{l+1})\le(1+C_1\rho)\sum\nolimits_{X_j\in\mathcal{G}(\mathrm{int}(\gamma^{l+1}))\setminus\mathcal{G}(\hat{S}_{l+1})}|\Gamma_j|_*.$$

Now it is not hard to see that (a)-(c) are satisfied for S_{l+1}.

After the last iteration step we define $S_*=S_n\cap\Omega_\rho$. Observe that by construction (see before (9.115)) each $\hat{P}_i$ satisfies $|\partial\hat{P}_i|_\infty\ge\rho^{q-2}$. Thus $n\le C_1\rho^{2-q}$ and then we obtain $\mathcal{H}^1(S_*)\le\|\Omega^H_y\|_*+C_1\rho$ since $n\rho^{q-1}\le C_1\rho$. Similarly as before, $S_*\cup\partial\Omega_\rho$ might not be connected in the components of $\Omega_\rho\setminus\hat{\Omega}$. Consequently, we proceed as in (I) (see construction below (9.117)) to find a set $S\supset S_*$ such that (i) still holds and $S\cup\partial\Omega_\rho$ is connected in the components of $\Omega_\rho\setminus\hat{\Omega}$. This gives (v). Moreover, (b) implies (ii) and similarly as in (I) also (iii) holds. (Here we do not have to consider generalized Jordan curves.) Finally, to see (iv) we use (c) and the fact that each $x\in\Omega_\rho\setminus\hat{\Omega}$ satisfies $\mathrm{dist}(x,\partial\hat{\Omega}\setminus\partial\Omega_\rho)\le C_1\rho^{q-2}$ by (9.115).

(III) We now finally treat the case that the components $(\hat{P}_i)_{i=0}^n$ may also contain other components of $\hat{\Omega}$. To simplify the exposition we assume that there is exactly one component, say $\hat{P}_0$, such that $\hat{P}_0\neq\mathrm{int}(\gamma^0)\cap\hat{\Omega}$. The general case follows by inductive application of the following arguments.

We proceed as in (II) (assuming we had $\hat{P}_0=\mathrm{int}(\gamma^0)\cap\hat{\Omega}$) and construct a set S' particularly satisfying (i),(iii),(v). We have to verify (ii) for $\hat{P}_0$ and find a set $S\supset S'$ such that (iv) is satisfied and (i),(iii),(v) still hold. By $(\hat{P}_{i_j})_j$ we denote the components with $\hat{P}_{i_j}\subset\mathrm{int}(\gamma_0)$. As (ii) holds for these components we find pairwise disjoint Jordan curves $\gamma_1,\ldots,\gamma_m$ with $\bigcup_j\hat{P}_{i_j}\subset\bigcup_{j=1}^m\mathrm{int}(\gamma_j)\subset\mathrm{int}(\gamma_0)$. Consequently, defining the generalized Jordan curve $\gamma=\bigcup_{j=0}^m\gamma_j$ we find $\hat{P}_0=\mathrm{int}(\gamma)\cap\hat{\Omega}$ which gives (ii).

Let $(Y_j)_j$ be the components of $\Omega_\rho\setminus\hat{\Omega}$ which are completely contained in $\mathrm{int}(\gamma_0)$. We observe that (iv) may be violated for $x\in Y^*:=\bigcup_j Y_j\setminus\bigcup_{j=1}^m\mathrm{int}(\gamma_j)$. We now proceed similarly as in (I) to obtain a set $R\subset Y^*$ such that $S:=S'\cup R$ is connected in the connected components of $\Omega_\rho\setminus\hat{\Omega}$ and $\mathrm{dist}(x,S)\le C_1\rho^{q-2}$ for all

$x \in \partial \hat{P}_1 \cap Y^*$. This implies (iii),(v) and together with (9.115) also (iv). Arguing similarly as in (II) we find that (i) is still satisfied since the sum in (9.118)(a) does not run over the components contained in Y^*. $\square$

We finally can give the proof of Theorem 6.1.1 by constructing an extension $\hat{y}$ of the function $\tilde{y}$. We briefly note that the function $\hat{y}$ has to be defined as an extension of the approximation and not of the original deformation y as only in this case we obtain the correct surface energy due to the higher regularity of the jump set of $\tilde{y}$ and the available trace estimates. Recall the definition of $E^\rho_\varepsilon(y,U)$ in (6.3), in particular $f^\rho_\varepsilon(x) = \min\{\frac{x}{\sqrt{\varepsilon}\rho}, 1\}$.

Proof of Theorem 6.1.1. Let $\Omega_y \subset \Omega_\rho$ with $\Omega_y \in \mathcal{V}^s$ and $\Omega^H_y \in \mathcal{V}^{3\varrho}$ with $\Omega^\circ_y \subset \Omega^H_y$ be given. Recall that $|\Omega \setminus \Omega_y| \le C_1\rho$. Let $\tilde{y} \in H^1(\Omega_y)$ be the approximation of $y \in SBV_M(\Omega)$ with $\|y - \tilde{y}\|^2_{L^2(\Omega_y)} + \|\nabla y - \nabla \tilde{y}\|^2_{L^2(\Omega_y)} \le C_1\varepsilon\rho$ and let $\hat{y} \in SBV_{cM}(\Omega^H_y, \mathbb{R}^2)$ be the extension of $\tilde{y}$ given by Corollary 9.2.6. Let $\hat{\Omega}$ be the set constructed in Lemma 9.5.1. We first consider the jumps of $\hat{y}$ in $(\Omega^H_y \cap \hat{\Omega})^\circ$. By (9.76) and Hölder's inequality we find

$$\begin{aligned}\Big(\int_{J_{\hat{y}} \cap (\Omega^H_y)^\circ} |[\hat{y}]| \, d\mathcal{H}^1\Big)^2 &\le \Big(\sum_{Q_t \subset \Omega^H_y} \int_{J_{\hat{y}} \cap \overline{Q_t}} |[\hat{y}]| \, d\mathcal{H}^1\Big)^2 \\ &\le \sum_{Q_t} |J_{\hat{y}} \cap \overline{Q_t}|_{\mathcal{H}} \cdot \sum_{Q_t} |J_{\hat{y}} \cap \overline{Q_t}|^{-1}_{\mathcal{H}} \Big(\int_{J_{\hat{y}} \cap \overline{Q_t}} |[\hat{y}]| \, d\mathcal{H}^1\Big)^2 \\ &\le C\mathcal{H}^1(J_{\hat{y}}) \cdot \sum_{Q_t} CC^2_\rho \varrho^2(\gamma(N_t) + \delta_4(N_t) + \epsilon|\partial W \cap N_t|_{\mathcal{H}}),\end{aligned}$$

where W as defined in (9.85), $N_t := N(Q_t) = \{x \in W : \operatorname{dist}(x, Q_t) \le C\rho^{q-1}\}$ and $\gamma(N_t) = \|\nabla \operatorname{dist}(\nabla\hat{y}, SO(2))\|^2_{L^2(W)}$, $\delta_4(N_t) = \sum_{i=1}^4 \|\nabla\hat{y} - \hat{R}_i\|^4_{L^4(W)}$ (recall (9.87)). As each $x \in \Omega$ is contained in at most $\sim \rho^{-2}$ different N_t we find by (9.84), (9.85), (9.87), (9.70) and the fact that $\epsilon = \hat{c}\rho^{-1}\varepsilon$

$$\Big(\int_{J_{\hat{y}} \cap (\Omega^H_y \cap \hat{\Omega})^\circ} |[\hat{y}]| \, d\mathcal{H}^1\Big)^2 \le C\rho^{-2} \cdot CC^2_\rho\varrho^2\epsilon \le C\varrho^2\rho^{-3}C^2_\rho\varepsilon.$$

(Note that in the general case the set W and the rigid motions $\hat{R}_i$ were defined differently (see e.g. (9.109)), but here and in the following we prefer to refer to the proof of Theorem 9.4.2 for the sake of simplicity.) By Remark 8.4.3, 8.5.8 we get for $q = q(h_*)$ sufficiently large

$$\int_{J_{\hat{y}} \cap (\Omega^H_y \cap \hat{\Omega})^\circ} |[\hat{y}]| \, d\mathcal{H}^1 \le CC_\rho\rho^{q-\frac{3}{2}}\sqrt{\varepsilon} = C\rho^{q-(\frac{3}{2}+z)}\sqrt{\varepsilon} \le \rho^2\sqrt{\varepsilon}.$$

Recalling that $f^\rho_\varepsilon(x) \le \rho^{-1}\frac{x}{\sqrt{\varepsilon}}$ for $x \ge 0$ we get

$$\int_{J_{\hat{y}} \cap (\Omega^H_y \cap \hat{\Omega})^\circ} f^\rho_\varepsilon(|[\hat{y}]|) \, d\mathcal{H}^1 \le \varepsilon^{-1/2}\rho^{-1} \int_{J_{\hat{y}} \cap (\Omega^H_y \cap \hat{\Omega})^\circ} |[\hat{y}]| \, d\mathcal{H}^1 \le \rho. \tag{9.119}$$

We now concern ourselves with the components of $\partial\hat{\Omega}$. Let Y_t be a connected component of $\Omega_\rho \setminus (\hat{\Omega} \cup S)$, where S is the set constructed in Lemma 9.5.1. Set $S_t = S \cap \overline{Y_t}$ and $\Gamma_t = \overline{Y_t} \cap \partial\hat{\Omega}$. We observe that by Lemma 9.5.1(ii),(iii) Γ_t is a Jordan curve if $\overline{Y_t} \cap \partial\Omega_\rho = \emptyset$.

Define $J = I^\varrho(\hat{\Omega})$ and for Γ_t we choose $J(\Gamma_t) \subset J$ such that $\overline{Q^\varrho(p)} \cap \Gamma_t \neq \emptyset$ for all $p \in J(\Gamma_t)$. We set $M(\Gamma_t) = \bigcup_{p \in J(\Gamma_t)} \overline{Q^\varrho(p)}$. For later purpose, for components with $|\Gamma_t|_\infty > 2\rho^{q-2}$ we introduce a finer partition of $M(\Gamma_t)$: Define $J(\Gamma_t) = I_1 \dot{\cup} \ldots \dot{\cup} I_n$ and the connected sets $B_i = \bigcup_{p \in I_i} \overline{Q^\varrho(p)}$ such that $\rho^{-2} \le \#I_i \le C\rho^{-2}$, $i = 1, \ldots, n$, for a constant $C \gg 1$. For $|\Gamma_t|_\infty \le 2\rho^{q-2}$ we let $I_1 = J(\Gamma_t)$. It is elementary to see that $n \le \max\{C|\Gamma_t|_\mathcal{H}\rho^{2-q}, 1\} \le C|\Gamma_t|_\mathcal{H}\rho^{-q}$, where we used $|\Gamma_t|_\mathcal{H} \ge C\rho^q$.

Consider $\bar{R}_j : \Omega_y^H \to SO(2)$ and $\bar{c}_j : \Omega_y^H \to \mathbb{R}^2$, $j = 1, \ldots, 4$, as given in (9.90). Recall the definition $\tilde{\Omega} = \hat{\Omega} \setminus \overline{Z} \subset \Omega_y^H$ before (9.115). We extend the function $\hat{y}$ to $\hat{\Omega}$ by setting $\hat{y} = \mathbf{id}$ on $\hat{\Omega} \setminus \tilde{\Omega}$ and likewise let $\bar{R}_j = \mathbf{Id}$, $\bar{c}_j = 0$ on $\hat{\Omega} \setminus \tilde{\Omega}$. (If $\overline{Z} \cap \Omega_y^H \neq \emptyset$, we redefine the function on this set.) Applying Corollary 9.2.6 on each $Q_j^{3\varrho}(p) \subset \hat{\Omega}$ with $Q_j^{3\varrho}(p) \cap M(\Gamma_t) \neq \emptyset$, we get

$$\begin{aligned} &\|\hat{y} - (\bar{R}_j \,\cdot + \bar{c}_j)\|^2_{L^2(B_i)} \le C\varrho^2 C_\rho^2 \cdot \rho^{-2}\rho^{q-1}\epsilon \cdot \#I_i = C\rho^{3q-6}C_\rho^2\varepsilon, \\ &\|\hat{y} - (\bar{R}_j \,\cdot + \bar{c}_j)\|^2_{L^1(\partial B_i)} \le C\rho^{3q-6}C_\rho^2\varepsilon, \end{aligned} \tag{9.120}$$

for $j = 1, \ldots, 4$ and $i = 1, \ldots, n$. Here we used $k = \rho^{q-1}$, $\epsilon = \hat{c}\varepsilon\rho^{-1}$ and the fact that each $N(Q_j^{3\varrho}(p))$ contains $\sim m^{-2} = \rho^{-2}$ different $Q^{3\varrho}(p) \subset \Omega_y^H$. The triangle inequality then yields

$$\|(\bar{R}_{j_1} \,\cdot + \bar{c}_{j_1}) - (\bar{R}_{j_2} \,\cdot + \bar{c}_{j_2})\|^2_{L^2(B_i)} \le C\rho^{3q-6}C_\rho^2\varepsilon$$

for $1 \le j_1, j_2 \le 4$ and $i = 1, \ldots, n$. The strategy will be to cover Y_t with n different rigid motions. We argue as in (9.91)f. and Lemma 7.1.3(ii) recalling Remark 7.1.4(iii) to get $\hat{R}_i \in SO(2)$, $\hat{c}_i \in \mathbb{R}^2$ such that

$$\|\hat{y} - (\hat{R}_i \,\cdot + \hat{c}_i)\|^2_{L^2(B_i)} \le C(\#I_i)^4\rho^{3q-6}C_\rho^2\varepsilon \le C\rho^{3q-14}C_\rho^2\varepsilon.$$

Here we used Hölder's inequality (cf. (7.19)). A similar argument shows that we even find

$$\sum\nolimits_{j=-1,0,1} \|\hat{y} - (\hat{R}_{i+j} \,\cdot + \hat{c}_{i+j})\|^2_{L^2(B_i)} \le C\rho^{3q-14}C_\rho^2\varepsilon \tag{9.121}$$

for $i = 1, \ldots, n$, where (in the case that Γ_t is a Jordan curve) we set $\hat{R}_{n+1} = \hat{R}_1$, $\hat{c}_{n+1} = \hat{c}_1$ and $\hat{R}_0 = \hat{R}_n$, $\hat{c}_0 = \hat{c}_n$. Without restriction recalling Remark 7.1.4(v) we can assume that $\hat{R}_i \in \mathrm{im}_{\bar{R}_4}(M(\Gamma_t)) \subset SO(2)$, where $\mathrm{im}_{\bar{R}_4}$ denotes the image of the function $\bar{R}_4$. For shorthand let $\bar{R} = \bar{R}_4$ and $\bar{c} = \bar{c}_4$. By (9.120) and (9.121) we get

$$\sum\nolimits_{j=-1,0,1} \|(\hat{R}_{i+j} \,\cdot + \hat{c}_{i+j}) - (\bar{R} \,\cdot + \bar{c})\|^2_{L^2(B_i)} \le C\rho^{3q-14}C_\rho^2\varepsilon. \tag{9.122}$$

Using Hölder's inequality and passing to the trace (argue as in (7.10)ff. on each $Q^{3\varrho}(p)$) we obtain for all $i = 1, \dots, n$

$$\begin{aligned}\sum\nolimits_{j=-1,0,1} &\|(\hat{R}_{i+j} \cdot + \hat{c}_{i+j}) - (\bar{R} \cdot + \bar{c})\|^2_{L^1(B_i \cap \Gamma_t)} \\ &\le C \sum\nolimits_{j=-1,0,1} |B_i \cap \Gamma_t|_{\mathcal{H}} \|(\hat{R}_{i+j} \cdot + \hat{c}_{i+j}) - (\bar{R} \cdot + \bar{c})\|^2_{L^2(B_i \cap \Gamma_t)} \\ &\le C \varrho \rho^{-2} \cdot \varrho^{-1} \rho^{3q-14} C^2_\rho \varepsilon \le C \rho^{3q-16} C^2_\rho \varepsilon.\end{aligned}$$

Together with (9.120) this implies

$$\sum\nolimits_{j=-1,0,1} \|\hat{y} - (\hat{R}_{i+j} \cdot + \hat{c}_{i+j})\|^2_{L^1(B_i \cap \Gamma_t)} \le C \rho^{3q-16} C^2_\rho \varepsilon.$$

This and the fact that $n \le C |\Gamma_t|_{\mathcal{H}} \rho^{-q}$ yield

$$H_1 := \sum\nolimits_i \sum\nolimits_{j=-1,0,1} \|\hat{y} - (\hat{R}_{i+j} \cdot + \hat{c}_{i+j})\|_{L^1(B_i \cap \Gamma_t)} \le C |\Gamma_t|_{\mathcal{H}} \rho^{\frac{q}{2}-8} C_\rho \sqrt{\varepsilon}. \quad (9.123)$$

For the difference of the rigid motions we get by the triangle inequality and (9.121)

$$\sum\nolimits_{j_1, j_2 = -1,0,1} \|(\hat{R}_{i+j_1} \cdot + \hat{c}_{i+j_1}) - (\hat{R}_{i+j_2} \cdot + \hat{c}_{i+j_2})\|^2_{L^2(B_i)} \le C \rho^{3q-14} C^2_\rho \varepsilon.$$

Let $\tilde{B}_i = \{x \in \Omega : \operatorname{dist}(x, B_i) \le \bar{C} \rho^{q-2}\}$. Arguing similarly as in (7.14) it is not hard to see that

$$\begin{aligned}\sum\nolimits_{j_1, j_2 = -1,0,1} &\|(\hat{R}_{i+j_1} \cdot + \hat{c}_{i+j_1}) - (\hat{R}_{i+j_2} \cdot + \hat{c}_{i+j_2})\|^2_{L^2(\tilde{B}_i)} \\ &\le C (\rho^{-2})^2 \cdot \rho^{-4} \cdot \rho^{3q-14} C^2_\rho \varepsilon \le C \rho^{3q-22} C^2_\rho \varepsilon\end{aligned} \quad (9.124)$$

as $\frac{|\tilde{B}_i|}{|B_i|} \le C\rho^{-4}$ and $\frac{|\partial \tilde{B}_i|_\infty}{|\partial B_i|_\infty} \le C \rho^{-2}$. Define $\tilde{I}_i = I^\varrho(\tilde{B}_i)$. Again using Hölder's inequality, passing from the traces to a bulk integral and recalling $n \le C|\Gamma_t|_{\mathcal{H}} \rho^{-q}$, $\# \tilde{I}_i \le C \rho^{-4}$ we derive (let $\cdot = (\hat{R}_{i+j_1} \cdot + \hat{c}_{i+j_1}) - (\hat{R}_{i+j_2} \cdot + \hat{c}_{i+j_2})$ for shorthand)

$$\begin{aligned} H_2 &:= \sum\nolimits_i \sum\nolimits_{p \in \tilde{I}_i} \sum\nolimits_{j_1, j_2 = -1,0,1} \| \cdot \|_{L^1(\partial Q^\varrho(p))} \\ &\le C \sum\nolimits_i \sum\nolimits_{p \in \tilde{I}_i} \sum\nolimits_{j_1, j_2 = -1,0,1} \varrho^{1/2} \| \cdot \|_{L^2(\partial Q^\varrho(p))} \\ &\le C \sum\nolimits_i (\#\tilde{I}_i)^{\frac{1}{2}} \Big(\sum\nolimits_{p \in \tilde{I}_i} \sum\nolimits_{j_1, j_2 = -1,0,1} \varrho \| \cdot \|^2_{L^2(\partial Q^\varrho(p))} \Big)^{1/2} \\ &\le C \sum\nolimits_i \rho^{-2} \Big(\sum\nolimits_{j_1, j_2 = -1,0,1} \| \cdot \|^2_{L^2(\tilde{B}_i)} \Big)^{1/2} \le C |\Gamma_t|_{\mathcal{H}} \rho^{\frac{q}{2}-13} C_\rho \sqrt{\varepsilon}. \end{aligned} \quad (9.125)$$

By $(T_j)_j$ we denote the connected components of $Q^\varrho(p) \setminus (\hat{\Omega} \cup S)$ for all $Q^\varrho(p)$ with $Q^\varrho(p) \cap Y_t \neq \emptyset$. We now choose suitable rigid motions: Observe that $\operatorname{dist}(\Gamma_t \cup \partial \Omega_\rho, x) \le C_1 \rho^{q-2}$ for all $x \in Y_t$ by Lemma 9.5.1(iv) and the fact that Y_t is a connected component of $\Omega_\rho \setminus (\hat{\Omega} \cup S)$. Therefore, for every T_j with $\mathrm{dist}(T_j, \partial\Omega_\rho) \gg \rho^{q-2}$ we find some (possibly non unique) B_{i_j} with $\mathrm{dist}(T_j, B_{i_j}) \le C\rho^{q-2}$. In particular, we get $T_j \subset \tilde{B}_{i_j}$ choosing $\bar{C}$ in the definition of $\tilde{B}_i$ large enough. We define

$$\hat{y}(x) = \hat{R}_{i_j}\, x + \hat{c}_{i_j} \quad \text{for } x \in T_j \cap Y_t \cap \Omega_{2\rho} \tag{9.126}$$

for all j and note that we have found an extension $\hat{y}$ to $Y_t \cap \Omega_{2\rho}$. (If $Y_t \cap \Omega^H_y \neq \emptyset$, we redefine the function on this set.) Taking Lemma 9.5.1(v) into account the choice of B_{i_j} can be done in a way that for neighboring sets T_1, T_2 with $\overline{T}_1 \cap \overline{T}_2 \neq \emptyset$ one has $i_1 - i_2 \in \{-1, 0, 1\}$ and that $\mathcal{H}^1(J_{\hat{y}} \cap Y_t) \le C_1 \mathcal{H}^1(\Gamma_t)$. Now by (9.123) and (9.125) it is not hard to see that

$$\int_{(J_{\hat{y}} \cap \overline{Y_t}) \setminus S} |[\hat{y}]|\, d\mathcal{H}^1 \le CH_1 + CH_2 \le C|\Gamma_t|_{\mathcal{H}} \rho^{\frac{q}{2}-13} C_\rho \sqrt{\varepsilon}.$$

Repeating the arguments for all components Y_t we obtain a configuration $\hat{y} \in SBV_{cM}(\Omega_\rho, \mathbb{R}^2)$ with $\hat{y} = \tilde{y}$ on $\Omega^*_y := \Omega_y \cap \tilde{\Omega}$, where by Lemma 9.5.1 we have $|\Omega \setminus \Omega^*_y| \le C_1\rho$. (With a slight abuse of notation we replace Ω^*_y by Ω_y in the assertion of Theorem 6.1.1.) Summing over all Y_t and recalling that $\mathcal{H}^1(\partial\hat{\Omega}) \le C_1$ by Lemma 9.5.1 we get

$$\sum\nolimits_t \int_{(J_{\hat{y}} \cap \overline{Y_t}) \setminus S} f^\rho_\varepsilon(|[\hat{y}]|)\, d\mathcal{H}^1 \le C\rho^{\frac{q}{2}-13} C_\rho \le \rho$$

for $q = q(h_*)$ sufficiently large. Together with (9.119), Lemma 9.5.1(i) and (9.83)(i) this implies

$$\int_{J_{\hat{y}}} f^\rho_\varepsilon(|[\hat{y}]|)\, d\mathcal{H}^1 \le \int_{J_{\hat{y}} \setminus S} f^\rho_\varepsilon(|[\hat{y}]|)\, d\mathcal{H}^1 + \mathcal{H}^1(S) \le (1 + C_1 h_*)\mathcal{H}^1(J_y) + C_1\rho.$$

Choosing $h_* = \rho$ we finally get

$$\int_{J_{\hat{y}}} f^\rho_\varepsilon(|[\hat{y}]|)\, d\mathcal{H}^1 \le \mathcal{H}^1(J_y) + C_1\rho. \tag{9.127}$$

We observe $\nabla\hat{y} \in SO(2)$ on $\Omega_\rho \setminus \Omega_y$ (see construction in Corollary 9.2.6, (9.126) and recall $\hat{y} = \mathbf{id}$ in $\hat{\Omega} \setminus \tilde{\Omega}$). As $\|\tilde{y} - y\|^2_{L^2(\Omega_y)} + \|\nabla\tilde{y} - \nabla y\|^2_{L^2(\Omega_y)} \le C_1\varepsilon\rho$ we obtain $E^\rho_\varepsilon(\hat{y}, \Omega_\rho) \le E_\varepsilon(y) + C_1\rho$ which gives (6.4). Here we used $\|\nabla\tilde{y}\|_\infty + \|\nabla y\|_\infty \le cM$ and the regularity of the stored energy density W.

Let $(P_j)_j$ be the connected components of $\Omega_\rho \setminus S$. By Lemma 9.5.1(ii),(iii) it is not hard to see that for every index j there is a (unique) connected component $\hat{P}_j$ of $\hat{\Omega}$ such that $\hat{P}_j \subset P_j$. Then there is either a connected component P^H_j of Ω^H_y such that $\hat{P}_j = P^H_j$ (see proof of Theorem 9.4.2) or $\hat{y} = \mathbf{id}$ on $\hat{P}_j$ (see construction before (9.120)). We now define (6.5) by $u(x) = \hat{y}(x) - (R_j\, x + c_j)$

for $x \in P_j$, where $R_j x + c_j$ is either the rigid motion on P_j^H given in Theorem 9.4.1 or $R_j = \mathbf{Id}$, $c_j = 0$, respectively. For later purpose, we note that for (9.127) we can also write

$$\sum\nolimits_j \tfrac{1}{2} P(P_j, \Omega_\rho) + \int_{J_{\hat{y}} \setminus \partial P} f_\varepsilon^\rho(|[\hat{y}]|)\, d\mathcal{H}^1 \le \mathcal{H}^1(J_y) + C_1 \rho, \tag{9.128}$$

where $\partial P = \bigcup_j \partial^* P_j$ and $P(P_j, \Omega)$ denotes the perimeter of P_j in Ω_ρ.

It remains to confirm (6.6). First, (i) follows by $\mathcal{H}^1(J_{\hat{y}} \cap (\Omega_y^H)^\circ) \le C_1$ (see (9.70) and (9.88)), $\mathcal{H}^1(\partial\hat{\Omega}) \le C_1$ (see Lemma 9.5.1) and the fact that the $\mathcal{H}^1$-measure of the jump set added in the construction of $\hat{y}$ (see (9.126)) is controlled by $\mathcal{H}^1(\partial\hat{\Omega})$ and $\mathcal{H}^1(S)$. In view of (9.83)(ii)-(iv) (see also (9.92)) the properties (ii)-(iv) already hold on the set $\hat{\Omega}$ for a sufficiently large constant $C(\rho, q) = C(\rho)$. (Recall $q = q(h_*)$ and the definition $h_* = \rho$. See also Remark 9.1.3.)

Recall that $\Omega_\rho \setminus \hat{\Omega} \subset \bigcup_t \overline{Y_t}$. Repeating the arguments leading to (9.92) we find by (9.122), (9.124) and (9.126)

$$\sum\nolimits_j \|\hat{y} - (R_j \,\cdot + c_j)\|^2_{L^2(P_j \setminus \hat{\Omega})} \le C(\rho)\varepsilon.$$

This gives (ii). Moreover, as on each $Q^\varrho(p) \subset P_j \setminus \hat{\Omega}$ we have $\nabla\hat{y} = R$ for some $R \in \mathrm{im}_{\bar{R}_4}(\hat{\Omega})$ (see construction before (9.122)) we get

$$\begin{aligned}\|\nabla\hat{y} - R_j\|^p_{L^p(P_j\setminus\hat{\Omega})} &\le C(\rho)\|\bar{R}_4 - R_j\|^p_{L^p(P_j\cap\hat{\Omega})}\\ &\le C(\rho)\Big(\|\nabla\hat{y} - \bar{R}_4\|^p_{L^p(P_j\cap\hat{\Omega})} + \|\nabla\hat{y} - R_j\|^p_{L^p(P_j\cap\hat{\Omega})}\Big)\end{aligned}$$

for $p = 2, 4$. By (9.90) and (9.92) this yields

$$\sum\nolimits_j \|\nabla\hat{y} - R_j\|^4_{L^4(P_j\setminus\hat{\Omega})} \le C(\rho)\varepsilon, \quad \sum\nolimits_j \|\nabla\hat{y} - R_j\|^2_{L^2(P_j\setminus\hat{\Omega})} \le C(\rho)\varepsilon^{1-\eta}.$$

This together with (9.23) gives (iii),(iv). □

Having completed the main rigidity result, we can now prove the linearized version. We may essentially follow the proof of Theorem 6.1.1 with some minor changes. The proof, however, is considerably simpler as a lot of estimates and lemmas can be skipped.

Proof of Theorem 6.1.3. We only give a short sketch of the proof. Define $y = \mathbf{id} + u$. As the approximation argument presented in the proof of Theorem 9.4.1 also holds in the SBD-setting, it again suffices to prove the result under the assumption that there is some $\tilde{\Omega}_u \in \mathcal{V}_\varepsilon^s$ such that $u|_{\tilde{\Omega}_u} \in H^1(\tilde{\Omega}_u)$. We skip Section 9.2.1 and always set $\hat{R}_i = \mathbf{Id}$ for $i = 1, \dots, 4$. Similarly as in Lemma 9.2.5 we find sets Ω_u, $\Omega_u^H \in \mathcal{V}_{9k}^{3\varrho}$ for $k = \rho^{q-1}$, $\varrho = \rho^q$, as well as mappings $\bar{A}_j : \Omega_u^H \to \mathbb{R}^{2\times 2}_{\mathrm{skew}}$ and $\bar{c}_j : \Omega_u^H \to \mathbb{R}^2$, which are constant on $Q_j^{3\varrho}(p)$, $p \in I_j^{3\varrho}(\Omega^{3k})$, such that

$$\begin{aligned}&(i)\ \ \|u - (\bar{A}_j \,\cdot + \bar{c}_j)\|^2_{L^2(\Omega_u)} \le CC_\rho^2 \varrho^2(\alpha + \epsilon\|W\|_*),\\ &(iii)\ \ \|(\bar{A}_{j_1} \,\cdot + \bar{c}_{j_1}) - (\bar{A}_{j_2} \,\cdot + \bar{c}_{j_2})\|^2_{L^2(\Omega_u^H)} \le CC_\rho^2 \varrho^2(\alpha + \epsilon\|W\|_*)\end{aligned}$$

for $j_1, j_2 = 1, \dots, 4$, $j = 1, \dots, 4$, where $\alpha = \|e(\nabla u)\|^2_{L^2(\tilde{\Omega}_u)}$ and $\epsilon = \hat{c}\rho^{-1}\varepsilon$. This can be established following the lines of the proof of Lemma 9.2.5 with the difference that in (9.58) we do not replace $\mathbf{Id} + A$ by a different rigid motion $\bar{R}$, but proceed with $\mathbf{Id} + A$. Analogously, we find an extension Ω_u^H as constructed in Corollary 9.2.6 and then we obtain the result up to a small set following the lines of Theorem 9.4.2. Finally, the jump set and the extension to Ω_ρ may be constructed as in Section 9.5. □

Chapter 10

Compactness and Γ-convergence

10.1 Compactness of rescaled configurations

This section is devoted to the proof of the main compactness result. For the compactness theorem in GSBD (see Theorem A.1.3) it is necessary that the integral for some integrand ψ with $\lim_{x\to\infty}\psi(x) = \infty$ is uniformly bounded. We first give a simple criterion for the existence of such a function which is, loosely speaking, based on the condition that the functions coincide in a certain sense on the bulk part of the domain.

Lemma 10.1.1. *For every increasing sequence $(b_i)_i \subset (0,\infty)$ with $b_i \to \infty$ there is an increasing concave function $\psi : [0,\infty) \to [0,\infty)$ with $\lim_{x\to\infty}\psi(x) = \infty$ and $\psi(b_i) \le 2^i$ for all $i \in \mathbb{N}$.*

Proof. Let $f : [0,\infty) \to [0,\infty)$ be the function with $f(0) = 0$, $f(b_i) = 2^i$ which is affine on each segment $[b_i, b_{i+1}]$. Clearly, f is increasing and satisfies $f(x) \to \infty$ for $x \to \infty$, but is possibly not concave. We now construct ψ and first let $\psi = f$ on $[0, b_1]$. Assume ψ has been defined on $[0, b_i]$ and satisfies $\psi(b_i) = f(b_i) = 2^i$. If $f'(b_i-) \ge f'(b_i+)$ we set $\psi = f$ on $[b_i, b_{i+1}]$. Here, $f'(x\pm)$ denote the one-sided limits of the derivative at point x. Otherwise, we let $\psi(x) = f(b_i) + f'(b_i-)(x - b_i)$ for $x \in [b_i, \bar{x}]$, where $\bar{x}$ is the smallest value larger than b_i such that $f(\bar{x}) = f(b_i) + f'(b_i-)(\bar{x} - b_i)$. If $\bar{x}$ does not exist we are done . If $\bar{x}$ exists we assume $\bar{x} \in (b_{j-1}, b_j]$ and define $\psi = f$ on $[\bar{x}, b_j]$ noting that $\psi'(\bar{x}-) \ge \psi'(\bar{x}+)$. We end up with an increasing concave function ψ with $\psi \le f$ and $\psi(x) \to \infty$ for $x \to \infty$, as desired.

Lemma 10.1.2. *Let $\Omega \subset \mathbb{R}^2$ and let $(y_l)_l \subset L^1(\Omega)$ be a sequence satisfying $|\Omega \setminus \bigcup_{n\in\mathbb{N}} \bigcap_{l\ge n}\{|y^n - y^l| \le 1\}| = 0$. Then there is a not relabeled subsequence such that*

$$\int_\Omega \psi(|y^l|) \le C$$

for a constant independent of l, where ψ is an increasing continuous function with $\lim_{x\to\infty}\psi(x) = +\infty$.

Proof. Define $C_l := \max_{1\le i\le l} \|y_i\|_{L^1(\Omega)}$ for all $l \in \mathbb{N}$. Let $A_n = \bigcap_{l\ge n}\{|y^n - y^l| \le 1\}$ and set $B_1 = A_1$ as well as $B_n = A_n \setminus \bigcup_{m=1}^{n-1} B_m$ for all $n \in \mathbb{N}$. The sets $(B_n)_n$ are pairwise disjoint with $\sum_n |B_n| = |\Omega|$. We choose $0 = n_1 < n_2 < \ldots$ such that $\sum_{1\le n\le n_i} \frac{|B_n|}{|\Omega|} \ge 1 - 4^{-i}$. We let $B^i = \bigcup_{n=n_i+1}^{n_{i+1}} B_n$ and observe $|B^i| \le 4^{-i}|\Omega|$.

We pass to the subsequence of $(n_i)_i \subset \mathbb{N}$ and choose $E^i \supset B^i$ such that $|E^i| = 4^{-i}|\Omega|$. Let $b_i = \frac{C_{n_{i+1}}}{|E^i|} + 2 = 4^i \frac{C_{n_{i+1}}}{|\Omega|} + 2$ for $i \in \mathbb{N}$ and note that $(b_i)_i$ is increasing with $b_i \to \infty$. By Lemma 10.1.1 we get an increasing concave function $\psi : [0,\infty) \to [0,\infty)$ with $\lim_{x\to\infty} \psi(x) = \infty$ and $\psi(b_i) \le 2^i$ for all $i \in \mathbb{N}$. Clearly, ψ is also continuous.

For $\hat{B}^i := \Omega \setminus \bigcup_{n=1}^{n_i} B_n$ we have $|\hat{B}^i| \le 4^{-i}|\Omega|$ and choose $\hat{E}^i \supset \hat{B}^i$ with $|\hat{E}^i| = 4^{-i}|\Omega|$. We then obtain $\frac{C_{n_i}}{|\hat{E}^i|} = 4^i \frac{C_{n_i}}{|\Omega|} \le b_i$. Now let $l = n_i$. Using Jensen's inequality, the definition of the sets B^i, $\|y_l\|_{L^1(\Omega)} \le C_l$ and the monotonicity of ψ we compute

$$\begin{aligned}
\int_\Omega \psi(|y^l|) &= \sum\nolimits_{1\le j\le i-1} \int_{B^j} \psi(|y^l|) + \int_{\hat{B}^i} \psi(|y^l|) \\
&= \sum\nolimits_{1\le j\le i-1} \int_{B^j} \psi(|y^{n_{j+1}}| + 2) + \int_{\hat{B}^i} \psi(|y^l|) \\
&\le \sum\nolimits_{1\le j\le i-1} |E^j| \psi\Big(\fint_{E^j} |y^{n_{j+1}}| + 2\Big) + |\hat{E}^i| \psi\Big(\fint_{\hat{E}^i} |y^l|\Big) \\
&\le \sum\nolimits_{1\le j\le i-1} 4^{-j}|\Omega| 2^j + 4^{-i}|\Omega| 2^i \le |\Omega| \sum\nolimits_{j\in\mathbb{N}} 2^{-j}.
\end{aligned} \tag{10.1}$$

As the estimate is independent of $l \in (n_i)_i$, this yields $\int_\Omega \psi(|y^l|) \le C$ uniformly in l, as desired. □

Now we are in a position to give the proof of the main compactness result. In the first part we show that (6.10), (6.11) and (6.12) hold.

Proof of Theorem 6.2.1, part 1. Let $(\varepsilon_k)_k$ be an arbitrary null sequence. Let $y_k \in SBV_M(\Omega)$ with $E_{\varepsilon_k}(y_k) \le C$ be given. The fact that $W(G) \ge c\,\mathrm{dist}^2(G, SO(2))$ for all $G \in \mathbb{R}^{2\times 2}$ implies $\|\mathrm{dist}(\nabla y_k, SO(2))\|^2_{L^2(\Omega)} \le C\varepsilon_k$ for a constant independent of ε_k. Moreover, we have $\mathcal{H}^1(J_{y_k}) \le C$ for all $k \in \mathbb{N}$.

Choose $\rho_0 > 0$ and let $\rho_l = 2^{-3l}\rho_0$ for all $l \in \mathbb{N}$. By Theorem 6.1.1 we find modifications $y^l_k \in SBV_{cM}(\Omega, \mathbb{R}^2)$ with $E^{\rho_l}_{\varepsilon_k}(y^l_k, \Omega_{\rho_l}) \le E_{\varepsilon_k}(y_k) + C\rho_l$ and

$$\|y^l_k - y_k\|^2_{L^2(\Omega^l_k)} + \|\nabla y^l_k - \nabla y_k\|^2_{L^2(\Omega^l_k)} \le C\varepsilon_k \rho_l, \tag{10.2}$$

where $\Omega^l_k := \Omega_{y^l_k}$ with $|\Omega \setminus \Omega^l_k| \le C\rho_l$. We further get Caccioppoli partitions $(P^{k,l}_j)_j$ of Ω_{ρ_l} with $\sum_j P(P^{k,l}_j, \Omega_{\rho_l}) \le C$ and corresponding piecewise rigid motions $T^l_k(x) := \sum_j (R^{k,l}_j\, x + c^{k,l}_j)\chi_{P^{k,l}_j}(x)$ such that the functions $v^l_k : \Omega \to \mathbb{R}^2$ defined by

$$v^l_k(x) = \begin{cases} \frac{1}{\sqrt{\varepsilon_k}} (R^{k,l}_j)^T \big(y^l_k(x) - (R^{k,l}_j\, x + c^{k,l}_j)\big) & \text{for } x \in P^{k,l}_j,\ j \in \mathbb{N}, \\ 0 & \text{else,} \end{cases} \tag{10.3}$$

satisfy by (6.6)

$$\mathcal{H}^1(J_{v_k^l}) \le C, \quad \|v_k^l\|_{L^2(\Omega)} + \|e(\nabla v_k^l)\|_{L^2(\Omega)} \le \hat{C}_l, \quad \|\nabla v_k^l\|^2_{L^2(\Omega)} \le \hat{C}_l \varepsilon_k^{-\eta} \quad (10.4)$$

for some $\hat{C}_l = \hat{C}(\rho_l) > 0$ and $\eta > 0$ small. Observe that $|c_j^{k,l}| \le cM$ for a universal constant as $\|y_k^l\|_\infty \le cM$ for all $k \in \mathbb{N}$. Clearly, each partition may be extended to Ω by adding the element $\Omega \setminus \Omega_{\rho_l}$ and $\sum_j P(P_j^{k,l}, \Omega) \le C$ is still satisfied as $\mathcal{H}^1(\partial\Omega_{\rho_l}) \le C\mathcal{H}^1(\partial\Omega)$.

Using a diagonal argument we get a (not relabeled) subsequence of $(\varepsilon_k)_k$ such that by Theorem A.1.3 for every $l \in \mathbb{N}$ we find a function $v^l \in GSBD^2(\Omega)$ with

$$v_k^l \to v^l \text{ a.e. in } \Omega \quad \text{and} \quad e(\nabla v_k^l) \rightharpoonup e(\nabla v^l) \text{ weakly in } L^2(\Omega, \mathbb{R}^{2\times 2}_{\mathrm{sym}}) \quad (10.5)$$

for $k \to \infty$. Moreover, by the compactness result for piecewise constant functions (see Theorem A.2.3) we obtain an (ordered) partition $(P_j^l)_j$ of Ω with $\sum_j P(P_j^l, \Omega) \le C$ and a piecewise rigid motion $T^l(x) := \sum_j (R_j^l\, x + c_j^l)\chi_{P_j^l}(x)$ such that for all $l \in \mathbb{N}$ letting $k \to \infty$ we obtain (again up to a subsequence) $R_j^{k,l}\chi_{P_j^{k,l}} \to R_j^l \chi_{P_j^l}$ and $c_j^{k,l}\chi_{P_j^{k,l}} \to c_j^l \chi_{P_j^l}$ in measure for all $j \in \mathbb{N}$. This also implies

$$\sum_j |P_j^{k,l} \triangle P_j^l| + \|T_k^l - T^l\|_{L^2(\Omega)} + \|\nabla T_k^l - \nabla T^l\|_{L^2(\Omega)} \to 0 \quad (10.6)$$

for $k \to \infty$, where $\triangle$ again denotes the symmetric difference of two sets. We now show that

$$\|v^l\|_{L^1(\Omega)} \le C\|v^l\|_{L^2(\Omega)} \le \hat{C}_l, \quad \mathcal{H}^1(J_{v^l}) \le C, \quad \|e(\nabla v^l)\|^2_{L^2(\Omega)} \le C. \quad (10.7)$$

The first two claims follow directly from (10.4) and (A.5). To see the third estimate we let $\chi_k^l(x) := \chi_{[0,\varepsilon_k^{-1/8}]}(|\nabla v_k^l(x)|)$ and recall that by the linearization formula (9.23) we have $\mathrm{dist}^2(G, SO(2)) = |\bar{e}_R(G)|^2 + \omega(G-R)$ with $\sup\{|G|^{-3}\omega(G) : |G| \le 1\} \le C$ and $\omega(RG) = \omega(G)$ for $G \in \mathbb{R}^{2\times 2}$, $R \in SO(2)$. We compute

$$\begin{aligned} C \ge E_{\varepsilon_k}^{\rho_l}(y_k^l, \Omega_{\rho_l}) &\ge \frac{C}{\varepsilon_k}\int_{\Omega_{\rho_l}} \mathrm{dist}^2(\nabla y_k^l, SO(2)) \\ &\ge \frac{C}{\varepsilon_k}\sum_j \int_{P_j^{k,l}} \chi_k^l\Big(|\bar{e}_{R_j^{k,l}}(\nabla y_k^l)|^2 + \omega(\nabla y_k^l - R_j^{k,l})\Big) \\ &= C\int_\Omega \chi_k^l\Big(|e(\nabla v_k^l)|^2 + \frac{1}{\varepsilon_k}\omega(\sqrt{\varepsilon_k}\nabla v_k^l)\Big). \end{aligned} \quad (10.8)$$

The second term of the integral can be estimated by

$$\int_\Omega \chi_k^l(x)\frac{1}{\varepsilon_k}\omega(\sqrt{\varepsilon_k}\nabla v_k^l) = \int_\Omega \chi_k^l(x)\sqrt{\varepsilon_k}|\nabla v_k^l|^3 \frac{\omega(\sqrt{\varepsilon_k}\nabla v_k^l)}{|\sqrt{\varepsilon_k}\nabla v_k^l|^3} \le C\varepsilon_k^{\frac{1}{8}} \to 0. \quad (10.9)$$

As $e(\nabla v_k^l) \rightharpoonup e(\nabla v^l)$ weakly in L^2 and $\chi_k^l \to 1$ boundedly in measure on Ω by (10.4) for η sufficiently small, it follows $\chi_k^l e(\nabla v_k^l) \rightharpoonup e(\nabla v^l)$ weakly in $L^2(\Omega)$. By lower semicontinuity we obtain $\|e(\nabla v^l)\|^2_{L^2(\Omega)} \le C$ for a constant independent of ρ_l which concludes (10.7).

We now want to pass to the limit $l \to \infty$. Similarly as in the argumentation leading to (10.6), by the compactness result for piecewise constant functions (see Theorem A.2.3) we find a partition $(P_j)_j$ of Ω and a piecewise rigid motion $T(x) := \sum_j (R_j\, x + c_j)\chi_{P_j}(x)$ such that for a suitable (not relabeled) subsequence $R_j^l \chi_{P_j^l} \to R_j \chi_{P_j}$, $c_j^l \chi_{P_j^l} \to c_j \chi_{P_j}$ in measure for all $j \in \mathbb{N}$ and thus

$$\sum\nolimits_j |P_j^l \triangle P_j| + \|T^l - T\|_{L^2(\Omega)} + \|\nabla T^l - \nabla T\|_{L^2(\Omega)} \to 0 \tag{10.10}$$

for $l \to \infty$. Recalling (10.6) and using a diagonal argument we can choose a (not relabeled) subsequence of $(\rho_l)_l$ and afterwards of $(\varepsilon_k)_k$ such that for all l we have

$$\sum\nolimits_j |P_j^l \triangle P_j| \le 2^{-l}, \quad \sum\nolimits_j |P_j^{k,l} \triangle P_j^l| \le 2^{-l} \ \text{ for all } \ k \ge l. \tag{10.11}$$

We see that the compactness result in GSBD cannot be applied directly on the sequence $(v^l)_l$ as the L^2 bound in (10.7) depends on ρ_l. We now show that by choosing the rigid motions on the elements of the partitions appropriately (see (10.3)) we can construct the sequence $(v^l)_l$ such that we obtain

$$\Big|\Omega \setminus \bigcup\nolimits_{n\in\mathbb{N}} \bigcap\nolimits_{m\ge n} \{|v^n - v^m| \le 1\}\Big| = 0 \tag{10.12}$$

and thus Lemma 10.1.2 is applicable.

We fix $k \in \mathbb{N}$ and describe an iterative procedure to redefine $R_j^{k,l}, c_j^{k,l}$ for all $l, j \in \mathbb{N}$. Let v_k^1 as defined in (10.3) and assume $\hat{R}_j^{k,l}, \hat{c}_j^{k,l}$ have been chosen for all $j \in \mathbb{N}$ (which possibly differ from $R_j^{k,l}, c_j^{k,l}$) such that (10.4) still holds possibly passing to a larger constant $\hat{C}_l$. Fix some $P_j^{k,l+1}$, $j \in \mathbb{N}$, and recall that $|P_j^{k,l+1}| \ge C(\rho_{l+1})$ as $P_j^{k,l+1}$ contains squares of size $\sim \rho_{l+1}^q$ (see the proof of Theorem 6.1.1.) Define $D^{l+1} = P_j^{k,l+1} \cap \Omega_k^{l+1}$ and let $D_i^l = P_{j_i}^{k,l} \cap \Omega_k^l$ be the components with $P_{j_i}^{k,l} \cap P_j^{k,l+1} \neq \emptyset$ for $i = 1, \ldots, n$. Without restriction assume that $P_{j_1}^{k,l} \cap P_j^{k,l+1}$ has largest Lebesgue measure. If $|P_{j_1}^{k,l} \cap P_j^{k,l+1}| > 2|P_j^{k,l+1} \setminus (\Omega_k^l \cap \Omega_k^{l+1})|$, we define

$$\hat{R}_j^{k,l+1} = \hat{R}_{j_1}^{k,l}, \quad \hat{c}_j^{k,l+1} = \hat{c}_{j_1}^{k,l} \quad \text{on} \quad P_j^{k,l+1}.$$

Otherwise we set $\hat{R}_j^{k,l+1} = R_j^{k,l+1}$ and $\hat{c}_j^{k,l+1} = c_j^{k,l+1}$. In the first case we then obtain $|D_1^l \cap D^{l+1}| \ge \frac{1}{2}|P_{j_1}^{k,l} \cap P_j^{k,l+1}| \ge C(\rho_{l+1})$ and thus for $p = 2, 4$ we get by $\|\nabla y_k^l\|_\infty, \|\nabla y_k^{l+1}\|_\infty \le cM$

$$\begin{aligned}|\hat{R}_j^{k,l+1} - R_j^{k,l+1}|^p \le C(\rho_{l+1})\Big(&\|\nabla y_k^l - \hat{R}_{j_1}^{k,l}\|^p_{L^p(D_1^l)} + \|\nabla y_k^l - \nabla y_k^{l+1}\|^2_{L^2(D_1^l \cap D^{l+1})} \\ &+ \|\nabla y_k^{l+1} - R_j^{k,l+1}\|^p_{L^p(D^{l+1})}\Big).\end{aligned}$$

The calculation may be repeated to estimate the difference of the rigid motions. Summing over all components and recalling (10.2), (10.4) (for l) as well as the estimates for the original rigid motions (for $l+1$, see (9.92)) we find that (10.4) still holds possibly passing to larger constants.

We define $A_{k,l} = \bigcap_{n \le m \le l} \{|v_k^m - v_k^n| \le \frac{1}{2}\}$ for all $n \in \mathbb{N}$ and $n \le l \le k$. If we show

$$|\Omega \setminus A_{k,l}| \le C 2^{-n}, \tag{10.13}$$

then (10.12) follows. Indeed, for given $l \ge n$ we can choose $K = K(l) \ge l$ so large that $|\{|v_K^m - v^m| > \frac{1}{4}\}| \le 2^{-m}$ for all $n \le m \le l$ since $v_k^m \to v^m$ in measure for $k \to \infty$. This implies

$$|\Omega \setminus \bigcap\nolimits_{n \le m \le l} \{|v^m - v^n| \le 1\}| \le |\Omega \setminus A_{K,l}| + \sum\nolimits_{n \le m \le l} |\{|v_K^m - v^m| > \tfrac{1}{4}\}| \le C 2^{-n}.$$

Passing to the limit $l \to \infty$ we find $|\Omega \setminus \bigcap_{m \ge n} \{|v^m - v^n| \le 1\}| \le C 2^{-n}$ and taking the union over all $n \in \mathbb{N}$ we derive (10.12).

To show (10.13) we proceed in two steps. Employing the redefinition of the piecewise rigid motions we first show that the set where $T_k^m, n \le m \le l$, differ is small. Afterwards we use (10.2) to find that the set where $y_k^m, n \le m \le l$, differ is small. We define $B_{k,l} = \bigcap_{n \le m \le l} \{T_k^m = T_k^n\}$ and prove that

$$|\Omega \setminus B_{k,l}| \le C 2^{-n} \tag{10.14}$$

for all $k \ge l \ge n$. To this end, consider $\{T_k^m = T_k^{m+1}\}$ for $n \le m \le l-1$ and first note that by (10.11) we have $\sum_j |P_j^{k,m+1} \triangle P_j^{k,m}| \le 3 \cdot 2^{-m}$. Define $J_1 \subset \mathbb{N}$ such that $|P_j^{k,m} \cap P_j^{k,m+1}| \le 2|P_j^{k,m+1} \setminus (\Omega_k^m \cap \Omega_k^{m+1})|$ for all $j \in J_1$ and let $J_2 \subset \mathbb{N} \setminus J_1$ such that $|P_j^{k,m+1} \cap P_j^{k,m}| > \frac{1}{2}|P_j^{k,m+1}|$ for all $j \in J_2$. Observe that $|P_j^{k,m+1}| \le 2|P_j^{k,m+1} \setminus P_j^{k,m}|$ for $j \in J_3 := \mathbb{N} \setminus (J_1 \cup J_2)$. Due to the above construction of the rigid motions we obtain $\{T_k^m = T_k^{m+1}\} \supset \bigcup_{j \in J_2} (P_j^{k,m+1} \cap P_j^{k,m})$ and therefore recalling $|\Omega \setminus (\Omega_k^m \cap \Omega_k^{m+1})| \le C 2^{-3m}$

$$\begin{aligned} |\Omega \setminus \{T_k^m = T_k^{m+1}\}| &\le \sum\nolimits_{j \in J_2} |P_j^{k,m+1} \setminus P_j^{k,m}| + \sum\nolimits_{j \in J_1 \cup J_3} |P_j^{k,m+1}| \\ &\le \sum\nolimits_{j \in J_2} |P_j^{k,m+1} \setminus P_j^{k,m}| + \sum\nolimits_{j \in J_1 \cup J_3} 2|P_j^{k,m+1} \setminus P_j^{k,m}| \\ &\quad + \sum\nolimits_{j \in J_1} 2|P_j^{k,m+1} \setminus (\Omega_k^m \cap \Omega_k^{m+1})| \le C 2^{-m}. \end{aligned}$$

Summing over $n \le m \le l-1$ we establish (10.14). Now recalling (10.2), (10.14), $|\Omega \setminus \Omega_k^l| \le C\rho_l$ and the fact that $(\rho_l)_l \subset (2^{-3l}\rho_0)_l$ we find

$$|\Omega \setminus A_{k,l}| \le |\Omega \setminus B_{k,l}| + \sum\nolimits_{n \le m \le l-1} |\{|y_k^{m+1} - y_k^m| > 2^{-m-1}\sqrt{\varepsilon_k}\}| \le C 2^{-n}$$

for all $k \ge l \ge n$, as desired.

By (10.7) and (10.12) we can apply Lemma 10.1.2 on the sequence $(v^l)_l$. We employ Theorem A.1.3 and obtain a function $v \in GSBD(\Omega)$ and a further not relabeled subsequence with $v^l \to v$ a.e in Ω and $e(\nabla v^l) \rightharpoonup e(\nabla v)$ weakly in $L^2(\Omega, \mathbb{R}^{2\times 2}_{\mathrm{sym}})$.

We now select a suitable diagonal sequence such that (6.11) and (6.12) hold. Observe that by (10.8), (10.9) the functions $\hat{v}^l_k := \chi^l_k v^l_k$ fulfill $\|e(\nabla \hat{v}^l_k)\|_{L^2(\Omega)} \le C$ and $\|\nabla \hat{v}^l_k\|_\infty \le \varepsilon_k^{-1/8}$ for a constant independent of $k, l \in \mathbb{N}$. As weak convergence in L^2 is metrizable on bounded sets and convergence in measure is metrizable (take $(f,g) \mapsto \int_\Omega \min\{|f-g|, 1\}$) we can apply a diagonal sequence argument and find a not relabeled subsequence $(y_n)_n$ and a corresponding diagonal sequence $(w_n)_{n\in\mathbb{N}} \subset (\hat{v}^l_k)_{k,l}$ with corresponding partitions $(P^n_j)_j$ and piecewise rigid motions $(T_n)_n$ such that by (10.5), (10.6) and (10.10)

$$
\begin{aligned}
& w_n \to v \ \text{ in measure on } \Omega, && e(\nabla w_n) \rightharpoonup e(\nabla v) \text{ weakly in } L^2(\Omega, \mathbb{R}^{2\times 2}_{\mathrm{sym}}),\\
& T_n \to T \text{ in } L^2(\Omega), && \nabla T_n \to \nabla T \text{ in } L^2(\Omega),\\
& \chi_{P^n_j} \to \chi_{P_j} \ \text{ in measure on } \Omega \ \text{ for all } j \in \mathbb{N},
\end{aligned}
$$

for $n \to \infty$. Up to a further subsequence we can assume $w_n \to v$ a.e. and $\nabla T_n \to \nabla T$ a.e. Finally, define $u_n = \nabla T_n w_n$ for all $n \in \mathbb{N}$ and $u = \nabla T v$ and observe that (6.11), (6.12) hold. Moreover, by (10.2), (10.3), the fact that $\|\nabla u_n\|_\infty \le \varepsilon_n^{-1/8}$, $\|\nabla y_n\|_\infty \le cM$ and the regularity of W it is not hard to see that

$$
\frac{1}{\varepsilon_n}\int_\Omega W(\mathbf{Id} + \sqrt{\varepsilon}_n \nabla T_n^T \nabla u_n) \le \frac{1}{\varepsilon_n}\int_\Omega W(\nabla y_n) + o(1).
$$

This gives (6.10)(ii). As $\chi^l_k \to 1$ boundedly in measure on Ω and $|\Omega \setminus \Omega^l_k| \to 0$ for $k, l \to \infty$ we also get (6.10)(i) recalling (10.2), (10.3) and possibly passing to a further subsequence. □

It remains to show (6.13).

Proof of Theorem 6.2.1, part 2. The sets $J^c_v := \{x \in J_v : [v](x) = c\}$ for $c \in B_1(0) \setminus \{0\}$ are pairwise disjoint with $\mathcal{H}^1$-σ finite union, i.e. $\mathcal{H}^1(J^c_v) = 0$ up to at most countable values of c. Consequently, we can choose a sequence (c_j) with $0 \le |c_j| < \frac{1}{2}$ such that $\mathcal{H}^1(J^{c_i - c_j}_v) = 0$ for $i \neq j$. Replacing v by $\tilde{v} = v + \sum_j c_j \chi_{P_j}$ we thus obtain $\mathcal{H}^1(\partial P \setminus J_{\tilde{v}}) = 0$ (recall that $\partial P = \bigcup_j \partial^* P_j$, where ∂^* denotes the essential boundary.) We first show that it suffices to prove

$$
\liminf_{k\to\infty} \sum\nolimits_j \mathcal{H}^1(J_{y_k}) \ge \mathcal{H}^1(J_{\tilde{v}}). \tag{10.15}
$$

To see this we have to show $\mathcal{H}^1(J_{\tilde{v}}) = \mathcal{H}^1(J_u \setminus \partial P) + \frac{1}{2}\sum_j P(P_j, \Omega)$. By (A.9) we obtain $2\mathcal{H}^1(J_{\tilde{v}} \cap \partial P) = 2\mathcal{H}^1(\bigcup_j \partial^* P_j \cap \Omega) = \sum_j P(P_j, \Omega)$ and thus $\mathcal{H}^1(J_{\tilde{v}}) = \mathcal{H}^1(J_{\tilde{v}} \setminus \partial P) + \frac{1}{2}\sum_j P(P_j, \Omega) = \mathcal{H}^1(J_u \setminus \partial P) + \frac{1}{2}\sum_j P(P_j, \Omega)$, as desired.

We now show (10.15) in two steps first passing to the limit $k \to \infty$ and then letting $l \to \infty$. We replace v^l_k by $\tilde{v}^l_k = v^l_k + \sum_j c_j \chi_{P^{k,l}_j}$ and v^l by $\tilde{v}^l = v^l + \sum_j c_j \chi_{P^l_j}$

noting that $\tilde{v}^l_k \to \tilde{v}^l$ for $k \to \infty$ and $\tilde{v}^l \to \tilde{v}$ for $l \to \infty$ in the sense of (A.5). In the following we write $J^l_k = J_{\tilde{v}^l_k} \cap \Omega_{\rho_l}$ for shorthand. Recalling (10.3) we obtain by (9.128)

$$\begin{aligned}\mathcal{H}^1(J_{y_k}) &\geq \int_{J^l_k \setminus \partial P^{k,l}} g_{\rho_l}(|[\tilde{v}^l_k]|)\, d\mathcal{H}^1 + \frac{1}{2}\sum\nolimits_j P(P^{k,l}_j, \Omega_{\rho_l}) - C\rho_l \\ &\geq \int_{J^l_k} g_{\rho_l}(\,[\tilde{v}^l_k]|)\, d\mathcal{H}^1 - C\rho_l\end{aligned} \tag{10.16}$$

where $\partial P^{k,l} = \bigcup_j \partial^* P^{k,l}_j$ and $g_{\rho_l}(x) = \min\{\frac{x}{\rho_l}, 1\}$. Here we note that the passage from v^l_k to $\tilde{v}^l_k$ does not affect the estimate. We cannot directly apply lower semicontinuity results for GSBD functions due to the involved function g_{ρ_l}. We therefore pass to the limit $k \to \infty$ on one-dimensional sections.

Let $\sigma > 0$ and $\hat{\mu}^{\sigma,\nu}_{v^l}$ be the measure given in (A.6) inserting the concave function g_σ for θ and let $\hat{\mu}^{0,\nu}_{v^l}$ be the measure when setting $\theta \equiv 1$. By Lemma A.1.4 we have

$$\hat{\mu}^{\sigma,\nu}_{\tilde{v}^l}(U) \leq \liminf_{k\to\infty} \hat{\mu}^{\sigma,\nu}_{\tilde{v}^l_k}(U)$$

for all $\sigma \geq 0$, $\nu \in S^1$ and for every open set $U \subset \Omega$. Let $\kappa_1 = \int_{S^1} |\xi \cdot \nu|\, d\mathcal{H}^1(\nu)$ for some $\xi \in S^1$ which clearly does not depend on the particular choice of ξ. Using Fatou's lemma and (A.6) we compute for l sufficiently large

$$\begin{aligned}\liminf_{k\to\infty} \mathcal{H}^1(J_{y_k}) + C\rho_l &\geq \liminf_{k\to\infty} \int_{J^l_k} g_\sigma(|[\tilde{v}^l_k]|)\, d\mathcal{H}^1 \\ &\geq \kappa_1^{-1} \int_{S^1} \liminf_{k\to\infty} \int_{J^l_k} g_\sigma(|[\tilde{v}^l_k](x)|) |\xi_{\tilde{v}^l_k}(x) \cdot \nu|\, d\mathcal{H}^1(x)\, d\mathcal{H}^1(\nu) \\ &\geq \kappa_1^{-1} \int_{S^1} \liminf_{k\to\infty} \hat{\mu}^{\sigma,\nu}_{\tilde{v}^l_k}(\Omega_{\rho_l})\, d\mathcal{H}^1(\nu) \geq \kappa_1^{-1} \int_{S^1} \hat{\mu}^{\sigma,\nu}_{\tilde{v}^l}(\Omega_{\rho_l})\, d\mathcal{H}^1(\nu).\end{aligned}$$

We pass to the limit $l \to \infty$ (i.e. $\rho_l \to 0$) and obtain by the dominated convergence theorem

$$\liminf_{k\to\infty} \mathcal{H}^1(J_{y_k}) \geq \kappa_1^{-1} \int_{S^1} \hat{\mu}^{\sigma,\nu}_{\tilde{v}}(\Omega)\, d\mathcal{H}^1(\nu).$$

Recall that $g_\sigma \to 1$ pointwise for $\sigma \to 0$. Now letting $\sigma \to 0$ we obtain by the dominated convergence theorem

$$\begin{aligned}\liminf_{k\to\infty} \mathcal{H}^1(J_{y_k}) &\geq \kappa_1^{-1} \int_{S^1} \hat{\mu}^{0,\nu}_{\tilde{v}}(\Omega)\, d\mathcal{H}^1(\nu) \\ &= \kappa_1^{-1} \int_{S^1} \int_{J_{\tilde{v}}} |\xi_{\tilde{v}}(x) \cdot \nu|\, d\mathcal{H}^1(x)\, d\mathcal{H}^1(\nu) = \mathcal{H}^1(J_{\tilde{v}}).\end{aligned}$$

This gives (10.15) and completes the proof. □

The proof of Corollary 6.1.2 is now straightforward.

Proof of Corollary 6.1.2 . Let $y \in SBV(\Omega)$ with $\mathcal{H}^1(J_y) < \infty$ as well as $\int_\Omega \mathrm{dist}^2(\nabla y, SO(2)) = 0$ be given. Define an arbitrary null sequence $(\varepsilon_k)_k$ and the sequence $y_k = y$ for all $k \in \mathbb{N}$. Applying Theorem 6.2.1 we obtain a subsequence and configurations u_k converging almost everywhere by (6.12). Moreover, we obtain piecewise rigid motions T, T_k such that $T_k \to T$, $\nabla T_k \to \nabla T$ in $L^2(\Omega)$ by (6.11) and $y_k - T_k \to 0$ a.e. for $k \to \infty$ by (6.10)(i). This implies $y = T$ is a piecewise rigid motion, as desired. □

10.2 Admissible & coarsest partitions and limiting configurations

We now prove Theorem 6.2.4 and begin with some preliminary observations. In the following let $(y_k)_k$ be a (sub-)sequence as considered in Theorem 6.2.1. Recall Definition 6.2.3. For notational convenience we will drop the dependence of $(y_k)_k$ in the sets $\mathcal{Z}_P, \mathcal{Z}_u, \mathcal{Z}_T$. Moreover, recall the definition of the set of piecewise infinitesimal rigid motions $\mathcal{A}((P_j)_j)$ in (6.9). We introduce a partial order on the admissible partitions $\mathcal{Z}_P$: Given two partitions $\mathcal{P}^1 := (P_j^1)_j, \mathcal{P}^2 := (P_j^2)_j$ in $\mathcal{Z}_P$ we say $\mathcal{P}^1 \geq \mathcal{P}^2$ if $\mathcal{P}^1$ is subordinated to $\mathcal{P}^2$, i.e. $\bigcup_j \partial^* P_j^1 \subset \bigcup_j \partial^* P_j^2$ up to an $\mathcal{H}^1$-negligible set. We observe that if $\mathcal{P}^1 \geq \mathcal{P}^2$ and $\mathcal{P}^2 \geq \mathcal{P}^1$, abbreviated by $\mathcal{P}^1 = \mathcal{P}^2$ hereafter, then the Caccioppoli partitions coincide: After a possible reordering of the sets we find $|P_j^1 \triangle P_j^2| = 0$ for all $j \in \mathbb{N}$.

We begin with the observation that the piecewise rigid motion is uniquely determined in the limit.

Lemma 10.2.1. *Let $(y_k)_k$ be a (sub-)sequence as considered in Theorem 6.2.1. Then there is a unique $T \in \mathcal{Z}_T(P)$ for all $\mathcal{P} \in \mathcal{Z}_P$.*

Proof. Assume there are $\mathcal{P}$, $\hat{\mathcal{P}} \in \mathcal{Z}_P$ and $T \in \mathcal{Z}_T(\mathcal{P})$, $\hat{T} \in \mathcal{Z}_T(\hat{\mathcal{P}})$. Let $(u_k, \mathcal{P}^k, T_k), (\hat{u}_k, \hat{\mathcal{P}}^k, \hat{T}_k) \in \mathcal{D}$ for $k \in \mathbb{N}$ be the triples according to Definition 6.2.3(ii). As $u_k - \hat{u}_k - (\frac{1}{\sqrt{\varepsilon_k}}(T_k - \hat{T}_k)) \to 0$ a.e. by (6.10)(i) and $u_k - \hat{u}_k$ converges pointwise a.e. (and the limits lie in $\mathbb{R}$ a.e.) by (6.12) we get $T_k - \hat{T}_k \to 0$ pointwise almost everywhere. This implies $T = \hat{T}$, as desired. □

From now on T will always denote the rigid motion given by Lemma 10.2.1. We state a lemma giving an equivalent characterization of the coarsest partition (recall Definition 6.2.3(iv)).

Lemma 10.2.2. *Let $(y_k)_k$ be a (sub-)sequence as considered in Theorem 6.2.1. Then $\mathcal{P} \in \mathcal{Z}_P$ is coarsest if and only if it is a maximal element in the partial order $(\mathcal{Z}_P, \geq)$, i.e. $\hat{\mathcal{P}} \geq \mathcal{P}$ implies $\hat{\mathcal{P}} = \mathcal{P}$.*

Proof. (1) Assume $\mathcal{P} = (P_j)_j$ was not coarsest. According to Definition 6.2.3 let be $(u_k, \mathcal{P}^k, T_k) \in \mathcal{D}$ and u be given such that $(u, \mathcal{P}, T) \in \mathcal{D}_\infty$ and (6.10)-(6.13) hold. Without restriction possibly passing to a subsequence we assume that $\frac{1}{\sqrt{\varepsilon_k}}\big(|R_1^k - R_2^k| + |c_1^k - c_2^k|\big) \le C$ for all $k \in \mathbb{N}$ (cf. (6.14)). By (9.23) we obtain $A^k \in \mathbb{R}^{2\times 2}_{\rm skew}$ with $|A^k| \le C$ such that $R_1^k - R_2^k = R_1^k(\mathbf{Id} - (R_1^k)^T R_2^k) = R_1^k(\sqrt{\varepsilon_k}A^k + O(\varepsilon_k))$. Passing to a (not relabeled) subsequence we then obtain

$$\begin{aligned} S(x) &:= \lim_{k\to\infty} \frac{1}{\sqrt{\varepsilon_k}}\big((R_1^k - R_2^k)\,x + c_1^k - c_2^k\big) \\ &= \lim_{k\to\infty} \frac{1}{\sqrt{\varepsilon_k}}\big(\sqrt{\varepsilon_k}R_1^kA^k\,x + c_1^k - c_2^k\big) + O(\sqrt{\varepsilon_k}) = RA\,x + c \end{aligned} \tag{10.17}$$

for some $A \in \mathbb{R}^{2\times 2}_{\rm skew}$, $c \in \mathbb{R}^2$ and $R = \lim_{k\to\infty} R_1^k$. We now define $\hat{\mathcal{P}}^k$, $\hat{\mathcal{P}}$, $\hat{T}_k$, $\hat{u}_k, \hat{u}$ as follows. Let $\hat{P}_1^k = P_1^k \cup P_2^k$, $\hat{P}_2^k = \emptyset$, $\hat{P}_j^k = P_j^k$ for $j \ge 3$ and likewise for the limiting partition $\hat{\mathcal{P}}$. Let $\hat{T}_k(x) = R_1^k x + c_1^k$ for $x \in \hat{P}_1^k$ and $\hat{T}_k(x) = T_k(x)$ for $x \in \Omega \setminus \hat{P}_1^k$. It is elementary to see that (6.11) holds as $|R_1^k - R_2^k| + |c_1^k - c_2^k| \to 0$ for $k \to \infty$. Furthermore, we let

$$\hat{u}_k = u_k + \frac{1}{\sqrt{\varepsilon_k}}\big((R_1^k - R_2^k)\,x + c_1^k - c_2^k\big)\chi_{P_2^k}$$

and $\hat{u} = u + (RA\,x + c)\chi_{P_2}$ (see (10.17)). Then (6.10)(i) clearly holds as $\hat{T}_k - T_k = (R_1^k - R_2^k)\,x + c_1^k - c_2^k\big)\chi_{P_2^k}$. It is not hard so see that $\mathcal{H}^1(J_u \setminus \partial P) + \frac{1}{2}\sum_j P(P_j, \Omega) \ge \mathcal{H}^1(J_{\hat{u}} \setminus \partial \hat{P}) + \frac{1}{2}\sum_j P(\hat{P}_j, \Omega)$, where $\hat{P} = \bigcup_j \partial^* \hat{P}_j$ and thus also (6.13) is satisfied. It remains to verify (6.12) and (6.10)(ii). First, (6.12)(iii) is obvious and (6.12)(i) follows from (10.17) and the definition of $\hat{u}$. To see (6.12)(ii) we use $R_1^k = R_2^k + \sqrt{\varepsilon_k}R_1^kA^k + O(\varepsilon_k)$, $|A^k| \le C$ and observe

$$\begin{aligned} \chi_{\hat{P}_1^k} e((R_1^k)^T \nabla \hat{u}_k) &= \sum\nolimits_{j=1,2} \chi_{P_j^k} e((R_1^k)^T \nabla u_k) + \chi_{P_2^k} e((R_1^k)^T R_1^k A^k) + O(\sqrt{\varepsilon_k}) \\ &= \sum\nolimits_{j=1,2} \chi_{P_j^k} e((R_j^k)^T \nabla u_k) + O(\sqrt{\varepsilon_k}) \\ &\rightharpoonup \sum\nolimits_{j=1,2} \chi_{P_j} e(R_j^T \nabla u) = \chi_{\hat{P}_1} e(R^T \nabla \hat{u}) \end{aligned}$$

weakly in $L^2(\Omega, \mathbb{R}^{2\times 2}_{\rm sym})$. Finally, to establish (6.10)(ii) we find by $\|\nabla u_k\|_{L^\infty(\Omega)} \le C\varepsilon_k^{-1/8}$

$$\nabla \hat{T}_k^T \nabla \hat{u}_k = (R_1^k)^T \nabla u_k + A^k + O(\sqrt{\varepsilon_k}) = (R_2^k)^T \nabla u_k + A^k + O(\varepsilon_k^{3/8}) \tag{10.18}$$

a.e. in P_2^k. Observe that $W(G) = \frac{1}{2}Q(e(G - \mathbf{Id})) + \omega(G - \mathbf{Id})$ with $\sup\{|F|^{-3}\omega(F) : |F| \le 1\} \le C$ and $\omega(RG) = \omega(G)$ for $G \in \mathbb{R}^{2\times 2}$, $R \in SO(2)$ by the assumptions on W, where $Q = D^2W(\mathbf{Id})$. Thus, we obtain by (10.18)

$$\begin{aligned} \frac{1}{\varepsilon_k}\int_{P_2^k} W(\mathbf{Id} + \sqrt{\varepsilon_k}\nabla \hat{T}_k^T \nabla \hat{u}_k) &= \int_{P_2^k} \Big(\frac{1}{2}Q(e(\nabla \hat{T}_k^T \nabla \hat{u}_k)) + \frac{1}{\varepsilon_k}\omega(\sqrt{\varepsilon_k}\nabla \hat{T}_k^T \nabla \hat{u}_k)\Big) \\ &= \int_{P_2^k} \Big(\frac{1}{2}Q(e(\nabla T_k^T \nabla u_k)) + \frac{\omega(\sqrt{\varepsilon_k}\nabla \hat{u}_k)}{\varepsilon_k}\Big) + O(\varepsilon_k^{\frac{3}{4}}) \end{aligned}$$

and likewise

$$\frac{1}{\varepsilon_k}\int_{P_2^k} W(\mathbf{Id}+\sqrt{\varepsilon_k}\nabla T_k^T\nabla u_k) = \int_{P_2^k}\Big(\frac{1}{2}Q(e(\nabla T_k^T\nabla u_k)) + \frac{1}{\varepsilon_k}\omega(\sqrt{\varepsilon_k}\nabla u_k)\Big).$$

In both estimates the second terms converge to 0 arguing as in (10.9) and using $\|\nabla u_k\|_\infty, \|\nabla \hat{u}_k\|_\infty \le C\varepsilon_k^{-1/8}$. Consequently, we get

$$\frac{1}{\varepsilon_k}\int_{P_2^k} W(\mathbf{Id}+\sqrt{\varepsilon_k}\nabla \hat{T}_k^T\nabla \hat{u}_k) = \frac{1}{\varepsilon_k}\int_{P_2^k} W(\mathbf{Id}+\sqrt{\varepsilon_k}\nabla T_k^T\nabla u_k) + o(1) \quad (10.19)$$

for $\varepsilon_k \to 0$. Thus, $\hat{\mathcal{P}}$ is an admissible partition and thus $\mathcal{P}$ is not maximal.

(2) Conversely, assume that $\mathcal{P} = (P_j)_j$ was not maximal, i.e. we find $\hat{\mathcal{P}} = (\hat{P}_j)_j$ with $\hat{\mathcal{P}} \ge \mathcal{P}$, $\hat{\mathcal{P}} \neq \mathcal{P}$, i.e. $\bigcup_j \partial^* P_j$ and $\bigcup_j \partial^* \hat{P}_j$ differ by a set of positive $\mathcal{H}^1$-measure. We may assume without restriction that $P_1 \cap \hat{P}_1$ and $P_2 \cap \hat{P}_1$ have positive $\mathcal{L}^2$-measure. According to Definition 6.2.3(i) let $(u_k, \mathcal{P}^k, T_k), (\hat{u}_k, \hat{\mathcal{P}}^k, \hat{T}_k) \in \mathcal{D}$ and $u, \hat{u}$ be given such that $(u, \mathcal{P}, T), (\hat{u}, \hat{\mathcal{P}}, T) \in \mathcal{D}_\infty$ and (6.10)-(6.13) hold. As by (6.12) u_k and $\hat{u}_k$ convergence pointwise a.e., by (6.10) also $\frac{1}{\sqrt{\varepsilon_k}}(T_k - \hat{T}_k)$ converges pointwise a.e. (and the limits lie in $\mathbb{R}$ a.e.). But this implies $\frac{1}{\sqrt{\varepsilon_k}}\big(|R_j^k - \hat{R}_1^k| + |c_j^k - \hat{c}_1^k|\big) \le C$ for $j = 1, 2$ and $k \in \mathbb{N}$. Then the triangle inequality shows that (6.14) is violated and thus $\mathcal{P}$ is not a coarsest partition. □

The alternative characterization now directly implies that there is at most one coarsest partition.

Lemma 10.2.3. *Let $(y_k)_k$ be a (sub-)sequence as considered in Theorem 6.2.1. Then there is at most one maximal element in $(\mathcal{Z}_P, \ge)$.*

Proof. Assume there are two maximal elements $\mathcal{P}^1, \mathcal{P}^2 \in \mathcal{Z}_P$ with $\mathcal{P}^1 \neq \mathcal{P}^2$. As before, without restriction we may assume that $P_1^1 \cap P_1^2$ and $P_2^1 \cap P_1^2$ have positive $\mathcal{L}^2$-measure. We proceed as in the proof of Lemma 10.2.2(2) to see that the partition $\mathcal{P}^1$ is not coarsest and thus not a maximal element in $(\mathcal{Z}_P, \ge)$. □

We now analyze the admissible configurations if the partitions are given. Recall that T is uniquely determined by Lemma 10.2.5.

Lemma 10.2.4. *Let $(y_k)_k$ be a (sub-)sequence as considered in Theorem 6.2.1 and $\mathcal{P}, \hat{\mathcal{P}} \in \mathcal{Z}_P$ such that $\mathcal{P} \le \hat{\mathcal{P}}$. Let $\hat{u} \in \mathcal{Z}_u(\hat{\mathcal{P}})$. Then $\mathcal{Z}_u(\mathcal{P}) = \hat{u} + \nabla T \mathcal{A}(\mathcal{P})$.*

Proof. (1) To see $\mathcal{Z}_u(\mathcal{P}) \subset \hat{u} + \nabla T\mathcal{A}(\mathcal{P})$ we have to show that $u - \hat{u} \in \nabla T\mathcal{A}(\mathcal{P})$ for all $u \in \mathcal{Z}_u(\mathcal{P})$. To this end, consider $P_{j_1} \in \mathcal{P}$, $\hat{P}_{j_2} \in \hat{\mathcal{P}}$ such that $|P_{j_1} \setminus \hat{P}_{j_2}| = 0$. Let u_k, $\hat{u}_k$ and T_k, $\hat{T}_k$ be given according to Definition 6.2.3. As $u_k - \hat{u}_k$ and thus $\frac{1}{\sqrt{\varepsilon_k}}(T_k - \hat{T}_k)$ converge pointwise a.e. we find $|R_{j_1}^k - \hat{R}_{j_2}^k| + |c_{j_1}^k - \hat{c}_{j_2}^k| \le C\sqrt{\varepsilon_k}$. Repeating the argument in (10.17) we derive $u(x) - \hat{u}(x) = \lim_{k\to\infty} u_k(x) - \hat{u}_k(x) = \lim_{k\to\infty} \frac{1}{\sqrt{\varepsilon_k}}(T_k(x) - \hat{T}_k(x)) = \nabla T(x)(A\,x + c)$ for a.e. $x \in P_{j_1}$ for some $A \in \mathbb{R}^{2\times 2}_{\text{skew}}$, $c \in \mathbb{R}^2$.

(2) Conversely, to see $\mathcal{Z}_u(\mathcal{P}) \supset \hat{u} + \nabla T \mathcal{A}(\mathcal{P})$ we first consider the special case $\mathcal{P} = \hat{\mathcal{P}} = (P_h)_h$. Let $\bar{u} \in \mathcal{Z}_u(\mathcal{P})$ and $A(x) = \sum_h (A_h x + c_h)\chi_{P_h}$ be given. We have to show that $u := \bar{u} + \nabla T A \in \mathcal{Z}_u(\mathcal{P})$.

We first note that $\mathcal{H}^1(J_u \setminus \partial P) = \mathcal{H}^1(J_{\bar{u}} \setminus \partial P)$ and thus (6.13) is satisfied. According to Definition 6.2.3(iii) let $(\bar{u}_k, \mathcal{P}^k, \bar{T}_k) \in \mathcal{D}$ be given such that (6.10)-(6.13) hold. Assume that $\bar{T}_k$ has the form $\bar{T}_k(x) = \bar{R}^k_j x + \bar{c}^k_j$ for $x \in P^k_j$. Now choose R^k_j such that $|R^k_j - \bar{R}^k_j(\mathbf{Id} + \sqrt{\varepsilon_k} A_j)| = \operatorname{dist}(\bar{R}^k_j(\mathbf{Id} + \sqrt{\varepsilon_k}A_j), SO(2))$ and let $c^k_j = \bar{c}^k_j + \sqrt{\varepsilon_k} R^k_j c_j$. Define

$$T_k(x) = \sum\nolimits_j (R^k_j x + c^k_j)\chi_{P^k_j}$$

as well as $u_k = \bar{u}_k + \frac{1}{\sqrt{\varepsilon_k}}(T_k - \bar{T}_k)$. Clearly, this implies (6.10)(i). By (9.23) we have $R^k_j = \bar{R}^k_j + \sqrt{\varepsilon_k} R^k_j A_j + \omega_{j,k}$ with $\varepsilon_k^{-\frac{1}{2}}|\omega_{j,k}| \to 0$ for all $j \in \mathbb{N}$. Moreover, we find

$$T_k = \bar{T}_k + \sum\nolimits_j \big(\sqrt{\varepsilon_k} R^k_j A_j x + \omega_{j,k} x + \sqrt{\varepsilon_k} R^k_j c_j\big)\chi_{P^k_j} \to T$$

in measure for $k \to \infty$. Then it is not hard to see that $T_k \to T$ and $\nabla T_k \to \nabla T$ in L^2 which gives (6.11). Likewise, we obtain

$$\begin{aligned} u_k - \bar{u}_k = \frac{1}{\sqrt{\varepsilon_k}}(T_k - \bar{T}_k) &= \sum\nolimits_j \Big(R^k_j(A_j x + c_j) + \frac{1}{\sqrt{\varepsilon_k}}\omega_{j,k} x\Big)\chi_{P^k_j} \\ &\to \nabla T \sum\nolimits_j (A_j x + c_j)\chi_{P_j} = \nabla T A \end{aligned}$$

pointwise a.e. which implies $u_k \to \bar{u} + \nabla T A$ and shows (6.12)(i). We observe that there are decreasing sets V_k with $|V_k| \to 0$ for $k \to \infty$ such that $\|\nabla A\|_{L^\infty(\Omega \setminus V_k)} \le C\varepsilon_k^{-1/8}$ and $\|\sum_j \chi_{P^k_j} \varepsilon_k^{-1/2}\omega_{j,k}\|_{L^\infty(\Omega\setminus V_k)} \to 0$ for $k \to \infty$. Consequently, we obtain

$$\|\nabla u_k - \nabla \bar{u}_k\|_{L^\infty(\Omega\setminus V_k)} \le \|\nabla A\|_{L^\infty(\Omega\setminus V_k)} + \|\sum\nolimits_j \chi_{P^k_j}\varepsilon_k^{-1/2}\omega_{j,k}\|^2_{L^\infty(\Omega\setminus V_k)} \le C\varepsilon_k^{-1/8}$$

and therefore, replacing u_k by $\chi_{\Omega\setminus V_k} u_k$ we find that (6.10)(i) still holds and (6.12)(iii) is fulfilled. Arguing similarly as in (10.18) and taking $\|\nabla \bar{u}_k\|_{L^\infty(\Omega\setminus V_k)} + \|\nabla A\|_{L^\infty(\Omega\setminus V_k)} \le C\varepsilon_k^{-1/8}$ we find

$$\begin{aligned} (R^k_j)^T \nabla u_k(x) &= (R^k_j)^T \nabla \bar{u}_k(x) + A_j + (R^k_j)^T \varepsilon_k^{-1/2} w_{j,k} \\ &= (\bar{R}^k_j)^T \nabla \bar{u}_k(x) + O(\varepsilon_k^{1/4}) + A_j + (R^k_j)^T \varepsilon_k^{-1/2} w_{j,k} \end{aligned}$$

for a.e. $x \in P^k_j \setminus V_k$. Thus, also (6.12)(ii) follows from the fact that (6.12)(ii) holds for the sequence $\bar{u}_k$ and

$$\begin{aligned} &\sum\nolimits_j \int_{P^k_j \setminus V_k} |e\big((R^k_j)^T \nabla u_k\big) - e\big((\bar{R}^k_j)^T \nabla \bar{u}_k\big)|^2 \\ &\qquad\qquad \le C\|\sum\nolimits_j \chi_{P^k_j}\varepsilon_k^{-1/2}\omega_{j,k}\|^2_{L^\infty(\Omega\setminus V_k)} + C\varepsilon_k^{1/2} \to 0. \end{aligned}$$

Finally, the above estimates together with a similar argumentation as in (10.19) yield (6.10)(ii).

In the general case we have to show $u := \hat{u} + \nabla T A \in \mathcal{Z}_u(\mathcal{P})$ for given $\hat{u} \in \mathcal{Z}_u(\hat{\mathcal{P}})$, $\hat{\mathcal{P}} \geq \mathcal{P}$ and $A \in \mathcal{A}(\mathcal{P})$. As $\mathcal{P} \in \mathcal{Z}_P$ we find some $\bar{u} \in \mathcal{Z}_u(\mathcal{P})$ which by (1) satisfies $\bar{u} - \hat{u} = \nabla T \bar{A}$ for $\bar{A} \in \mathcal{A}(\mathcal{P})$. Consequently, we get $u = \bar{u} + \nabla T(A - \bar{A})$ and by the special case in (2) we know that $u \in \mathcal{Z}_u(\mathcal{P})$, as desired. □

To guarantee existence of coarsest partitions we show that each totally ordered subset has upper bounds such that afterwards we may apply Zorn's lemma.

Lemma 10.2.5. *Let $(y_k)_k$ be a (sub-)sequence as considered in Theorem 6.2.1. Let I be an arbitrary index set and let $\{\mathcal{P}_i = (P_{i,j})_j : i \in I\} \subset \mathcal{Z}_P$ be a totally ordered subset, i.e. for each $i_1, i_2 \in I$ we have $\mathcal{P}_{i_1} \leq \mathcal{P}_{i_2}$ or $\mathcal{P}_{i_2} \leq \mathcal{P}_{i_1}$. Then there is a partition $\mathcal{P} \in \mathcal{Z}_P$ with $\mathcal{P}_i \leq \mathcal{P}$ for all $i \in I$.*

Proof. To prove the existence of an upper bound we first show that it suffices to consider a suitable countable subset of $\{\mathcal{P}_i : i \in I\}$. For notational convenience we write $i_1 \leq i_2$ for $i_1, i_2 \in I$ if $\mathcal{P}_{i_1} \leq \mathcal{P}_{i_2}$. Choose an arbitrary $i_0 \in I$ and note that it suffices to find an upper bound for all $i \geq i_0$. We observe that for each $i \geq i_0$ we have $\bigcup_j \partial^* P_{i,j} \subset \bigcup_j \partial^* P_{i_0,j}$ (up to an $\mathcal{H}^1$-negligible set). Thus, for each $k \in \mathbb{N}$ there are (coarsened) partitions $\mathcal{P}_i^k = (P_{i,j}^k)_j$ with $\bigcup_j \partial^* P^k_{i,j} = \bigcup_j \partial^* P_{i,j} \setminus \big(\bigcup_{j \geq k} \partial^* P_{i_0,j} \setminus \partial^*(\bigcup_{j\geq k} P_{i_0,j})\big)$ up to $\mathcal{H}^1$-negligible sets for all $i \geq i_0$. Observe that typically $\mathcal{P}_i^k$ are not elements of $\{\mathcal{P}_i : i \in I\}$, but satisfy

$$\mathcal{H}^1\Big(\bigcup\nolimits_j \partial^* P_{i,j} \setminus \bigcup\nolimits_j \partial^* P^k_{i,j}\Big) \leq \omega(k)$$

with $\omega(k) \to 0$ for $k \to \infty$. After identifying partitions whose boundaries only differ by $\mathcal{H}^1$-negligible sets we find that each $\{\mathcal{P}_i^k : i \geq i_0\}$ contains only a finite number of different elements and therefore contains a maximal element $\mathcal{P}^k = (P_j^k)_j$. Now we can choose $i_0 \leq i_1 \leq i_2 \leq \ldots$ such that $\mathcal{P}^k = \mathcal{P}^k_{i_k}$ for $k \in \mathbb{N}$. It now suffices to construct an upper bound $\mathcal{P} = (P_j)_j \in \mathcal{Z}_P$ with $\mathcal{P} \geq \mathcal{P}_{i_k}$ for all $k \in \mathbb{N}$. Indeed, we then obtain for all $i \geq i_0$

$$\begin{aligned}
\mathcal{H}^1\Big(\bigcup\nolimits_j \partial^* P_j \setminus \bigcup\nolimits_j \partial^* P_{i,j}\Big) &\leq \mathcal{H}^1\Big(\bigcup\nolimits_j \partial^* P_j \setminus \bigcup\nolimits_j \partial^* P^k_{i,j}\Big)\\
&\leq \mathcal{H}^1\Big(\bigcup\nolimits_j \partial^* P_j \setminus \bigcup\nolimits_j \partial^* P^k_{j}\Big)\\
&\leq \mathcal{H}^1\Big(\bigcup\nolimits_j \partial^* P_j \setminus \bigcup\nolimits_j \partial^* P_{i_k,j}\Big) + \omega(k) = \omega(k)
\end{aligned}$$

and as $k \in \mathbb{N}$ was arbitrary, we derive $\bigcup_j \partial^* P_j \subset \bigcup_j \partial^* P_{i,j}$, as desired.

Now consider the totally ordered sequence of partitions $(\mathcal{P}_{i_k})_k$. For notational convenience we will denote the sequence by $(\mathcal{P}_i)_{i \in \mathbb{N}}$ in the following. By the compactness theorem for Caccioppoli partitions (see Theorem A.2.2) we get an (ordered) Caccioppoli partition $\mathcal{P} = (P_j)_j$ such that $\chi_{P_{i,j}} \to \chi_{P_j}$

in measure for $i \to \infty$ for all $j \in \mathbb{N}$. This also implies $\mathcal{H}^1(\bigcup_j \partial^* P_j \setminus E) \le \liminf_{i\to\infty} \mathcal{H}^1(\bigcup_j \partial^* P_{i,j} \setminus E)$ for every $\mathcal{H}^1$-measurable set with $\mathcal{H}^1(E) < \infty$ (see e.g. [35, Theorem 2.8]). Consequently, we apply $\bigcup_j \partial^* P_{i,j} \subset \bigcup_j \partial^* P_{k,j}$ for $i \ge k$ to derive $\mathcal{H}^1(\bigcup_j \partial^* P_j \setminus \bigcup_j \partial^* P_{k,j}) = 0$ for all $k \in \mathbb{N}$. This implies $\mathcal{P} \ge \mathcal{P}_k$ for all $k \in \mathbb{N}$ and therefore it suffices to show that $\mathcal{P} \in \mathcal{Z}_P$. To this end, we will construct partitions $\mathcal{P}^n$, rigid motions $T_n \in \mathcal{R}(\mathcal{P}^n)$ and a limiting function u by a diagonal sequence argument.

For all $i \in \mathbb{N}$, according to Definition 6.2.3(i), we find $(u_i^k, \mathcal{P}_i^k, T_i^k) \in \mathcal{D}$ and a sequence of admissible limiting configurations $u_i \in \mathcal{Z}_u(\mathcal{P}_i)$ such that (6.10)-(6.13) hold as $k \to \infty$. The strategy is to select u_i in a suitable way such that we find a limiting configuration $u \in GSBD(\Omega)$ with

$$\begin{aligned} &u_i \to u \ \text{ a.e.},\\ &e(\nabla T^T \nabla u_i) \rightharpoonup e(\nabla T^T \nabla u),\\ &\liminf_{i\to\infty} \mathcal{H}^1(J_{u_i}) \ge \mathcal{H}^1(J_u). \end{aligned} \tag{10.20}$$

Then we can choose a diagonal sequence $(\bar{u}_n) := (u_n^{k(n)})_n$ converging to the triple $(u, \mathcal{P}, T)$ in the sense of (6.10)-(6.13). Indeed, $k(n)$ can be selected such that letting $\bar{\mathcal{P}}^n = (\bar{P}_j^n)_j = \mathcal{P}_n^{k(n)}$ and $\bar{T}_n = T_n^{k(n)} \in \mathcal{R}(\mathcal{P}^n)$ we find $\chi_{\bar{P}_j^n} \to \chi_{P_j}$ in measure for all $j \in \mathbb{N}$ and $\bar{T}_n \to T$, $\nabla \bar{T}_n \to \nabla T$ in $L^2(\Omega)$ which gives (6.11). Moreover, possibly passing to a further subsequence this can be done in a way that $\bar{u}_n \to u$ a.e., $\bar{u}_n - \varepsilon_n^{-1/2}(y_n - \bar{T}_n) \to 0$ a.e. and therefore also (6.10), (6.12)(i) hold.

Likewise, (6.12)(ii) can be achieved by (10.20) and the fact the weak convergence is metrizable as $\|e(\nabla(T_i^k)^T \nabla u_i^k)\|_{L^2(\Omega)} \le C$ for a constant independent of k, i. The last property follows from the construction of $\hat{v}_k^l$ in the proof of Theorem 6.2.1 (see (10.8), (10.9)). Moreover, (6.12)(iii) and (6.10)(ii) directly follow from the corresponding estimates for the functions u_i^k. Finally, to see (6.13) it suffices to prove

$$\liminf_{i\to\infty} \Big(\mathcal{H}^1(J_{u_i} \setminus \partial P_i) + \frac{1}{2}\sum\nolimits_j P(P_{i,j}, \Omega)\Big) \ge \Big(\mathcal{H}^1(J_u \setminus \partial P) + \frac{1}{2}\sum\nolimits_j P(P_j, \Omega)\Big),$$

where $\partial P_i = \bigcup_j \partial^* P_{i,j}$. This can be derived arguing as in (10.15): We may consider an infinitesimal perturbation of the form $\tilde{u}_i = u_i + \sum_j c_j \chi_{P_{i,j}}$, $\tilde{u} = u + \sum_j c_j \chi_{P_j}$ with c_j small such that $\mathcal{H}^1(\partial P_i \setminus J_{\tilde{u}_i}) = \mathcal{H}^1(\partial P \setminus J_{\tilde{u}}) = 0$ and the convergence in (10.20) still holds after replacing u_i, u by $\tilde{u}_i$, $\tilde{u}$, respectively. Then the claim follows from (10.20). Consequently, $\mathcal{P} \in \mathcal{Z}_P$ due to Definition 6.2.3(i).

It suffices to show (10.20). Clearly, we have $\|e(\nabla T^T \nabla u_i)\|^2_{L^2(\Omega)} \le C$ and $\mathcal{H}^1(J_{u_i}) \le C$ for a constant independent of $i \in \mathbb{N}$. This follows by a lower semicontinuity argument taking (6.13) and $\|e(\nabla(T_n^k)^T \nabla u_n^k)\|_{L^2(\Omega)} \le C$ into account. Consequently, in order to apply Theorem A.1.3 we have to find an increasing continuous function $\psi : [0, \infty) \to [0, \infty)$ with $\lim_{x\to\infty} \psi(x) = +\infty$ such that $\int_\Omega \psi(|u_i|) \le C$.

We proceed similarly as in the proof of Theorem 6.2.1 and define u_i iteratively. Choose $u_1 \in \mathcal{Z}_u(\mathcal{P}_1)$ arbitrarily. Given u_i we define u_{i+1} as follows. We recall $\bigcup_j \partial^* P_{i_2,j} \subset \bigcup_j \partial^* P_{i_1,j}$ for $i_1 \le i_2$. Consider some $P_{i+1,j}$ and choose $l_{1,j} < l_{2,j} < \ldots$ such that $P_{i+1,j} = \bigcup_{k=1}^\infty P_{i,l_{k,j}}$ up to an $\mathcal{L}^2$- negligible set (observe that the union may also be finite). Without restriction assume that $P_{i,l_{1,j}}$ has largest Lebesgue measure. Choose some $\tilde{u}_{i+1} \in \mathcal{Z}_u(\mathcal{P}_{i+1})$. By Lemma 10.2.4 for $\mathcal{P} = \mathcal{P}_i, \hat{\mathcal{P}} = \mathcal{P}_{i+1}$ we get $(u_i - \tilde{u}_{i+1})\chi_{P_{i+1,j}} = \sum_{k=1}^\infty (A_{l_{k,j}}\, x + c_{l_{k,j}})\chi_{P_{i,l_{k,j}}}$ for $A_{l_{k,j}} \in \mathbb{R}^{2\times 2}_{\text{skew}}$, $c_{l_{k,j}} \in \mathbb{R}^2$. Now define

$$u_{i+1}(x) = \tilde{u}_{i+1}(x) + A_{l_{1,j}}\, x + c_{l_{1,j}}$$

for $x \in P_{i+1,j}$ and observe that $u_i = u_{i+1}$ on $P_{i,l_{1,j}}$. Proceeding in this way on all $P_{i+1,j}$ we find some $\tilde{A}^{i+1} \in \mathcal{A}(\mathcal{P}_{i+1})$ such that $u_{i+1} := \tilde{u}_{i+1} + \tilde{A}^{i+1} \in \mathcal{Z}_u(\mathcal{P}_{i+1})$ applying Lemma 10.2.4 for $\mathcal{P} = \hat{\mathcal{P}} = \mathcal{P}_{i+1}$. Moreover, there is a corresponding $A^i \in \mathcal{A}(\mathcal{P}_i)$ such that $u_{i+1} = u_i + A^i$ with $A^i = 0$ on $\bigcup_j P_{i,l_{1,j}}$.

We now show that $\sum_{i\in\mathbb{N}} |A^i| < +\infty$ a.e. To see this, we recall that $\chi_{P_{i,j}} \to \chi_{P_j}$ in measure for all $j \in \mathbb{N}$. Consequently, as due to the total order of the partitions the sets $P_{i,j}$ are increasing for fixed $j \in \mathbb{N}$, the construction of the functions $(u_i)_i$ implies $A^i = 0$ on $P_{i,j}$ for i so large that $|P_{i,j}| > \frac{1}{2}|P_j|$. Thus, for a.e. $x \in P_j$ the sum $\sum_{i\ge 1} |A^i(x)|$ is a finite sum and therefore finite. Taking the union over all $j \in \mathbb{N}$ we obtain $\sum_{i\in\mathbb{N}} |A^i| < +\infty$ a.e.

Consequently, there is an increasing continuous function $\psi : [0,\infty) \to [0,\infty)$ with $\lim_{x\to\infty} \psi(x) = \infty$ such that $\|\psi(|u_1| + \sum_{i\in\mathbb{N}} |A^i|)\|_{L^1(\Omega)} < \infty$. Using the definition $u_{i+1} = u_i + A^i$ and the monotonicity of ψ we find $\|\psi(|u_i|)\|_{L^1(\Omega)} \le \|\psi(|u_1| + \sum_{k\in\mathbb{N}} |A^k|)\|_{L^1(\Omega)} < \infty$ for all $i \in \mathbb{N}$, as desired. □

After these preparatory lemmas we are finally in a position to prove Theorem 6.2.4.

Proof of Theorem 6.2.4. First, (i) follows from Lemma 10.2.1. The uniqueness of the coarsest partition is a consequence of Lemma 10.2.3 and Lemma 10.2.2. We obtain existence by Zorn's lemma: As $(\mathcal{Z}_P, \ge)$ is a partial order and every chain has an upper bound by Lemma 10.2.5, there exists a maximal element $\bar{\mathcal{P}} \in \mathcal{Z}_P$. Lemma 10.2.2 shows that $\bar{\mathcal{P}}$ is a coarsest partition which gives (ii). Finally, assertion (iii), namely $\mathcal{Z}_u(\bar{\mathcal{P}}) = v + \nabla T \mathcal{A}(\bar{\mathcal{P}})$ for some $v \in \mathcal{Z}_u(\bar{\mathcal{P}})$, follows from Lemma 10.2.4 for the choice $\mathcal{P} = \hat{\mathcal{P}} = \bar{\mathcal{P}}$. □

10.3 Derivation of linearized models via Γ-convergence

This section is devoted to the proof of Theorem 6.3.1.

Proof of Theorem 6.3.1. (i) Thanks to the preparations in the last section the lower bound is almost immediate. Let $(u, \mathcal{P}, T) \in \mathcal{D}_\infty$ be given as well as a sequence $(y_k)_k \subset SBV_M(\Omega)$ with corresponding $(u_k, \mathcal{P}^k, T_k) \in \mathcal{D}$ such that (6.10)-(6.13) hold. By (6.13) it suffices to show that

$$\liminf_{k\to\infty} \frac{1}{\varepsilon_k}\int_\Omega W(\nabla y_k) \geq \int_\Omega \frac{1}{2}Q(e(\nabla T^T \nabla u)).$$

We proceed as in (10.8): Recall that $W(G) = \frac{1}{2}Q(e(G - \mathbf{Id})) + \omega(G - \mathbf{Id})$ with $\sup\{|F|^{-3}\omega(F) : |F| \leq 1\} \leq C$ by the assumptions on W, where $Q = D^2W(\mathbf{Id})$. We compute by (6.10)(ii)

$$\begin{aligned}\frac{1}{\varepsilon_k}\int_\Omega W(\nabla y_k) &\geq \frac{1}{\varepsilon_k}\int_\Omega W(\mathbf{Id} + \sqrt{\varepsilon_k}\nabla T_k^T \nabla u_k) + o(1)\\ &= \int_\Omega \frac{1}{2}\Big(Q(e(\nabla T_k^T\nabla u_k)) + \frac{1}{\varepsilon_k}\omega(\sqrt{\varepsilon_k}\nabla T_k^T \nabla u_k)\Big) + o(1)\end{aligned}$$

as $k \to \infty$. The second term converges to 0 arguing as in (10.9) and using $\|\nabla u_k\|_\infty \leq C\varepsilon_k^{-1/8}$ (see (6.12)). As $e(\nabla T_k^T \nabla u_k) \rightharpoonup e(\nabla T^T \nabla u)$ weakly in $L^2(\Omega, \mathbb{R}^{2\times 2}_{\mathrm{sym}})$ by (6.12)(ii) and Q is convex we conclude

$$\liminf_{k\to\infty} \frac{1}{\varepsilon_k}\int_\Omega W(\nabla y_k) \geq \int_\Omega \frac{1}{2}Q(e(\nabla T^T \nabla u)),$$

as desired.

(ii) By a general density result in the theory of Γ-convergence together with Theorem A.1.6, Theorem A.1.8 and the fact that the limiting functional $E(u, \mathcal{P}, T)$ is continuous in u with respect to the convergence given in Theorem A.1.6 and Theorem A.1.8 it suffices to provide recovery sequences for $u \in \mathcal{W}(\Omega)$. Moreover, as in the proof of Theorem 6.2.1 we may assume that $\mathcal{H}^1(\partial P \setminus J_u) = 0$ up to an infinitesimal small perturbation of u (a similar argument was used in the proof of Lemma 10.2.5). Let $(u, \mathcal{P}, T) \in \mathcal{D}_\infty$ and $\varepsilon_k \to 0$ be given. Define $y_k(x) = Tx + \sqrt{\varepsilon_k}u(x)$ for all $x \in \Omega$. It is not hard to see that $(y_k)_k \subset SBV_M(\Omega)$ for ε_k small enough. Moreover, define $\mathcal{P}^k = \mathcal{P}$, $T_k(x) = Tx$ and $u_k = \frac{1}{\sqrt{\varepsilon_k}}(y_k - T_k) \equiv u$ for all $k \in \mathbb{N}$. Then (6.10),(6.11), (6.13) and the first two parts of (6.12) hold trivially. To see the (6.12)(iii) it suffices to note that $\|\nabla u_k\|_\infty = \|\nabla u\|_\infty \leq C \leq C\varepsilon_k^{-1/8}$.

We finally confirm $\lim_{k\to\infty} E_{\varepsilon_k}(y_k) = E(u, \mathcal{P}, T)$. As for all $k \in \mathbb{N}$ we have $\mathcal{H}^1(J_{u_k}) = \frac{1}{2}\sum_j P(P_j, \Omega) + \mathcal{H}^1(J_u \setminus \partial P)$, it suffices to show $\lim_{k\to\infty} \frac{1}{\varepsilon_k}\int_\Omega W(\nabla y_k) = \int_\Omega \frac{1}{2}Q(e(\nabla T^T \nabla u))$. Using again that $W(G) = \frac{1}{2}Q(e(G - \mathbf{Id})) + \omega(G - \mathbf{Id})$ we compute

$$\begin{aligned}\frac{1}{\varepsilon_k}\int_\Omega W(\nabla y_k) &= \frac{1}{\varepsilon_k}\int_\Omega W(\nabla T_k^T \nabla y_k)\\ &= \int_\Omega \Big(\frac{1}{2}Q(e(\nabla T_k^T \nabla u_k)) + \frac{1}{\varepsilon_k}\omega(\sqrt{\varepsilon_k}\nabla T_k^T \nabla u_k)\Big)\\ &= \int_\Omega \frac{1}{2}Q(e(\nabla T^T \nabla u)) + O(\sqrt{\varepsilon_k}) \to \int_\Omega \frac{1}{2}Q(e(\nabla T^T \nabla u)).\end{aligned}$$

This finishes the proof. □

Remark 10.3.1. Due to the assumptions in the density result of Theorem A.1.8 we have to suppose that $u \in L^2(\Omega)$ in Theorem 6.3.1(ii). A possible strategy to drop this additional assumption is to show that each limiting configuration u given by Theorem 6.2.1 can be approximated in the sense of (6.10)-(6.13) by a sequence $(v^l)_l \subset GSBD(\Omega) \cap L^2(\Omega)$ such that $E(u, \mathcal{P}, T) = \lim_{l\to\infty} E(v^l, \mathcal{P}, T)$. A natural candidate seems to be the sequence $(v^l)_l$ given in the proof of Theorem 6.2.1, but the verification of the convergence of the surface energy appears to be a subtle problem.

The proof of Corollary 6.3.2 is now straightforward.
Proof of Corollary 6.3.2. To see the liminf-inequality assume without restriction that $E_{\varepsilon_k}(y_{\varepsilon_k}) \le C$ and $y_{\varepsilon_k} \to y$ in L^1 for $k \to \infty$. By (6.10)(i), (6.11) we obtain $y = T$ for some $T \in \mathcal{R}(\mathcal{P})$ for a Caccioppoli partition $\mathcal{P}$. Moreover, Theorem 6.3.1 yields $\liminf_{k\to\infty} E_{\varepsilon_k}(y_k) \ge E_{\rm seg}(y)$. A recovery sequence is obviously given by $y_k = y$ for all $k \in \mathbb{N}$. □

10.4 Application: Cleavage laws

We are finally in a position to prove the cleavage law in Theorem 6.3.3.
Proof of Theorem 6.3.3. The proof is very similar to the proof of Theorem 1.6.5 and we only indicate the necessery changes. Let $(y_{\varepsilon_k})_k$ be a sequence of almost minimizers. Passing to a suitable subsequence, by Theorem 6.2.1 we obtain a triple $(u_k, \mathcal{P}^k, T_k) \in \mathcal{D}$ and a limiting triple $(u, \mathcal{P}, T) \in \mathcal{D}_\infty$ such that (6.10)-(6.13) hold and

$$E(u, \mathcal{P}, T) \le \liminf_{\varepsilon\to 0} \inf\{E_\varepsilon(y) : y \in \mathcal{A}(a_\varepsilon)\}$$

by Theorem 6.3.1(i). Due to the boundary conditions it is not hard to see that on each component $P_j \in \mathcal{P}$ we find $A_j \in \mathbb{R}^{2\times 2}_{\rm skew}$ and $c_j \in \mathbb{R}^2$ such that

$$\begin{aligned} u_1(x) &= \lim_{k\to\infty} \varepsilon_k^{-1/2}(\mathbf{e}_1 \cdot (\mathbf{Id} - R^k_j)\,x - \mathbf{e}_1 \cdot c^k_j + a_\varepsilon x_1) \\ &= \mathbf{e}_1 \cdot A_j\,x + \mathbf{e}_1 \cdot c_j + a x_1 \end{aligned} \tag{10.21}$$

for a.e. $x \in \Omega'$ with $x_1 < 0$ or $x_1 > l$ and $x \in P_j$. In particular, this implies

$$u_1(x_1, x_2) - u_1(\hat{x}_1, x_2) = |x_1 - \hat{x}_1| a \tag{10.22}$$

for a.e. $x \in \Omega'$ with $\hat{x}_1 < 0$, $x_1 > l$ and $(x_1, x_2), (\hat{x}_1, x_2) \in P_j$.

We first derive the limiting minimal energy and postpone the characterization of the sequence of almost minimizers to the end of the proof. The argument in (10.21) shows that $\nabla T = \mathbf{Id}$ on P_j if $|P_j \cap \{x : x_1 < 0 \text{ or } x_1 > l\}| > 0$. As in the proof of Lemma 10.2.5 (cf. also proof of Theorem 6.3.1(ii)) we may assume that $\mathcal{H}^1(\partial P \setminus J_{u_1}) = 0$ after a possible infinitesimal perturbation. Consequently, it is not restrictive to assume $\nabla T^T \nabla u = \nabla u$ a.e. Indeed, we may replace u by $\nabla T u$ in a component P_j which does not intersect the boundaries without changing the energy. By (6.16), a slicing argument in GSBD (see Theorem A.1.5 or [34, Section 8,9]) and the fact that $\inf\{Q(F) : \mathbf{e}_1^T F \mathbf{e}_1 = a\} = \alpha a^2$ (see Section 6.3) we obtain

$$
\begin{aligned}
E(u, \mathcal{P}, T) &\geq \int_{\Omega'} \frac{1}{2} Q(e(\nabla u)) + \int_{J_u} |\nu_u \cdot \mathbf{e}_1| d\mathcal{H}^1 + \mathcal{E}(u) \\
&\geq \int_0^1 \Big(\int_0^l \frac{\alpha}{2} (\mathbf{e}_1^T \nabla u(x) \mathbf{e}_1)^2 \, dx_1 + S^{x_2}(u) \Big) \, dx_2 + \mathcal{E}(u),
\end{aligned}
$$

where S^{x_2} denotes the number of jumps of u_1 on a slice $(-\eta, l+\eta) \times \{x_2\}$ and $\mathcal{E}(u) = \int_{J_u} (1 - |\nu_u \cdot \mathbf{e}_1|) d\mathcal{H}^1$. If $S^{x_2} \geq 1$ the inner integral is bounded from below by 1. By the structure theorem for Caccioppoli partitions (see Theorem A.2.1) we find that $((-\eta, 0) \cup (l, l+\eta)) \times \{x_2\} \subset P_j$ for some $j \in \mathbb{N}$ for $\mathcal{H}^1$-a.e. x_2 with $S^{x_2} = 0$ and then arguing as in (5.18) we derive that the term is bounded from below by $\frac{1}{2}\alpha l a^2$ due to the boundary conditions (10.22). This implies $E(u) \geq \min\{\frac{1}{2}\alpha l a^2, 1\}$.

Otherwise, it is not hard to see that the configurations $y^{\mathrm{el}}_{\varepsilon_k} = x + F^{a_{\varepsilon_k}} x$ for $x \in \Omega'$ satisfy $E_{\varepsilon_k}(y^{\mathrm{el}}_{\varepsilon_k}) \to \frac{1}{2}\alpha l a^2$ for $k \to \infty$. Likewise, we get $E_{\varepsilon_k}(y^{\mathrm{cr}}_{\varepsilon_k}) = 1$ for all $k \in \mathbb{N}$, where $y^{\mathrm{cr}}_{\varepsilon_k}(x) = x\chi_{x_1 < \frac{1}{2}} + (x + (l a_{\varepsilon_k}, 0))\chi_{x_1 > \frac{1}{2}}$ for $x \in \Omega$ and $y^{\mathrm{cr}}_{\varepsilon_k} = (x_1(1 + a_{\varepsilon_k}), x_2)$ for $x \in \Omega' \setminus \Omega$. This completes (6.17).

It remains to characterize the sequences of almost minimizers. Let u be a minimizer of E under the boundary conditions (10.21). Let first $|a| < a_{\mathrm{crit}}$. Repeating the arguments in the proof of Theorem 1.6.5(i) we find that $u \in H^1(\Omega')$ with

$$
u(x) = F^a x + A\,x + c
$$

for $x \in \Omega$ and suitable $A \in \mathbb{R}^{2\times 2}_{\mathrm{skew}}$, $c \in \mathbb{R}^2$, where the matrix A appears as in contrast to the proof of Theorem 1.6.5 we cannot derive $\mathbf{e}_1^T \nabla u \mathbf{e}_2 = 0$, but only $\mathbf{e}_1^T \nabla u \mathbf{e}_2 + \mathbf{e}_2^T \nabla u \mathbf{e}_1 = 0$ (cf. (10.21)). In particular, this implies $\mathcal{P}$ consists only of $P_1 = \Omega'$ and thus by (10.21) we get $A = \lim_{k\to\infty} \varepsilon_k^{-1/2}(\mathbf{Id} - R_1^k)$ and $\mathbf{e}_1 \cdot c = -\lim_{k\to\infty} \varepsilon_k^{-1/2} \mathbf{e}_1 \cdot c_1^k$. Letting $s = \lim_{k\to\infty} \mathbf{e}_2 \cdot (\varepsilon_k^{-1/2} c_1^k + c)$ (which exists by (6.10)(i), (6.12)(i)), we now conclude by (6.10) (i) for a.e. $x \in \Omega$

$$
\begin{aligned}
&\lim_{k\to\infty} \varepsilon_k^{-1/2}(y_{\varepsilon_k}(x) - x) \\
&\quad = u(x) + \lim_{k\to\infty} \varepsilon_k^{-1/2}\big((R_1^k - \mathbf{Id})\,x + c_1^k\big) \qquad (10.23) \\
&\quad = u(x) - A\,x - c + (0, s) = (0, s) + F^a\,x.
\end{aligned}
$$

If $|a| > a_{\mathrm{crit}}$ we argue as in the proof of Theorem 1.6.5(ii) to find $p \in (0, l)$,

$A_i \in \mathbb{R}^{2\times 2}_{\text{skew}}$, $c_i \in \mathbb{R}^2$ for $i = 1, 2$ such that

$$u(x) = \begin{cases} A_1\, x + c_1 & \text{for } x_1 < p, \\ A_2\, x + c_2 & \text{for } x_1 > p, \end{cases}$$

where we used that $\mathcal{E}(u) = 0$ iff $\nu_u = \pm\mathbf{e}_1$ a.e. Indeed, the linearized rigidity estimate in [25] can also be applied in the GSBD-setting as it relies on a slicing argument and an approximation which is also available in the generalized framework (see [54, Section 3.3]). (The only difference is that the approximation does not converge in L^1 but only pointwise a.e. which does not affect the argument.) Now repeating the calculation in (10.23) for the sets $P_1 = \{x \in \Omega' : x_1 < p\}$ and $P_2 = \Omega' \setminus P_1$ we find $s, t \in \mathbb{R}$ such that for $x \in \Omega$ a.e.

$$\lim_{k\to\infty} \varepsilon_k^{-1/2}(y_{\varepsilon_k}(x) - x) = u(x) - (A_1\, x + c_1)\chi_{x_1<p}(x) + (A_2\, x + c_2)\chi_{x_1>p}(x)$$
$$+ (0, s)\chi_{x_1<p}(x) - ((la, t))\chi_{x_1>p}(x).$$

This finishes the proof. □

Appendix A

Functions of bounded variation and Caccioppoli partitions

A.1 (G)SBV and (G)SBD functions

In this section we collect the definitions and fundamental properties of the function spaces needed in this thesis.

Let $\Omega \subset \mathbb{R}^d$ open, bounded with Lipschitz boundary. Recall that the space $SBV(\Omega, \mathbb{R}^d)$, abbreviated as $SBV(\Omega)$ hereafter, of *special functions of bounded variation* consists of functions $y \in L^1(\Omega, \mathbb{R}^d)$ whose distributional derivative Dy is a finite Radon measure, which splits into an absolutely continuous part with density ∇y with respect to Lebesgue measure and a singular part $D^j y$ whose Cantor part vanishes and thus is of the form

$$D^j y = [y] \otimes \xi_y \mathcal{H}^{d-1} \lfloor J_y,$$

where $\mathcal{H}^{d-1}$ denotes the $(d-1)$-dimensional Hausdorff measure, J_y (the 'crack path') is an $\mathcal{H}^{d-1}$-rectifiable set in Ω, ξ_y is a normal of J_y and $[y] = y^+ - y^-$ (the 'crack opening') with $y^\pm$ being the one-sided limits of y at J_y. If in addition $\nabla y \in L^2(\Omega)$ and $\mathcal{H}^{d-1}(J_y) < \infty$, we write $y \in SBV^2(\Omega)$. See [6] for the basic properties of this function space.

Likewise, we say that a function $y \in L^1(\Omega, \mathbb{R}^d)$ is a *special function of bounded deformation* if the symmetrized distributional derivative $Ey := \frac{(Dy)^T + Dy}{2}$ is a finite $R^{d \times d}_{\text{sym}}$-valued Radon measure with vanishing Cantor part. It can be decomposed as

$$Ey = e(\nabla y)\mathcal{L}^d + E^j y = e(\nabla y)\mathcal{L}^d + [y] \odot \xi_y \mathcal{H}^{d-1}|_{J_y}, \tag{A.1}$$

where $e(\nabla y)$ is the absolutely continuous part of Ey with respect to the Lebesgue measure $\mathcal{L}^d$, $[y]$, ξ_y, J_y as before and $a \odot b = \frac{1}{2}(a \otimes b + b \otimes a)$. For basic properties of this function space we refer to [5, 8].

To treat variational problems as considered in Section 6 (see in particular (6.2)) the spaces $SBV(\Omega)$ and $SBD(\Omega)$ are not adequate due to the lacking L^∞-bound being essential in the compactness theorems. To overcome this difficulty the space of $GSBV(\Omega)$ was introduced consisting of all $\mathcal{L}^d$-measurable functions $u : \Omega \to \mathbb{R}^d$ such that for every $\phi \in C^1(\mathbb{R}^d)$ with the support of $\nabla\phi$ compact, the composition $\phi \circ u$ belongs to $SBV_{\mathrm{loc}}(\Omega)$ (see [38]). In this new setting one may obtain a more general compactness result (see [6, Theorem 4.36]). Unfortunately, this approach cannot be pursued in the framework of SBD functions as for a function $u \in SBD(\Omega)$ the composite $\phi \circ u$ typically does not lie in $SBD(\Omega)$. In [34], Dal Maso suggested another approach which is based on certain properties of one-dimensional slices.

First we have to introduce some notation. For every $\nu \in \mathbb{R}^d \setminus \{0\}$, for every $s \in \mathbb{R}^d$ and for every $B \subset \Omega$ we let

$$B^{\nu,s} = \{t \in \mathbb{R} : s + t\nu \in B\}. \tag{A.2}$$

Furthermore, define the hyperplane $\Pi^\nu = \{x \in \mathbb{R}^d : x \cdot \nu = 0\}$. Moreover, for every function $y : B \to \mathbb{R}^d$ we define the function $y^{\nu,s} : B^{\nu,s} \to \mathbb{R}^d$ by

$$y^{\nu,s}(t) = y(s + t\nu) \tag{A.3}$$

and $\hat{y}^{\nu,s} : B^{\nu,s} \to \mathbb{R}$ by $\hat{y}^{\nu,s}(t) = y(s+t\nu) \cdot \nu$. If $\hat{y}^{\nu,s} \in SBV(B^{\nu,s}, \mathbb{R})$ and $J_{\hat{y}^{\nu,s}}$ denotes the the *approximate jump set* we define

$$J^1_{\hat{y}^{\nu,s}} := \{x \in J_{\hat{y}^{\nu,s}} : |[\hat{y}^{\nu,s}](x)| \geq 1\}.$$

The space $GSBD(\Omega, \mathbb{R}^d)$ of *generalized functions of bounded deformation* is the space of all $\mathcal{L}^d$-measurable functions $y : \Omega \to \mathbb{R}^d$ with the following property: There exists a nonnegative bounded Radon measure λ on Ω such that for all $\nu \in S^{d-1} := \{x \in \mathbb{R}^d : |x| = 1\}$ we have that for $\mathcal{H}^{d-1}$-a.e. $s \in \Pi^\nu$ the function $\hat{y}^{\nu,s} = y^{\nu,s} \cdot \nu$ belongs to $SBV_{\mathrm{loc}}(\Omega^{\nu,s})$ and

$$\int_{\Pi^\nu} \Big(|D\hat{y}^{\nu,s}|(B^{\nu,s} \setminus J^1_{\hat{y}^{\nu,s}}) + \mathcal{H}^0(B^{\nu,s} \cap J^1_{\hat{y}^{\nu,s}}) \Big) \, d\mathcal{H}^{d-1}(s) \leq \lambda(B)$$

for all Borel sets $B \subset \Omega$. If in addition $e(\nabla y) \in L^2(\Omega)$ and $\mathcal{H}^1(J_y) < \infty$, we write $y \in GSBD^2(\Omega)$. We refer to [34] for basic properties of this space.

We recall fundamental compactness, slicing and approximation results.

Compactness and lower semicontinuity

We state a version of Ambrosio's compactness theorem in SBV adapted for our purposes (see e.g. [6]):

Theorem A.1.1. *Let $(y_k)_k$ be a sequence in $SBV(\Omega, \mathbb{R}^d)$ such that*

$$\int_\Omega |\nabla y_k(x)|^2 \, dx + \mathcal{H}^{d-1}(J_{y_k}) + \|y_k\|_\infty \leq C$$

for some constant C not depending on k. Then there is a subsequence (not relabeled) and a function $y \in SBV(\Omega, \mathbb{R}^d)$ such that $y_k \to y$ in $L^1(\Omega)$, and

$$\begin{aligned} &\nabla y_k \rightharpoonup \nabla y \text{ in } L^2(\Omega), \\ &D^j y_k \rightharpoonup^* D^j y \text{ as Radon measures.} \end{aligned} \tag{A.4}$$

An important subset of SBV is given by the indicator functions χ_W, where $W \subset \Omega$ is measurable with $\mathcal{H}^{d-1}(\partial^* W) < \infty$. Sets of this form are called *sets of finite perimeter* (cf. [6]). As a direct consequence of Theorem A.1.1 we get the following compactness result.

Theorem A.1.2. *Let $(W_k)_k \subset \Omega$ be a sequence of measurable sets satisfying $\mathcal{H}^{d-1}(\partial W_k) \leq C$ for some constant C independent of k. Then there is a subsequence (not relabeled) and a measurable set W such that $\chi_{W_k} \to \chi_W$ in measure for $k \to \infty$.*

We now state a compactness result for the generalized spaces. In [34, Theorem 11.3] we find the following theorem which we slightly adapt for our purposes.

Theorem A.1.3. *Let $(y_k)_k$ be a sequence in $GSBD(\Omega)$. Suppose that there exist a constant $M > 0$ and an increasing continuous functions $\psi : [0, \infty) \to [0, \infty)$ with $\lim_{x \to \infty} \psi(x) = +\infty$ such that*

$$\int_\Omega \psi(|y_k|) + \int_\Omega |e(\nabla y_k)|^2 + \mathcal{H}^1(J_{y_k}) \leq M$$

for every $k \in \mathbb{N}$. Then there exist a subsequence, still denoted by $(y_k)_k$, and a function $y \in GSBD(\Omega)$ such that

$$\begin{aligned} &y_k \to y \quad \text{pointwise a.e. in} \quad \Omega, \\ &e(\nabla y_k) \rightharpoonup e(\nabla y) \quad \text{weakly in } L^2(\Omega, \mathbb{R}^{2\times 2}_{\mathrm{sym}}), \\ &\liminf_{k \to \infty} \mathcal{H}^1(J_{y_k}) \geq \mathcal{H}^1(J_y). \end{aligned} \tag{A.5}$$

An analogous result holds in GSBV replacing $e(\nabla y_k), e(\nabla y)$ by $\nabla y_k, \nabla y$ (see [4, Theorem 2.2]). The lower semicontinuity result for the jump set can be generalized considering one-dimensional slices. For a concave function $\theta : (0, \infty) \to [0, 1]$ let

$$\hat{\mu}^\nu_y(B) := \int_{\Pi^\nu} \int_{B^{\nu,s} \cap J_{\hat{y}^{\nu,s}}} \theta(|[\hat{y}^{\nu,s}](t)|) \, d\mathcal{H}^0(t) \, d\mathcal{H}^{d-1}(s)$$

for all Borel sets $B \subset \Omega$.

Lemma A.1.4. *Let $(y_k)_k$ be a sequence in $GSBD(\Omega)$ converging to a function $y \in GSBD(\Omega)$ in the sense of* (A.5)*. Then*

$$\hat{\mu}_y^\nu(U) \le \liminf_{k\to\infty} \hat{\mu}_{y_k}^\nu(U)$$

for every $\nu \in S^{d-1}$ and for all open sets $U \subset \Omega$.

Proof. As $y_k \to y$ in the sense of (A.5) we may assume that $(y^{\nu,s})_k \to y^{\nu,s}$ in $GSBV(U^{\nu,s})$ for $\mathcal{H}^{d-1}$-a.e. $s \in U^\nu := \{s \in \Pi^\nu : U^{\nu,s} \neq \emptyset\}$. This is one of the essential steps in the proof of Theorem A.1.3 (cf. [34, Theorem 11.3] or [8, Theorem 1.1] for an elaborated proof in the SBD-setting). The desired claim now follows from the corresponding lower semicontinuity result for GSBV functions (see e.g. [6, Theorem 4.36]) and Fatou's lemma. □

We briefly note that using the area formula (see e.g. [6, Theorem 2.71, 2.90])) and fine properties of GSBD functions (see [34]), $\hat{\mu}_y^\nu(B)$ can be written equivalently as

$$\hat{\mu}_y^\nu(B) = \int_{J_y \cap B} \theta(|[y]\cdot\nu|)|\xi_y\cdot\nu|\, d\mathcal{H}^{d-1} \tag{A.6}$$

for all $\nu \in S^{d-1}$ and all Borel sets $B \subset \Omega$ (see also [34, Remark 9.3]).

Slicing

We briefly state the main slicing properties of SBV functions. For a proof we refer to [6, Section 3.11]. Recall definitions (A.2) and (A.3).

Theorem A.1.5. *Let $y \in SBV(\Omega, \mathbb{R}^d)$. For all $\nu \in S^{d-1}$ and $\mathcal{H}^{d-1}$-a.e. s in $\Pi^\nu = \{x : x\cdot\nu = 0\}$ the function $y^{\nu,s}$ belongs to $SBV(\Omega^{\nu,s}, \mathbb{R}^d)$. Moreover, one has*

$$\begin{aligned} &\nabla y(s+t\nu)\cdot\nu = (y^{\nu,s})'(t) \ \ \textit{for a.e. } t \in \Omega^{\nu,s},\\ &J_{y^{\nu,s}} = \{t \in \mathbb{R} : s+t\nu \in J_y\},\\ &\int_{\Pi^\nu} \# J_{y^{\nu,s}}\, d\mathcal{H}^{d-1}(s) = \int_{J_y} |\xi_y\cdot\nu|\, d\mathcal{H}^{d-1}. \end{aligned}$$

There is an analogous result in (G)SBD (see [34, Section 8,9]): Replace $\nabla(s+t\nu)\cdot\nu$ by $\nu^T e(\nabla y(s+t\nu))\nu$ and $y^{\nu,s}$ by $\hat{y}^{\nu,s}$.

Approximation results

Another basic tool we need are density results. We start with a result in the space of SBV^2 (see [31]). We define $\mathcal{W}(\Omega, \mathbb{R}^d)$ as the space of all SBV^2 functions y such that J_y is a finite, disjoint union of (d-1)-simplices and $y \in W^{k,\infty}(\Omega \setminus J_y, \mathbb{R}^d)$ for all k.

Theorem A.1.6. *The space $\mathcal{W}(\Omega, \mathbb{R}^d)$ is dense in $SBV^2(\Omega, \mathbb{R}^d)$ in the following sense: For every $y \in SBV^2(\Omega, \mathbb{R}^d)$ there exists a sequence $y_n \in \mathcal{W}(\Omega, \mathbb{R}^d)$ such that for $n \to \infty$*

$$
\begin{aligned}
&(i) \quad \|y_n - y\|_{L^1(\Omega)} \to 0, \\
&(ii) \quad \|\nabla y_n - \nabla y\|_{L^2(\Omega)} \to 0, \\
&(iii) \quad \mathcal{H}^1(J_{y_n}) \to \mathcal{H}^1(J_y).
\end{aligned}
$$

We now state a density result for SBV functions due to Cortesani and Toader being appropriate for anisotropic surface energies (see [32]). Moreover, a proof very similar to that of Proposition 2.5 in [51] shows that we may also impose suitable boundary conditions on the approximating sequence. Assume that $\tilde{\Omega} \supset \Omega$ is a bounded, open domain in $\mathbb{R}^d$ with Lipschitz boundary defining the Dirichlet boundary $\partial_D\Omega = \partial\Omega \cap \tilde{\Omega}$ of Ω. Moreover, let $\Omega_{D,\delta} := \{x \in \tilde{\Omega} : \mathrm{dist}(x, \partial_D\Omega) \le \delta\} \cup (\tilde{\Omega} \setminus \Omega)$ for $\delta > 0$.

Theorem A.1.7. *Let $g \in W^{1,\infty}(\tilde{\Omega})$. For every $u \in SBV^2(\tilde{\Omega}, \mathbb{R}^d) \cap L^\infty(\tilde{\Omega}, \mathbb{R}^d)$ with $u = g$ on $\tilde{\Omega} \setminus \Omega$, there exists a sequence u_n and a sequence of neighborhoods $U_n \subset \tilde{\Omega}$ of $\tilde{\Omega} \setminus \Omega$ such that $u_n = g$ on $\Omega_{D,\frac{1}{n}}$, $u_n \in W^{1,\infty}(U_n)$ and $u_n|_{V_n} \in \mathcal{W}(V_n, \mathbb{R}^2)$, where $V_n \subset \Omega$ is some neighborhood of $\Omega \setminus U_n$, such that $\|u_n\|_\infty \le \|u\|_\infty$ and*

(i) $u_n \to u$ *strongly in* $L^1(\tilde{\Omega}, \mathbb{R}^d)$, $\nabla u_n \to \nabla u$ *strongly in* $L^2(\tilde{\Omega}, \mathbb{R}^d)$,

(ii) $\limsup_{n\to\infty} \int_{J_{u_n}} \phi(\nu_{u_n}) d\mathcal{H}^1 \le \int_{J_u} \phi(\nu_u) d\mathcal{H}^1$ *for every upper semicontinuous function* $\phi : S^1 \to [0, \infty)$ *satisfying* $\phi(\nu) = \phi(-\nu)$ *for every* $\nu \in S^1$.

Recall that u is defined on $\tilde{\Omega}$ and thus it will be penalized in $\int_{J_u} \phi(\nu_u)\, d\mathcal{H}^1$ if u does not attain the boundary condition g on the Dirichlet boundary $\partial_D\Omega$. The following result proved in [54] together with Theorem A.1.6 provides a density result in GSBD.

Theorem A.1.8. *Let $u \in GSBD^2(\Omega, \mathbb{R}^d) \cap L^2(\Omega, \mathbb{R}^d)$. Then there exists a sequence $u_n \in SBV^2(\Omega, \mathbb{R}^d)$ such that each J_{u_n} is contained in the union of a finite number of closed connected pieces of C^1-surfaces, each u_n belongs to $W^{1,\infty}(\Omega \setminus J_{u_n}, \mathbb{R}^2)$ and the following properties hold:*

$$
\begin{aligned}
&(i) \quad \|u_n - u\|_{L^2(\Omega)} \to 0, \\
&(ii) \quad \|e(\nabla u_n) - e(\nabla u)\|_{L^2(\Omega)} \to 0, \\
&(iii) \quad \mathcal{H}^1(J_{u_n}) \to \mathcal{H}^1(J_u).
\end{aligned}
$$

The last result is an adaption of [26] to the GSBD-setting. In the proof of Theorem 9.4.1 we will draw ideas from [26] to establish a slightly different variant of the density result [31] where an L^∞-bound for the derivative is preserved.

Distance to weakly differentiable functions

The distance of an SBV function to Sobolev functions can be measured by the distribution $\operatorname{curl}\nabla y$ (see [25, Proposition 5.1]).

Theorem A.1.9. *Let $Q = (0,1)^d$. Let $y \in SBV_\infty(Q) := \{y \in SBV(Q,\mathbb{R}^d) : \|\nabla y\|_\infty < \infty, \ \ \mathcal{H}^{d-1}(J_y) < \infty\}$. Then $\mu_y := \operatorname{curl}\nabla y$ is a measure concentrated on J_y such that*

$$|\mu_y| \leq C\|\nabla y\|_\infty \mathcal{H}^{d-1}|_{J_y}.$$

Moreover, for $p < \frac{d}{d-1}$ there is a constant $C = C(p) > 0$ such that for all $y \in SBV_\infty(Q)$ there is a function $u \in H^1(Q,\mathbb{R}^d)$ such that

$$\|\nabla u - \nabla y\|_{L^p(Q)} \leq C|\mu_y|(Q) \leq C\|\nabla y\|_\infty \mathcal{H}^{d-1}(J_y).$$

A.2 Caccioppoli partitions

We first introduce the notions of perimeter and essential boundary. Consider $E \subset \mathbb{R}^d$ measurable and let

$$P(E,\Omega) = \sup\left\{\int_E \operatorname{div}(\varphi) : \varphi \in C^1_c(\Omega,\mathbb{R}^d), \|\varphi\|_\infty \leq 1\right\} \qquad \text{(A.7)}$$

be the *perimeter* of E in Ω (see [6, Section 3.3]). Moreover, we define the *essential boundary* by

$$\partial^* E = \mathbb{R}^d \setminus \bigcup\nolimits_{t=0,1} \left\{x \in \mathbb{R}^d : \lim\nolimits_{\varrho\downarrow 0} \frac{|E \cap B_\varrho(x)|}{|B_\varrho(x)|} = t\right\}. \qquad \text{(A.8)}$$

By [6, (3.62)] we have

$$P(E,\Omega) = \mathcal{H}^{d-1}(\Omega \cap \partial^* E).$$

We say a partition $\mathcal{P} = (P_j)_j$ of Ω is a *Caccioppoli partition* of Ω if $\sum_j P(P_j,\Omega) < +\infty$. We say a partition is *ordered* if $|P_i| \geq |P_j|$ for $i \leq j$. In the whole thesis we will always tacitly assume that partitions are ordered. Given a rectifiable set S we say that a Caccioppoli partition is *subordinated* to S if (up to an $\mathcal{H}^{d-1}$-negligible set) the essential boundary $\partial^* P_j$ of P_j is contained in S for every $j \in \mathbb{N}$.

The local structure of Caccioppoli partitions can be characterized as follows (see [6, Theorem 4.17]).

Theorem A.2.1. *Let $(P_j)_j$ be a Caccioppoli partition of Ω. Then*

$$\bigcup\nolimits_j (P_j)^1 \cup \bigcup\nolimits_{i\neq j} \partial^* P_i \cap \partial^* P_j$$

contains $\mathcal{H}^{d-1}$-almost all of Ω.

Here $(P)^1$ denote the points where P has density one (see [6, Definition 3.60]). Essentially, the theorem states that $\mathcal{H}^{d-1}$-a.e. point of Ω either belongs to exactly one element of the partition or to the intersection of exactly two sets $\partial^* P_i$, $\partial^* P_j$. In particular, the structure theorem implies (see [6, (4.24) and Theorem 4.23])

$$2\mathcal{H}^{d-1}\Big(\bigcup\nolimits_j \partial^* P_j \cap \Omega\Big) = \sum\nolimits_j P(P_j, \Omega). \tag{A.9}$$

We now state a compactness result for ordered Caccioppoli partitions (see [6, Theorem 4.19, Remark 4.20]).

Theorem A.2.2. *Let $\Omega \subset \mathbb{R}^d$ bounded, open with Lipschitz boundary. Let $\mathcal{P}_i = (P_{j,i})_j$, $i \in \mathbb{N}$, be a sequence of ordered Caccioppoli partitions of Ω with $\sup_i \sum_j P(P_{j,i}, \Omega) \le C$ independent of $i \in \mathbb{N}$. Then there exists a Caccioppoli partition $\mathcal{P} = (P_j)_j$ and a not relabeled subsequence such that $P_{j,i} \to P_j$ in measure for all $j \in \mathbb{N}$ as $i \to \infty$.*

Caccioppoli partitions are naturally associated to piecewise constant functions. We say $y : \Omega \to \mathbb{R}^d$ is *piecewiese constant in* Ω if there exists a Caccioppoli partition $(P_j)_j$ of Ω and a sequence $(t_j)_j \subset \mathbb{R}^d$ such that $y = \sum_j t_j \chi_{P_j}$. We close this section with a compactness result for piecewise constant functions (see [6, Theorem 4.25]).

Theorem A.2.3. *Let $\Omega \subset \mathbb{R}^d$ bounded, open with Lipschitz boundary. Let $(y_i)_i \subset SBV(\Omega, \mathbb{R}^d)$ be a sequence of piecewise constant functions such that $\sup_i(\|y_i\|_\infty + \mathcal{H}^{d-1}(J_{y_i})) \le C$ independent of $i \in \mathbb{N}$. Then there exists a not relabeled subsequence converging in measure to a piecewise constant function y.*

Appendix B

Rigidity and Korn-Poincaré's inequality

The following geometric rigidity result by Friesecke, James and Müller (see [49]) is fundamental in the thesis.

Theorem B.1. *Let $\Omega \subset \mathbb{R}^d$ a (connected) Lipschitz domain and $1 < p < \infty$. Then there exists a constant $C = C(\Omega, p)$ such that for any $y \in W^{1,p}(\Omega, \mathbb{R}^d)$ there is a rotation $R \in SO(d)$ such that*

$$\|\nabla y - R\|_{L^p(\Omega)} \le C \|\mathrm{dist}(\nabla y, SO(d))\|_{L^p(\Omega)} . \tag{B.1}$$

One ingredient in the proof is the following decomposition into a harmonic and a rest part.

Theorem B.2. *Let $\Omega \subset \mathbb{R}^2$ open and $1 < p < \infty$. There is a constant $C = C(p)$ such that all $y \in W^{1,p}(\Omega, \mathbb{R}^2)$ can be split into $y = w + z$, where w is a harmonic function and z satisfies*

$$\|\nabla y - \nabla w\|_{L^p(\Omega)} = \|\nabla z\|_{L^p(\Omega)} \le C\| \mathrm{dist}(\nabla y, SO(2))\|_{L^p(\Omega)}.$$

Note that the constant C is independent of the domain Ω. In higher dimensions one additional needs $\|\nabla y\|_\infty \le M$ for $M > 0$.

Proof. Following the singular-integral estimates in [30, Section 2.4] we find $\|\nabla z\|_{L^p(\Omega)} \le c\|\mathrm{cof}\nabla y - \nabla y\|_{L^p(\Omega)}$. The assertion follows from the fact that $|\mathrm{cof}A - A|^p \le C_p \,\mathrm{dist}^p(A, SO(2))$ for all $A \in \mathbb{R}^{2\times 2}$ (see also (3.11) in [49]). □

We state a Poincaré inequality in BV (see [6]).

Theorem B.3. *Let $\Omega \subset \mathbb{R}^d$ bounded, connected with Lipschitz boundary. Then there is a constant $C > 0$, which is invariant under rescaling of the domain, such that for all $u \in BV(\Omega, \mathbb{R}^d)$*

$$\|u - c\|_{L^{\frac{d}{d-1}}(\Omega)} \le C|Du|(\Omega)$$

for some $c \in \mathbb{R}^d$, where $|\cdot|$ denotes the total variation.

Finally, we recall a Korn-Poincaré inequality and a trace theorem in BD (see [21, 62]).

Theorem B.4. *Let $\Omega \subset \mathbb{R}^d$ bounded, connected with Lipschitz boundary and let $P : L^2(\Omega, \mathbb{R}^d) \to L^2(\Omega, \mathbb{R}^d)$ be a linear projection onto the space of infinitesimal rigid motions. Then there is a constant $C > 0$, which is invariant under rescaling of the domain, such that for all $u \in BD(\Omega, \mathbb{R}^d)$*

$$\|u - Pu\|_{L^{\frac{d}{d-1}}(\Omega)} \leq C|Eu|(\Omega),$$

where $Eu = \frac{Du^T + Du}{2}$ is the symmetrized distributional derivative.

Theorem B.5. *Let $\Omega \subset \mathbb{R}^2$ bounded, connected with Lipschitz boundary. There exists a constant $C > 0$ such that the trace mapping $\gamma : BD(\Omega, \mathbb{R}^2) \to L^1(\partial\Omega, \mathbb{R}^2)$ is well defined and satisfies the estimate*

$$\|\gamma u\|_{L^1(\partial\Omega)} \leq C\big(\|u\|_{L^1(\Omega)} + |Eu|(\Omega)\big)$$

for each $u \in BD(\Omega, \mathbb{R}^2)$.

For sets which are related through bi-Lipschitzian homeomorphisms with Lipschitz constants of both the homeomorphism itself and its inverse uniformly bounded the constants in Theorem B.1 and Theorem B.4 can be chosen independently of these sets, see e.g. [49].

Bibliography

[1] G. Alberti, C. Mantegazza. *A note on the theory of SBV functions.* Boll. Un. Mat. Ital. B(7) **11** (1989), 375–382.

[2] R. Alicandro, M. Focardi, M. S. Gelli. *Finite-difference approximation of energies in fracture mechanics.* Ann. Scuola Norm. Sup. **29** (2000), 671–709.

[3] L. Ambrosio. *A Compactness Theorem for a Special Class of Functions of Bounded Variation.* Boll. Un. Mat. Ital. **3-B** (1989), 857–881.

[4] L. Ambrosio. *Existence theory for a new class of variational problems.* Arch. Ration. Mech. Anal. **111** (1990), 291–322.

[5] L Ambrosio, A. Coscia, G. Dal Maso. *Fine properties of functions with bounded deformation.* Arch. Ration. Mech. Anal. **139** (1997), 201–238.

[6] L. Ambrosio, N. Fusco, D. Pallara. *Functions of bounded variation and free discontinuity problems.* Oxford University Press, Oxford 2000.

[7] L. Ambrosio, V. M. Tortorelli. *On the approximation of free discontinuity problems.* Boll. Un. Mat. Ital. B **7** (1992), 105–123.

[8] G. Bellettini, A. Coscia, G. Dal Maso. *Compactness and lower semicontinuity properties in $SBD(\Omega)$.* Math. Zl. **228** (1998), 337–351.

[9] X. Blanc, C. Le Bris, P.-L. Lions. *From molecular models to continuum mechanics.* Arch. Ration. Mech. Anal. **164** (2002), 341–381.

[10] B. Bourdin, G. A. Francfort, J. J. Marigo. *Numerical experiments in revisited brittle fracture.* J. Mech. Phys. Solids **48** (2000), 797–826.

[11] B. Bourdin, G. A. Francfort, J. J. Marigo. *The variational approach to fracture.* J. Elasticity **91** (2008), 5–148.

[12] A. Braides. *Γ-convergence for Beginners.* Oxford University Press, Oxford 2002.

[13] A. BRAIDES. *Non-local variational limits of discrete systems.* Commun. Contemp. Math. **2** (2000), 285–297.

[14] A. BRAIDES, M. CICALESE. *Surface energies in nonconvex discrete systems.* Math. Models Methods Appl. Sci. **17** (2007), 985–1037.

[15] A. BRAIDES, G. DAL MASO, A. GARRONI. *Variational formulation of softening phenomena in fracture mechanics. The one-dimensional case.* Arch. Ration. Mech. Anal. **146** (1999), 23–58.

[16] A. BRAIDES, M. S. GELLI. *Limits of discrete systems without convexity hypotheses.* Math. Mech. Solids **7** (2002), 41–66.

[17] A. BRAIDES, M. S. GELLI. *Limits of discrete systems with long-range interactions.* J. Convex Anal. **9** (2002), 363–399.

[18] A. BRAIDES, A. LEW, M. ORTIZ. *Effective cohesive behavior of layers of interatomic planes.* Arch. Ration. Mech. Anal. **180** (2006), 151–182.

[19] A. BRAIDES, M. SOLCI, E. VITALI. *A derivation of linear elastic energies from pair-interaction atomistic systems.* Netw. Heterog. Media **2** (2007), 551–567.

[20] J. BRAUN, B. SCHMIDT. *On the passage from atomistic systems to nonlinear elasticity theory for general multi-body potentials with p-growth.* Netw. Heterog. Media **4** (2013), 789–812.

[21] K. BREDIES. *Symmetric tensor fields of bounded deformation.* Ann. Mat. Pura Appl. **192** (2013), 815–851.

[22] G. BUTTAZZO. *Energies on BV and variational models in fracture mechanics.* Proceedings of "'Curvautre Flows and Related Topics"', Levico 27 June-2 July 1994, Gakkotosho, Tokyo 1995.

[23] M. CARRIERO, A. LEACI, AND F. TOMARELLI. *About Poincare inequalities for functions lacking summability.* Note Mat. **1** (2011), 67–84.

[24] A. CHAMBOLLE, S. CONTI, G. FRANCFORT. *Korn-Poincaré inequalities for functions with a small jump set.* 2014. <hal-01091710>

[25] A. CHAMBOLLE, A. GIACOMINI, M. PONSIGLIONE. *Piecewise rigidity.* J. Funct. Anal. Solids **244** (2007), 134–153.

[26] A. CHAMBOLLE. *An approximation result for special functions with bounded deformation..* J. Math. Pures Appl. **83** (2004), 929–954.

[27] S. Conti, G. Dolzmann, B. Kirchheim, S. Müller. *Sufficient conditions for the validity of the Cauchy-Born rule close to $SO(n)$.* J. Eur. Math. Soc. (JEMS) **8** (2006), 515–539.

[28] S. Conti, G. Dolzmann, S. Müller. *Korn's second inequality and geometric rigidity with mixed growth conditions.* Calc Var PDE **50** (2014), 437–454.

[29] S. Conti, D. Faraco, F. Maggi. *A new approach to counterexamples to L^1 estimates: Korn's inequality, geometric rigidity, and regularity for gradients of separately convex functions.* Arch. Rat. Mech. Anal. **175** (2005), 287–300.

[30] S. Conti, B. Schweizer. *Rigidity and Gamma convergence for solid-solid phase transitions with $SO(2)$-invariance.* Comm. Pure Appl. Math. **59** (2006), 830–868.

[31] G. Cortesani. *Strong approximation of $GSBV$ functions by piecewise smooth functions.* Ann. Univ. Ferrara Sez. **43** (1997), 27–49.

[32] G. Cortesani, R. Toader. *A density result in SBV with respect to non-isotropic energies.* Nonlinear Analysis **38** (1999), 585–604.

[33] G. Dal Maso. *An introduction to Γ-convergence.* Birkhäuser, Boston · Basel · Berlin 1993.

[34] G. Dal Maso. *Generalized functions of bounded deformation.* J. Eur. Math. Soc. **15** (2013), 1943–1997.

[35] G. Dal Maso, G. A. Francfort, R. Toader. *Quasistatic crack growth in nonlinear elasticity.* Arch. Ration. Mech. Anal. **176** (2005), 165–225.

[36] G. Dal Maso, M. Negri, D. Percivale. *Linearized elasticity as Γ-limit of finite elasticity.* Set-valued Anal. **10** (2002), 165–183.

[37] G. Dal Maso, R. Toader. *A model for quasi-static growth of brittle materials: Existence and approximation results.* Arch. Ration. Mech. Anal. **162** (2002), 101–135.

[38] E. De Giorgi, L. Ambrosio. *Un nuovo funzionale del calcolo delle variazioni.* Acc. Naz. Lincei, Rend. Cl. Sci. Fis. Mat. Natur. **82** (1988), 199–210.

[39] E. De Giorgi, M. Carriero, A. Leaci. *Existence theorem for a minimum problem with free discontinuity set.* Arch. Rational Mech. Anal. **108** (1989), 195–218.

[40] L. C Evans, R. F. Gariepy. *Measure theory and fine properties of functions.* CRC Press, Boca Raton · London · New York · Washington, D.C. 1992.

[41] H. Federer. *Geometric measure theory.* Springer, New York, 1969.

[42] M. Focardi, M. S. Gelli. *Approximation results by difference schemes of fracture energies: the vectorial case.* NoDEA Nonlinear Differential Equations Appl. Vol. 4 No. **4** (2003), 469–495.

[43] G. A. Francfort, C, J. Larsen. *Existence and convergence for quasi-static evolution in brittle fracture.* Comm. Pure Appl. Math. **56** (2003), 1465–1500.

[44] G. A. Francfort, J. J. Marigo. *Revisiting brittle fracture as an energy minimization problem.* J. Mech. Phys. Solids **46** (1998), 1319–1342.

[45] M. Friedrich. *From atomistic to continuum theory for brittle materials: A two-dimensional model problem.* Master's Thesis, Technische Universität München (2012).

[46] M. Friedrich, B. Schmidt. *An atomistic-to-continuum analysis of crystal cleavage in a two-dimensional model problem.* J. Nonlin. Sci. **24** (2014), 145–183.

[47] M. Friedrich, B. Schmidt. *An analysis of crystal cleavage in the passage from atomistic models to continuum theory.* Arch. Rational Mech. Anal., published online 2014, doi: 10.1007/s00205-014-0833-y.

[48] M. Friedrich, B. Schmidt. *On a discrete-to-continuum convergence result for a two dimensional brittle material in the small displacement regime.* Preprint, 2014.

[49] G. Friesecke, R. D. James, S. Müller. *A theorem on geometric rigidity and the derivation of nonlinear plate theory from three-dimensional elasticity.* Comm. Pure Appl. Math. **55** (2002), 1461–1506.

[50] G. Friesecke, F. Theil. *Validity and failure of the Cauchy-Born hypothesis in a two-dimensional mass spring lattice.* J. Nonlinear. Sci. **12** (2002), 445–478.

[51] A. Giacomini. *Ambrosio-Tortorelli approximation of quasi-static evolution of brittle fractures.* Calc. Var. Partial Differential Equations. **22** (2005), 129–172.

[52] A. A. Griffith. *The phenomena of rupture and flow in solids.* Philos. Trans. R. Soc. London **221** (1921), 163–198.

[53] R. L. Hayes, M. Ortiz, E. A. Carter *Universal binding-energy relation for crystals that accounts for surface relaxation.* Phys. Rev. B **69** (2004), 172104.

[54] F. Iurlano. *A density result for GSBD and its application to the approximation of brittle fracture energies.* Calc. Var. **51** (2014), 315–342.

[55] E. A. A. Jarvis, R. L. Hayes, E. A. Carter *Effects of Oxidation on the Nanoscale Mechanisms of Crack Formation in Aluminum.* ChemPhysChem **2** (2001), 55–59.

[56] J. Kristensen. *Lower semicontinuity in spaces of weakly differentiable functions.* Math. Ann. **313** (1999), 653–710.

[57] C. Mora-Corral. *Explicit energy-minimizers of incompressible elastic brittle bars under uniaxial extension.* C. R. Acad. Sci. Paris **348** (2010), 1045–1048.

[58] M. Negri. *Finite element approximation of the Griffith's model in fracture mechanics.* Numer. Math. **95** (2003), 653–687.

[59] M. Negri. *A discontinuous finite element approximation of free discontinuity probems.* Adv. Math. Sci. Appl. **15** (2005), 283–306.

[60] M. Negri, R. Toader. *Scaling in fracture mechanics by Bažant's law: from finite to linearized elasticity.* Preprint SISSA, Trieste, 2013.

[61] O. Nguyen and M. Ortiz. *Coarse-graining and renormalization of atomistic binding relations and universal macroscopic cohesive behavior.* J. Mech. Phys. Solids **50** (2002), 1727–1741.

[62] R. Temam. *Mathematical Problems in Plasticity.* Bordas, Paris 1985.

[63] B. Schmidt. *A derivation of continuum nonlinear plate theory from atomistic models.* SIAM Multiscale Model. Simul. **5** (2006), 664–694.

[64] B. Schmidt. *Linear Γ-limits of multiwell energies in nonlinear elasticity theory.* Continuum Mech. Thermodyn. **20** (2008) 375–396.

[65] B. Schmidt. *On the derivation of linear elasticity from atomistic models.* Netw. Heterog. Media **4** (2009), 789–812.

[66] B. Schmidt, F. Fraternali, M. Ortiz. *Eigenfracture: an eigendeformation approach to variational fracture.* SIAM Mult. Model. Simul. **7** (2009), 1237–1266.

[67] K. Zhang. *An approximation theorem for sequences of linear strains and its applications.* ESAIM Control Optim. Calc. Var. **10** (2004), 224–242.